THE PEDERSEN MEMORIAL ISSUE

ADVANCES IN INCLUSION SCIENCE

VOLUME 7

The titles published in this series are listed at the end of this volume.

The Pedersen Memorial Issue

Edited by

R. M. IZATT
and
J. S. BRADSHAW

Department of Chemistry,
Brigham Young University,
Provo, Utah, U.S.A.

Reprinted from
Journal of Inclusion Phenomena and Molecular Recognition in Chemistry,
Volume 12, Nos. 1–4 (1992)

SPRINGER SCIENCE+BUSINESS MEDIA, B.V.

Library of Congress Cataloging-in-Publication Data

The Pedersen memorial issue / edited by R.M. Izatt and J.S. Bradshaw.
p. cm. -- (Advances in inclusion science ; v. 7)
"Journal of inclusion phenomena and molecular recognition in chemistry, volume 12, nos. 1-4, January-April 1992"--P.
Includes index.
ISBN 978-94-010-5118-7 ISBN 978-94-011-2532-1 (eBook)
DOI 10.1007/978-94-011-2532-1
1. Crown ethers. 2. Cyclic compounds. 3. Pedersen, Charles J., 1904-1989. I. Izatt, Reed McNeil, 1926- . II. Bradshaw, J. S. III. Series.
QD305.E7P36 1992
547'.035--dc20 92-10361

ISBN 978-94-010-5118-7

JOURNAL OF INCLUSION PHENOMENA
AND MOLECULAR RECOGNITION IN CHEMISTRY

Volume 12 Nos. 1–4 January–April 1992

THE PEDERSEN MEMORIAL ISSUE

Guest Editor:

R. M. IZATT

Department of Chemistry, Brigham Young University, Provo, UT 84602, U.S.A.

Journal of Inclusion Phenomena and Molecular Recognition in Chemistry **12**: 1–6, 1992.

Foreword

Charles J. Pedersen (1904–1989), Nobel Laureate in Chemistry (1987)

This issue is dedicated to the memory of the late Charles J. Pedersen in recognition of his outstanding contribution to scientific research, culminating in his discovery of crown ethers and their remarkable cation complexing properties and his receipt of the 1987 Nobel Prize in Chemistry.

Charlie's origin and early years in Korea did not portend the creative work in chemistry which would characterize his later life. However, we can see in his early years the influence of his Norwegian father and Japanese mother who considered his formal education to be of utmost importance. At the age of eight, he was sent abroad to Japan for schooling, first at a convent school in Nagasaki, and two years later at a French–American preparatory school in Yokohama run by a Marianist order of Catholic priests and brothers. The latter group encouraged him to attend the order's University of Dayton in Ohio where he received a bachelors degree in chemical engineering.

Charlie's academic experiences, his employment with du Pont, and the creative spark which he manifested at an early stage of his scientific career are detailed in the paper in this issue by Herman Schroeder. Schroeder had a long-time association with Charlie at du Pont as a co-worker, supervisor, and friend. His recollections provide insight into Charlie's creative mind. In addition, they make it clear that a long period of creative work preceded the accidental discovery of the first synthetic crown ether. It is important to note that Charlie's mind was well prepared to recognize the importance of his discovery.

The field of macrocyclic chemistry, to a large degree, had its beginnings with Charlie's discovery. A first-person account of his discovery is given as the first paper in this issue. This account was prepared by him and was read at the 12th Symposium on Macrocyclic Chemistry in Hiroshima, Japan in 1987 by Herman Schroeder. The growth of this field since Charlie's first publication on the subject in 1967 has been enormous. This growth is evidenced in one segment of the field by the three-fold increase in the number of references in two *Chemical Reviews* articles on thermodynamic quantities associated with cation–macrocycle interaction authored by us in 1985 and 1991. Charlie lived to see much of this growth. He saw many of his own predictions of possible uses of crown ethers and related macrocycles realized.

Recognition for Charlie came late in his career. He found it satisfying to see so many capable scientists go in so many directions as they applied his discovery to a wide range of chemical and other fields. He made seminal contributions to the broad area known today as molecular recognition. His work illustrates how one individual can make an enormous difference in science. The effect of his life and work on those of us who contributed papers for this issue and on many others is appreciated and is acknowledged by several of the authors in their individual papers. It is entirely appropriate to honor his memory with this special issue.

Department of Chemistry,
Brigham Young University,
Provo, UT 84602, U.S.A.

R. M. IZATT

J. S. BRADSHAW

(a) Charles Pedersen on campus at the University of Dayton in the mid-1920's.

(b) Charles Pedersen in his laboratory at the Experimental Station in 1968.

(c) Charles J. Pedersen receiving the Noble Prize, 1987 (© Pressers Bild AB).

(d) Charles J. Pedersen, Nobel Laureate 1987 (Photographed by Lief Skoogfors).

Journal of Inclusion Phenomena and Molecular Recognition in Chemistry **12**: 7–10, 1992.

The Discovery of Crown Ethers*

CHARLES J. PEDERSEN

I regret that I have nothing new to add to this field in which so many able investigators have obtained interesting results extending the scope of my discovery. So I must resort to old material, perhaps so old that some have forgotten, and some of the younger of you never knew of it. Briefly, I will recount how I discovered the crown ethers and the fun I had doing so [1, 2].

In 1961, at Dr. H. E. Schroeder's suggestion, I began research on vanadium compounds. We hoped that an understanding of the effect of various ligands on the catalytic behavior of vanadium might lead to the development of superior catalysts for the polymerization of olefins. We also hoped to find means for suppressing the pro-oxidant catalytic action of residual vanadium in the resulting polymers. There was then little understanding of the effects of structure on contemporary vanadium catalysts. Most were discovered empirically by reactions of inorganic vanadium compounds, such as vanadium tetrachloride or vanadyl chloride ($VOCl_2$), with various aluminum alkyls.

I decided to study the effects of uni- and multidentate phenolic ligands on the catalytic properties of VO. To this end, I had synthesized the desired ligands up to and including the quadridentate. Now I was preparing the quinquedentate, bis[2-(*o*-hydroxyphenoxy)ethyl] ether **3** by reacting a catechol derivative with a protected hydroxy group **1** with bis(2-chlorethyl) ether in the presence of sodium hydroxide (Figure 1). The sample of the partly protected catechol **2** was known to be contaminated with about 10% unreacted catechol, but it was used without further purification.

A brownish goo was the product of the reaction, and in my initial attempt to recover the desired quinquedentate ligand, I obtained a colorless crystalline by-product in 0.4% yield. This fibrous substance had an intriguing combination of properties: (1) elementary analysis fitted benzo-9-crown-3, a possible product of the reaction of the contaminant catechol with bis(2-chlorethyl) ether; (2) it was little soluble in methanol but solubilized by sodium hydroxide; (3) its UV spectrum was that of a phenolic compound but without a hydroxyl group which would have explained the solubilization by sodium hydroxide; and (4) its UV spectrum was altered by the base in a way we had never seen before (Figure 2).

There was no known cause for the solubilization in methanol by sodium hydroxide without an appropriate functional group. To complete the picture, the product was submitted for molecular weight determination.

In the meantime, I was greatly excited to discover that the unknown was solubilized in methanol by any methanol-soluble salt of the alkali metals. Therefore,

* Reprinted with permission from *Current Topics in Macrocyclic Chemistry in Japan* E. Kimura (ed.), Hiroshima University School of Medicine, Hiroshima, 1987. The text was prepared by Charles J. Pedersen and was read by Dr. Herman Schroeder at the 12th International Symposium on Macrocyclic Chemistry, held in Hiroshima, Japan, 20–23 July, 1987.

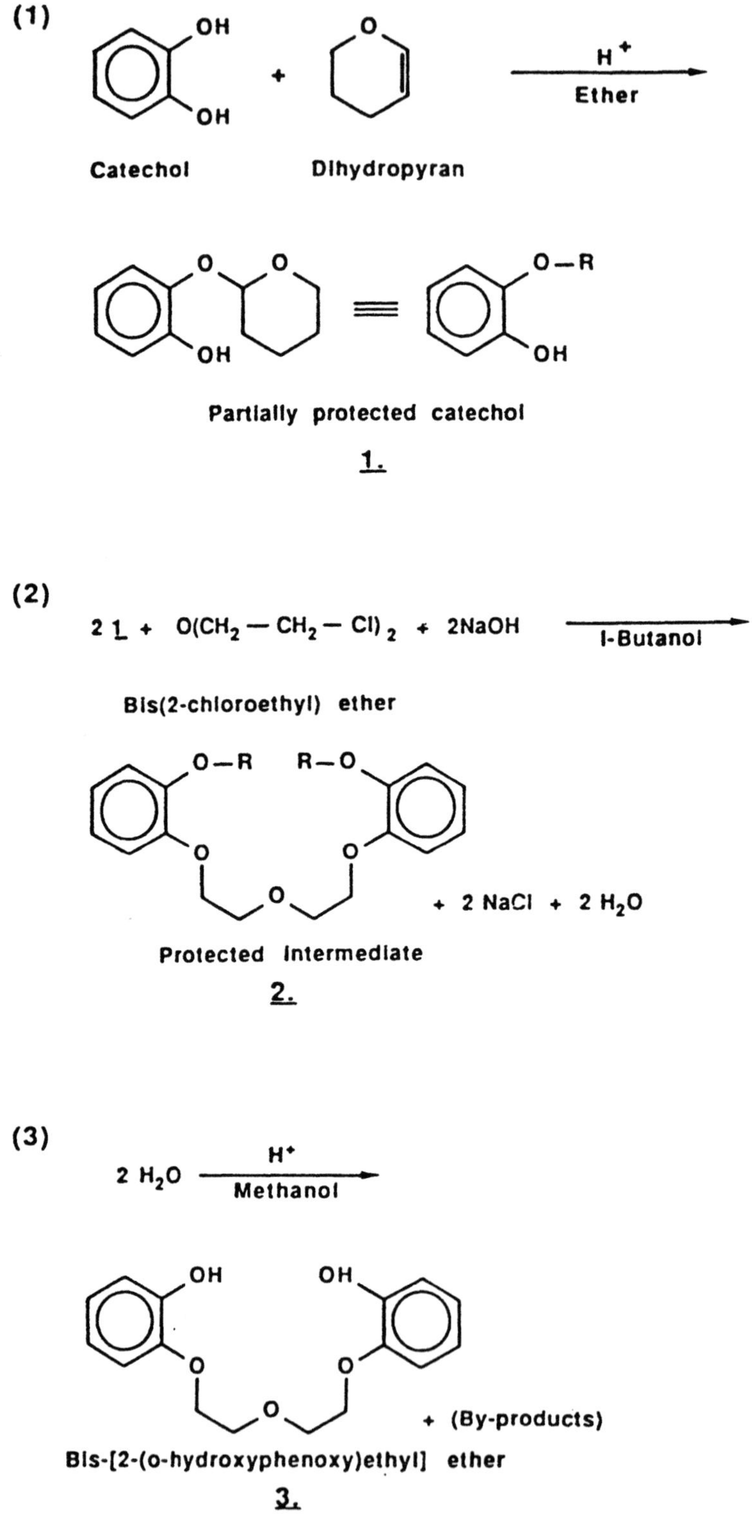

Fig. 1. Synthesis of bis-[2-*o*-hydroxyphenoxy)ethyl] ether.

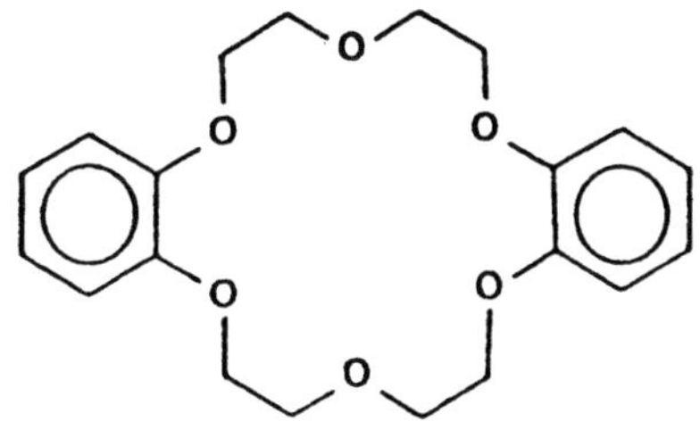

4

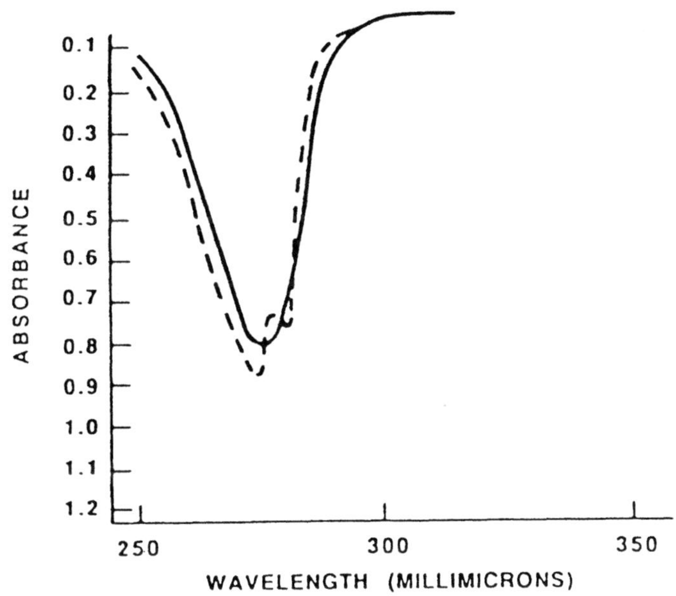

Fig. 2. Effect of NaOH on the ultraviolet spectrum of dibenzo-18-crown-6. **4**. (- - -, after addition of NaOH). Soluble sodium salts have the same effect.

solubilization by sodium hydroxide was due to the sodium cation and not to the base! Its molecular weight was found to be double that of benzo-9-crown-3, namely corresponding to dibenzo-18-crown-6, **4** in Figure 2. The mystery was solved; in my synthesis of the quinquedentate ligand, nature had lent a hand to produce a hexadentate ligand by reacting two molecules of the contaminant catechol with two molecules of bis(2-chlorethyl) ether and thus closing the ring.

Thus did I discover dibenzo-18-crown-6, the first crown ether and the first synthetic compound capable of complexing the alkali metal cations. So now, with the realization that I had something very unusual and with the utmost curiosity and anticipation, I devoted all my energies to the study of this fascinating class of ligands by synthesizing a great variety of macrocyclic polyethers and determining their interaction with inorganic cations. I was specially interested in the stability of the 'complexes' and the reason for their behavior. For example, I found that for maximum stability of its salt complex, each cation has an optimum size of the ring

Fig. 3. Courtauld model of dibenzo-18-crown-6, **4**.

of the polyether. A complex can form even if the fit is not the best by forming a sandwich complex consisting of two molecules of polyether per cation. The thermal stability of some salt complexes, e.g., KCNS is attested to by their having melting points higher than those of the components.

When their unique properties were realized, an exhilarating period of research was inaugurated. Every successful experiment produced a significantly novel result and led to new thoughts on what to synthesize and also as to the many potential uses of these extraordinary substances. I also derived great esthetic pleasure from the structure of the crown ethers as shown by the Courtauld model (Figure 3); so simple, elegant, and effective a means for the trapping of hitherto recalcitrant alkali cations. The crown nomenclature was created because the official names of crown ethers were so complex and hard for me to remember.

It might appear that the work on the crown ethers was mine alone save for the help of an able laboratory technician. True, the concepts and decisions were mine, but there was a host of indispensable collaborators who made all this possible. I would be more than remiss were I not to make the following acknowledgement; to the management of the Elastomer Chemicals Department of the du Pont Company for support in every way for almost a decade; to the Analytical Groups of the Company who made all their resources available; for consultations with the technical staff of the company, and the versatile skill of my laboratory technician.

And now, I place myself again in the kindly hands of my friend and mentor, Herman Schroeder, I cannot adequately express my gratitude to him for all of his actions on my behalf.

References

1. C. J. Pedersen: *J. Am. Chem. Soc.* **89,** 7017 (1967).
2. C. J. Pedersen: in *Synthetic Multidentate Macrocyclic Compounds*, R. M. Izatt and J. J. Christensen (eds.), Academic Press, New York, Ch 1, 1978.

Journal of Inclusion Phenomena and Molecular Recognition in Chemistry **12**: 11–21, 1992.

The Productive Scientific Career of Charles Pedersen* (October 3, 1904–October 26, 1989)

HERMAN E. SCHROEDER
4031 Kennett Pike, # 74, Greenville, DE 19807, U.S.A.

(Received: 11 March 1991; in final form: 25 March 1991)

Abstract. Charles Pedersen's career is reviewed from the time of his early life in Korea and Japan and scientific training in the United States to the present. His 42-year research career was practically and scientifically productive, leading finally to a share of the Nobel prize in chemistry for 1987. Commercially significant accomplishments included a large improvement in the yield of tetraethylead antiknock and discovery of the first oil soluble agents for inhibiting catalytic effects of copper in degradation of petroleum products and rubber. This led to a study of the interactions of metals with chelating agents and the resultant effects on oxidation reactions, thence to the behavior of peroxides and their reactions with substrates. He also discovered the antiknock activity of ferrocene. While studying the effects of ligands on the catalytic behavior of vanadium he discovered the crown compounds and their ability to include alkali metal ions. He devoted the last years of his career to elaborating their chemistry.

Key words. Charles Pedersen, 1987 Chemistry Nobel Prize, du Pont industrial chemist, oil soluble agents, ferrocene, crown compounds, alkali metal complexation.

1. Introduction

On October 14, 1987 it was revealed that a relatively unknown and unrecognized du Pont industrial chemist had been awarded a share of the 1987 Nobel prize in Chemistry for his "development and use of molecules with structure-specific interactions of high specificity". While this surprised many chemists even in the du Pont company, those aware of his pioneering work with 'crown ethers' had been awaiting such recognition. In 1967, he had startled the world of coordination chemists with a description of his studies at a conference in Japan and in a paper which Marshall Gates, Editor of the *Journal of the American Chemical Society*, said was "a monumental piece of work".

What sort of man was this who had made so 'lucky' a find at the age of 58, in an industrial laboratory of all places? I write this as one who has known Charlie Pedersen as friend, associate, and when I was Director of Research, as a treasured member of my staff. To understand him we need to know of his interesting personal background and its effect on his character. These events, together with a stimulating industrial environment, led to an unusually productive 42-year career in research with the du Pont company which gave a basis for his discovery of the crown ethers and his recognition of their significance. Fortune surely smiled on a prepared mind.

* Taken in part with permission from *Current Topics in Macrocyclic Chemistry in Japan*, E. Kimura (ed.), Hiroshima University School of Medicine, Hiroshima, 1987.

The breadth of Pedersen's interests and his creativity are revealed in 65 patents and about 28 papers, many of which are listed in the reference section. A partial review of his accomplishments shows how a capable and creative scientist can shape his career to fit his unique abilities, though in a practical environment devoted largely to the support and growth of existing businesses. It also shows the beneficial effects of proximity to the varied problems of a highly diversified chemical business in broadening Pedersen's interests and stimulating his natural ingenuity.

2. Early Life

Charles John Pedersen was born October 3, 1904 in Fusan, Korea and died at his home in Salem, New Jersey, October 26, 1989. His father was Brede Pedersen, an expatriate Norwegian engineer, and his mother Takino Yasui, a Japanese from Kyushu whose family had emigrated to Korea to trade in soybeans and silkworms. His sister, Astrid, who was five years older and became his youthful mentor, played a very important role in his young life. They lived at the Unsan gold mines in northwestern Korea where his father worked as a mechanical engineer. The 400 square mile American concession abutted the Yalu river in a primitive area where pony harnesses had bells to ward off the Siberian tigers which still roamed the countryside. In the bitter cold of winter nights, wolves devoured children and foxes slept against rooftop chimneys to keep warm.

Since the Unsan mines were in an American enclave, life there was as American as possible and much gentler than on the American frontier. Social life centered at the country club and the common language was English, even in the Pedersen household. Life at an isolated American outpost undoubtedly influenced the boy strongly and fostered his independence. The procedures and attitudes involved in operating the mine and gold refinery plus the availability of interesting minerals generated his interest in technology.

Since foreign-language schools did not then exist in Korea, the youth went to Japan at the age of eight, where he enrolled for two years in a convent school in Nagasaki. Two years later he moved to St. Joseph's College, a French preparatory school in Yokohama run by Marianist priests and brothers, where he completed his secondary education. He returned to his family in Korea only during the summer vacations, traveling by train and boat in the style of a European. With his father's encouragement he came to America for his college education in 1922 shortly before his 18th birthday and enrolled at the University of Dayton (Ohio). He became vice-president of his class and received a chemical engineering degree in 1926 and then an M.S. in organic chemistry in 1927 at Massachusetts Institute of Technology. His professor, James F. Norris, recognized the ability of this unassuming young man and tried to persuade him to seek a doctorate, but Pedersen no longer wished to burden his father financially and chose to go to work.

With Norris's help he obtained a position with the du Pont Company where he spent his entire career of 42 years, mostly at Jackson Laboratory, their most diversified applied research organization. It served businesses in dyestuffs, intermediates, refrigerants, and made products for the rubber and petroleum industries. Here he had a distinguished career in industrial research with many notable accomplishments of both a practical and a scientific nature, valuable contributions to chemistry and to our material well-being, all well before his discovery of the crown ethers.

3. The du Pont Years

He started in analytic work and after a brief indoctrination came under the wing of two mentors, his leader W. S. Calcott and his associate A. S. Carter. Like many who had been responsible for the young man earlier, they did all they could to guide and encourage him while at the same time becoming close friends. Calcott sensed Pedersen's unusual quality and chose to keep him in research rather than send him to manufacturing as was usually done with those who lacked a doctoral degree.

They also fostered some of his lifelong interests: love of the natural world, stamp collecting, and fishing which was originally inspired by his father's love for the sea. He learned the native flowers and where to find them, the birds, what they ate and where they lived, and their unusual traits. Though he became seasick at the slightest wiggle of a boat, he loved to fish and pursued it wherever he could, locally, in Wyoming and Alaska, and with me for several weeks in Iceland. In life, as in science, Pedersen has been a fine student. A voracious reader, he rapidly broadened his knowledge of science and of art, history and nature. He once read through much of the Encyclopedia Brittanica to satisfy his consuming desire for knowledge. He even read it during his lunch hour to pass the time more pleasantly.

Pedersen soon learned the nature of research and of relevant du Pont technology. He also developed a personal investigative style which was intuitive and unfettered by scientific dogma, and remarkably gifted. Though scientific in his methodology he was very much an industrial chemist. He wanted his research to be useful, to have practical and beneficial applications. A hands-on chemist, and experimentalist rather than a theoretician, he was a keen observer of what was happening with an uncanny ability to ignore confusing details and focus directly on his goal. He had an eye for fruitful problems and a flair for appraising situations quickly. His approach to a problem was surprising in its simplicity and often led directly to the answer. Little activity was apparent on a visit to his laboratory; yet with simple set-ups and measurements he produced a prodigious amount of work. It is hard to appreciate that the monumental studies described in the first publications on the crown compounds were not the work of a professor with a corps of scientists. It came solely from a lone research scientist, Charles Pedersen, with an able but technically untrained technician, Ted Malinowski. Pedersen's research managers soon learned how much he could accomplish so they loosed their reins and urged him to follow his own intuition. He rewarded them with a flood of interesting and valuable discoveries.

After several successes with a variety of problems, he was asked in 1932 to find out how to increase the yield of antiknock TEL (tetraethyl lead) in the plant process where consumption of sodium–lead alloy indicated more TEL had been produced than was isolated from the steam distillation. Pedersen noted that the reaction mass was a sludge of very fine lead particles and guessed that the TEL was so strongly adsorbed that it could not be recovered. This he proved by taking 10 mL of TEL, adding 100 g of the most finely powdered lead he could obtain, and steam distilling. He recovered only 10% of the TEL compared with 99% recovery in a similar experiment where TEL was distilled in the presence of *coarse* lead shot. He then added a non-foaming wetting agent to the experiment with the powder. A 15 minute distillation yielded the rest of the TEL! Thus he enabled a 15% increase in plant yield of a high volume process and made what was probably the most profitable discovery of his career [1]!

Next came a succession of important patents: corrosion inhibitors and evaporation retarders for ethanol anti-freeze [2–4] and a very effective stabilizer, rosin acid, for sodium hydrosulfite which was then very important as a reducing agent for vat dyes [5]. By then the plant people had such faith in him that they asked him to treat each batch as it was produced in the plant. In 1935, he made another important discovery. He found the first good agents for deactivating metals in petroleum products and rubber. Heavy metals, especially copper, were normal contaminants in gasoline, oils, and rubber and were known to catalyze their oxidation. Clues he uncovered in a search for hydrocarbon soluble agents to precipitate copper led him to disalicylal ethylene diimine **1** but it was high melting and almost insoluble (<0.01%) in gasoline. So he made the 1,2-propylene diamine analog **2**. The melting point fell to 45°C and the solubility increased 100 fold, more than enough to make it practical. Disalicylal propylene diamine proved so effective in countering the degradative effects of metals it was soon used in almost all gasoline and much rubber. It is still important today [6–9].

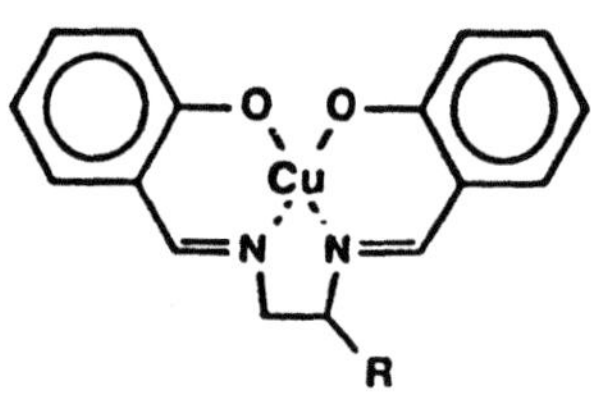

1., R = H; 2., R = CH_3

The metal deactivator discovery reshaped his career. It turned his attention toward peroxide and radical behavior, the catalytic effects of metal ions, and especially the behavior of metals in complexes [10]. With loving care he varied structures of ligands to find the features which led to the strongest complexes and the most effective deactivation of transition metals. Over the next ten years he and his group studied oxidation of hydrocarbons with and without inhibitors and deactivators. He also examined the corrosive action of the oils on metals while seeking means of blocking those processes. He himself became inventor or coinventor of about thirty patents covering anti-oxidants, stabilizers and inhibitors, and chelating compounds. He also sought stable synthetic lubricants which could better resist heat and oxidation.

During World War II he experienced none of the difficulties endured by the Nisei on the west coast. Although carrying a Norwegian passport, he had never been to Norway to confirm it, so he was officially classified a stateless alien which kept him from research on certain problems. He continued to work in du Pont throughout World War II on urgent war-related problems especially on the oxidation of lubricating oils. As an acknowledged expert in the stabilization of petroleum products he was of value to the government and the armed forces as a war-time consultant and was well liked and highly regarded.

Shortly after the war he married Susan J. Ault. Sue and Charlie became devoted companions until her death in 1983. Concurrently, in recognition of his research talent, he was appointed Research Associate, the top scientific position then available in du Pont. This freed him from the supervisory responsibilities he never liked and allowed him to practice research as he wished, with full responsibility for his own programs. With this new freedom he returned to his studies of the mechanisms of oxidation and the various reaction products of peroxides.

During this period, I got to know Charlie Pedersen very well as we were among the co-founders of an evening seminar group wherein we discussed science and exchanged ideas. This gave me a fine chance to appreciate his abilities; so when I was promoted to Laboratory management I asked for responsibility for him. Thus began a close technical and personal association. I did two things for him. I was the first in his management to discuss chemistry with him, not only his work but also unusual science that either of us encountered. That stimulated us both and broadened his perspective. Second, I urged him to publish more. This led to twenty papers while we were together compared with four in the previous 24 years. Despite the change of focus toward science and publication, the flow of patents continued unabated, twenty-nine for the 19 years we were together compared with thirty-six for the earlier 24 years. This is testimony to what a good scientist can achieve if given encouragement and freedom to pursue his own ideas.

For a while Pedersen continued his studies of the pro-oxidant activity of metal chelates and of oxidation and peroxides [11–13]. Then came new developments. The first resulted from curiosity about the redox behavior of iron and cobalt phthalocyanines which did not appear to be shared by the copper, nickel or metal-free types. On closer scrutiny, he found that the purportedly stable types readily formed solvent-soluble adducts **3** with peroxides. Subsequent reduction, chemical or photochemical, restored the original pigments **4** [14]. He also showed that these reactions could be employed to produce brilliant fast blue and green colors on textiles or to generate photographic images. Similar conversions could be effected by careful treatment with chlorine in methanol to form an adduct from which the pigment was easily reconstituted.

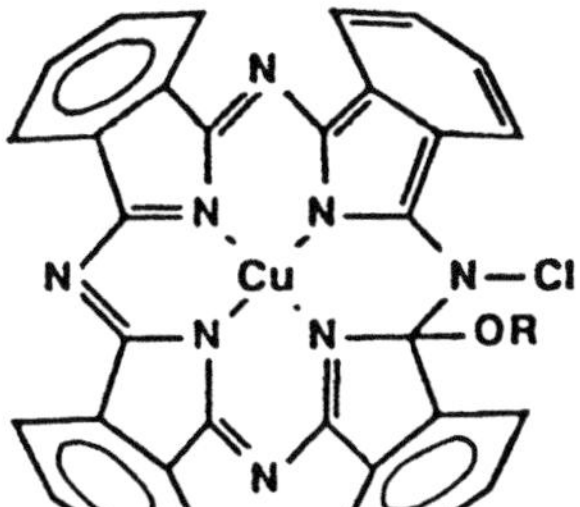

3.

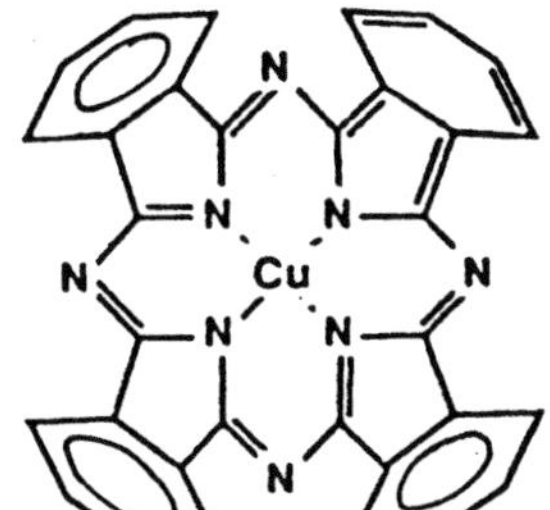

4.

Another interesting development arose because he wondered what happened to the *N,N'*-disubstituted *p*-phenylene diamine antioxidants when consumed during their protection of a substrate. This led to a new class of *N,N'*-disubstituted *p*-quinonediimine-*N,N'*-dioxides **5** [15]. He prepared many of these substances and found they had unusual properties. In their capacities of polymerization inhibitors and antioxidants they appear to react with hydrocarbon radicals and not the peroxy intermediates which generate them. Also when exposed to light (300–450 nm) they decompose rapidly and quantitatively into *N*-substituted *p*-quinoneimine-*N*-oxides **6** and azo compounds [16]. Further irradiation with light of 20 nm shorter wavelength converts these *N*-oxides to the *p*-quinone and azo compounds. Mixtures of azo compounds are obtained from unsymmetric *N,N'*-dioxides.

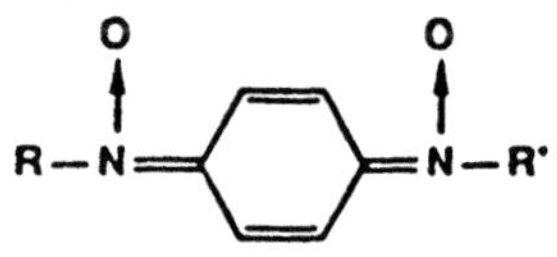

5.

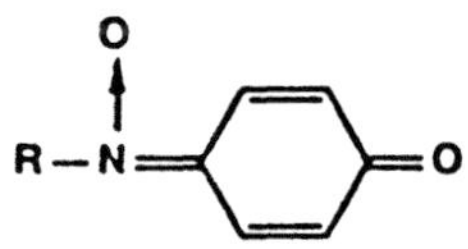

6.

One afternoon in December 1951, Charlie came bustling into my office all excited. He had just read an interesting note in *Nature* about Kealy and Pauson's discovery of an unusual soluble dicyclopentadienyl iron complex later called ferrocene. He felt that they had not fully grasped the nature of the substance nor the significance of their discovery. He described his own conception of what they had, which later proved to be correct. On the strength of Pedersen's enthusiasm, I invited Pauson to visit our Laboratory to discuss his work and there we made arrangements to help Pauson file an appropriate patent case. We also secured a license so that we could use the invention in some of the areas Charlie visualized as of practical interest.

He went right to work on ferrocene and quickly made two important discoveries. He found that dicyclopentadienyl iron compounds were excellent combustion control agents, specially valuable because they were both hydrocarbon soluble and volatile [17]. Of most importance, ferrocene functioned as an anti-knock agent when added in very small amounts to gasoline. This was very exciting as these compounds were the first really different highly effective anti-knock agents since the discovery of tetraethyl lead and promised great advantages because they would avoid the toxicity problems encountered with lead derivatives. Charlie quickly filed a patent application on the discovery (March 19, 1952) [18]. In view of its importance, the application was immediately placed under a U.S. government secrecy order. This was rescinded in 1960 but by then larger scale practical tests had showed that the ferric oxide formed on combustion of ferrocene–gasoline compositions caused excessive engine wear. Nevertheless, the patent proved to be valuable as we were able to trade rights for a patent covering a new method for handling sodium–lead alloy which was worth over a million dollars to us.

His other discovery also related to the extraordinary catalytic effect of ferrocene in promoting combustion. Small amounts (0.1%) stop the formation of soot or smoke in a hydrocarbon flame by catalysis of combustion. When a hydrocarbon fuel containing ferrocene was mixed with fuming nitric acid the mixture ignited spontaneously [19]. Such a hypergolic mixture was of potential value as a rocket fuel so it was of great interest to the U.S. government.

In 1957, our Organic Chemicals Department was divided to form a new Elastomer Chemicals Department with Carter and me as Research Directors. With our strong encouragement Pedersen chose to join us. Although we never tried to influence his programs, he somehow seemed to feel he should work on subjects he deemed of interest to us. This included elastomers, elastomer chemicals, and isocyanate and urethane chemistry. He soon made an important contribution in a well studied area. We were trying to find ways of increasing the use of neoprene in light colored stocks, particularly for tire side walls where its ozone and age resistance are outstanding. But, since white neoprene stocks discolored on light exposure, its use was largely restricted to black stocks. Pedersen showed by the simplest of experiments (exposing neoprene to light in the presence and absence of water and air) that the goal was attainable, but with practical difficulties. On photolysis, neoprene liberates hydrogen chloride which in turn catalyzes color-forming condensation reactions involving the aldehyde and ketone products of oxidation. If water or bases are present to dilute or neutralize the acid, no color is formed. Unfortunately, impractically large amounts of basic oxides would be required to protect the neoprene completely over the life of a tire.

His next endeavors involved the chemistry of peroxycarbamates **7** which he synthesized by reaction of hydroperoxides with isocyanates or carbamoyl chlorides. They were found to decompose smoothly both homolytically and by intramolecular

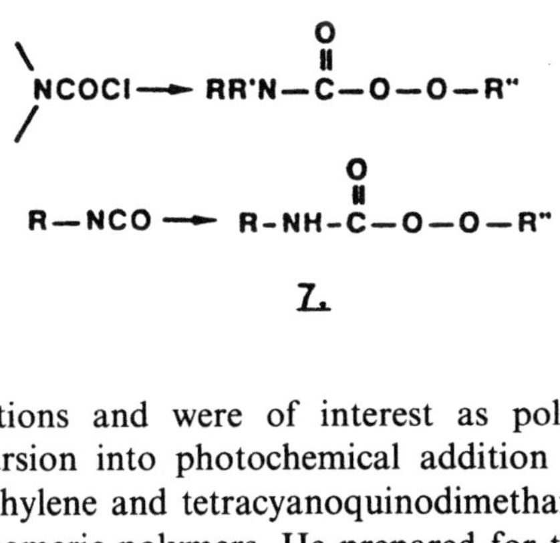

7

reactions and were of interest as polymerisation catalysts [20]. After a brief excursion into photochemical addition reactions of tetrahydrofuran to tetracyanoethylene and tetracyanoquinodimethan [21] **8**, **9**, he shifted his attention toward elastomeric polymers. He prepared for the first time novel functionally substituted ethylene/propylene copolymers with pendant amino and sulfide groups which had many very interesting properties [22–23].

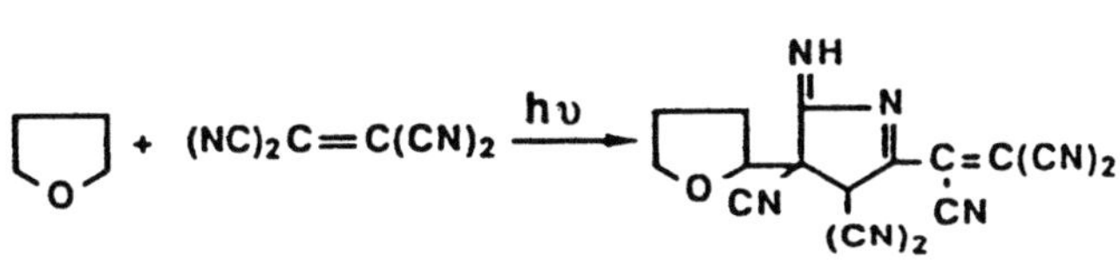

8

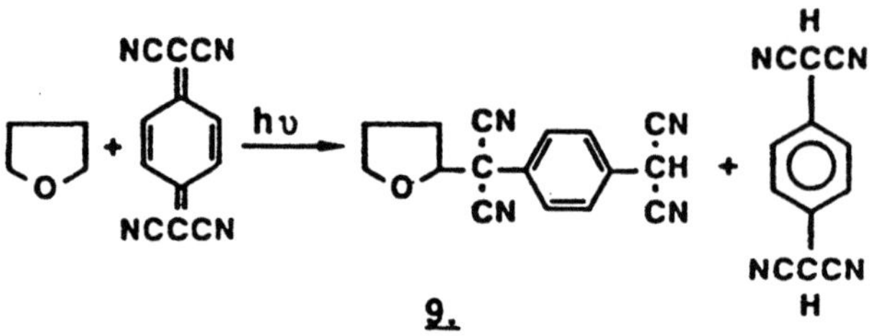

9.

Despite the obvious practical possibilities of these studies, I felt that this simple chemistry was not the best use of Pedersen's unusual capabilities, so I encouraged him to return to his beloved coordination chemistry, still of great interest to him. To convince him that there were good subjects of importance to us in that field I suggested that the chemistry of transition elements like vanadium might be worthy of attention. We were then using vanadium in a coordination catalyst system to make a very important polymer whose microstructure depended on the vanadium ligands. When polymerization was completed, we used an acid extraction to remove the vanadium residues from the elastomer since vanadium catalyzed its oxidative degradation. So I asked him if he would like to study vanadium to see if he could develop an understanding of why it was so fine a polymerization catalyst in one oxidation state and so strong an oxidation catalyst in another. I also wondered whether he could think of any substances from his earlier work with metal deactivators which might deactivate the vanadium so that the costly extraction would not be necessary.

Pedersen jumped in enthusiastically. He suggested that it would be interesting to study polydentate oxy structures derived from various polyhydric phenols since these had not been examined. A few months later he told me he enjoyed the catalyst ligand chemistry but that the various complexing agents all *enhanced* the pro-oxidative behavior of vanadium except for tetrasalicylal pentaerythrityltetraamine which was not practical for manufacture. He also planned synthesis of a new pentadentate ligand for possible use as a catalyst [24]. (It later turned out to be ineffective.)

4. Crown Ether Discovery

This is when the lightening struck! From one of his reactions he isolated the first of the *Crown Ethers*, which he later called dibenzo-18-crown-6, **10**. His first person account of his discovery and his reaction thereto written for the 1987 Symposium

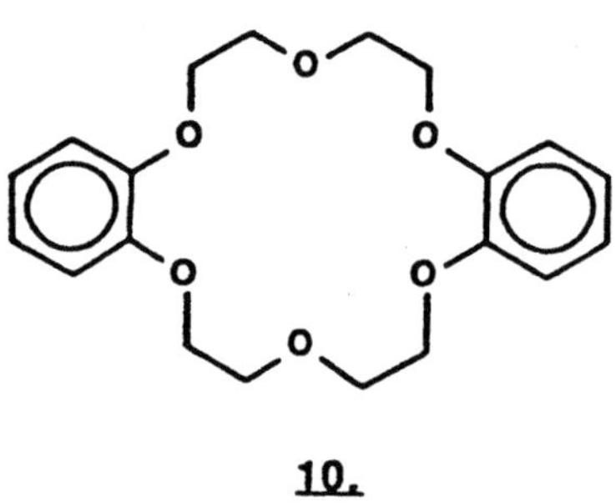

10.

on Macrocyclic Chemistry commemorating the 20th anniversary of his first publication on 'crown' ethers is found elsewhere in this volume [25]. His account is truly a remarkable story! Intuitively and almost immediately he recognized the significance of his find and he pursued it the way a bird dog goes after his game.

Pedersen first described his research at the International Conference on Coordination Chemistry in Nikko, Japan in September 1967 (see paper in this issue by D. Busch) and in his beautiful 1967 note and paper in the *Journal of the American Chemical Society* [26] which describes about 50 crown ethers with from 9- to 60-membered oligoether rings. This was an amazing accomplishment for a single chemist assisted by one technician. Charlie's description of events leading to the discovery has been published [27, 28].

Reaction to his work was fast and enthusiastic. *Angewandte Chemie* noted it in the hot news section of their July 7, 1967 edition and in early 1968 it won the American Chemical Society Delaware Section award for best paper of 1967. Now everyone wanted a talk or paper from him! Knowing full well the unusual nature of his discovery Pedersen spent the rest of his career with du Pont on the chemistry of the crown ethers until he retired in 1969. After that, at the invitation of the late Sir Ronald Nyholm, he spent three months collaborating with Professor Mary Truter (University College, London). Their work was concerned mainly with the crystal structures of crown ethers and their metal derivatives [29].

His own further studies greatly broadened the scope of the discovery and included syntheses of new crown ethers, factors governing inclusion of other metal ions, products with sulfur and nitrogen in the rings, extensions of the preparation and characterization of crystalline salt complexes, ionic complexes, and unusual thiourea complexes. He made many suggestions for uses of the crown ethers, especially for catalysis and separations, which have proved to be prophetic. He was fully aware of the potential of these substances for carrying various metal ions into organic media to effect reactions. They are described in six papers and ten patents [30–37] and in the publications of some of his colleagues. In the course of this work he also studied reactions of crown ethers as organic chemicals and their utility as catalysts and as agents for effecting ionic reactions in organic media.

While the complexation of transition-metal cations is a commonplace phenomenon, practically no alkali-metal ion complexes were known before the advent of the crown compounds. Thus, the discovery of a large variety of easily synthesized alkali-cation-complexing agents represented a real breakthrough. This complexing property is far more than a laboratory curiosity; in fact, it is the key to the importance of the crown ethers. The complexes are soluble in a variety of organic solvents in which the uncomplexed salts are at best sparingly soluble. Moreover, the inter-ionic forces are considerably reduced by the oligoether ring which surrounds the cation, so that there is far less ion pairing and 'naked cations' are available in much higher concentration for reaction.

5. Impact of Crown Ether Discovery on the Scientific World

The major impact of Pedersen's discovery is shown by the frequency of citation of his crown ether papers and even more strikingly by the many new directions of research it has opened up in which many groups have been engaged. These include:

cryptates, J. M. Lehn; host–guest chemistry, D. J. Cram; naked-ion chemistry, C. L. Liotta; phase-transfer catalysis, C. L. Liotta; solubilization of alkali metals, J. L. Dye; ion pairing in non-polar solvents, J. Šmid; X-ray structure of complexes, M. J. Truter; crown ethers as ionphores, G. Eisenman; and many others.

As Professor D. J. Cram described it [38]:

Pedersen's discoveries illustrate organic systhesis and the organic chemist at their best. Using a simple Williamson ether synthetic procedure and simple starting materials, he synthesized in good yields a new family of cyclic polyethers. He noted that these 'crown compounds' complex and lipophilize metal guest cations whose diameters are similar to the diameters of the host crown compounds. He attributed the good yields of the crown compounds in part to templating effects during the ring closures. He varied the ring sizes over a wide range by appropriate selection of starting materials and the orders in which the molecular parts are assembled. The anions of salts lipophilized by the crown compounds have in many cases never been brought into non-polar organic media before, and show greatly enhanced reactivity over what they have shown in polar media.

Charles Pedersen's discovery of the crown compounds has had a major impact in a variety of investigative fields: inorganic chemistry (the crowns are ligand assemblies); organic chemistry (the crowns are commercial synthetic aids); physical organic chemistry (crown compounds greatly affect ion pair organization); biological chemistry (crowns are important ion transport agents); and analytical chemistry (analytic reagents and ion selective electrodes are based on crown compounds). Pedersen's discoveries were not the usual result of the efforts of a group of competitors that he led or beat into print. Nor are they those of a large group of scientists under a director; they were his alone.

Professor D. J. Clam added:

He had no competition – he was alone – he was original – he did something important – he knew it was important – and he knew what to do with his discovery... It is my belief that Pedersen's demonstration of the feasibility for synthesizing multiheteromacrocycles is leading to the development of a whole new synthetic field of organic host–guest chemistry.

The man who did all this was not after riches, glory, or status. He loved the thrills and problems of research and in his heart he most wanted his work to be worthwhile and so recognized. We could not find so deserving nor more modest and unassuming a person to honor than my dear friend and associate Charles Pedersen.

References

1. F. B. Downing, A. E. Parmalee and C. J. Pedersen: *U.S. Patent* 2,004,160, June 11, 1935.
2. F. B. Downing and C. J. Pedersen: *U.S. Patent* 2,008,753, July 23, 1935; U.S. Patent 2,087,103, July 13, 1937.
3. R. G. Clarkson and C. J. Pedersen: *U.S. Patent* 2,054,282, Sept. 15, 1936.
4. L. Spiegler and C. J. Pedersen: *U.S. Patent* 2,087,098, July 13, 1937.
5. F. B. Downing and C. J. Pedersen: *U.S. Patent* 2,121,397, June 21, 1938.
6. F. B. Downing and C. J. Pedersen: *U.S. Patent* 2,181,121, Nov. 28, 1939.
7. C. J. Pedersen: *Oil Gas Journal* p. 97, July 27, 1939.
8. C. J. Pedersen: *Ind. Eng. Chem.* **41**, 824 (1949).
9. C. J. Pedersen: 'Pro-oxidant Catalytic Activity of Metal Chelates', *Delaware Chemical Symposium*, Dec. 1, 1948.
10. C. J. Pedersen: 'Mechanism of Decomposition of Perbenzoic Acid compared with Dibenzoyl Peroxide', *Delware Chemical Symposium*, Jan. 21, 1950.
11. C. J. Pedersen: 'Antioxidants', *Encyclopedia Britanica*, 1953.
12. C. J. Pedersen, *Symposium on Chelate Chemistry*, Centenary celebration of Brooklyn Polytechnic Institute, New York, N.Y. published in *Advances in Chelate Chemistry*, p. 113 (1954).

13. C. J. Pedersen: *J. Org. Chem.* **22**, 127 (1957); C. J. Pedersen: *U.S. Patent* 2,662,895, December 15, 1953; C. J. Pedersen: *U.S. Patent* 2,662,896, December 15, 1953; C. J. Pedersen: *U.S. Patent* 2,662,897, December 15, 1953; C. J. Pedersen: *U.S. Patent* 2,681,347, June 15, 1954; C. J. Pedersen: *U.S. Patent* 2,741,531, May 10, 1956; C. J. Pedersen: *U.S. Patent* 2,831,805, April 22, 1958.
14. C. J. Pedersen: *Ind. Eng. Chem.* **48**, 1881 (1956).
15. C. J. Pedersen: *J. Am. Chem. Soc.* **79**, 2295 (1957).
16. C. J. Pedersen: *J. Am. Chem. Soc.* **79**, 5014 (1957); C. J. Pedersen: *U.S. Patent* 2,681,918, June 22, 1954; C. J. Pedersen: *U.S. Patent* 2,741,625, April 10, 1956; C. J. Pedersen: *U.S. Patent* 2,831,805, April 22, 1958.
17. C. J. Pedersen: *U.S. Patent* 2,867,516, January 6, 1959.
18. C. J. Pedersen: *U.S. Patent* 3.341.311, September 12, 1967.
19. C. J. Pedersen: *U.S. Patent* 3.038,299, June 12, 1962; C. J. Pedersen: *U.S. Patent*, 3,038,300, June 12, 1962.
20. C. J. Pedersen: *J. Org. Chem.* **23**, 252 (1958); C. J. Pedersen: *J. Org. Chem.* **23**, 255 (1958).
21. J. Diekamnn and C. J. Pedersen: *J. Org. Chem.* **28**, 2879 (1963).
22. C. J. Pedersen: *U.S. Patent* 3,232,914, February 1, 1966.
23. C. J. Pedersen: *U.S. Patent* 3.320,214, May 16, 1967.
24. C. J. Pedersen: *U.S. Patent* 3.361,778, January 2, 1968.
25. C. J. Pedersen: *J. Incl. Phenom. Mol. Recognit.* **12,** 7 (this issue).
26. C. J. Pedersen: *J. Am. Chem. Soc.* **89**, 2495 (1967); C. J. Pedersen: *J. Am. Chem. Soc.* **89**, 7017 (1967).
27. C. J. Pedersen: *Aldrichimica Acta* **1**(4), 1 (1971).
28. C. J. Pedersen: in *Synthetic Multidentate Macrocyclic Compounds*, Eds. R. M. Izatt and J. J. Christensen, Academic Press, New York, 1978, Ch. 1.
29. M. R. Truter and C. J. Pedersen: *Endeavor* **111**, 142, (1971).
30. C. J. Pedersen: *Fed. Proc., Fed. Am. Soc. Exp. Biol.* **27**, 1305 (1968).
31. C. J. Pedersen: *J. Am. Chem. Soc.* **92**, 386 (1970).
32. C. J. Pedersen: *J. Am. Chem. Soc.* **92**, 391 (1970).
33. C. J. Pedersen: *J. Org. Chem.* **36**, 254 (1971).
34. C. J. Pedersen: *J. Org. Chem.* **36**, 1690 (1971).
35. C. J. Pedersen and H. K. Frensdorff: *Angew. Chem.* **84**, 16 (1972); C. J. Pedersen and H. K. Frensdorff: *Angew. Chem. (Int. Ed. Engl.)* **11**, 16 (1972).
36. C. J. Pedersen: *Org. Syn.* **52**, 66 (1972).
37. C. J. Pedersen: *U.S. Patent* 3,562,295, February 9, 1971; C. J. Pedersen: *U.S. Patent* 3,622,577, November 23, 1971; C. J. Pedersen: *U.S. Patent* 3,686,225, August 22, 1972; C. J. Pedersen: *U.S. Patent* 3,687,978, August 19, 1972; C. J. Pedersen and M. Bromels: *U.S. Patent* 3,847,949, November 12, 1974; C. J. Pedersen: *U.S. Patent* 3,856,813, December 24, 1974; C. J. Pedersen: *U.S. Patent* 3,873,569, March 25, 1975; C. J. Pedersen: *U.S. Patent* 3,987,061, October 19, 1976; C. J. Pedersen: *U.S. Patent* 3,998,838, December 21, 1976; C. J. Pedersen and M. Bromels: *U.S. Patent* 4,031,111, June 21, 1977.
38. D. J. Cram, personal communication.

Journal of Inclusion Phenomena and Molecular Recognition in Chemistry **12**: 23–53, 1992.

Free Energy Perturbation Simulations of Cation Binding to Valinomycin#

GEORGE EISENMAN*
Department of Physiology, UCLA Medical School, Los Angeles, CA 90024-1751, U.S.A.

OSVALDO ALVAREZ
Department of Biology, Faculty of Sciences, University of Chile, Santiago, Chile.

JOHAN AQVIST
Department of Molecular Biology, Uppsala Biomedical Center, S-75124 Uppsala, Sweden.

(Received: 27 September 1990; in final form: 11 December 1990)

Abstract. Experimental values of the free energies of cation binding to the cyclic depsipeptide molecule, valinomycin, obtained from Pedersen-type salt extraction measurements, provide data against which it is possible to test the adequacy of the procedures and force fields of the molecular dynamics algorithms, MOLARIS and GROMOS. These data are then used to assess appropriate values for the partial charges of the ester carbonyl oxygen and carbon. Valinomycin was chosen because it has only one kind of ion-binding ligand and because the cation is sufficiently enfolded by the molecule in the ion-complexes that the overall size and shape of the complex is virtually the same regardless of the species of cation bound. For such an 'isosteric complex', the experimentally measured selectivities are sufficiently similar in a wide variety of solvent environments that the *differences* in free energies measured between the different ion-valinomycin complexes by two-phase salt extraction experiments into dichloromethane can be taken as equivalent to the differences in free energies *in vacuo*. These *differences* were therefore compared with those computed for ion-valinomycin complexation *in vacuo* by Free Energy Perturbation/Molecular Dynamics (FEP/MD) simulations using the MOLARIS and GROMOS programs. Starting with a set of Lennard-Jones 6–12 parameters for the monovalent cations assessed for aqueous solution we explored the effect of varying the partial charges of the ester carbonyl ligands on binding free energy differences (i.e. the selectivity) among Na, K, Rb, and Cs. The computed selectivity was found to depend strongly on the value of partial charge, following a typical 'Eisenman Selectivity Pattern' in which the correct selectivity sequence and magnitude occurred only over a very narrow range of partial charge (around 0.33 and 0.6 for the standard carbonyls of MOLARIS and GROMOS, respectively). Using MOLARIS we explored the effect of varying the size of the ester carbonyl ligands by comparing the standard carbonyl of MOLARIS with the somewhat smaller carbonyls of GROMOS and found an equally satisfactory ability to reproduce the experimental data with a partial charge value of 0.41. These results validate the use of both the MOLARIS and GROMOS force fields as starting points for quantitative calculations of ion-binding in more complex molecules (e.g., ion-binding sites and channels in proteins).

Key words. Molecular dynamics, free energy perturbation simulations, ion-binding ion-complexation, ion-selectivity, ion-carbonyl bond angles and distances, valinomycin selectivity, MOLARIS, GROMOS, Pedersen two-phase salt-extraction, Eisenman selectivity sequences, Lennard-Jones parameters for ions, isostericity.

This paper is dedicated to the memory of the late Dr C. J. Pedersen.
* Author for correspondence.

1. Introduction

Charles Pedersen discovered the first synthetic neutral compounds, the crown ethers, that formed stable complexes with alkali metal cations [1]. The importance of these molecules as model compounds for ion carriers in membranes, and their fundamental similarity to the neutral antibiotic ion carriers like valinomycin and monactin was immediately recognized [2]. Indeed, the first author had the pleasure in 1968 of inviting Pedersen to present his work to the American Physiological Society Symposium on 'Biological and Artificial Membranes' [3]. Pedersen's pioneering work provided a crucial system in which to confront classical chemical concepts and measurements with those of the newly opened field of ion permeation in lipid bilayer membranes. It played a central role in establishing a rigorous understanding of the chemistry underlying the effects of macrocyclic carrier molecules on bilayer membranes [4–6] and ultimately proved to be central in clarifying a number of fundamental assumptions regarding the stoichiometry of membrane carriers [7]. Understanding how ions selectively interact with his molecules is still fundamental to understanding ion binding to the peptide backbone, a central feature of ion binding sites in protein enzymes and channels.

It seems particularly appropriate to dedicate the present paper to Charlie Pedersen, since at the aforementioned Symposium Pedersen described the two-phase salt extraction procedure for measuring free energies of ion binding [3] which will be used here, some 22 years later, to measure experimental free energies of ion-binding to valinomycin for comparison with those computed by free energy perturbation (FEP) simulations. These experimental data enable a test to be made of the validity of the procedures and force fields of molecular dynamics for simulating selective ion binding to a peptide-like molecule. These data also provide a basis for assigning appropriate values for the partial charges of the carbonyl group.

The present contribution represents a small step directed towards a central problem in molecular biophysics, namely how to assess the energetics of specific interactions between ions and proteins. These energies determine the affinities and rates of ion binding, unbinding and permeation for protein channels as well as for ion-activated enzymes. Recent advances in computational power available on laboratory minisupercomputers together with great progress in developing appropriate force fields and algorithms for the computational chemistry of proteins [8–12] offer the promise of being able to deduce details of channel permeation from structural knowledge, and vice versa. Examples of channels for which calculations have been performed are the gramicidin channel [13, and references therein], the acetylcholine receptor channel [14–16] and certain channel-like structures in the protein coat of icosahedral viruses [17–18]. Relevant calculations of the affinity of crown ethers and other small ion-binding molecules [19–21], of specific ion binding to proteins [22], and to specific ion effects on catalysis [23] are also encouragingly reasonable.

Despite the apparent adequacy of the force fields used, as judged by ability to generate correct protein structures [24–27], the use of molecular dynamics simulations to calculate the specific interactions of ions with protein binding sites is still in an early stage. In particular, no algorithm has yet been tested for any molecule whose ion-binding selectivity and three-dimensional structure are *both* known. Indeed, only very recently has a reliable set of Lennard-Jones 6–12 ion-water interaction parameters been derived for the Group Ia and IIa cations from free energy simulations [28] in bulk water. This set of parameters was derived by performing free energy perturbation simulations at 298 K of the process of charging an ion in aqueous solution. The model system consisted of an ion, described as a point charge imbedded in Lennard-Jones spheres characterized by the A and B parameters, surrounded by a surface constrained drop of explicit (flexible or rigid) SPC water molecules. The oxygen Lennard-Jones A and B parameters were 793.3 and 25.01. The computations produced the observed absolute hydration free energies and the radial distribution function of water around the ions. The A and B parameters for the monovalent cations are listed in Section 4.2 'General Procedure'. These parameters are the starting point for the present study of ion binding to a prototype ion-binding ligand, the ester carbonyl group of the backbone of the cyclic depsipeptide ion-carrying molecule, valinomycin.

Such carbonyls are important ligands in many cation binding sites in proteins [29–31]. Indeed, backbone amide carbonyls are the sole ligands in the five-fold ion binding site of the satellite tobacco necrosis virus [32, 33] and the gramicidin channel [34, 35]. We show here that the experimental selectivity data and structure for valinomycin can indeed be described by molecular dynamics procedures such as the MOLARIS force field [9, 36, 37], provided the carbonyl partial charges are appropriately parametrized. Our results suggest that the partial charge parameters computed here can serve as a useful starting point for refining previous computations of free energies of ion interactions with the amide carbonyl ligands in more complex protein structures such as the Gramicidin channel [13] and the five-fold Ca binding site in the satellite tobacco necrosis virus [18, 22, 38]. When appropriate account is taken of variable stoichiometry, structure, and hydration, quantitative calculations for ion complexes with Pedersen's crown compounds should also become feasible.

This contribution has been adapted from a paper presented to the Royal Society of Chemistry Faraday Symposium No. 26 'Molecular Transport in Confined Regions and Membranes', University of Oxford, England, Dec. 17–18, 1990; and the analysis of structure and selectivity has been considerably extended.

2. Valinomycin Structure

The primary structure of valinomycin is

cyclo-(D-HYL-D-VAL-L-LAC-L-VAL-)$_3$

where D-DYL is D-hydroxyisovaleric acid, L-LAC is L-lactic acid, and D-VAL and L-VAL are D- and L- valine, respectively. The atom types, standard partial charges, and Lennard-Jones A and B parameters used in MOLARIS are given in Figure 1.

In low dielectric constant solvents, as well as in the crystalline cation complexes, the valinomycin molecule wraps around a cation to form a bracelet shaped structure in which the ion is liganded by six ester carbonyl oxygens and which is stabilized by six intramolecular H-bonds between all the —NH groups and the

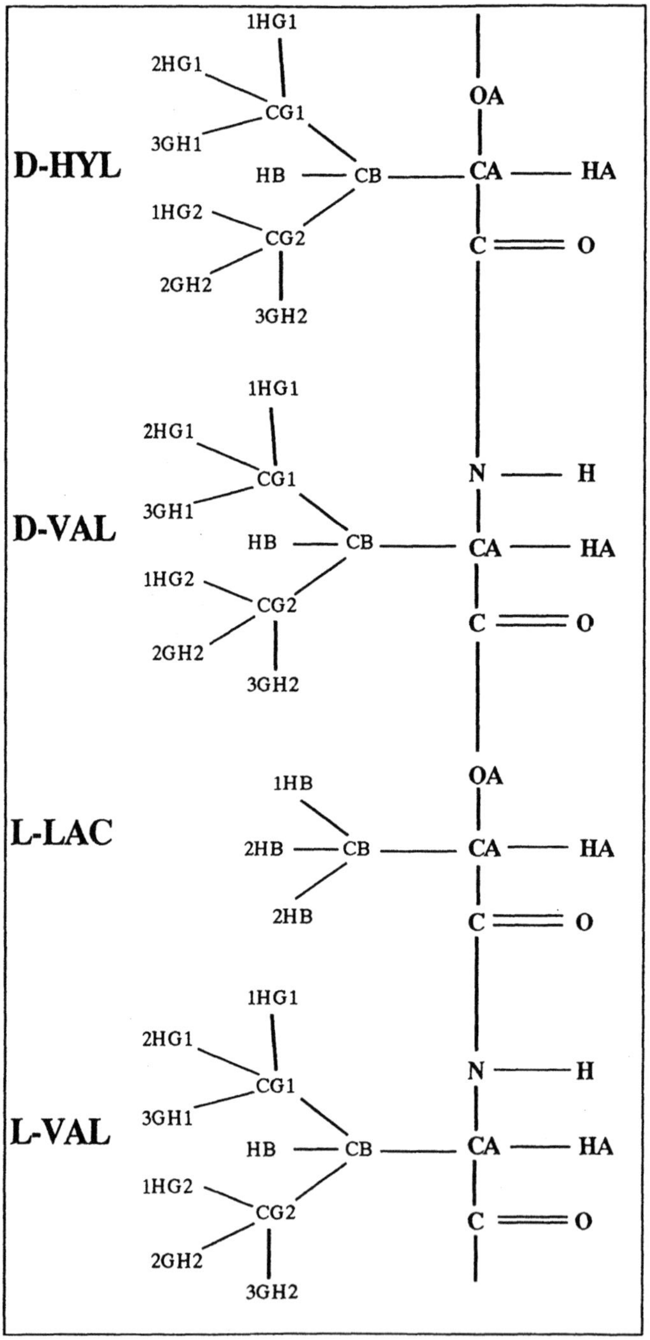

OA (Hyl and Lac)
Partial charge -0.400
A 793.30 B 25.00

CA (Hyl and Lac)
Partial charge 0.300
A 1956.00 B 32.00

HA (Hyl and Lac)
Partial charge 0.100
A 22.00 B 1.00

C
Partial charge 0.550
A 1956.00 B 32.00

O
Partial charge -0.550
A 793.30 B 25.00

N (VAL)
Partial charge -0.400
A 900.00 B 42.00

H (Val)
Partial charge 0.400
A 4.00 B 0.00

CA (Val)
Partial charge -0.097
A 1956.00 B 32.00

HA (Val)
Partial charge 0.097
A 22.00 B 1.00

CB (Hyl, Val)
Partial charge -0.097
A 1956.00 B 32.00

CB(Lac) CG1, CG2
Partial charge -0.291
A 1956.00 B 32.00

1HB, 2HB, 3HB, 1HG1, 2HG1, 3HG1, 1HG2, 2HG2, 3HG2
Partial charge 0.097
A 22.00 B 1.00

Fig. 1. Valinomycin chemical structure and atom types. The Lennard-Jones 6–12 A and B parameters and partial charges used in Amino88 amino acid library for MOLARIS are indicated at the right.

amide carbonyl oxygens (see Ovchinnikov *et al.* [39] for review of the extensive conformation studies). One face of the structure is formed by D-HYL, the other by L-LAC. The three-dimensional structure of the K complex in the bis-(Valinomycin) potassium tri-iodide penta-iodide crystal has been determined (including hydrogens by Neupert-Laves and Dobler [40]) and is available as the Cambridge Data Bank structure VALINK. Structures without H atoms are also available for the Cs picrate complex and the Na picrate monohydrate *m*-xylene solvate complex (Cambridge codes: DOWDAU [41] and BINFIN [42], respectively).

The crystallographic coordinates of the three-dimensional structure of VALINK were converted using the ability of the Biodesign BIOGRAF program to read a Cambridge format file, make the connections of the atom using geometrical rules, and create an intermediate file in which the atoms are segregated by type, rather than by residue, but in which their connections are given explicitly. To make the final Brookhaven format file for input to MOLARIS, the identity of each atom in the intermediate file was deduced from the connections, the atoms were grouped by residues and the ring was closed by hand-connecting the carbonyl carbon of VAL-12 with the hydroxyl oxygen of HYL-1. For computations all the iodide data were deleted.

3. Calculation of Free Energies for Binding Ions to Valinomycin *in vacuo*

Well-defined free energy *differences* among the alkali metal cations in their interactions with valinomycin can be measured in several ways: (1) by two-phase salt-extraction equilibria [3–6]; (2) by ion complexation equilibria in a single phase [43]; (3) by effects on the electrical conductance and permeability of lipid bilayers [44, 5, 6]; and (4) by the electrical potentials of thick ion-selective electrodes [45]. The free energy differences are directly related to the ratios between species I^+ and J^+ of their two-phase equilibrium constants (K_I/K_J), their one-phase complexation constants (K_{Is}/K_{Js}), their membrane permeabilities (P_I/P_J), their membrane conductances (G_I/G_J), and their electrode selectivities (S_I/S_J).

3.1. THEORETICAL EXPECTATIONS FOR ISOTERIC COMPLEXES

The following approximate identity is theoretically expected [4–6] to relate to the above ratios for 'isosteric complexes' in which the overall size, shape, and charge distribution of the complex are the same regardless of the species of ion complexed:

$$K_I/K_J = K_{Is}/K_{Js} = P_I/P_J = G_I/G_J = S_I/S_J \qquad (1)$$

Equation (1) is exact for exactly isosteric complexes; and we will present evidence below that it holds surprisingly well for valinomycin. The results presented here show that valinomycin-ion complexes are exactly isosteric for K and all larger ions, and quite isosteric for Na. The importance of this is that the free energy difference calculated between any pair of these ions by any of these measurements should depend only on the energy of complex formation *in vacuo* relative to the experimentally known free energies of ionic hydration. This enables us to use experimentally measured selectivity data to check the energies calculated by molecular dynamics (MD) simulations using the free energy perturbation (FEP) procedure.

3.2. EXPERIMENTAL RESULTS

We summarize here some results of a study of ion-valinomycin complexation (Gomez-Lojero and Eisenman, to be published) which are more extensive and precise than those previously reported [5, 6]. Figure 2 shows an example of typical

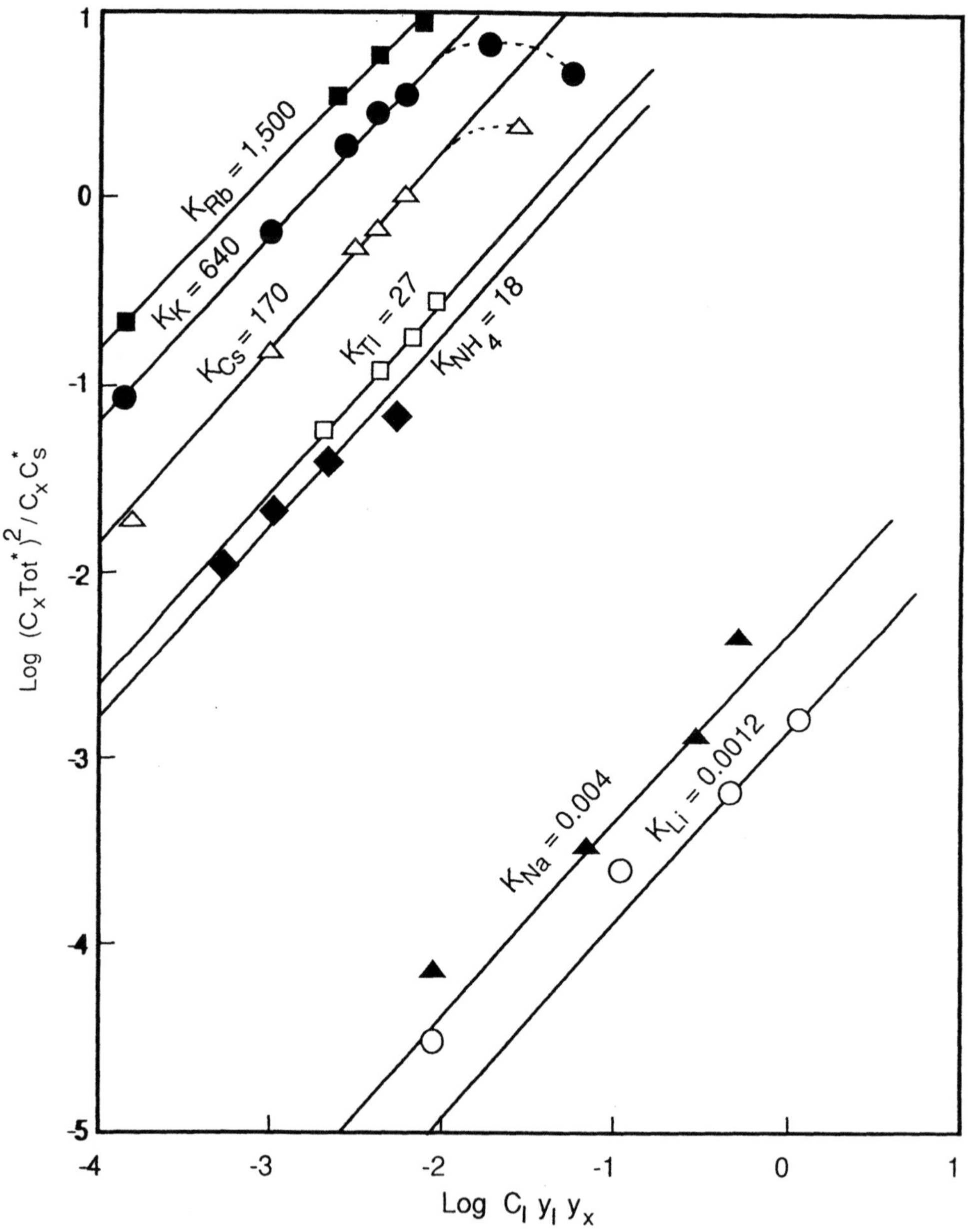

Fig. 2. Equilibrium extraction of 2,4-dinitrophenolates of the indicated cations by valinomycin into dichloromethane. The data are plotted in logarithmic form according to the expectations of Equation (50) of Ref. [4] in the manner of Figure 7 of that reference. The ordinate is dimensionless; the concentrations are in moles/liter and y's are molar activity coefficients. The slope of the lines = 1.

data for the extraction by valinomycin into dichloromethane for the 2,4-DNP salts and demonstrates that the behavior is consistent with the chemistry of forming 1:1 ion-complexes in this solvent uncomplicated by ion-pairing with the 2,4-DNP anion. The behavior is seen to be as regular as that previously found for the macrotetralide actins [4].

Figure 3 shows, for all cations except Li^+, that the equilibrium constant ratio K_I/K_J is independent of the anion, a finding which is only expected for isosteric complexes. This indicates that only the Li^+ complex is seriously non-isosteric, which is probably a consequence of the ability of the small Li ion to move away from the center of the complex toward its surface, as well as the possibility that a water molecule might be included in the complex. There is evidence in *n*-hexane that Li, and to a lesser extent Na, could move toward the surface of the complex under the influence of an adjacent anion to form a 'tight' ion pair. In *n*-hexane a blue shift in the picrate UV absorbance spectrum has been found for Li and Na (Gomez-Lojero and Eisenman, unpublished results) despite the absence of any such shift for the complexes of the larger ions, which are strictly isosteric even in *n*-hexane. Clearly, Li (and to a lesser extent, Na) are not as well screened from the picrate as the larger cations. For this reason, and because there is room within the valinomycin-Li complex for a water molecule, we will exclude Li from the quantitative comparisons.

Figure 4 illustrates the effect of varying the dielectric constant of the solvent (from a value of less than 2 in *n*-hexane to more than 9 in dichloromethane) on the salt extraction equilibrium constants. Despite the huge dependence of K_I on the dielectric constant, the independence of the ratios between ions is apparent from the parallelism of the lines. This behavior indicates that the selectivity (i.e. the difference in free energies) is completely independent of the dielectric constant.

This conclusion is further supported by the results of Figure 5, which show a very close agreement between the ratios of heterogeneous ion-complexation constants (K_I/K_K) measured by two-phase salt extraction equilibria into dichloromethane and the ratios of single-phase ion-complexation constants (K_{Is}/K_{Ks}) measured in methanol by Grell *et al.* [43]. Measurements for Li^+ complexation in methanol do not exist, and only for Na^+ is there any detectable deviation from the straight line expected for strict isostericity between these two extremely different solvent environments.

These results are summarized in Tables I and II. Table I compares the equilibrium constants measured for two-phase salt extraction into three different solvent mixtures: pure dichloromethane, a 64% *n*-hexane–36% dichloromethane mixture, and pure *n*-hexane. It can be seen in this table that, although the absolute value of the heterogeneous equilibrium constant decreases by a factor of 10 000 000 on lowering the dielectric constant (cf. the value of K_K of 52 000 in CH_2Cl_2 vs. 0.00044 in *n*-hexane), the ratios K_IK_{Cs} is remarkably species independent, as can be seen by the constancy of values of the K_K/K_{Cs} and K_{Rb}/K_{Cs} ratios, which vary by less than 25%.

Table II extracts the free energy differences for exchanging an ion for K^+ relative to the aqueous reference state. Column 1 summarizes the equilibrium constants reported previously for dichloromethane [5]; while column 2 summarizes the more accurate ones from the first row of Table I. These values are used to calculate the

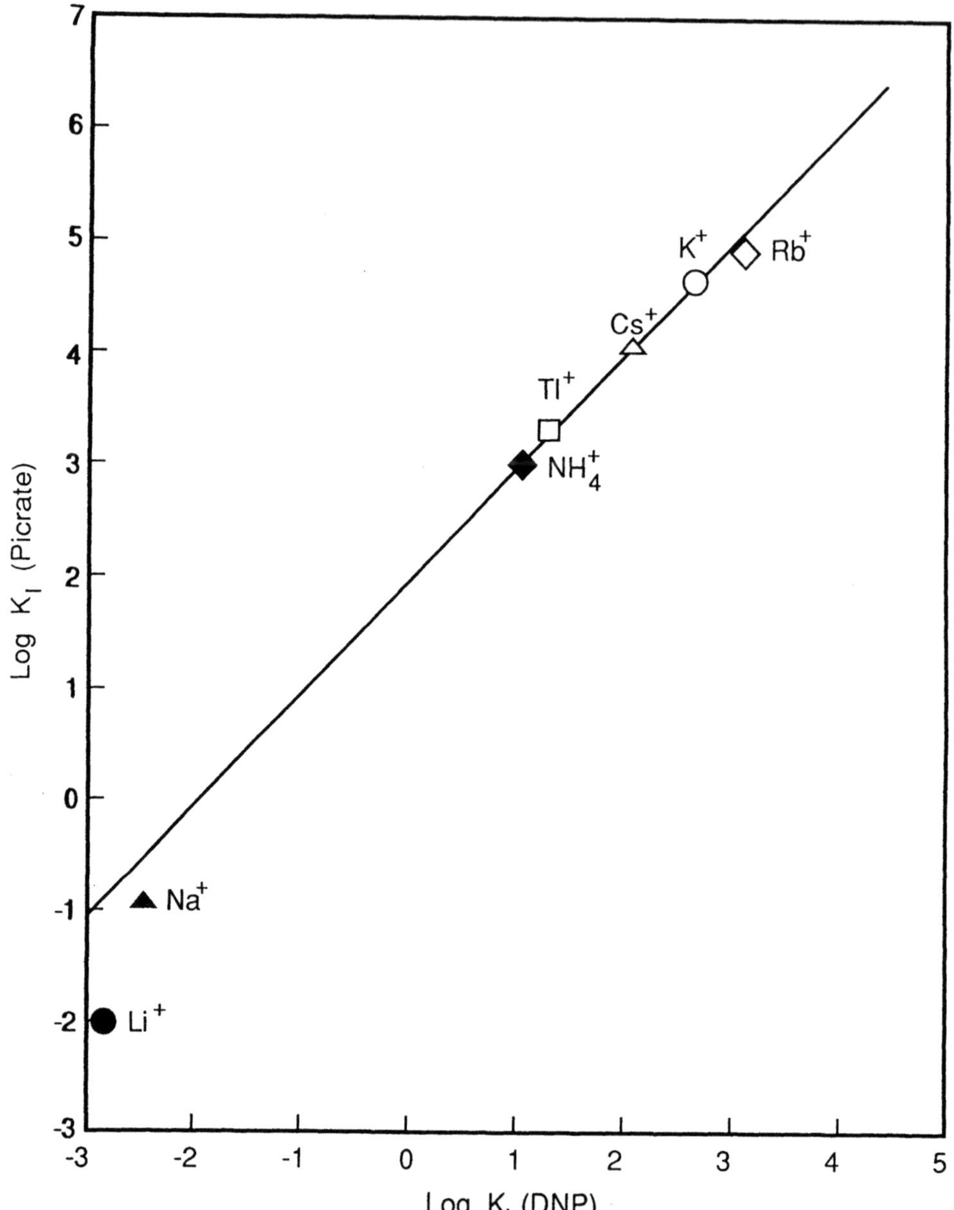

Fig. 3. Correlation between absolute values of salt extraction equilibrium constants for valinomycin in dichloromethane for picrates (ordinate) vs 2,4-dinitrophenolates (abscissa). Unpublished data of Gomez-Lojero and Eisenman. The line is drawn with a slope of 1.

ratios K_I/K_K in the third column. The last column summarizes the free energy differences corresponding to these. We will use these values as a measure of the difference in free energy of the reaction to form the complex *in vacuo* relative to the experimental free energies of hydration. The validity of this rests only on the assumption that the energy of interaction of the complex with any medium external to it is independent of the species of ion complexed (i.e. is independent of the

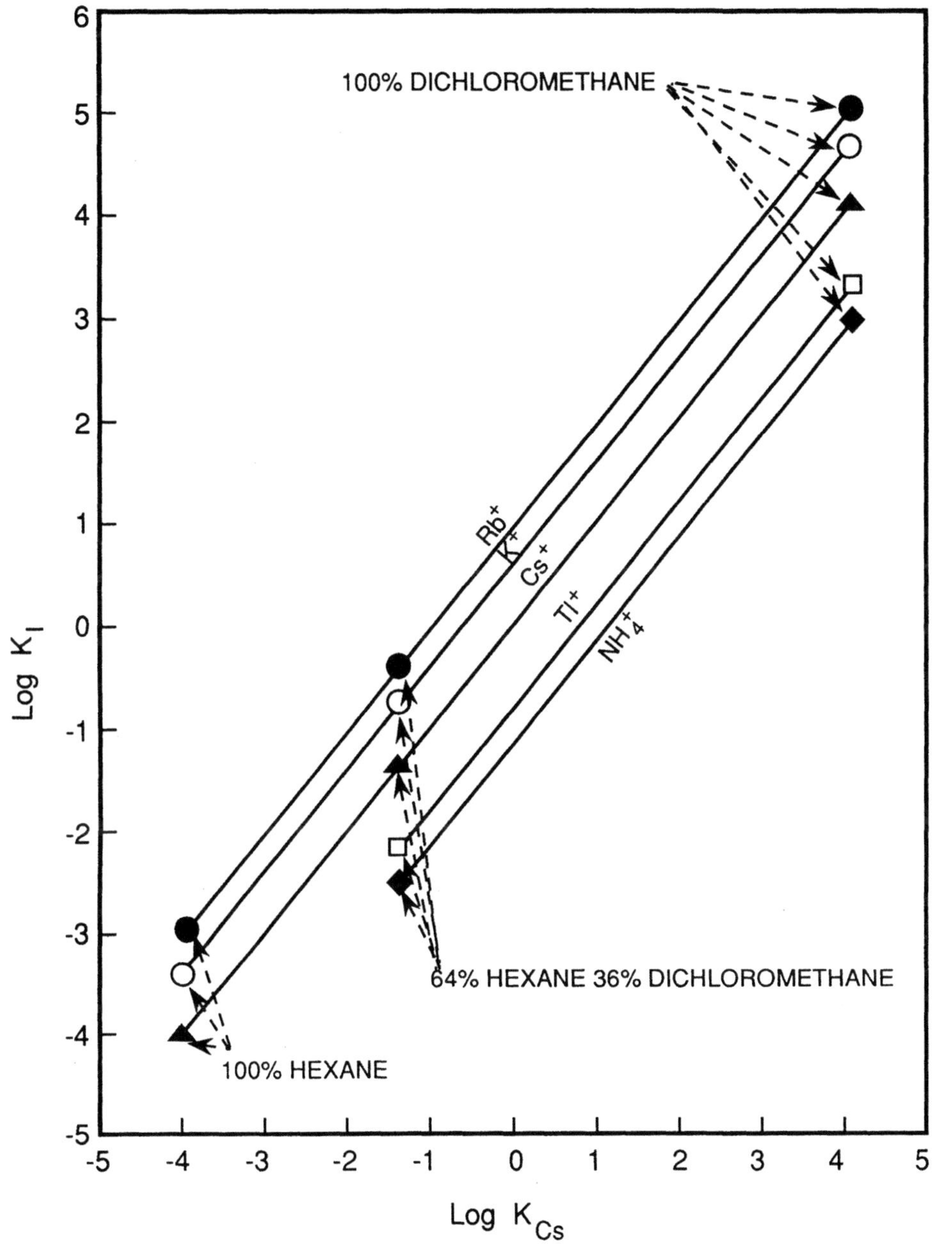

Fig. 4. Demonstration of the independence of the ratios of two-phase equilibrium constants of the nature and dielectric constant of the solvent as this is varied from hexane to dichloromethane. The effect of varying the solvent on the values of $\log_{10} K_I$ are plotted against $\log_{10} K_{Cs}$. Unpublished data of Gomez-Lojero and Eisenman.

dielectric constant of the solvent (and even of its chemical nature). This is so because when computing the free energy differences, the conformations of the uncomplexed valinomycin cancel out, as do the energies of interaction of isosteric complexes with the medium [6].

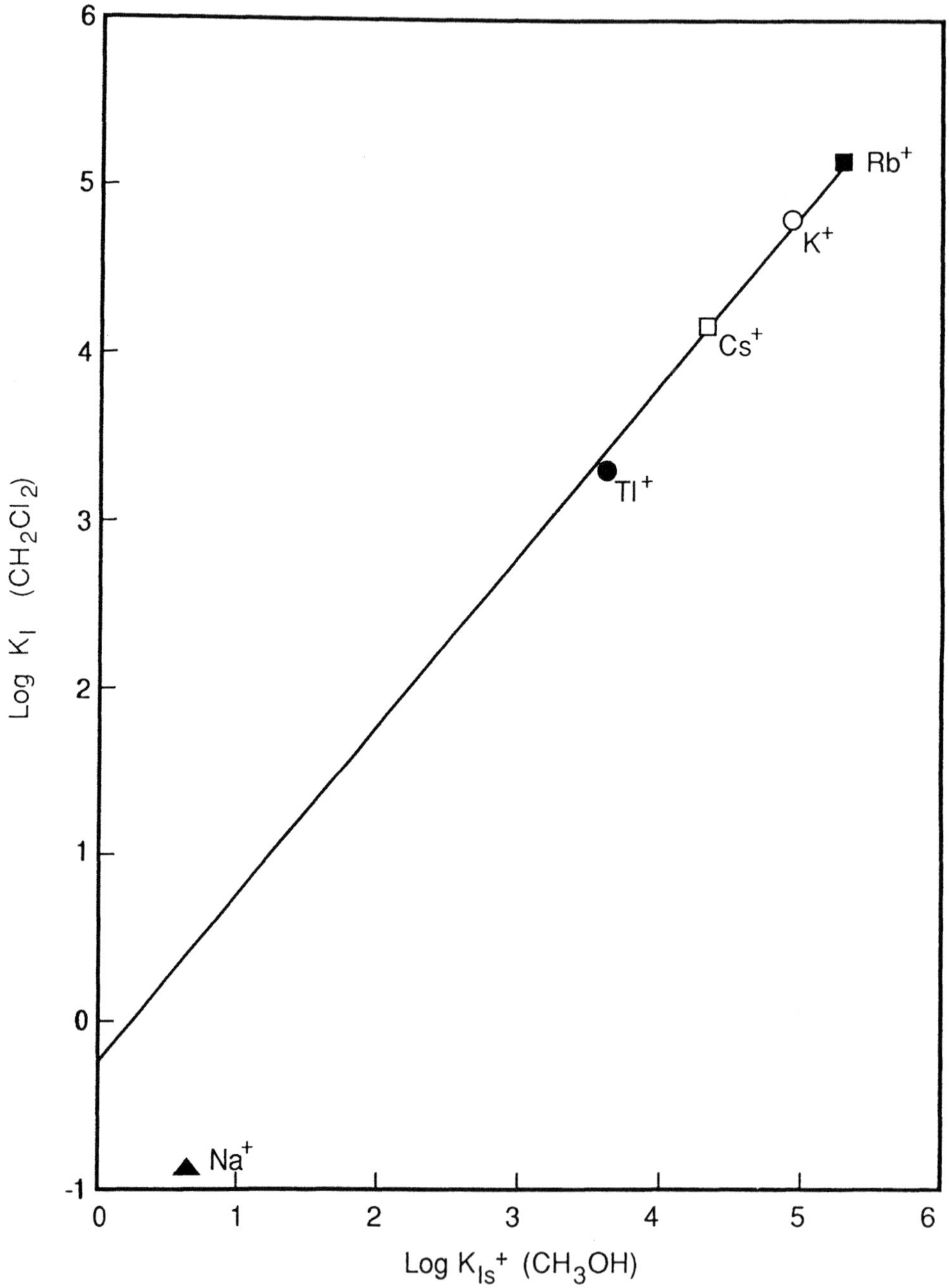

Fig. 5. Correlation between salt extraction constants into dichloromethane (ordinate) from Table I and stability constants in methanol (abscissa) [41]. Only Na^+ deviates from the line of slope 1 expected for perfect isostericity.

Table I. Equilibrium constants of picrate with valinomycin

Equilibrium Constants	Solvent	Li	Na	K	Rb	Cs	NH_4	Tl
K_I	CH_2Cl_2	0.0098	0.13	52 000	110 000	14 500	1200	2400
	64% hexane–36% CH_2Cl_2			0.14	0.25	0.031	0.0037	0.0064
	hexane			0.00044	0.001	0.0001		
K_I/K_{Cs}	CH_2Cl_2	0.00000067	0.000009	3.54	7.6	1	0.082	0.165
	64% hexane–36% CH_2Cl_2			4.5	8	1	0.12	0.2
	hexane			4.4	10	1		

Table II. Salt extraction equilibrium constants (K_I), their ratios (K_I/K_K), and the differences in free energies of exchanging $((\Delta F_{exch})(I_K))$ an ion for K^+ (relative to the aqueous standard state)

	K_I^a	K_I	K_I/K_K	$\Delta F_{exch}(I_K)$
Li	0.07	0.0098	0.00000019	+9.17 kcal
Na	0.18	0.13	0.0000025	+7.64
K	10 400	52 000	1.0	0
Rb	20 500	110 000	2.147	−0.45
Cs	6450	14 500	0.2825	+0.75
NH_4	2000	1200	0.0232	+2.23
Tl	—	2400	0.0466	+1.82

[a]Eisenman *et al.* [5]

Notice in Table II that the selectivity sequence is Rb > Cs > K ≫ Na > Li, which is Eisenman selectivity sequence III [46, 47], and that there is a large, 7.64 kcal, difference in selectivity favoring K over Na. Satisfying these two observations is the task of any theoretical algorithm that is to be relied upon for inferring ion-protein interactions. We will test the MOLARIS [36] and GROMOS [25] algorithms here.

3.3 OTHER SIGNS OF ISOSTERICITY

Table III compares the ratios of salt extraction equilibrium constants from Table II with selectivity ratios determined in four other measurements: complexation constants in methanol [43], permeability ratios in bilayers (Krasne and Eisenman, unpublished data), conductance ratios in bilayers [44], and bulk phase electrode selectivities [45]. All measurements agree that the selectivity sequence is Rb > K > Cs ≫ Na > Li; and the agreement can be seen in Table III to be satisfactorily quantitative, given the technical difficulty of measuring very small values of permeability and conductance ratios in bilayers. As discussed elsewhere [5, 6] such agreement is only expected for isosteric complexes.

Table III. Comparisons of ratios of salt extraction equilibrium constants (K_I/K_K), complexation constants (K_{Is}/K_{Ks}) in methanol, permeability ratios (P_I/P_K) in bilayers, conductance ratios (G_I/G_K) in bilayers, and selectivity ratios (S_I/S_K) in ion selective electrodes

	K_I/K_K	K_{Is}/K_{Ks}^a	p_I/p_K^b	G_I/G_K^c	S_I/S_K^d
Li	0.00000019	—	0.0000047	<0.005	<0.00021
Na	0.0000025	0.00016	0.000036	<0.006	<0.00026
K	1.0	1.0	1.0	1.0	1.0
Rb	2.147	2.17	1.8	1.5	1.9
Cs	0.2825	0.267	0.76	0.25	0.48

[a]Grell, *et al.* [43], Table VII.
[b]Unpublished data of Krasne and Eisenman for glycerol monooleate bilayers.
[c]Mueller and Rudin [44], at 35°C and 0.05 M salt.
[d]Pioda *et al.* [45] in diphenyl ether.

3.4. FREE ENERGY DIFFERENCES OF VALINOMYCIN–ION COMPLEX *IN VACUO*

The differences in free energy of binding of the various ions to valinomycin *in vacuo* can be calculated from ΔF_{exch} in Table II by subtracting from these the experimentally known differences in free energy of hydration [48]. The free energy of exchange, ΔF_{exch}, relative to Cs^+ from Table II are:

Li^+	8.42 kcal/M
Na^+	6.89
K^+	−0.75
Rb^+	−1.20
Cs^+	0

The free energies of hydration and the binding energy differences *in vacuo* are:

	$\Delta F_{hydration}$[48]	$\Delta F_{binding}$
Li	−54.3	−45.88
Na	−30.4	−23.31
K	−12.8	−13.55
Rb	−7.7	−8.9
Cs	0.0	0.0

The absolute values of these latter energies are seen to be monotonically increasing functions of decreasing ionic radius. The selectivity implicit in these will be examined in a later section.

4. Computations

4.1 MOLECULAR DYNAMICS/FREE ENERGY PERTURBATION SIMULATIONS

Since the criterion of isostericity is fulfilled for the alkali ion-valinomycin complexes, one only needs to calculate the difference in free energy between the various ion complexes *in vacuo* to obtain the differential binding free energies and thus the selectivity properties of the host. Our strategy in all the reported calculations was therefore to start from the Na-valinomycin complex and in two separate runs slowly 'mutate' Na into K, Rb, and Cs in one run and into Li in another. (Additionally, we started from the Cs–valinomycin complex and slowly mutated Cs into Rb, K, and Na in one run and started from the Li complex and slowly mutated into the Na). During the process of mutation, the structure was allowed to rearrange and the overall free energy change of the system was recorded. All computations were carried out using the MOLARIS program [36] with its standard force fields and parameters (Amino88 library), except as specifically noted, to perform MD/FEP computations [49, 37]. The MOLARIS code was provided by Arieh Warshel and run on an Ardent Titan minisupercomputer.

4.2. GENERAL PROCEDURE

The crystallographic structure for the K^+–valinomycin complex of Neupert-Laves and Dobler [40] was first energy-minimized by 1 ps of molecular dynamics at 5 K with K^+ bound in order to remove possible bad contacts. The resulting atomic coordinates were used as the starting point for subsequent calculations and called the 'annealed structure'.

The free energy change of replacing one ion by another at 100 K was calculated by slowly 'mutating' the Lennard-Jones 6–12 parameters of one ion to another. In forward mutations from smaller to larger sized ions, K^+ in the annealed structure was replaced by Na^+ and subjected to 3 ps of molecular dynamics at 100 K to bring it into the Na^+ form. The Lennard-Jones parameters of the ion were then gradually 'mutated' from those for Na^+ to those for Cs^+ in 35 steps (the spacing between these steps was chosen for most efficient weighting of the energies), passing through K and Rb in the process. The reverse procedure started with replacing Cs^+ for K^+ in the annealed structure and performing 3 ps of molecular dynamics at 100 K to bring the complex into the Cs^+ form. In this way it is possible to obtain the free energy change for each of the ions (Na, K, Rb, and Cs) in one run since one passes through the intermediate sized ions on the way from the smaller to the larger and vice versa. K^+ corresponds to an ion that is 54% of the way from Na^+ to Cs^+; and Rb^+ to an ion that is 74% of the way from Na^+ to Cs^+. The reversibility of this procedure was verified and can be seen in Figure 6 and by comparing the top and bottom parts of Figure 7. A similar procedure was followed for the Na^+ to Li^+ and Li^+ to Na^+ mutations. The Lennard-Jones 6–12 parameters used for the ions [28] are given below.

	A_I	B_I
Li^+	30.2	2.66
Na^+	154.0	4.00
K^+	589.2	4.42
Rb^+	925.0	4.74
Cs^+	1830.0	5.60

At each of the 35 steps of a complete FEP/MD mutation a 1 ps trajectory was calculated, of which the last 0.6 ps were used for evaluating the overall free energy difference. The total free energy corresponding to a given mutation was obtained as the sum over the individual contributions from the 35 steps. For each step the free energy of 'moving' to the next point (i.e. of changing the effective MD potential by a small amount) was calculated as the Boltzmann average of a corresponding small perturbation to the potential energy function. Overall, a mutation run in 35 steps took about 33 hours of CPU time on one processor of an Ardent Titan minisupercomputer (viz. about 30 days to construct Figure 8). The total free energy change includes the cost of rearranging the host molecule and the non-bonded Coulomb and van der Waals interactions between the ion and the host. The present potential energy function does not include induced dipoles.

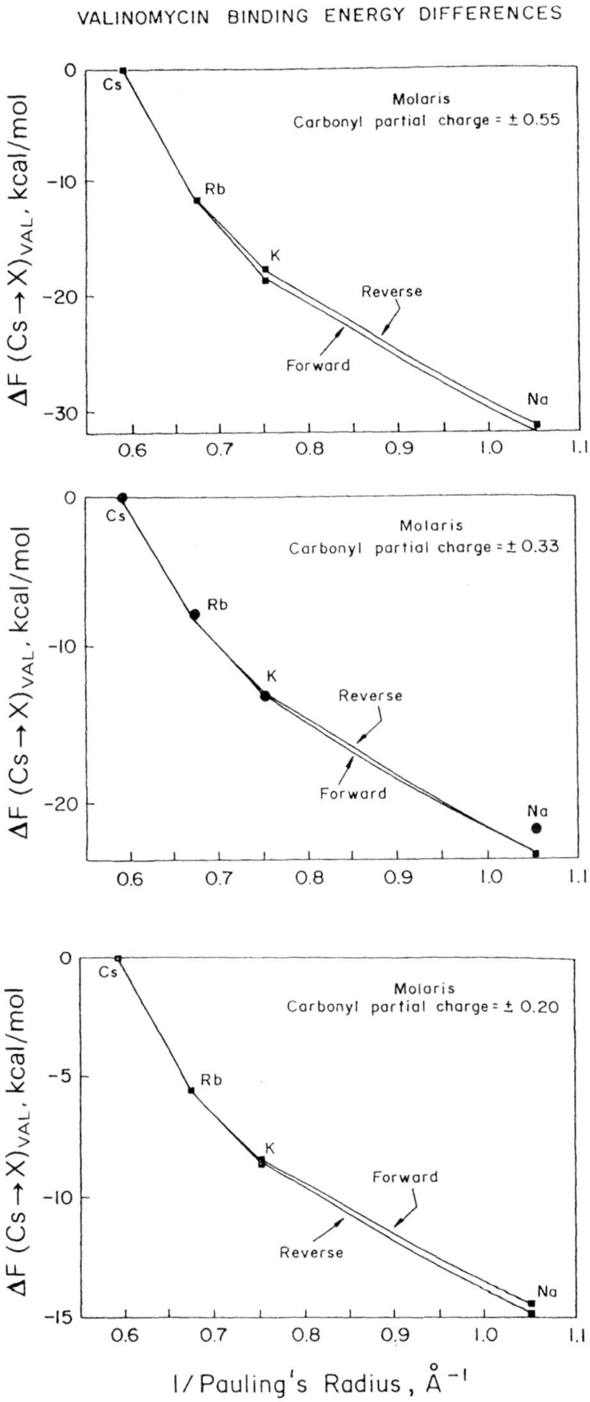

Fig. 6. Reversibility of FEP mutations in the forward (from Na to Cs) and backward (from Cs to Na) directions. Results for simulations for 3 different values of ester carbonyl partial charge are shown: $+/-0.55$, the default value for MOLARIS, at the top; $+/-0.20$ at the bottom; and $+/-0.33$, in the middle which gives the best agreement with the experimental values from salt extraction equilibria (indicated by experimental data points, the large dots in the middle figure).

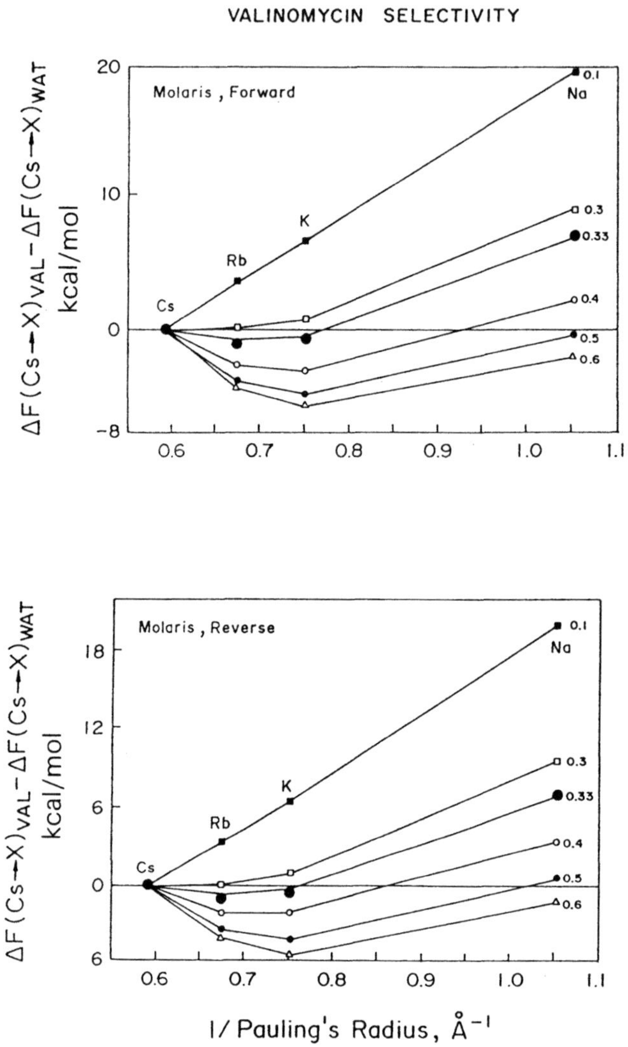

Fig. 7. Calculated values of free energy selectivity are plotted for the indicated values of ester carbonyl partial charge as a function of the reciprocal of the Pauling crystal radius for forward mutations (top) and reverse mutations (bottom). Selectivity is defined as the free energy difference between the ion–valinomycin complex and the Cs^+–valinomycin complex minus the difference in the experimental hydration free energy of the ion relative to Cs^+. Increasing selectivity is plotted downward. Experimental values are shown by the data points. Note the successive shift of selectivity optima from Cs^+ at a partial charge of 0.1, to Rb^+ at a partial charge of 0.33, and to K^+ at a partial charge of 0.5 or greater. This pattern is recognizable as having a 'Coulomb topology' [50]. The larger dots are experimental data.

4.3. CONSTRAINTS

We applied a very weak harmonic constraint of 0.1 kcal/$Å^2$ to the valinomycin atoms, which restrains them to their crystallographic structure. This constraint was included in order to compensate for the missing van der Waals' interaction with a surrounding (non-polar) solvent and because of uncertainties with respect to the torsional potentials. The magnitude of this constraint is so small that it is not likely to have any influence on the present results and conclusions (for an atom that moves 1.0 Å the energy cost would only be 0.05 kcal/mol). We have recently reduced this constraint to zero in computations using GROMOS (see Section 6.2 of the Discussion) and have obtained results compatible with the results to be presented below.

4.4. RESULTS

4.4.1. *Reversibility*

Figure 6 demonstrates the reversibility of the MD/FEP procedure. The difference in free energy between Cs and the indicated cations *in vacuo* is plotted for the 'forward' mutation from Na to Cs and the 'reverse' mutation from Cs to Na as a function of the reciprocal of the Pauling crystal radius. The reversibility is shown for three different values of partial charge of the six ester carbonyl ion binding ligands. The top is the (high) default value (+0.55 for C, −0.55 for O) of the carbonyl partial charge for the standard (Amino88) parameters of MOLARIS. The bottom figure is for an extremely low value of partial charge (+/−0.20). The middle is for a partial charge of +/−0.33 which we will show below reproduces the experimental data for valinomycin quite well.

4.4.2. *Comparison of Computed Free Energies in vacuo with Experimental Values*

The differences in free energy of binding of the various ions to valinomycin *in vacuo* were calculated in Section 3.3. These 'experimental' *in vacuo* binding energies can now be directly compared with the *in vacuo* calculations by plotting the experimental values and noting how well these (solid dots) agree with the free energies for a partial charge of +/−0.33 in the middle part of Figure 6.

4.4.3. *Selectivity: Comparison of Computated with Experimental Values*

Figure 7 shows the selectivity calculated by the MD/FEP procedure for the 'forward' mutation (top) and the 'reverse' mutation (bottom) for the indicated values of partial charge of the carbonyl ligand. For comparison, the experimental values of ΔF_{exch} recalculated from Table II relative to Cs^+ are plotted as large dots. It is apparent that the computed and experimental values agree quite satisfactorily for a value of partial charge of +/−0.33 and also that the minimum in the selectivity curve occurs at Rb for this value.

An alternative comparison is presented in Figure 8 which shows more clearly how the two experimental constraints on selectivity can be imposed to define quite narrowly the range of permissible values of carbonyl partial charge. It will be recalled that the first experimental constraint is that the selectivity sequence must be Rb > Kb > Cs ≫ Na > Li. The top part of Figure 8 plots the free energy change for mutating a Na complex first into a K complex, then into a Rb complex, and finally into a Cs complex. The bottom part of Figure 8 plots the reverse process, mutating from larger to smaller ion. The Roman numerals indicate the ranges of carbonyl

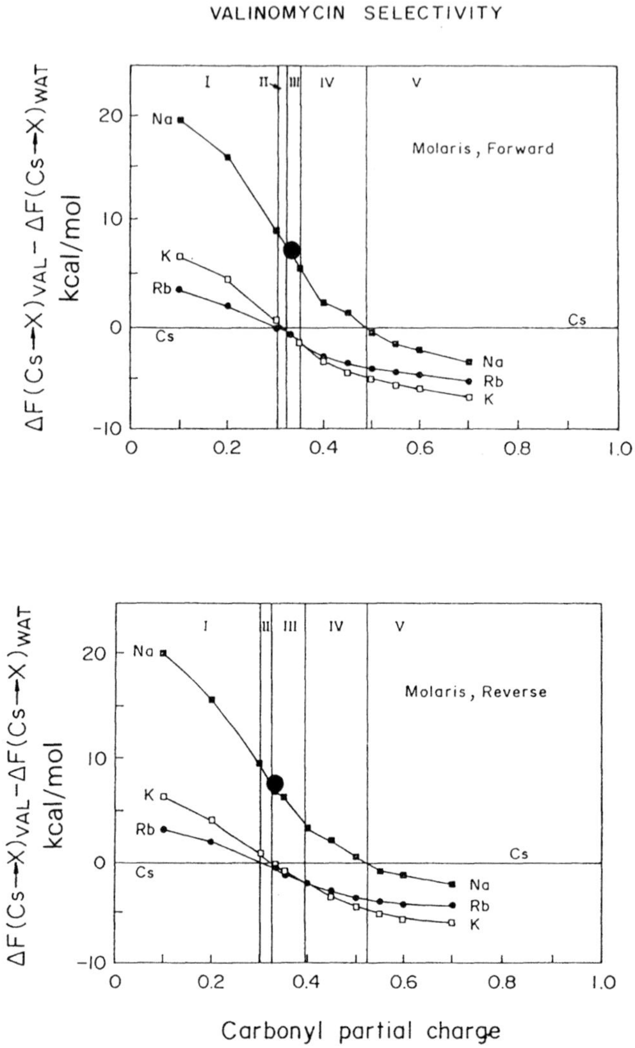

Fig. 8. Calculated values of free energy selectivity, defined as in Figure 7, are plotted as a function of the ester carbonyl partial charge for the indicated ions. Increasing selectivity is plotted downward. The intersections between the isotherms for each cation mark shifts in selectivity sequence, with the Roman numerals indicating the regions in which particular Eisenman sequences are seen. Experimental salt extraction data points for Na^+ (relative to Cs^+) are also plotted as a large dot at a partial charge of 0.33 for comparison with the Na^+ isotherm.

partial charge in which the Eisenman selectivity sequences corresponding to these numbers occur. They correspond to the regions between intersections of the isotherms for each ion.

It can be seen for both directions that there is a relatively narrow range of carbonyl partial charge, labelled 'III', in which the selectivity sequence Rb > K > Cs > Na is observed. This occurs in the region in which the Rb isotherm is the lowest, and, moreover, in the region where the K isotherm lies between the Rb and Cs isotherms (see below for more discussion of the selectivity pattern). If we now apply the additional constraint that the free energy difference between Na and Cs must be 6.89 kcal/mol in favor of Cs we find that this indeed is satisfied within the sequence III region (compare experimental data point) with the Na isotherm and, moreover, that this occurs for a partial charge of 0.33 for both the forward and reverse mutations. In this way we have parameterized the ester carbonyl partial charge at a value of $+/-0.33$ for the standard Lennard-Jones 6–12 parameters of MOLARIS (see Figure 1) for the carbonyl oxygen and the alkali cations. In a subsequent section we will parameterize the partial change for the somewhat smaller carbonyl oxygen used by GROMOS [25].

The degree of quantitative agreement between computation and experimental data can be seen for each ion in Figure 7; but it is worth emphasizing the data plotted in Figure 9 which shows for both the forward and reverse mutations the very satisfactory agreement between experimental free energies (large filled circles) and the computed ones for a carbonyl particle charge of 0.33.

Finally, for completeness we show in Figure 10 the results for computations for Li in the reverse mutation (from Na to Li). As mentioned before, we will not try to compare these calculations with experimental data because it is highly likely that the Li^+ complex contains a water molecule and also because there are indications that 'tight' ion pairing between the Li and picrate ions make it difficult to interpret the salt extraction data. Briefly, whereas the computations indicate that Li should be highly disfavored to Na, the experiments indicate that the disfavoring is not as great. Such a difference between computed expectations and the observations of Tables II and III makes sense and probably can even be resolved quantitatively by extending the computations to allow a water molecule to enter the Li^+ complex.

4.4.4. *Parameterization of the C=O Partial Charge*

To our knowledge this constitutes the first test of a molecular dynamics simulations to reproduce *quantitatively* the free energies underlying ion-protein binding selectivity. It gives a strong indication that the MOLARIS force field, parameters, and procedure are adequate for ion-protein interactions. Of course, this has been tested only for the calculation of a free energy *difference* and needs to be extended to the calculation of the absolute values of free energy, which are known for the case of the ion-complexation constants measured by Grell *et al.* [43] for valinomycin in methanol (recall Table III), to which we plan to extend our computations. The partial charge value of 0.33 is much smaller than that used in MOLARIS for its standard amide carbonyl (0.55). It should be noted that the large value of the standard partial charge has been imposed to produce the correct H-bonding energies on a solely electrostatic basis with the Lennard-Jones parameters for the atoms used by MOLARIS. An alternative variable in parameterizing the C=O ligands is the size of the oxygen (i.e. an alternative Lennard-Jones 6–12 potential); for the effects of altering the partial charge can be produced to some extent by altering the 6–12 potentials. Since the well-established GROMOS [25] algorithm uses a different set of 6–12 potentials and partial charges for carbonyl oxygens than does MOLARIS, we will now examine this question by using GROMOS Lennard-Jones parameters for the ligand carbonyls.

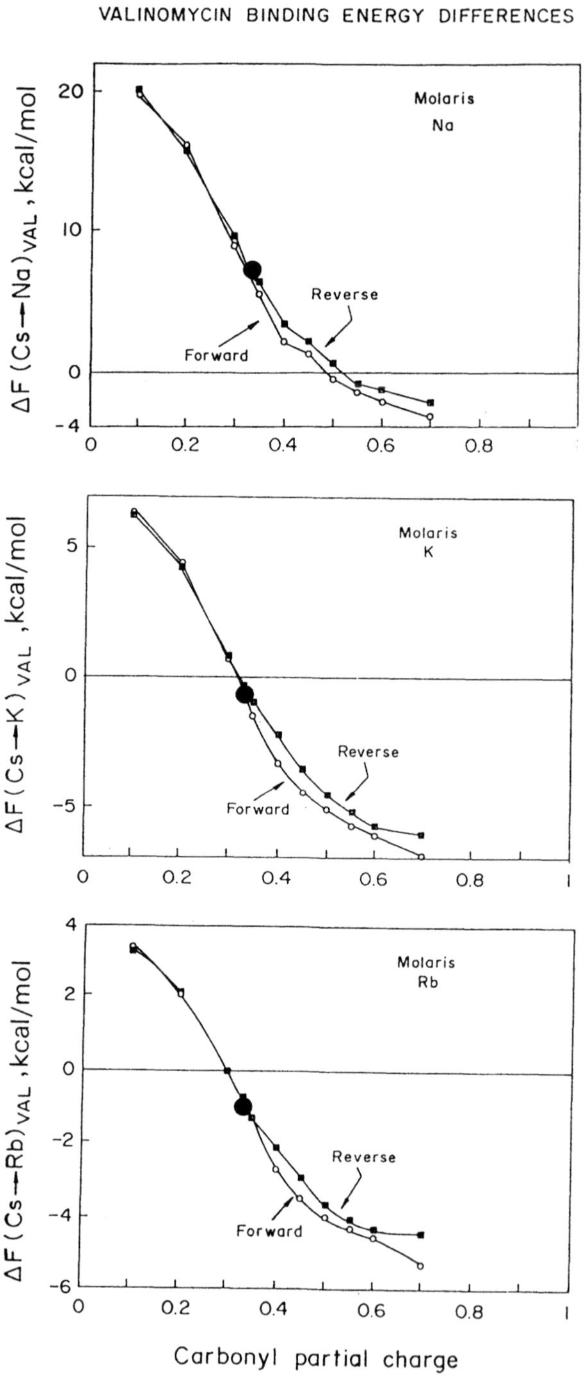

Fig. 9. Demonstration of quantitative agreement between calculated and experimentally measured selectivity for a carbonyl partial charge of 0.33 for both the forward (open circles) and reverse (filled boxes) mutations. Top: Na relative to Cs; middle: K relative to Cs; bottom: Rb relative to Cs. The large filled circles represent the experimental data.

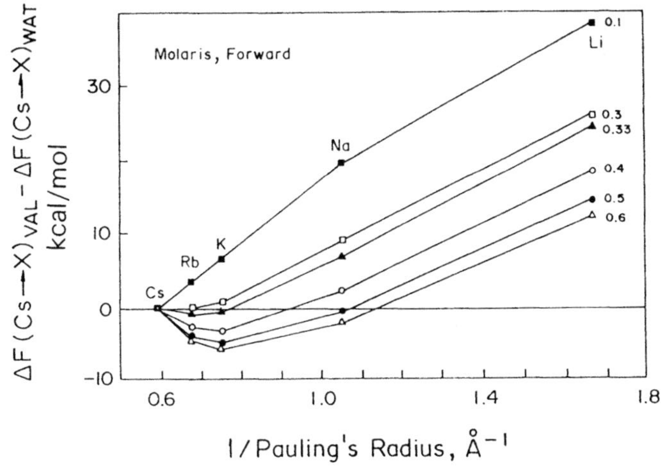

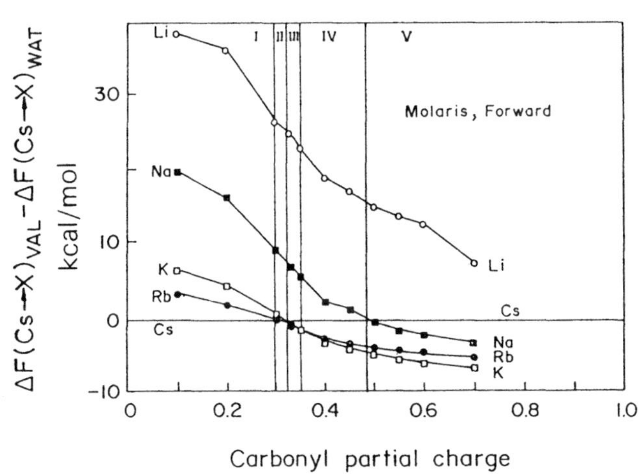

Fig. 10. Values of selectivity including Li calculated by MOLARIS for the forward mutations (Li to Na combined with Na to Cs). *Top*: plotted in the manner of Figure 7; *Bottom*: plotted in the manner of Figure 8.

4.5. AN ALTERNATIVE PARAMETERIZATION OF CARBONYL OXYGENS FOR THE MOLARIS FORCE FIELD USING GROMOS LENNARD-JONES PARAMETERS FOR THESE OXYGENS

The GROMOS program uses somewhat different Lennard-Jones 6–12 parameters for the ester carbonyl oxygens (A = 550.0, B = 23.25) than MOLARIS, defining a somewhat smaller oxygen. The valinomycin molecule offers an opportunity to compare MOLARIS with GROMOS oxygens using the MOLARIS force field. Figure 11 summarizes such computations using MOLARIS for the forward mutation using GROMOS Lennard-Jones parameters for the ester carbonyl oxygens. Figure 11 compares the experimental data with the selectivity curve for the indicated values of partial charge, including the standard GROMOS partial charge of 0.383. It is apparent that a satisfactory representation of the experimental data

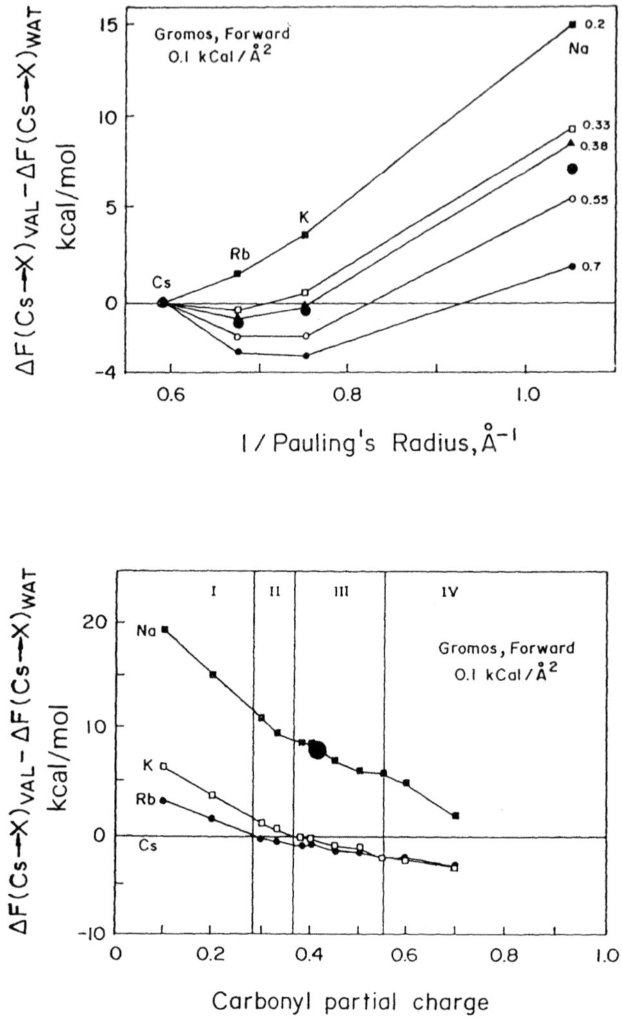

Fig. 11. Selectivity of MOLARIS using GROMOS Lennard-Jones 6–12 parameters for the ester carbonyl. *Top*: plotted in the manner of Figure 7; *Bottom*: plotted in the manner of Figure 8. Notice that the best agreement between experimental data points (large dots) and theoretical expectations would occur for a partial change around 0.41, slightly higher than the standard GROMOS value of 0.383.

can indeed be obtained for a value of partial charge around 0.41, slightly higher than the standard value for a GROMOS amide carbonyl. This is a somewhat surprising finding when combined with the conclusion that the partial charge that gives the correct selectivity is 0.33 for the *larger* GROMOS carbonyl; for intuitively one would have expected it to be *smaller* to get a comparable electric field at the surface of the oxygen. There is a simple, and highly instructive, explanation for this which will be given in the next section.

At the present stage of this work it is not possible to decide (from energetics alone) whether GROMOS or MOLARIS parameters are better to use for ions. Structurally, the ion-ligand distances and efficiency with which the carbonyl bonds points at the ion are summarized in Table IV, which will be discussed below.

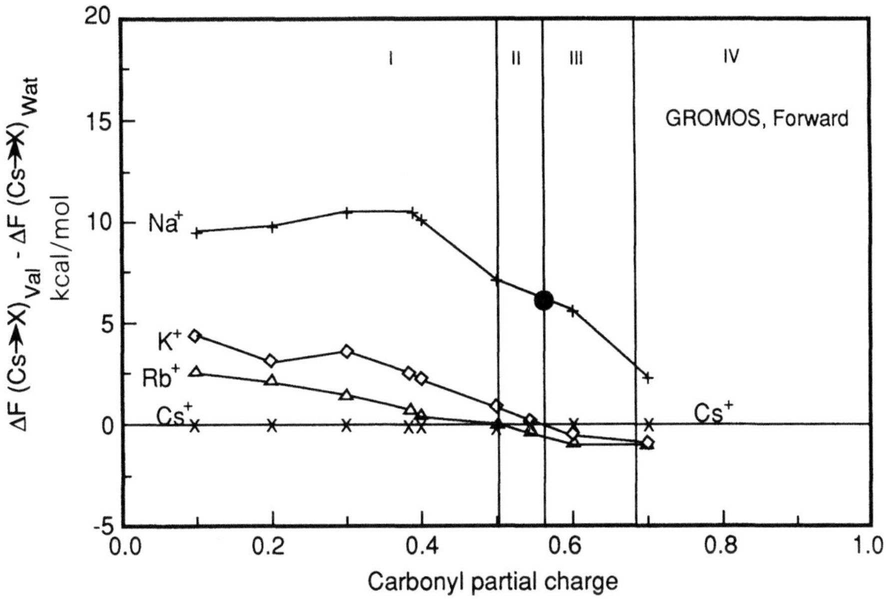

Fig. 12. Selectivity of GROMOS with extended atoms and standard parameters. The forward reaction is plotted in the manner of Figure 7. Notice that the best agreement between experimental data point (large dot) and theoretical expectations would occur for a partial charge around 0.55, somewhat higher than the standard GROMOS value of 0.383.

4.6. RECENT SIMULATIONS WITH THE GROMOS FORCE FIELD

We have very recently repeated these computations using the GROMOS force field [25] (with extended atoms) and parameters. The results are shown in Figure 12 for the forward mutation and should be compared with those of Figures 11 and 8. The overall behavior is quite similar; although the pattern of inversion of selectivity is shifted even further to the right than in Figure 11 relative to the standard MOLARIS results of Figure 8 (for GROMOS the best fit to the experimental data is seen to occur for a partial charge around 0.55).

Since FEP/MD simulations with both the MOLARIS and GROMOS force fields can successfully account for the experimental data, the use of both MOLARIS and GROMOS algorithms for evaluating ion binding to proteins is validated.

5. Structure

For the MOLARIS force field with both MOLARIS and GROMOS carbonyls we have compared the original crystallographic structure for K^+ with the energy minimized structure for K^+ snapshot generated in the molecular dynamics mutation from Na^+ to K^+. We find the backbones to be virtually identical with only slight differences between the side chains. This indicates that the host molecule neither flies apart nor collapses unreasonably. The energy minimized structures of snapshots from the FEP dynamics runs for K^+ give virtually identical results to the X-ray structure for K^+.

The energy minimized structures of snapshots for the different cations show systematic puckering of the backbone and systematic changes in the ion-oxygen

Table IV. Distances and angles between ions and ligands

		MOLARIS Carbonyl pt.chg: 0.33		GROMOS Carbonyl pt.chg: 0.41	
		Distance	Angle	Distance	Angle
Na	1	2.635	152.127	2.678	163.095
	2	2.667	157.923	2.533	166.404
	3	2.647	153.306	2.494	162.839
	4	2.686	150.817	2.731	159.840
	5	2.674	152.851	2.636	161.270
	6	2.726	155.836	3.028	160.981
K	1	2.843	147.370	2.791	157.227
	2	2.830	154.825	2.873	161.522
	3	2.805	148.518	2.787	156.973
	4	2.854	140.773	2.819	155.388
	5	2.837	147.296	2.876	157.144
	6	2.811	151.635	2.827	160.237
Rb	1	2.910	141.460	2.852	156.396
	2	2.949	155.306	2.871	159.763
	3	2.885	148.755	2.905	156.403
	4	2.931	130.010	2.862	154.552
	5	2.951	145.195	2.886	154.627
	6	2.889	149.032	2.903	159.399
Cs	1	3.051	139.631	2.973	152.615
	2	3.069	149.785	2.961	157.226
	3	3.150	110.842	3.055	155.481
	4	3.123	119.677	2.961	152.117
	5	3.120	116.349	2.988	150.566
	6	3.076	150.008	3.028	157.844

Distances (in ångstrom units) are between the ion center and the carbonyl oxygen center. The angle represents the angle between the carbonyl oxygen, the ion, and the carbonyl carbon. An angle of 180 degrees indicates that the C=0 bond points directly at the ion.

distances and angles summarized in Table IV for both MOLARIS and GROMOS carbonyl oxygens. Figure 13 compares the ion-oxygen distances for the MOLARIS carbonyls with those expected from Pauling crystal radii (straight line), the agreement is seen to be excellent for Rb^+ and Cs^+, indicating that the Lennard-Jones parameters used are appropriate. However, the agreement is less satisfactory for the smaller cations, the structure being apparently too rigid to collapse more completely around K^+ and Na^+. That these ions do not find an off-center energy minimum is surprising, though a tendency to do so does seem to exist with the smaller GROMOS carbonyls, as can be seen from Table IV. Possibly in the energy minimization procedure near absolute zero temperature we are finding the minimum potential energy conformation but missing the existence of an ensemble of 3 or 6 (symmetrical) local minima around the periphery of the cavity where the smaller ions at room temperature can transiently find minima at the closest distance of approach expected from their van der Waals sizes. Perhaps evidence in support

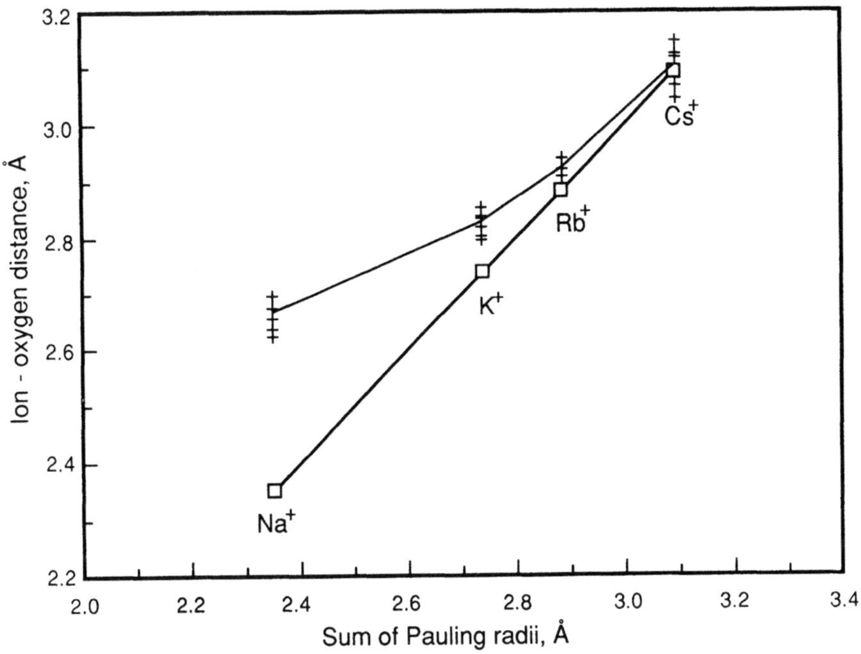

Fig. 13. Ion-carbonyl ligand distances observed in the energy minimized (40 ps at 1K) coordinates of the 'snapshots' from the FEP/MD simulations at 100 K (plotted as plus symbols) compared to the values expected according to the Pauling crystal radii (plotted as squares).

of this notion is contained in the ion-oxygen distances in the non-energy minimized snapshot for MOLARIS carbonyls with Na^+, where the distances are: 2.675, 2.714, 2.733, 2.993, 3.009, 3.075. In this transient structure, the shortest potassium–oxygen distance is quite close to the value of 2.75 Å expected from the sum of the Pauling crystal radii. This contrasts with the average distance of 2.82 Å calculated for the energy-minimized structure and plotted in Figure 13. That a low-temperature energy minimized structure might not show local minima accessible only at higher temperatures has been discussed recently by Van Gunstaren *et al.* [52].

From these results it is clear that, although 'field strength' considerations alone clearly can account for the observed selectivity sequence (see Section 6.1 of the Discussion), part of the high discrimination against Na by valinomycin is probably played by the steric inability of the molecule to achieve tight coordination without paying a substantial price in deformation energy.

5.1. ION–CARBONYL BOND ANGLES

Table IV presents for each of the cations the carbon-oxygen ion angle (in degrees) found for the MOLARIS force field with the MOLARIS and GROMOS carbonyls. This table enables one to compare the distances and angles calculated for the larger MOLARIS carbonyls with those computed for the smaller GROMOS carbonyls. The most striking observation is that the GROMOS carbonyl angle is closer to the 180 degrees expected for a carbonyl group pointing directly at the ion, an angle that maximizes the cation-oxygen attraction and minimizes the carbon-carbon repulsion. The difference between GROMOS and MOLARIS carbonyls becomes increasingly pronounced as cation size increases (notice that for Cs three angles are less than 120 degrees for MOLARIS carbonyls, while none is less than 150 degrees for GROMOS carbonyls). This indicates that the bulkier MOLARIS ligands cannot point as favorably at the larger cations. This results in an essentially steric (bad orientation) discrimination against the larger cations for the larger MOLARIS carbonyls that disfavors these relative to the smaller ones and leads to the paradoxical crossovers, mentioned above, in selectivity sequence at lower partial charge values for the larger MOLARIS carbonyls than for the smaller GROMOS carbonyls (compare Figures 11 and 10). This has interesting implications for the interpretation of selectivity in terms of both steric and field strength contributions, as discussed in Section 6.1 and elsewhere [56].

6. Discussion

6.1. THE SELECTIVITY PATTERN

It is clear for both MOLARIS carbonyls (Figure 8) and GROMOS carbonyls (Figure 11) that the selectivity isotherms pass through the so-called Eisenman selectivity sequences [46, 47, 50] as a function of increasing partial charge. The lowest 'field strength' sequence I is seen at the extreme left. Here the Cs isotherm is the lowest (most preferred); and the sequence of selectivity is Cs > Rb > K > Na. Looking at the 'forward' mutations and proceeding to the right, the first change of sequence occurs when the partial charge is increased to about 0.3, at which point the Rb isotherm crosses the Cs isotherm; so that the sequence becomes Rb > Cs > K > Na, which is Eisenman sequence II. Proceeding further to the right, the K isotherm crosses the Cs isotherm when the partial charge is increased to about 0.32 for the MOLARIS carbonyl and 0.36 with the GROMOS carbonyl. The sequence is now Rb > K > Cs > Na, which is Eisenman sequence III, and is the one characteristic of valinomycin.

This sequence exists over a relatively narrow range for MOLARIS carbonyls and a wider range of GROMOS carbonyls since the K isotherm crosses the Rb isotherm to produce sequence IV (K > Rb > Cs > Na) at a partial charge of 0.34 for MOLARIS and 0.54 for GROMOS. Finally, at a partial charge of 0.48 for the MOLARIS carbonyl and at 0.8 (extrapolated) for the GROMOS carbonyl the Na isotherm crosses the Cs isotherm resulting in sequence V (K > Rb > Na > Cs) for this and higher values of partial charge. As noted above this is paradoxical in that the smaller carbonyl oxygen appears to require a larger partial charge to produce an equivalent 'field strength' [46, 47] effect. This illustrates a limitation of interpreting observations from real protein structures in terms of oversimplified models, as discussed elsewhere [56]. In particular, the achievement of a 'higher field strength' Eisenman sequence for a given value of partial charge by the *larger* carbonyl ligand is a consequence of discrimination *against* the larger ions rather than a favoring of the smaller ones! This explains the counter-intuitive finding noted in a preceding section that the partial charge that gives the experimentally correct selectivity is *larger* (sic) for the *smaller* GROMOS carbonyl instead of being smaller.

The same pattern of selectivity inversions is seen for the reverse mutation, though the range of partial charge over which sequence III is observed is somewhat broader. Nevertheless, when the second constraint of the proper value for the Na–Cs free energy difference is applied (see solid experimental data point), the best agreement is constrained to a partial charge of 0.33, just as in the case of the forward mutation.

The selectivity pattern can also be seen on plots of selectivity vs ion size like Figure 7. In the case of such plots, the selectivity optimum occurs at the size of the ion that is preferred [50].

6.2. PREVIOUS CALCULATIONS OF VALINOMYCIN SELECTIVITY

Theoretical calculations of the ion interactions with valinomycin were first carried out by A. Pullman *et al.* [51]. These successfully reproduced the selectivity but were based on a frozen structure (the crystallographic structure for the K-valinomycin complex) which was kept the same for all ions while the intramolecular interaction energy was computed by optimizing the position of the cation inside the (rigid) cavity. Such calculations are interesting but do not reflect the ability of valinomycin to adjust its structure to relieve local strains. This is not a trivial matter. In calculations on gramidicin [6] and on viral ion binding sites [4, 16] it was found that using a fixed structure causes a large component of the selectivity to be attributed to steric effects (e.g. van der Waals repulsion) which should become completely negligible in a structure that can relax.

Comparing Pullman's minimum ion-ligand distances with Pauling radii, it is apparent that all her ions are too small, being approximately shifted by one ion towards Li. In particular, her Rb radius corresponds to that observed for K. Therefore, Rb will be the ion that fits best in the K crystal structure; and the correct selectivity apparently emerges as a consequence of using too small ions in a rigid structure, a situation not too dissimilar to the conclusions previously reached in primitive modelling of a frozen viral selectivity filter [18].

It will be interesting to compare the energies of the optimized valinomycin and viral structures with those of the frozen structures. The subject of selectivity in 'frozen' vs. rearrangeable structures has been discussed in general terms [13, 22, 38, 53] and is examined in detail elsewhere [56] for valinomycin.

Another example of the importance of extending selectivity modelling to include the energetics of structural rearrangement has been given in Section 6.1. Here the paradoxical finding of an effectively higher 'field strength' for a given partial charge imbedded in a larger carbonyl oxygen than in a smaller one was rationalized by recognizing a 'steric' contribution.

6.3. SOME COMMENTS ON ISOSTERICITY

A few comments are now in order on the question of whether or not isostericity can strictly exist in a protein when replacing one ion by another and on the related question of whether or not a strictly isomorphous replacement is ever possible between ions. There is a large body of spectroscopic data germane to this (see pp. 127–131 of [39] and 19–24 of [54] for discussion and references); but to discuss these exceeds the scope of the present paper. From the preceding Section 5 it should be apparent that, despite the apparently isosteric behavior of the experimental data, valinomycin *does* change its structure slightly to accommodate the different sized ions; and this has important energetic consequences. A useful analogy can be made with a system of springs. The van der Waals repulsion term represents a very stiff spring (12th power spatial dependence) while the other interactions between protein atoms represent much weaker springs. In a strictly frozen system in which an ion replaces another isomorphously the only species dependent contribution to selectivity comes from van der Waals interactions between the ion and the ligands (see [53, 56] for details). In such a system selectivity has a purely steric origin. But we would argue that this can never happen in a real protein since small readjustments will relieve the van der Waals strain, essentially distributing it over a somewhat larger region. For example consider what actually occurs when an ion that fits easily into the natural cavity of valinomycin (e.g. Na^+) is replaced by a larger ion (e.g. Cs^+). In a rigid structure this would imply a large van der Waals repulsive energy; but what actually happens [56] is that the van der Waals strain is released by compressing the weaker springs that constitute the bond and non-bond interactions between all the other atoms of the protein (bond stretching, bond-angle bending, torsional-angle contributions, and intramolecular charge-charge and VDW interactions). In this way the repulsive charge effect of differing ion size is distributed over a larger group of atoms than the immediate neighbors, which of course implies that the replacement cannot be strictly isomorphous. The experimentally observed 'isostericity' in this case is therefore only approximate and signifies that the changes in structure are relatively localized so that the overall shape and size of the complex are not altered significantly thereby. In the literal sense, strict isostericity between ions and strictly isomorphous replacements can never occur in a protein!

7. Summary and Conclusions

Rationale. Studies of two-phase salt extraction equilibria underlying the ability of valinomycin to extract the picrates and dinitrophenolates of Cs^+, Rb^+, K^+, Na^+, Li^+, Tl^+, and NH_4^+ from aqueous solutions into organic phases (dichloromethane, n-hexane, and their mixtures) have shown that the valinomycin–ion complexes of most of these ions are quite isosteric. For isosteric complexes the *differences* in free energy between the different ions are expected to be independent of the solvent in which the complex exists. Such free energy *differences* correspond to *ratios* which are the usual measures of selectivity. Experimentally it is found that the ratios of two-phase equilibrium constants relative to Cs^+ agree closely with the ratios of the homogeneous equilibrium constants for ion-complex formation in methanol, as well as with ratios of conductance and permeabilities in lipid bilayers, and even with selectivity coefficients of bulk ion-selective electrodes. These observations imply that isostericity of ion-valinomycin complexes both for media ranging in polarity from hexane to methanol. This is strictly true for K^+, Rb^+, Cs^+, Tl^+, and NH_4^+, approximately true for Na^+, but breaks down for Li^+. The property of isostericity enables meaningful theoretical calculations to be carried out *in vacuo* since the *differences* in calculated free energy can be directly compared with those measured experimentally, regardless of the media in which they were measured. This provides an opportunity to test the GROMOS and MOLARIS force fields and parameters against experimental data.

Measurements. Precise ion selectivity ratios were measured by two-phase salt extraction into dichloromethane and compared with ratios measured for bilayers and for the homogeneous complexation reaction in methanol. All observed ratios are found to exist in the Eisenman selectivity sequence III: Rb > K > Cs ⩾ Na > Li. The 2-phase salt extraction measurements into dichloromethane yield the most accurate free energy differences. These are (in kcal/mol referred to Cs in the aqueous reference state): Rb: −1.2; K: −0.75; Cs: 0.0; Na: +6.89; Li: +8.42.

Simulations. These values were compared with those calculated *in vacuo* by FEP/MD simulations for the valinomycin-alkali cation complexes carried out for both the MOLARIS and GROMOS force fields starting with the crystallographic atomic coordinates of the K–valinomycin complex using Lennard-Jones 6–12 parameters for the ions derived from aqueous solution data. On the one hand, for the standard partial charges and Lennard-Jones parameters of MOLARIS, the calculations predict the selectivity sequence K > Rb > Na > Cs > Li. This is Eisenman sequence V, which would be expected for a binding site having a field strength stronger than that of valinomycin. On the other hand, for the standard partial charges and Lennard-Jones parameters of GROMOS the predicted selectivity sequence is Cs > Rb > K > Na > Li. This is the lyotropic sequence (Eisenman sequence I) and would be expected of a binding site having a field strength weaker than that of valinomycin.

Parameterization of ester carbonyl and comparison of simulations with experiments. Despite these differences we show that it is possible with both GROMOS and MOLARIS force fields to fit the experimental data both as to sequence (Rb > K > Cs ⩾ Na > Li) and magnitude (particularly the large preference of K, Rb, and Cs relative to Na (6.89 kcal/mol Cs preference over Na)). This is accomplished if one parameterizes the partial charges of the ester carbonyl ligands appropriately. When this is done, FEP/MD simulations using either the MOLARIS or GROMOS force fields give quantitative fits to the experimental data. We conclude that these two fields are valid as appropriate procedures for computing ion binding selectivity in proteins.

Acknowledgement

The research was supported by USPHS Grant GM 24749, by the Swedish Natural Science Research Council (NFR), and by FONDECYT GRANT 1112-1989. We thank Emily Maverick for the Cambridge Data Base structures for valinomycin, Rich Little for help in programming and analyzing ligand distances and bond angles, and Fred Lee for help in customizing MOLARIS and in preparing the library entries for lactate and hydroxyisovalerate.

References

1. C. J. Pedersen: *J. Am. Chem. Soc.* **89**, 7017 (1967).
2. G. Eisenman, S. Ciani and G. Szabo: *Federation Proceedings* **27**, 1289 (1968).

3. C. J. Pedersen: *Federation Proceedings* **27**, 1305 (1968).
4. G. Eisenman, S. Ciani and G. Szabo: *J. Membr. Biol.* **1**, 294 (1969).
5. G. Eisenman, G. Szabo, S. G. A. McLaughlin and S. M. Ciani: in *Symposium on Molecular Mechanics of Antibiotic Action on Protein Biosynthesis and Membranes*, eds. E. Munoz, F. G. Ferrandiz and D. Vasquez, Elsevier, Amsterdam, London, New York (1972), pp. 545–602.
6. G. Eisenman, G. Szabo, S. Ciani, S. McLaughlin and S. Krasne: in *Progress in Surface and Membrane Science* **6**, 139 (1973) Academic Press, New York.
7. S. McLaughlin, G. Szabo, S. Ciani, and G. Eisenman: *J. Membr. Biol.* **9**, 3 (1972).
8. J. A. McCammon, P. G. Wolynes and M. Karplus: *Biochemistry* **18**, 927 (1979).
9. A. Warshel, and J. Aqvist: in *Theoretical Biochemistry and Molecular Biophysics*, eds. D. L. Beveridge and R. Lavery, Adenine Press (1990), pp. 1–18.
10. B. R. Brooks, R. E. Brucoleri, B. D. Olafson, D. J. States, S. Swaminathan and M. Karplus: *J. Comp. Chem.* **4**, 187 (1983).
11. P. K. Weiner and P. A. Kollman: *J. Comp. Chem.* **2**, 287 (1981).
12. W. F. van Gunsteren and P. K. Weiner: *Computer Simulation of Biomolecular Systems*, Escom, Leiden (1989).
13. J. Åqvist and A. Warshel: *Biophys. J.* **56**, 171 (1989).
14. S. Furois-Corbin and A. Pullman: in *Transport Through Membranes: Carriers, Channels and Pumps*, eds. A. Pullman, J. Jortner and B. Pullman, Kluwer Academic Publishers, Dordrecht, Boston, London (1988), pp. 337–357.
15. S. Furous-Corbin and A. Pullman: *Fed. Eur. Biochem. Soc.* **252,** 63 (1989).
16. G. Eisenman, A. Villarroel, M. Montal and O. Alvarez; in *Progress in Cell Research*, eds. J. M. Ritchie, P. J. Magestretti and L. Bolis, Elsevier (1990), Vol. 1, pp. 195–211.
17. A. M. Silva, R. E. Cachau and D. J. Goldstein: *Biophys. J.* **52**, 595 (1987).
18. G. Eisenman, A. Oberhauser and F. Bezanilla: in *Transport Through Membranes: Carriers, Channels and Pumps*, eds. A. Pullman, J. Jortner, B. Pullman, Kluwer Academic Publ. Dordrecht, Boston, London (1988), pp. 27–50.
19. F. Sussman and Harel Weinstein: *Proc. Nat. Acad. Sci. USA* **86**, 7880 (1989).
20. G. Wipff, and J. M. Wurtz: in *Transport Through Membranes: Carriers, Channels and Pumps*, eds. A. Pullman, J. Jortner, B. Pullman, Kluwer Academic Publishers, Dordrecht, Boston, London (1988), pp. 1–26.
21. P. D. J. Grootenhuis and P. A. Kollman: *J. Am. Chem. Soc.* **111**, 4046 (1989).
22. G. Eisenman and O. Alvarez: in *Ca Transport and Homeostasis*, ed. D. Pansu, NATO ASI Series, vol H48, Springer Verlag, Berlin, New York, Tokyo (1990), p. 283.
23. J. Åqvist and A. Warshel: *J. Am. Chem. Soc.* **112**, 2860 (1990).
24. S. J. Weiner, P. A. Kollman, D. T. Nguyen and D. A. Case: *J. Comp. Chem.* **7**, 230 (1986).
25. W. F. Van Gunsteren, H. J. C. Berendsen, J. Hermans, W. G. J. Hol and J. P. M. Postma: *Proc. Natl. Acad. Sci. USA* **80**, 4315 (1983).
26. J. Åqvist, W. F. van Gunsteren, M. Leijonmarck and O. Tapia: *J. Mol. Biol.* **183**, 461 (1985).
27. A. T. Brunger, R. L. Campbell, G. M. Clore, A. M. Gronenborn, M. Karplus, G. A. Petsko and M. M. Teeter: *Science* **235**, 1049 (1987).
28. J. Åqvist: *J. Phys. Chem.* **94**, 8021 (1990).
29. R. H. Kretzinger: *CRC Crit. Rev. Biochem.* **8**, 119 (1980).
30. G. Eisenman and J. A. Dani: *Ann. Rev. Biophys. Biophys. Chem.* **16**, 205 (1987).
31. O. Herzberg and M. N. G. James: *Biochem.* **24**, 5289 (1985).
32. T. A. Jones and L. Liljas: *J. Mol. Biol.* **177**, 735 (1984).
33. I. Montelius, L. Liljas and T. Unge: *J. Mol. Biol.* **201**, 353 (1988).
34. D. W. Urry: *Proc. Natl. Acad. Sci. U.S.A.* **68**, 672 (1971).
35. B. A. Wallace and K. Ravikumar: *Science* **241**, 182 (1988).
36. A. Warshel and S. Creighton: in *Computer Simulation of Biomolecular Systems*, eds. W. F. van Gunsteren and P. K. Weiner, Escom, Leiden (1989), pp. 120–137.
37. A. Warshel, F. Sussman and G. King: *Biochemistry* **25**, 8368 (1985).
38. G. Eisenman and A. Villarroel: in *Monovalent Cations in Biological Systems*, ed. C. A. Pasternak, CRC Press Publishers. Boca Raton, FL (1989), pp. 1–29.
39. Y. A. Ovchinnikov, V. T. Ivanov and A. M. Shkrob: *BBA Library*, Elsevier Scientific Publ. Co., Amsterdam, Oxford, New York (1974), Vol. 12 pp. 1–464.

40. Neupert-Laves and M. Dobler: *Helv. Chim. Acta* **58**, 432 (1975).
41. L. K. Steinrauf and K. Folting: *Isr. J. Chem.* **24**, 290 (1984).
42. L. K. Steinrauf, J. A. Hamilton, and M. N. Sabesan: *J. Am. Chem. Soc.* **104,** 4085 (1982).
43. E. Grell, T. Funck and F. Eggers: in *Membranes*, ed. G. Eisenman, Dekker, New York (1975), Vol. 3. Ch. 1. pp. 1–126.
44. P. Mueller and D. O. Rudin: *Biochem. Biophys. Res. Commun.* **26**, 398 (1967).
45. L. A. R. Pioda, V. Stankova and W. Simon: *Anal. Letters.* **2**, 665 (1969).
46. G. Eisenman: in *Symposium on Membrane Transport and Metabolism*, eds. A. Kleinzeller and A. Kotyk, Academic Press, New York (1961), pp. 163–179.
47. G. Eisenman: *Biophys. J.*, **2**, (2), 259 (1962).
48. J. Burgess: *Metal ions in Solution.* Ellis Horwood Ltd. Chicester, England (1977) p. 186.
49. R. W. Zwansig: *J. Chem. Phys.* **22,** 1420 (1954).
50. G. Eisenman and R. Horn: *J. Membr. Biol.* **76**, 197 (1983).
51. N. Gresh, C. Etchebest, O. de la luz Rojas and A. Pullman: *Int. J. Quantum Chem*: *Quantum Biol. Symp.* **8**, 109 (1981).
52. W. F. Van Gunstaren, P. Gros, and A. E. Torda: *Abstr. 10th Int. Biophys. Congress* (1990), p. 11.
53. G. Eisenman, and O. Alvarez: *J. Membr. Biol.* **119**, 109 (1991).
54. K. Easwaran: in *Ion Treatment Through Membranes*, ed. K. Yagi and B. Pullman, Academic Press, Tokyo, New York (1987) pp. 17–39.
55. G. Eisenman: *Excerpta Medica, Int. Congr. Ser.* No. 87, 489 (1965).
56. G. Eisenman, and O. Alvarez: in *Membrane Structure and Function*: *The State of the Art*, K. R. K. Easwaran and B. P. Gaber, eds. Adenine Press, New York (1991). In Press.

Journal of Inclusion Phenomena and Molecular Recognition in Chemistry **12**: 55–74, 1992.

An Approach to Channel Type Molecular Structures. 1. Synthesis of *Bouquet*-Shaped Molecules Based on an [18]-O_6 Polyether Macrocycle

LUDOVIC JULLIEN and JEAN-MARIE LEHN*
Laboratoire de Chimie des Interactions Moléculaires, Collège de France, 11, Place Marcelin Berthelot, F-75005 Paris, France.

(Received: 31 January 1991; in final form: 5 March 1991)

Abstract. An approach to a transmembrane cation channel is described. It is based on the grafting of oxygen bearing side-chains on a macrocyclic polyether unit that constitutes the organizing core. The resulting species has a structure of overall *bouquet* shape. The synthesis of such a molecule $\mathscr{B}_M^O$, **14a** is described, together with that of its analogue bearing polymethylene side-chains $\mathscr{B}_M^C$, **14b**. The physicochemical properties of these molecules indicate that they possess the features expected on the basis of their structure.

Key words. Polyether macrocycle, molecular channel, molecular bundle.

1. Introduction

Communication between two regions of space separated by an impermeable barrier (e.g. a membrane) is a crucial event in the transfer of information and of signal generation in molecular and supramolecular systems. Such communication may be established and controlled via specialized structures capable of handling selectively ionic and molecular substrates. Ion transport through biological membranes is an essential phenomenon in living cells, for processes as different as energy metabolism, transmission of nerve impulses, visual transduction, etc. (1, 2a). Its study forms a basic area of supramolecular chemistry [3], involving the design of artificial transport processes and effectors [2, 3]. Two types of transport mechanisms can be distinguished: shuttle (or free carrier) and channel processes [1, 2].

In the first case, an ion binds to a free carrier species on one side of the membrane forming a complex which diffuses through the bilayer; the ion is released on one of the two faces of the membrane depending on the different gradients present in the system, making the carrier available for a new cycle. The shuttle approach, involving comparatively small and simple carrier molecules has been extensively investigated in the last twenty years [1, 2, 4, 5]. Numerous examples have been described involving both natural (e.g. valinomycin [4], monensin A [5], etc.) and artificial (crown ethers [6], cryptands [2b]) carrier molecules.

In the second mechanism, an ion crosses the membrane directly through a specialized transmembrane structure, such as that provided by gramicidin A [7] or alamethicin [8]. Much less work has been done on the design of artificial channel

This paper is dedicated to the memory of the late Dr C. J. Pedersen.

* Author for correspondence

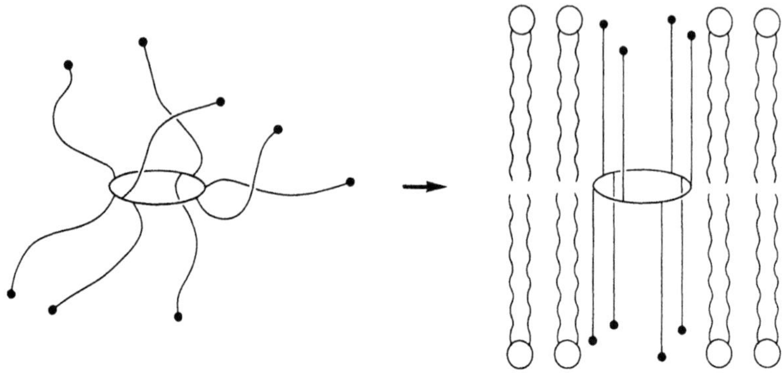

Fig. 1. The *Bouquet* approach: the functional structure might be attained by favorable organisation in a bilayer membrane of a molecule composed of flexible chains grafted on a circular core.

type species. However, several approaches have been explored [9], ranging from highly specialized structures in a well-defined conformation in the membrane to the creation of a defect in the organized bilayer and going through the generation of tubular mesophase [10a] or the face to face connection of several macrocyclic units [10b] as suggested by a solid state model [10c].

We report here an approach to the design of artificial ion channels based on the grafting of a bundle of amphiphilic chains on a central ring [11]. In appropriate anisotropic conditions (lipidic membrane of artificial vesicles for instance), these molecules might organize in a pseudo-cylindrical *structure* which may evoke the shape of a *bouquet* (or a sheaf) (Figure 1). *Functionally*, such a species represents a **ch**annel based on an organized b**undle** of chains, and may be described as a *chundle* [11]. This paper describes the synthesis and some physical properties of two molecules of *bouquet* structure $\mathscr{B}$ based on a tetrasubstituted macrocyclic polyether core [12] derived from Pedersen's 18-crown-6 parent compound [6a]. Ion binding studies, incorporation into bilayer membranes and ion-transport experiments will be reported elsewhere.

2. Experimental

2.1. GENERAL PROCEDURES

Microanalyses were performed by the 'Service Central d'Analyses du CNRS' (Vernaison) or by the 'Service de Microanalyses de l'Université P. et M. Curie' (Paris). Melting points were recorded on a Perkin-Elmer DSC 2 Differential Scanning Calorimeter. ^{1}H-NMR and ^{13}C-NMR spectra were recorded on an AM 200 SY Bruker spectrometer at room temperature unless otherwise indicated. Chemical shifts are given in ppm with protonated solvent as internal reference (^{1}H-NMR: $CHCl_3$ in $CDCl_3$ 7.26 ppm, CHD_2OD in CD_3OD 3.30 ppm, HOD in D_2O 4.70 ppm; ^{13}C-NMR: $^{13}CDCl_3$ in $CDCl_3$ 76.9 ppm, $^{13}CD_3OD$ in CD_3OD 49.0 ppm). Mass spectra (FAB positive) were performed by the 'Service de Spectrométrie de Masse du CNRS' (Vernaison). IR spectra and rotatory powers were respectively recorded on a Perkin-Elmer 297 spectrometer (film on KBr) and on a

Perkin-Elmer 241 polarimeter. Column chromatography was performed on silica gel 60 (0.040–0.063 mm) Merck. Fluorescent silica plates (analytical or preparative) Merck or Macherey-Nagel, or type E alumina Merck were used for thin layer chromatography. Analytical plates were revealed by UV (254 nm), iodine (I_2) or sulfuric acid solution (5%). Anhydrous solvents (SDS), stored on molecular sieves (3–4 Å) were used as received, except methylene chloride which was distilled from calcium hydride before use. All the catalytic hydrogenations were performed at a pressure of one bar.

Due to the physical state (waxes), satisfactory analyses were not obtained for the final products **13a**, **13b**, **14a** and **16a**. IR and ^{1}H-NMR spectroscopy respectively showed the presence of water and of residual solvent.

2.2. SYNTHETIC PROCEDURES

2.2.1. *Ethyl* 11-*Hydroxy*-3,6,9-*trioxaundecanoate* **1**

Ethyl diazoacetate (15 mL; 0.14 mol) was slowly added at room temperature to a stirred solution of triethyleneglycol (130 g; 0.87 mol) and trifluoride boron etherate (3 drops) in dry CH_2Cl_2 (200 ml). Abundant nitrogen formation occurred immediately. After stirring at r.t. for 7 h, the organic solution was washed (water, 100 mL; saturated NaCl, 2 × 50 ml), dried (Na_2SO_4), evaporated to dryness and distilled (bp: 145°C; 1–1.5 mm Hg) to give **1** as a colourless liquid (23.7 g; 70%). ^{1}H-NMR ($CDCl_3$): 4.22 (*q*, 7Hz, CH_2CH_3); 4.16 (*s*, OCH_2CO); 3.70 (*m*, 6 OCH_2); 2.43 (*s*, OH); 1.29 (*t*, 7Hz, CH_3). ^{13}C-NMR ($CDCl_3$): 169.8 (CO); 72.0 (CH_2CH_2OH); 70.2, 70.0, 69.7 (OCH_2); 68.0 (OCH_2CO); 61.0 (CH_2CH_2OH); 60.1 (CH_2CH_3); 13.6 (CH_3). TLC: SiO_2; CH_2Cl_2/MeOH (9/1); $R_f = 0.55$ (I_2). IR (cm^{-1}): 3360 (OH); 1760 (CO). *Anal. calc.* for $C_{10}H_{20}O_6$ (236) C 50.8; H 8.5; *found* C 51.5; H 8.75.

2.2.2. *Ethyl* 11-*Tosyloxy*-3,6,9-*trioxaundecanoate* **2**

Tosyl chloride (14.5 g; 77 mmol) was slowly added into a solution of **1** (15 g; 64 mmol) in pyridine (75 mL) cooled to 0°C. After one night at 4°C, the organic solution was poured onto crushed ice (100 g), then extracted with ether (150 mL). The organic phase was washed (3 M HCl, 500 mL; saturated NaCl, 50 mL), dried (Na_2SO_4) and evaporated to dryness yielding **2** as a light-yellow liquid used without further purification (22.3 g; 90%). ^{1}H-NMR ($CDCl_3$): 7.79 (*d*, 8Hz, 2H arom); 7.34 (*d*, 8Hz, 2H arom); 4.21 (*q*, 7Hz, CH_2CH_3); 4.16 (*m*, CH_2OTs); 4.13 (*s*, OCH_2CO); 3.72–3.59 (*m*, 5 OCH_2); 2.45 (*s*, CH_3 arom); 1.28 (*t*, 7Hz, CH_2CH_3). ^{13}C-NMR ($CDCl_3$): 169.9 (CO); 144.4, 132.9, 129.4, 127.5 (arom); 70.5, 70.3, 70.2, 68.9 (OCH_2); 68.3 (OCH_2CO); 60.3 (CH_2CH_3); 21.2 (CH_3 arom); 13.8 (CH_3). TLC: SiO_2; CH_2Cl_2/MeOH (95/5); $R_f = 0.63$ (UV, I_2). IR (cm^{-1}): 1760 (CO). *Anal. calc.* for $C_{17}H_{26}O_8S$ (390) C 52.3; H 6.7; *found* C 52.0; H 6.8.

2.2.3. *Ethyl* 11-*Azido*-3,6,9-*tiroxaundecanoate* **3**

A solution of **2** (25.8 g; 66 mmol) and NaN_3 (9.0 g; 138 mmol) in dry DMF (100 mL) was heated at 60°C for 20 h. The solvent was removed under vacuum and the residue was taken up in CH_2Cl_2 (125 mL), washed (water, 50 mL), dried (Na_2SO_4) and evaporated to dryness to yield **3** as a light-yellow liquid used without further purification (16.8 g; 97%). ^{1}H-NMR ($CDCl_3$): 4.22 (*q*, 7Hz, CH_2CH_3); 4.15 (*s*, OCH_2CO); 3.8–3.6 (*m*, 5 OCH_2), 3.39 (*t*, 5 Hz, CH_2N_3); 1.29 (*t*, 7Hz, CH_3). ^{13}C-NMR ($CDCl_3$): 169.9 (CO); 70.2, 70.2, 69.6 (OCH_2); 68.2 (OCH_2CO); 60.3 (CH_2CH_3); 50.2 (CH_2N_3); 13.8 (CH_3). TLC: SiO_2; CH_2Cl_2/MeOH (9/1); $R_f = 0.82$ (I_2). IR (cm^{-1}): 2120 (N_3); 1760 (CO). MS; $(M + H)^+ = 262$.

2.2.4. *Ethyl* 11-*Amino*-3,6,9-*trioxaundecanoate* **4a**

3 (5.20 g; 20 mmol) was catalytically hydrogenated on Pd/C 10% (700 mg) in dry CH_2Cl_2(100 mL) at r.t. for 6 h. After filtration of the catalyst and evaporation of the solvent, the amine **4a** was isolated as a liquid (4.59 g; 98%). ^{1}H-NMR ($CDCl_3$): 4.21 (*q*, 7Hz, CH_2CH_3); 4.15 (*s*, OCH_2CO); 3.8–3.6 (*m*, 5 OCH_2); 3.52 (*t*, 5 Hz, CH_2NH_2); 1.9 (broad *s*; NH_2); 1.29 (*t*, 7 Hz, CH_3). ^{13}C-NMR ($CDCl_3$): 170.0 (CO); 70.5, 70.3, 69.9, 69.6 (OCH_2); 68.3 (OCH_2CO); 60.4 (CH_2CH_3); 50.4 (CH_2NH_2); 13.8 (CH_3). TLC: SiO_2; CH_2Cl_2/MeOH (9/1); $R_f = 0.1$–0.3 (I_2). IR (cm^{-1}): 3400 (broad; NH_2); 1760 (CO). MS: $(M + H)^+ = 236$.

2.2.5. *N-Benzyloxycarbonyl-N-butylglycine* **5**

Benzylchloroformate (1.71 g; 10 mmol) and 4 M NaOH solution (2.5 mL; 10 mmol) were simultaneously and separately added dropwise at 0°C over a period of 30 min to a mixture of butylglycine (1.67 g; 10 mmol) and 2 M NaOH solution (10 mL; 20 mmol). After further stirring at 0°C for 20 min, unreacted products were extracted with ether (10 mL). The aqueous phase was acidified up to pH = 1 with 1 M HCl and then extracted twice with ether (50 mL; 25 mL). The combined organic phases were washed (saturated NaCl; 20 mL), dried (Na_2SO_4) and evaporated to dryness to yield **5** as a viscous colourless liquid (1.95 g; 74%). The proton NMR spectrum shows the presence of two amide rotamers. ^{1}H-NMR ($CDCl_3$): 7.35–7.30 (*m*, 5H arom); 5.17 and 5.13 (2*s*, OCH_2); 4.06 and 4.01 (2*s*, NCH_2CO); 3.34 (2*t*, CH_2N); 1.50 (2*tt*, CH_2CH_2N); 1.30 (2*tq*; CH_2CH_3); 0.92 and 0.89 (2*t*, CH_3). ^{13}C-NMR ($CDCl_3$): 174.2 (CO); 136.4, 128.3, 128.1, 127.9, (arom); 67.5 (OCH_2); 48.5 (several peaks; CH_2N and NCH_2CO); 30.2 (several peaks; CH_2CH_2N); 19.8 (CH_2CH_3); 13.6 (CH_3). TLC: SiO_2; CH_2Cl_2/MeOH (95/5); $R_f = 0.40$ (UV, I_2).

2.2.6. *Preparation of* **6a** and **6b**

DCC was slowly added to a mixture of **5**, **4** (either the free base or its chlorhydrate), DMAP and dry CH_2Cl_2 at 0°C. After stirring at r.t. overnight, the suspension was filtered and the residual solid washed with CH_2Cl_2 (20 mL). The combined organic phases were evaporated, taken up in Et_2O (250 mL) and filtered. The filtrate was washed (1 M HCl, 2 × 50 mL; saturated NaCl, 50 mL), dried (Na_2SO_4) and evaporated to dryness. **6a** and **6b** were purified by column chromatography (SiO_2; 100–150 g; gradient of acetone in CH_2Cl_2).

6a. From **5** (4.46 g; 17 mmol); **4a** (4.00 g; 17 mmol); DMAP (2.20 g; 18 mmol); CH_2Cl_2 (60 ml); DCC (3.70 g; 18 mmol). Yield: 2.91 g (36%). NMR 2 conformers: ^1H ($CDCl_3$): 7.36 (*m*, 5H arom); 6.59 and 6.40 (broad *s*, 0.55 and 0.45 NH); 5.14 (*s*, OCH_2Ph); 4.20 (*q*, 7 Hz, OCH_2CH_3); 4.13 (*s*, OCH_2CO); 3.91 (*s*, NCH_2CO); 3.8–3.4 (*m*, CH_2N and 5 OCH_2); 3.32 (*t*, 8Hz, $CH_2CH_2CH_2N$); 1.52 (*tt*; $J_1 \approx J_2 = 8$ Hz; $CH_2CH_2CH_2N$); 1.27 (*t*; 7 Hz; OCH_2CH_3); 1.24 (*m*; $CH_2CH_2CH_2N$); 0.92 (*t*; CH_3). ^{13}C ($CDCl_3$): 170.1 (COO); 169.0 (CON); 156 (broad, OCON), 136.3, 128.3, 127.8, 127.6 (arom); 70.7, 70.3, 70.1, 69.5, 68.5, 65.6 (OCH_2CO and 5 OCH_2); 67.3 (OCH_2Ph); 60.6 (OCH_2CH_3); 51.1 (NCH_2CO); 48.3 ($CH_2CH_2CH_2N$); 39.0 (OCH_2CH_2N); 33.8 ($CH_2CH_2CH_2N$); 29.9 (CH_2CH_3); 14.0 (OCH_2CH_3); 13.6 (CH_3). TLC: SiO_2; CH_2Cl_2/Acetone (6/4); $R_f = 0.70$ (UV, I_2). IR (cm^{-1}): 3340 (NH); 1755 (COO); 1710 (OCON); 1675 (CON). *Anal. calc.* for $C_{24}H_{38}N_2O_8$ (482) C 59.75 H 7.9 N 5.8; *found*: C 59.0 H 8.1 N 6.0. MS: $(M+H)^+ = 483$.

6b. From **5** (4.67 g; 18 mmol); **4b** chlorhydrate (4.68 g; 18 mmol); DMAP (4.40 g; 36 mmol); CH_2Cl_2 (100 ml); DCC (4.12 g; 20 mmol). Yield: 6.35 g (74%). NMR 2 conformers: ^1H ($CDCl_3$): 7.28 (*m*, 5H arom); 6.33 and 6.01 (broad *s*, 0.55 and 0.45 NH); 5.09 (*s*, OCH_2Ph); 4.06 (*q*, 7 Hz, OCH_2); 3.83 (*s*, NCH_2CO); 3.27 (*t*, 7 Hz, CH_2N); 3.14 (*dt*, $J_1 \approx J_2 = 6$ Hz, CH_2N); 2.23 (*t*, 7Hz, CH_2CO); 1.6–1.1 (*m*, 10 CH_2 and OCH_2CH_3); 0.85 (*t*; 7 Hz; CH_3). ^{13}C ($CDCl_3$): 173.4 (COO); 168.9 (CON); 136.1, 128.1, 127.8, 127.6 (arom); 67.3 (OCH_2Ph); 59.8 (OCH_2); 51.4 (NCH_2CO); 48.3 (CH_2N); 39.1 (CH_2N); 34.1 (CH_2COO); 29.2–28.8 (several peaks), 26.6, 24.7, 19.6 (CH_2); 14.0 (OCH_2CH_3); 13.5 (CH_3). TLC: SiO_2; CH_2Cl_2/MeOH (95/5); $R_f = 0.73$ (UV, I_2, H_2SO_4). *Anal. calc.* for $C_{27}H_{44}N_2O_5$ (476) C 68.1 H 9.2; *found*; C 67.9 H 9.4.

2.2.7. *Preparation of Ethyl* 11-(*N-butylglycylamino*)-3,6,9-*trioxaundecanoate* **7a** *and Ethyl* 11-(*N-butylglycyl amino*)-*undecanoate* **7b**

Hydrogenolysis of **6a** or **6b** on Pd/C 10% in CH_2Cl_2 at r.t. for 2 h gave **7a** and **7b** respectively. The catalyst was filtered off to give a solution which, after evaporation to dryness, yielded a colourless viscous liquid.

7a. 6a (543 mg; 1.2 mmol); Pd/C 10% (200 mg); CH_2Cl_2 (25 mL). Yield: 355 mg (92%). ^1H-NMR ($CDCl_3$): 7.64 (broad *s*, NH); 4.21 (*q*, 7 Hz, OCH_2CH_3); 4.14 (*s*, OCH_2CO); 3.8–3.4 (*m*, 5 OCH_2 and NCH_2); 3.32 (*s*, NCH_2CO); 2.80 (broad *s*, 1H); 2.63 (*t*, 7 Hz, $CH_2CH_2CH_2N$); 1.6–1.3 (*m*, $CH_2CH_2CH_2N$); 1.28 (*t*, 7 Hz, OCH_2CH_3); 0.91 (*t*, 7 Hz, CH_3). ^{13}C-NMR ($CDCl_3$): 170.3 (COO); 170.0 (CON); 70.6, 70.3, 70.2, 70.0, 69.5, 68.4 (OCH_2); 60.4 (OCH_2CH_3); 51.6 (NCH_2CO); 49.2 ($CH_2CH_2CH_2N$); 38.6 (CH_2N); 31.1 ($CH_2CH_2CH_2N$); 19.9 (CH_2CH_3; 13.9 (OCH_2CH_3); 13.5 (CH_3). TLC: Al_2O_3; CH_2Cl_2/MeOH (95/5); $R_f = 0.8$ (I_2).

7b. same as **7a** on a 13 mmol scale. Yield: 94%. ^1H-NMR ($CDCl_3$): 7.57 (*t*, 6 Hz, NH); 4.37 (broad *s*, NH); 3.93 (*q*, 7 Hz, OCH_2); 3.23 (*s*, NCH_2CO); 3.06 (*dt*, $J_1 = 6$ Hz and $J_2 = 7$ Hz, CH_2N); 2.52 (*t*, 7 Hz, $CH_2CH_2CH_2N$); 2.23 (*t*, 7Hz, CH_2CO); 1.5–1.0 (*m*, 10 CH_2); 1.07 (*t*, 7 Hz, OCH_2CH_3); 0.74 (*t*, 7 Hz, CH_3). ^{13}C-NMR ($CDCl_3$): 173.1 (COO); 169.4 (CON); 59.5 (OCH_2); 51.1 (NCH_2CO); 48.9 ($CH_2CH_2CH_2N$); 38.6 (CH_2N); 33.8 (CH_2CO); 30.6, 29.0, 28.9, 28.8, 28.7, 28.6, 26.4, 24.4, 19.7 (CH_2); 13.7 (OCH_2CH_3); 13.3 (CH_3). TLC: SiO_2; CH_2Cl_2/acetone (1/1); $R_f = 0.60$ (I_2).

2.2.8. *Preparation of* **9a** *and* **9b**

5-Nitro-1,3-benzenedicarbonyl chloride **8** was freshly prepared from the corresponding diacid by refluxing in excess thionyl chloride containing one drop of DMF for 2 h. After cooling to r.t. and evaporation of the solvent, the residue was taken up in benzene and evaporated twice, dissolved in dry CH_2Cl_2 and cooled to 0°C. To this solution, a solution of the appropriate amine in CH_2Cl_2 and TEA (freshly distilled; excess) was added dropwise at 0°C. The mixture was allowed to cool and was stirred at r.t. overnight. The organic solution was washed (1 M HCl; saturated NaCl), dried (Na_2SO_4) and evaporated to dryness. The purification of **9a** and **9b** was achieved by column chromatography (SiO_2; gradient of acetone in CH_2Cl_2).

9a. 5-Nitro-1,3-benzenedicarboxylic acid (500 mg; 2.4 mmol); $SOCl_2$ (3 ml); **7a** (1.81 g; 5.2 mmol); CH_2Cl_2 (25 mL); TEA (1 ml); 1 M HCl (20 mL); NaCl (10 mL); SiO_2 (100 g). Viscous liquid. Yield: 1.72 g (80%). ^{1}H-NMR ($C_2D_2Cl_4$; 375 K): 8.33 (*s*, 2H arom); 7.88 (*s*, 1H arom); 6.42 (broad *s*, 2NH); 4.22 (*q*, 7 Hz, 2 OCH_2CH_3); 4.11 (*s*, 2 OCH_2CO); 4.04 (broad *s*, 2 NCH_2CO); 3.8–3.55 (*m*, 10 OCH_2); 3.46 (*t*, 6Hz, 2 $CH_2CH_2CH_2N$); 3.42 (broad *t*, 2 NCH_2); 1.62 (*tt*, $J_1 \approx J_2 = 6$ Hz, 2 $CH_2CH_2CH_2N$); 1.4–1.2 (*m*, 2 CH_2CH_3); 1.30 (*t*, 7 Hz, 2 OCH_2CH_3); 0.90 (*t*, 7 Hz, 2 CH_3). ^{13}C ($CDCl_3$): 170.2 (COO); 168.6 (CON); 137.2, 114.1 (arom); 70.7, 70.4, 69.9, 69.5, 68.5 (OCH_2); 60.7 (OCH_2CH_3); 50 ($CH_2CH_2CH_2N$); 39.1 (CH_2N); 30 ($CH_2CH_2CH_2N$); 19.7 (CH_2CH_3); 14.0 (OCH_2CH_3); 13.5 (CH_3). TLC: SiO_2; CH_2Cl_2/MeOH (9/1); $R_f = 0.62$ (UV and I_2). *Anal. calc.* for $C_{40}H_{65}N_5O_{16}$ (871) C 55.1 H 7.5 N 8.0; *found*: C 54.4 H 7.7 N 8.1. MS: $(M+H)^+ = 873$; $(M+Na)^+ = 895$.

9b. Same preparation as for **9a** on a 16 mmol scale. Solid. Yield: 59%. ^{1}H-NMR ($CDCl_3$): 8.3–8.1 (*m*, 2H arom); 7.85–7.6 (*m*, 1H arom); 7.05–6.75 (*m*, 2NH); 3.94 (*q*, 7 Hz, 2 OCH_2CH_3); 3.9 and 3.7 (2 broad *s*, 2 NCH_2CO); 3.45–2.95 (*m*, 4 CH_2N); 2.12 (*t*, 7Hz, 2 CH_2CO); 1.55–0.9 (*m*, 20 CH_2); 1.08 (*t*, 7 Hz, 2 OCH_2CH_3); 0.9–0.6 (*m*, 2 CH_3). ^{13}C-NMR ($CDCl_3$): 173.3 (COO); 168.5, 167.4 (CON); 147.4, 137.7, 130.9, 122.3 (arom); 59.6 (OCH_2CH_3); 53.1, 51.7, 50.3, 49.1 (broad peaks; NCH_2CO and NCH_2); 39.3 (NCH_2); 33.9 (CH_2CO); 29.95, 29.1, 29.0, 28.9, 28.7, 28.6, 26.8, 24.8, 21.8, 19.4 (CH_2); 13.75 (OCH_2CH_3); 13.5, 13.1 (CH_3). M.P. 35.3°C and 87.4°C. TLC: SiO_2; CH_2Cl_2/acetone (8/2); $R_f = 0.80$ (UV, I_2). IR (cm^{-1}): 3280, 3100 (NH); 1740 (CO). *Anal. calc.* for $C_{46}H_{77}N_5O_{10}$ (859) C 64.3 H 9.0 N 8.15; *found*: C 64.4 H 9.05 N 8.2. MS: $(M+H)^+ = 861$.

2.2.9. *Preparation of* **10a** *and* **10b**

The nitro derivatives **9a** and **9b** were hydrogenated on Pd/C 10% in dry CH_2Cl at r.t. for 6 h. After filtration of the catalyst, the organic solution was evaporated to dryness to yield viscous colourless liquids.

10a. 9a (1.25 g; 1.4 mmol); Pd/C 10% (200 mg); CH_2Cl_2 (50 ml). Yield: 1.12 g (91%). ^{1}H-NMR ($CDCl_3$): 7.5–6.5 (*m*, 3 H arom and 2 NH); 4.16 (*q*, 7 Hz, 2 OCH_2CH_3); 4.1–2.6 (*m*, 12 OCH_2 and 6 CH_2N); 1.6–1.0 (*m*, 4 CH_2); 1.24 (*t*, 7 Hz, 2 OCH_2CH_3); 1.0–0.7 (*m*, 2 CH_3). ^{13}C-NMR ($CDCl_3$): 172.0 (CON); 170.8 (COO); 169.0 (CON); 147.8, 137.4, 113.8 (arom); 71.0, 70.5, 69.8, 68.7 (OCH_2); 61.3 (OCH_2CH_3); 52.5, 50.8, 49.3, 46.6, 39.5 (CH_2N); 30.7, 29.3 ($CH_2CH_2CH_2N$);

20.5, 20.0 (CH_2CH_3); 14.5 (OCH_2CH_3); 14.0 (2; CH_3). TLC: SiO_2; CH_2Cl_2/MeOH (9/1); $R_f = 0.46$ (UV, I_2). MS: $(M+H)^+ = 843$.

10b. Same preparation as for **10a** on a 0.45 mmol scale. Yield: 93%. ^{1}H-NMR ($CDCl_3$): 7.3 and 6.85–6.6 (*m*, 3 H arom and 2 NH); 4.10 (*q*, 7 Hz, 2 OCH_2CH_3); 4.2–3.8 (*m*, 2 NCH_2CO); 3.6–3.1 (*m*, 4 CH_2N); 2.29 (*t*, 6 Hz, 2 CH_2CO); 1.7–1.1 (*m*, 20 CH_2); 1.24 (*t*, 7 Hz, 2 OCH_2CH_3); 1.1–0.75 (*m*, 2 CH_3). TLC: SiO_2; CH_2Cl_2/acetone (6/4); $R_f = 0.29$ (UV, I_2). MS: $(M+H)^+ = 831$.

2.2.10. *Preparation of* **13a** $\mathscr{B}^{O}_{M}$ *COOEt and* **13b** $\mathscr{B}^{C}_{M}$ *COOEt*

A mixture of [18]-O_6 tetracarboxylic acid **11** [12] (75 mg; 0.17 mmol; Merck), phosphorus pentachloride (150 mg, large excess), in freshly distilled CH_2Cl_2 (5 ml) was stirred at r.t. overnight. A further amount of phosphorus pentachloride (145 mg) was added and the mixture stirred for 1 h. After solvent evaporation, the residue was taken up twice in anhydrous benzene and the solution was evaporated under vacuum at 40°C to give the tetraacyl chloride **12** [12]. A solution of that chloride and **10a** or **10b** (1 mmol) in anhydrous dimethylacetamide (7 mL) was cooled to 0°C. 4-Dimethylaminopyridine (423 mg; large excess) was slowly added and the mixture was stirred at r.t. for 24 h. After evaporation of the solvent under 1 mm-vacuum, the residue was taken up in CH_2Cl_2 (50 mL). The organic solution was washed (1 M HCl, 25 mL; saturated NaCl, 25 mL), dried (Na_2SO_4). After evaporation under vacuum and careful column chromatography (SiO_2; 120 g; gradient of MeOH in CH_2Cl_2) a colourless glassy wax was obtained.

13a. Yield: 87%. ^{1}H-NMR ($C_2D_2Cl_4$; 394 K): 10.1 (broad *s*, NH); 8.0 (broad *s*, 2 H arom); 7.3 (broad *s*, 1 H arom); 6.6 (broad *s*, 2 NH); 4.6 (broad *s*, CH); 4.24 (*q*, 7 Hz, 2 OCH_2CH_3); 4.13 (*s*, 2 OCH_2CO); 4.03 (broad *s*, 2 NCH_2CO); 4.0 and 3.8–3.3 (*m*, 10 OCH_2 and 4 CH_2N); 1.60 (*tt*, 2 $CH_2CH_2CH_2N$); 1.3 (*m*, 2 CH_2CH_3); 1.32 (*t*, 7 Hz, 2 OCH_2CH_3); 0.89 (*t*, 7 Hz, 2 CH_3). ^{13}C-NMR: ($C_2D_2Cl_4$; 383 K): 171.1 (CON); 170.4 (COO); 168.8 (CON); 138.3, 137.6, 121.8, 121.0 (arom); 81.5 (CH); 69.2 (OCH_2 crown); 71.4, 70.9, 70.8, 70.6, 69.1 (OCH_2); 60.9 (OCH_2CH_3); 51.2 (NCH_2CO); 49.4 ($CH_2CH_2CH_2N$); 39.9 (CH_2N); 30.4 ($CH_2CH_2CH_2N$); 20.2 (CH_2CH_3); 14.4 (OCH_2CH_3); 13.7 (CH_3). TLC: SiO_2; CH_2Cl_2/MeOH (9/1); $R_f = 0.45$; grafted C18 SiO_2 (Whatman LKC18F); MeOH/H_2O (95/5); $R_f = 0.28$ (UV and I_2). MS for $(M+Na)^+$: maximum calculated peak: 3759; maximum experimental peak: 3759.8; peaks distribution shape in agreement with calculated data. $[\alpha]^{25}_{589} = +15°$; $[\alpha]^{25}_{578} = +17°$; $[\alpha^{25}_{546} = +19°$ ($c = 1.06$; EtOH 95%).

13b. Yield: 71%. ^{1}H-NMR ($C_2D_2Cl_4$; 375 K): 10.05 (broad *s*, NH); 7.95 (broad *s*, 2 H arom); 7.25 (broad *s*, 1 H arom); 6.5 (broad *s*, 2 NH); 4.6 (broad *s*, CH); 4.10 (*q*, 7 Hz, 2 OCH_2CH_3); 4.0 (m, 2 OCH_2 and 6 CH_2N); 2.29 (*t*, 6 Hz, 2 CH_2CO); 1.7–1.2 (*m*, 20 CH_2); 1.24 (*t*, 7 Hz, 2 OCH_2CH_3); 0.87 (*t*, 2 CH_3). ^{13}C-NMR: ($C_2D_2Cl_4$; 375K): 174.2 (COO); 171.1, 168.8, 168.4 (CON); 138.2, 137.3, 121.8, 120.8 (arom); 81.3 (CH); 71.5, 68.9 (OCH_2 crown); 60.2 (OCH_2CH_3); 51.4 (NCH_2CO); 49.3 ($CH_2CH_2CH_2N$); 40.1 (CH_2N); 34.7, 30.3, 29.85, 29.7, 29.6, 29.5, 29.4, 27.2, 25.2, 20.2 (CH_2); 14.5 (OCH_2CH_3); 13.8 (CH_3). TLC: SiO_2; CH_2Cl_2/MeOH (9/1); $R_f = 0.65$ (UV and I_2). MS for $(M+Na)^+$: maximum calculated peak: 3711; maximum experimental peak: 3711.4; peaks distribution shape in agreement with calculated data. $[\alpha]^{25}_{589} = +12°$; $[\alpha]^{25}_{578} = +13°$; $[\alpha]^{25}_{546} = +15°$ ($c = 1.71$; EtOH 95%).

2.2.11. *Preparation of* **14a** $\mathcal{B}^O_M$ *COONa and* **14b** $\mathcal{B}^C_M$ *COONa*

A mixture of **13a** or **13b** (32 mg; 9 μmol), 1 M aqueous NaOH (170 μL; 170 μmol) and distilled MeOH (5 mL) was stirred at r.t. for 2–4 h (the course of the reaction was followed by the disappearance of the COOEt signal in the ^{1}H-NMR spectra; the reaction was quicker with **13a** than with **13b**). Solid CO_2 was then added, the solution was evaporated to dryness and the solid residue was taken up in absolute EtOH. The suspension was weakly bath-sonicated for one minute and filtered on millipore. The resulting solution was evaporated to dryness to yield **14a** and **14b** as glassy waxes, which were used without further purification. They were stored as MeOH or EtOH solutions at 4°C.

14a. ^{1}H-NMR (D_2O; 373K; ethanol as the internal reference; CH_3 signal; 1.14 ppm): 7.73 (*s*, 2 H arom); 7.28 (*s*, 1 H arom); 4.65 (*s*, CH); 4.08 (*s*, 2 NCH_2CO); 3.9–3.5 (*m*, 2 OCH_2 crown); 3.7 – 3.5 (*m*, 10 OCH_2); 3.5–3.2 (2*t*, 4 CH_2N); 1.49 (*m*, 2 CH_2); 1.12 (*m*, 2 CH_2CH_3); 0.74 (*t*, 7Hz, 2 CH_3). ^{13}C-NMR (D_2O; 373K; internal reference ethanol: CH_3 signal: 17.8 ppm): 177.8 (COO); 172.6, 170.6 (CON); 137.8, 121.3 (arom); 71.2; 70.4; 70.3; 69.65 (OCH_2); 50.2 (large; NCH_2): 39.9 (CH_2N); 30.1 (CH_2); 19.9 (CH_2CH_3); 13.5 (CH_3).

14b. ^{1}H-NMR (CD_3OD): 8.1–7.8 (*m*, 2 H arom); 7.4–7.2 (*m*, 1 H arom); 6.75 (broad *s*, 2 NH); 4.65 (broad *s*, CH); 4.3–3.0 (*m*, 2 OCH_2 and 6 CH_2N); 2.30 (*t*, 7 Hz, 2 CH_2CO); 1.8–1.3 (*m*, 20 CH_2); 1.0 and 0.8 (2 broad *t*, 2 CH_3). ^{13}C-NMR (CD_3OD): 176.0 (COO); 172.85, 170.2 (2; CON); 140, 138.4, 121.7 (arom); 81.85 (CH); 71.7, 69.8 (OCH_2); 53.1, 51.7, 40.6 (CH_2N); 34.8, 31.4, 30.4, 28.1, 26.0, 21.2, 20.9 (CH_2); 14.2 (CH_3).

2.2.12. *Preparation of the Model Compounds* **15a, 15b, 16a** *and* **16b**

15a. 5-Nitro-1,3-benzenedicarbonyl chloride was reacted with diethylamine under the same conditions as for the preparation of **9a** and **9b** to yield 5-nitro-*N,N'*-diethyl-1,3-benzenedicarboxamide (93%). This compound was catalytically hydrogenated as described for **10a** and **10b** to give 5-amino-*N,N'*-diethyl-1,3-benzenedicarboxamide (71%) which was then condensed with cyclohexylcarbonyl chloride as described for **13a** and **13b** to give **15a** (76%). ^{1}H-NMR ($CDCl_3$): 9.19 (*s*, NH); 7.64 (*s*, 2H arom); 6.95 (*s*, 1H arom); 3.46 (*m*, 2 NCH_2); 3.21 (*m*, 2 NCH_2); 2.22 (*m*, CH); 1.9–1.0 (*m*, 5 CH_2 and 4 CH_3). ^{13}C-NMR ($CDCl_3$): 175.1, 170.1 (CON); 139.2, 137.5, 116.7, 116.3 (arom); 45.8 (CH); 43.3 and 42.7 (NCH_2); 29.3 and 25.5 (CH_2); 13.9 and 12.8 (CH_3). *Anal. calc.* for $C_{23}H_{235}N_3O_3$ (401) C 68.8 H 8.7 N 10.5; *found*: C 68.0 H 8.6 N 10.3.

15b was prepared by esterification of 5-aminoisophthalic acid in acidic ethanol followed by condensation of the resulting diethylester with cyclohexylcarbonyl chloride as described for **13a** and **13b** (90%). ^{1}H-NMR ($CDCl_3$): 8.42 (*s*, NH); 8.40 (*s*, 2 H arom); 7.36 (*s*, 1 H arom); 4.40 (*q*, 7 Hz, 2 OCH_2); 2.27 (*m*, CH); 2.0–1.2 (*m*, 5 CH_2); 1.41 (*t*, 7 Hz, 2 CH_3). M. P. 162°C. *Anal. calc.* for $C_{19}H_{25}NO_5$ (347) C 65.7 H 7.2; *found*: C 65.6 H 7.3.

16a was prepared by reaction of 5-amino-*N*,*N'*-diethyl-1,3-benzenedicarboxamide with the tetracarbonyl chloride **12** as described for **13a** and **13b** (50%). ^{1}H-NMR ($CDCl_3$): 10.5 (broad *s*, NH); 7.86 (broad *s*, 2H arom); 7.05 (broad *s*, 1H arom); 4.59 (broad *s*, CH); 4.0–3.0 (*m*, 2 OCH_2 and 4 NCH_2); 1.4–0.8 (*m*, 4 CH_3). ^{13}C-NMR ($CDCl_3$): 169.4, 167.8 (CON); 137.7, 119.6, 116.6 (arom); 79.8 (CH); 70.1 and 67.9 (OCH_2); 43.1 and 39.1 (NCH_2); 13.9 and 12.5 (CH_3).

16b was prepared by reaction of diethyl 5- aminoisophthalic diester with the tetracarbonyl chloride **12** as described for **13a** and **13b** (35%; not optimized). ^{1}H-NMR ($CDCl_3$): 9.25 (*s*, NH); 8.30 (*s*, 2 H arom); 8.23 (*s*, 1 H arom); 4.66 (*s*, CH); 4.33 (*q*, 7 Hz, 2 OCH_2CH_3); 4.2–3.6 (*m*, 2 OCH_2); 1.38 (*t*, 7 Hz, 2 CH_3). ^{13}C-NMR ($CDCl_3$): 168.2, 165.1 (CO); 137.5, 131.0, 125.8, 124.1 (arom); 80.9 (CH); 69.8, 69.3 (OCH_2); 61.2 (OCH_2CH_3); 13.7 (CH_3). *Anal. calc.* for $C_{64}H_{76}N_4O_{26}$ (1316) C 58.3 H 5.8; *found*: C 57.5 H 5.9.

2.3. SPECTROSCOPIC MEASUREMENTS

Solvents were of spectroscopic grade. 1-pentanol and 1-octanol were distilled twice on activated charcoal before use. Solutions were prepared from a stock solution in CH_2Cl_2 or $CHCl_3$. Aliquots of mother solution were put in volumetric flasks. After complete evaporation of the volatile solvent, the appropriate solvent was added.

UV spectra were recorded on a Perkin-Elmer 554 spectrometer. Fluorescence spectra were measured with an AMINCO SPF 500 spectrometer coupled to a Kontron PSI 80 computer for data acquisition and were corrected for the residual solvent fluorescence. Slit widths were 4 nm (excitation) and 8 nm (emission). In all cases, absorbance at the excitation wavelength was less than 0.1. CD spectra were recorded on a Mark V Jobin-Yvon spectrometer and corrected for solvent residual absorption. In all cases, absorbance was less than 1.4.

3. Results

3.1. DESIGN OF THE MOLECULAR BOUQUETS $\mathcal{B}_M$

The design of the *bouquet* molecules was inspired at least in part by the geometrical and structural features of the natural peptides gramicidin A [7] and alamethicin [8]. Included in natural or artificial bilayers, they display their activity under the form of a cylinder with a length of about 30 Å and a diameter of a few Å (3–7 Å). Moreover, they present two gradients of increasing hydrophobicity: (i) from the interior to the exterior and, (ii) from the top to the middle of the cylinder. In addition to these essential structural criteria, a UV absorbing chromophore was introduced in the core of the *bouquet* molecules, in order to provide a spectroscopic label facilitating the interpretation of the eventual ionic conduction results. It may allow us to probe their incorporation and to study their possible orientations and conformations in bilayer membranes, so as to be able to draw structure-property relationships.

We have synthesised two series of molecules exhibiting these required characteristics. The first $\mathcal{B}_M$ is built up on an [18]-O_6 macrocyclic crown ether core and is the object of the present paper. The second $\mathscr{B}_{CD}$ is elaborated from a β-cyclodextrin derivative and will be described elsewhere. In view of comparative studies, two analogs were synthesised in each series: (i) the first, denoted $\mathscr{B}^{O}$, bears polyoxyethylenic chains and contains the oxygen sites necessary for the binding of cations along the channel in a functional ionophore, (ii) the second, denoted $\mathscr{B}^{c}$, is the alkyl analogue, which should display a similar structural behaviour but much weaker interactions with ions. Further criteria such as synthetic accessibility, chemical and structural characterization, means of visualization, postulated interactions with ions, geometric features in isotropic and anisotropic media, led finally to the two molecules $\mathscr{B}^{o}_{M}$ and $\mathscr{B}^{c}_{M}$ (**14a** and **14b**; Scheme 3) as first approaches to the *bouquet* type structures and to potential *chundle* type ion transport functions.

3.2. SYNTHESIS OF THE $\mathscr{B}_{M}$ MOLECULES

The linear aminoester chain **4a** was obtained in four steps starting from triethyleneglycol (Schemes 1 and 2). The condensation of ethyl diazoacetate in the

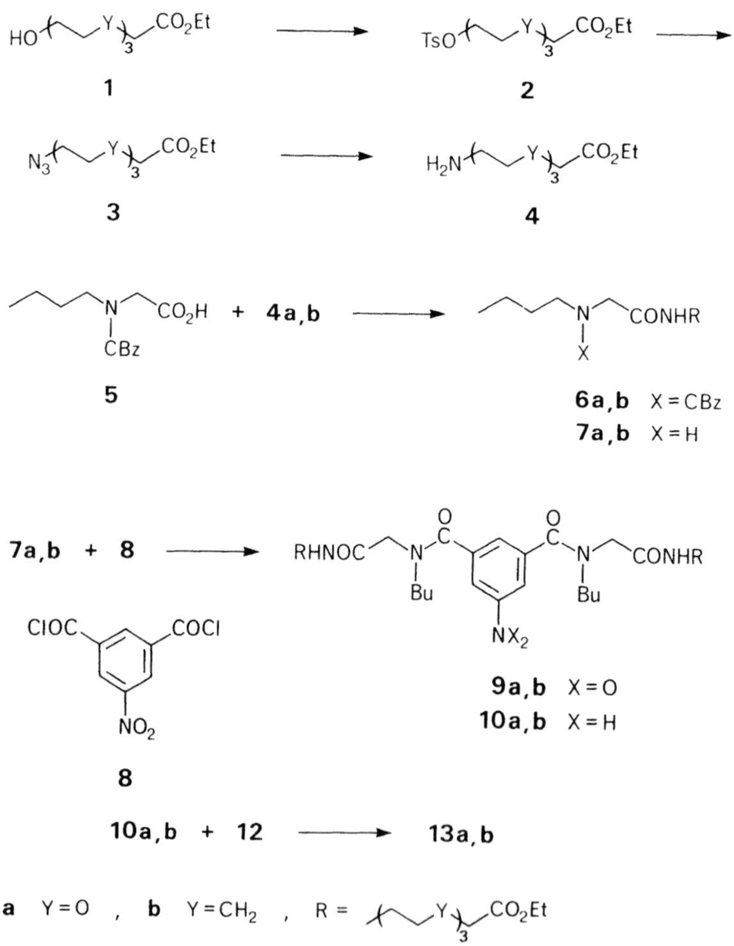

Scheme 1

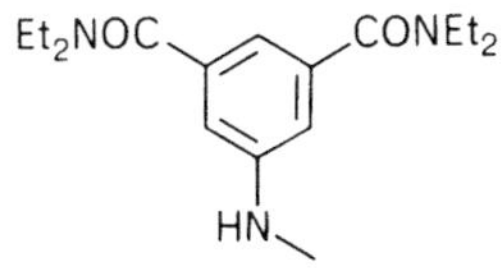

a

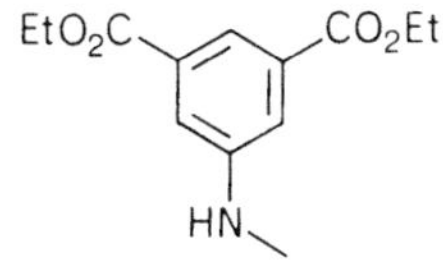

b

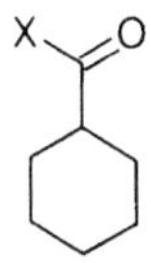

15a X = a
15b X = b

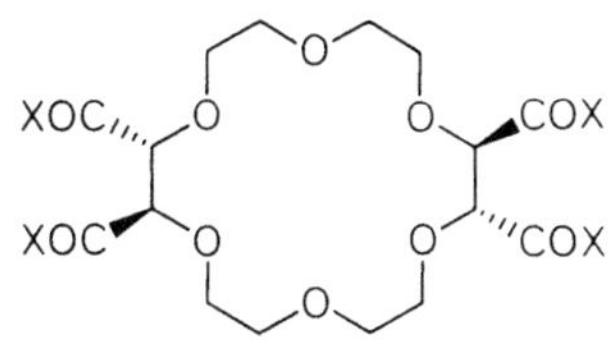

11 X = OH
12 X = Cl
16a X = a
16b X = b

Scheme 2

presence of BF_3 etherate [13] gave the hydroxyester **1** in 70% yield. **1** was then transformed into the aminoester **4a** via the tosylate **2** (TsCl, Pyr, 4°C) and the azide **3** (NaN_3, DMF, 60°C) which was hydrogenated to the amine **4a** (H_2, Pd/C 10%, CH_2Cl_2, RT). **4b** was obtained from the commercial aminoacid by esterification in acidified absolute ethanol [14]. The aminoesters **4a** and **4b** were coupled with carbobenzyloxybutylglycine **5** (obtained from *n*-butylglycine [15] and benzylchloroformate in the usual way [15]) using dicyclohexylcarbodiimide (DCC) and 4-dimethylaminopyridine (DMAP) to give **6a** and **6b**. Removal of the protecting groups by hydrogenolysis and condensation of the amines **7a** and **7b** with 5-nitro-1,3-benzenedicarbonyl chloride **8** yielded **9a** and **9b** (TEA, CH_2Cl_2, RT). After catalytic hydrogenation of the nitro group, the anilines **10a** and **10b** were grafted on the (+)-tartrocrown tetracarbonyl tetrachloride **12** [12] to give **13a** ($\mathscr{B}^{o}_{M}$ COOEt) and **13b** ($\mathscr{B}^{c}_{B}$ COOEt), respectively. The final saponification of the terminal ester groups to **14a** ($\mathscr{B}^{o}_{m}$ COONa) and **14b** ($\mathscr{B}^{c}_{M}$ COONa) was achieved by sodium hydroxide in methanol at room temperature. Polarimetric measurements were used to ensure that no epimerization had occured during saponification.

3.3. PROPERTIES OF THE *BOUQUET* MOLECULES $\mathscr{B}_M$

The properties of the $\mathscr{B}_M$ *bouquets* in solution were investigated mainly in order to prepare the further steps of this work i.e. membrane incorporation and ion transport studies.

The first features examined are related to the molecular geometry of the $\mathscr{B}_M$ species and its modification according to the medium. The aromatic chromophores were used as visualisation and potential conformational probes of the *bouquets*. UV spectra are affected by solvent effects, aggregation phenomena and conformational perturbation. In order to analyse eventual changes of UV spectra in membrane

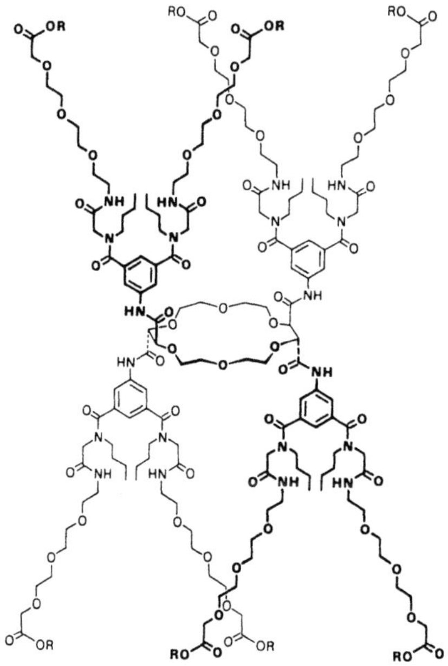

13a R = Et
14a R = Na

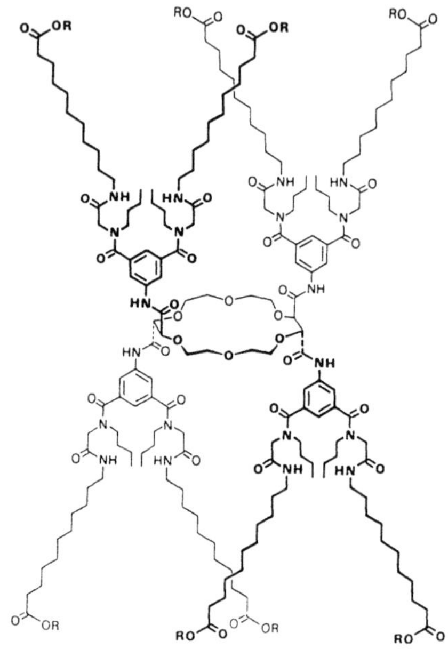

13b R = Et
14b R = Na

Scheme 3

medium relative to solvent, spectra were collected in several solvents displaying a wide range of liquid state characteristics: refraction index (from MeOH to DMSO), lipophilicity (from light alcohols, MeOH or EtOH, to higher homologs such as pentanol or octanol).

The second type of properties to be studied will deal with the interaction of the $\mathscr{B}_M$ *bouquets* with cations in an isotropic solvent. In view of their potential ionophore properties, it is necessary to investigate their possible associations with cations, though the molecular geometry may not be the same in an isotropic and in a membrane medium. Using the esters, **13a** and **13b**, or the model compounds **16a** and **16b**, allows the study of the interactions of cations mainly with both the core and the chains or the core alone of the *bouquets*. These results will be reported elsewhere.

3.3.1. *Spectroscopic Properties*

Electronic spectroscopy studies were performed on **13a**, **13b**, **14a** and **14b**. For comparison purposes, reference compounds containing the basic chromophore, **15a** and **15b**, as well as models devoid of chains, **16a** and **16b**, were also examined (Scheme 2) (Tables I and II). Figure 2 shows typical UV absorption and fluorescence emission spectra of these molecules. The UV spectra display a maximum with a strong molecular extinction coefficient ($\lambda_M = 210-225$ nm) and a shoulder ($\lambda_s = 240-245$ nm). When excited at λ_s, these molecules fluoresce weakly at about 360 nm. **13a**, **13b**, **14a**, **14b**, **16a** and **16b** being chiral and bearing strongly absorbing

Table I. UV absorption characteristics of **13a**, **13b**, **14a**, **14b**, **15a**, **15b**, **16a** and **16b** in several solvents

Solvent	n_D^{25} [a]	Compound	λ_M(nm)	ε_M	λ_s(nm)	ε_s
Methanol	1.326	**13a**	212	160 000	242	105 000
		13b	212	85 000	240	105 000
		15a	214	37 000	244	26 000
		15b	223	48 000	254	17 000
		16a	214	110 000	240	76 000
		16b	220	120 000	260	40 000
Ethanol 95%	1.359	**14a**	216	130 000		
		14b	210[c]	130 000		
		16b	220	120 000	260	35 000
n-Pentanol	1.408	**13a**	212	111 000	242	80 000
		13b	220	400 000		
		15a	215	29 000	240	25 000
		15b	225	41 000	256	12 500
		16a	216	91 000	240	67 000
		16b	222	100 000	260	35 000
n-Octanol	1.427	**13a**	220	330 000		
		13b	216	138 000	240	92 000
		15a	217	25 000	240	19 000
		15b	225	33 000	256	8 900
		16a	216	93 000	240	69 000
		16b	223	100 000	260	26 000
Chloroform	1.444	**16b**			260	29 000
DMSO	1.479[b]	**16b**			260	18 000

[a] Refraction index at 25°C.
[b] At 20°C
[c] Not a maximum.

Table II. Fluorescence emission characteristics of **13a**, **15b** and **16b** in different solvents

Solvent	Compound	$\lambda_{excitation}$ (nm)	$\lambda_{emission}$ (nm)
Methanol	**13a**	270	364
	15b	260	361
	16b	260	356
n-Pentanol	**13a**	270	362
	15b	260	358
	16b	260	355
n-Octanol	**13a**	270	360
	15b	260	357
	16b	260	351

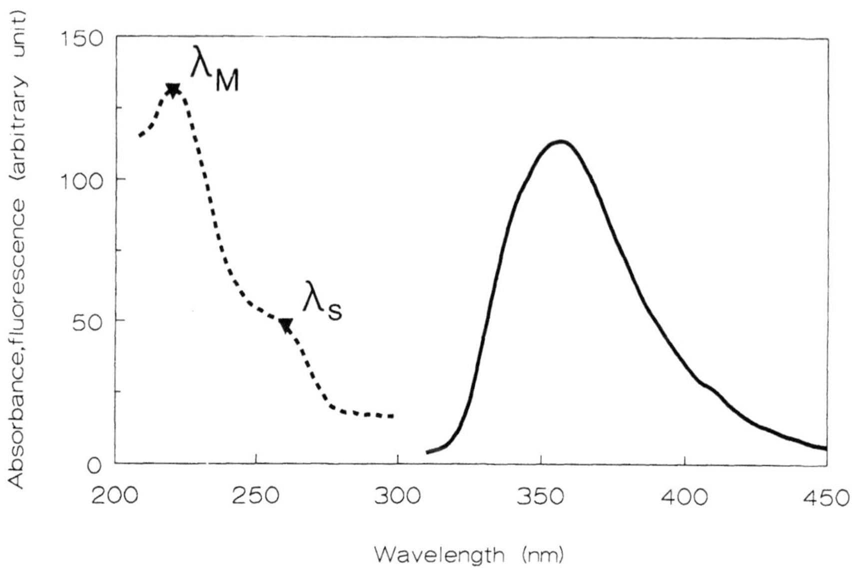

Fig. 2. UV absorption (- - -) and fluorescence emission (——) spectra (excitation at 260 nm) of **16b** in MeOH. λ_M and λ_s indicate the wavelengths mentioned in the text.

chromophores, some circular dichroism spectra were also recorded. They display an intramolecular excitonic coupling. The circular dichroism spectra show a negative $\Delta\varepsilon$ wave at long wavelength. Figure 3 presents a typical spectrum and Table III summarizes the results.

3.3.2. *^{1}H-NMR and ^{13}C-NMR Investigations*

The 200 MHz proton NMR spectrum of **13a** shows broad peaks at room temperature and is not informative. It is much better resolved at 394 K as shown in Figure 4.

The 50.3 MHz ^{13}C-NMR spectrum of **13a** at 298 K (Figure 5, bottom) displays broadened peaks as well as a doubling of the signals of the CH_2 carbons in α and β positions to N of the tertiary amide functions (in the 30 and 50 ppm regions). On heating to 383 K (Figure 5, top) sharp resonances are obtained and the doubled signals coalesce into narrow singlets.

Measurements of ^{13}C-NMR relaxation times are of general use to assess the degree of local and overall molecular mobility [17]. Because of the complexity of the ^{13}C-NMR spectra of **13a** and **13b** at room temperature, the longitudinal relaxation times T_1 of all the macrocycle-carbon atoms were determined on **16b** (Table IV). When corrected for the number of hydrogen atoms carried by each carbon atom, all the positions on the ring display the same mobility behavior. For **13a** it was possible to determine the ^{13}C relaxation times at room temperature for the carbon atoms of the side chains. They increase from the aromatic carbons ($NT_1 \approx 0.19$ s) towards the end of the chain: $NT_1 \approx 0.3$–0.6 s (CH_2 polyoxy), 1.8 s (CH_2 ester), 6 s (CH_3 ester) ($N =$ number of hydrogens on the carbon atom). A similar behaviour was found for **13b**.

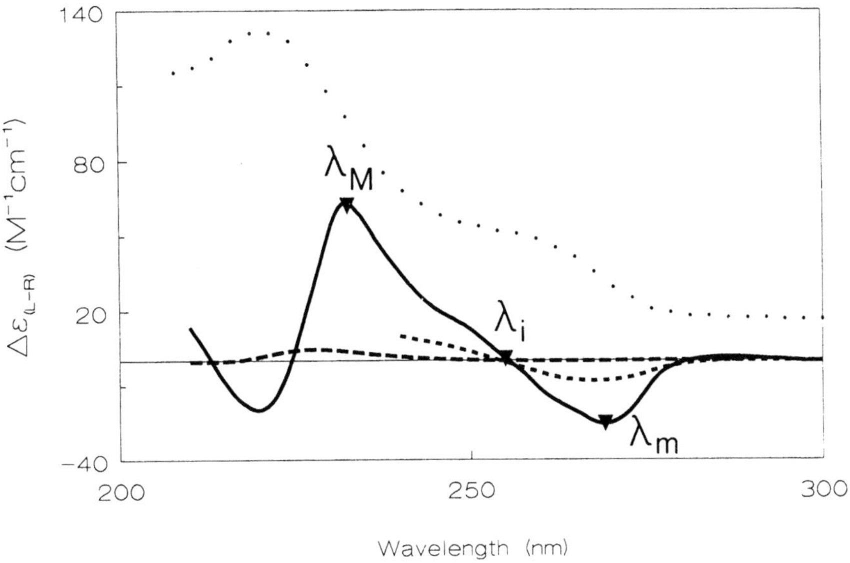

Fig. 3. Circular dichroism spectra in MeOH (——), *n*-pentanol (- - -), *n*-octanol (– – –) and UV absorption spectrum of **16b** in MeOH (. . . .). λ_M, λ_i and λ_m indicate the wavelengths mentioned in the text.

Table III. Cicular dichroism characteristics of **13a**, **13b**, **14a** and **16b** in different solvents

Solvent	Compound	λ_M(nm)	λ_i(nm)	λ_m(nm)
Methanol	**13a**	246	267	273
	13b	249	268	291
	14a	253	270	278
	16b	232	256	269
n-Pentanol	**13a**	—	266	272
	13b	—	273	288
	14a	—	245	264
	16b	—	254	267
n-Octanol	**13b**	249	—	—
	13b	246	276	290
	14a	—	252	267
	16b	228	253	—

Table IV. ^{13}C-NMR longitudinal relaxation times T_1 of the carbon atoms of the macrocycle in **16b**

Carbon	δ (ppm)	T_1(s)	Standard deviation	NT_1(s)
CH	80.1	0.11	0.03	0.11
CH_2	70.3	0.06	0.06	0.12
CH_2	68.0	0.065	0.03	0.13

100 mg of **16b** in 0.5 mL of $CDCl_3$ at r.t. (without degassing). δ = ^{13}C chemical shifts; N = number of hydrogens bound to the carbon atom.

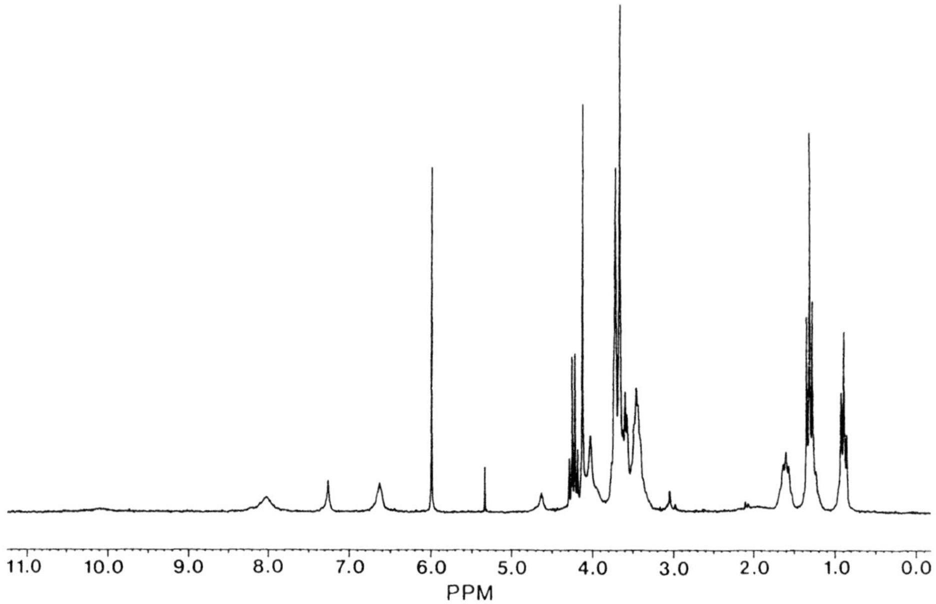

Fig. 4. 200 MHz proton NMR spectrum of the *bouquet* molecule **13a** at 394 K in $C_2D_2Cl_4$ (solvent signal at 6.0 p.p.m.); see experimental section for peak assignment.

4. Discussion

4.1. SPECTROSCOPIC PROPERTIES

The UV spectra in homogeneous medium clearly show the absence of a marked solvatochromic effect in the $\mathscr{B}_M$ molecules **13a** and **13b** investigated (little or no variation of the λ_{max} in the solvents studied). Thus a modification of medium polarity does not affect their UV spectra. As a consequence, any change which might take place on membrane incorporation would have to arise from other factors such as aggregation phenomenon or conformational effects.

4.2. MOLECULAR GEOMETRY OF THE BOUQUET MOLECULES $\mathscr{B}_M$

The structural features of the *bouquet* molecule $\mathscr{B}^{\circ}_M$ **14a** correspond to those desired for *chundle* type cation transport functions. The oxygen containing chains and core should be able to bind metal ions. With the side-chains in their extended state, the molecule is long enough (about 45–50 Å) to span a bilayer membrane. The terminal carboxylate groups provide the charged sites required for anchorage at the water/membrane interfaces and transmembrane orientation. The rigid phenyl groups may be expected to hinder back-bending of the chains and the *n*-butyl groups confer lipophilicity facilitating membrane inclusion.

The relaxation times of all carbon atoms of the macrocycle in **16b** are found to be similar (Table IV). This indicates that the crown ring is rigid on this time scale, and that all positions play the same role. The local dynamical properties are not significantly perturbed by the attached substituents. Although these results do not

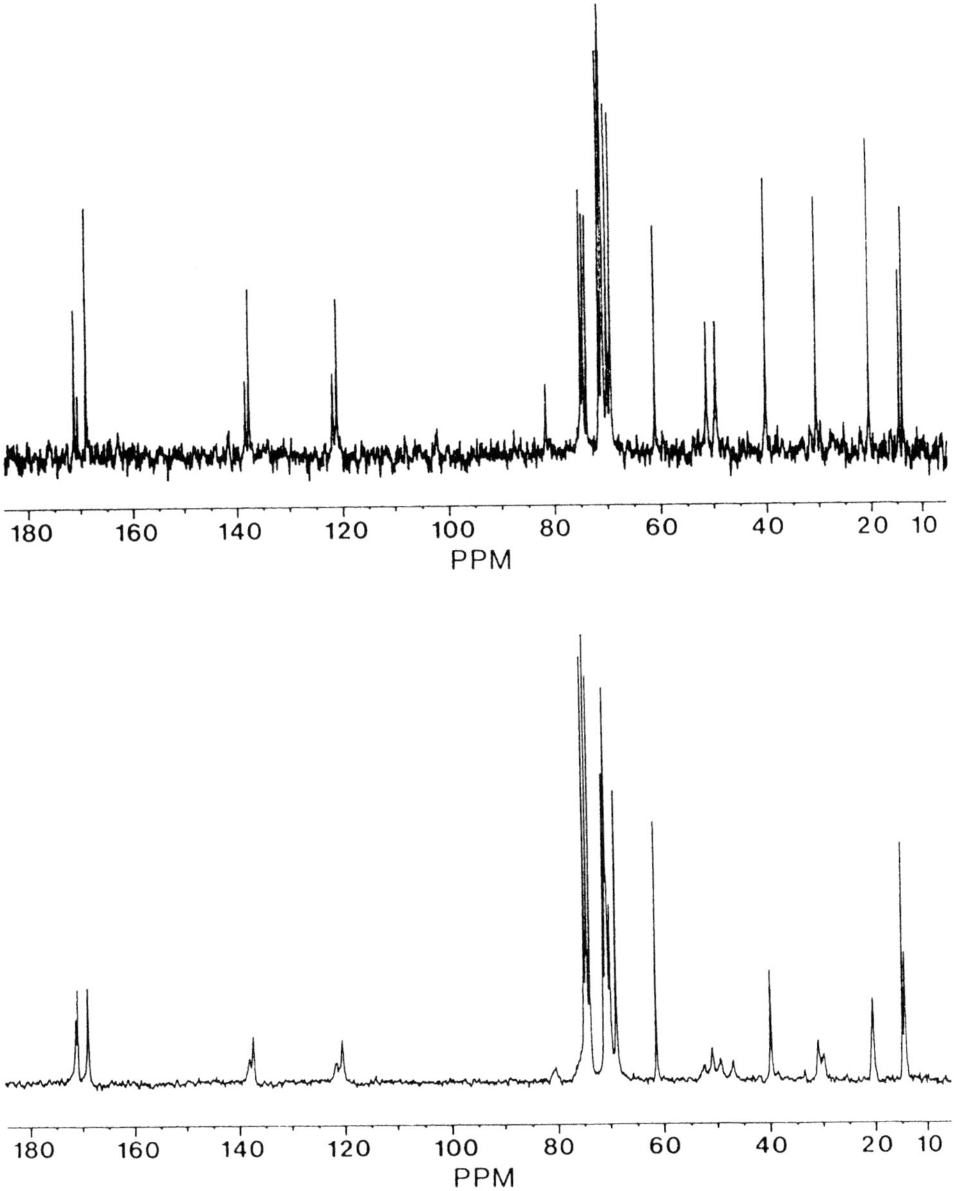

Fig. 5. 50.3 MHz carbon-13 NMR spectra of the *bouquet* molecule **13a** at 298 K (bottom) and at 383 K (top) in $C_2D_2Cl_4$ (solvent signal at 74.3 p.p.m.); see experimental section for peak assignment.

give information about the exact shape of the macrocycle, they agree with an axial symmetry, as in a circular core, but do not exclude other shapes. Indeed the parent [18]-O_6 ring displays structural flexibility, adopting different conformations [18]. It does not itself have a circular shape (D_{2d}) in the crystal [19], but takes it up in aqueous solution [20]. In the crystal structures of complexes of derivatives of the basic unit **11** the macrocycle is of circular shape, but this may well be a result of the complexation [10c, 21, 22]. The marked increase in ^{13}C relaxation times along the side chains agrees with the expected increase in local motions due to their flexibility.

The close similarity between UV and fluorescence emission spectra for all the molecules, including the models containing the elementary chromophores **15a** and **15b**, excludes the presence of a strong interaction between the four chains of $\mathscr{B}_M$ bouquets. This, together with the ^{13}C-NMR results, suggests that the ring is not a collapsed form under these conditions.

The circular dichroism spectra are only slightly informative. The $\Delta\varepsilon$ values are linearly related to the molecular masses. The qualitative similarity of all the recorded spectra in MeOH points to the similarity of the ring-chromophore environments in **13a**, **13b**, **14a** and **16b** in this solvent. The modifications observed for **14a** support this assumption; indeed, among the different molecules studied and in these poor solvents for ions and polyoxy chains, **14a**, which is charged and bears polyoxy elements, is the most likely to display a constraint, and a conformational change.

In order to specify the orientation of the chains with respect to the macrocycle, the determination of the dihedral angle in the tartaric units was attempted, but even for **16b** a precise value of the vicinal coupling constant could not be obtained from examination of the ^{13}C satellite of the CH resonance. The observed width at half-height (4 Hz) is in agreement with the amide groups on the ring being approximately in a *trans*-diaxial conformation. This is in line with earlier NMR studies [12] as well as crystal structure determinations [23] on derivatives of the tartocrown **11** bearing CONHR amide groups, which indicated that these side chains are oriented perpendicularly to the mean plane of the macrocycle.

The ^{13}C-NMR spectral data for **13a** and their temperature dependent features (Figure 5; see above) indicate (1) that the expected two rotamers around each tertiary amide CO–N bond are present and, (2) that rotation around these bonds is slow at 298 K and fast at 383 K, on the NMR time scale, in agreement with the well-known NMR features of such functional groups. In both rotamers the chains linked to the tertiary amide nitrogens have a similar overall orientation with respect to the axis of the macrocyclic core. Structures **13** and **14** represent just one of the 27 rotameric forms (three forms for each of the four side chains with D_2 symmetry of the core macrocycle). Medium and membrane incorporation effects (Figure 1) will affect the population of each form and the orientation of the lateral polyoxy and polymethylene chains.

5. Conclusions

We have synthetised two molecules $\mathscr{B}_M^o$ and $\mathscr{B}_M^c$ in order to investigate an approach to transmembrane structures. In particular, $\mathscr{B}_M^o$ displays favorable characteristics as a potential ionic channel; its structural and geometrical properties should allow its incorporation into phospholipid bilayer membranes with preservation of the geometry of the macrocycle core (existence of a central cavity and preferential diaxial orientation of the lateral carboxy groups). Whether these features will allow the functioning of the $\mathscr{B}_M^o$ bouquet molecule as an ion channel, will depend on its ion binding properties [23] and its incorporation into bilayer membranes. Such studies will be described elsewhere.

Acknowledgments

We thank L. Lacombe, T. Lazrak, L. Leroy and J. Pouget for their help in performing some measurements during the course of this work.

References

1. P. Läuger: *Angew. Chem. Int. Ed. Engl.* **24,** 905 (1985).
2. (a) G. Spach (Ed.): *Physical Chemistry of Transmembranar Ion Motions*, Elsevier, Amsterdam (1983). (b) J.-M. Lehn: *ibid*, p. 181.
3. J.-M. Lehn: *Angew. Chem. Int. Ed. Engl.* **27,** 89 (1988).
4. (a) Yu. A. Ovchinnikov, V. T. Ivanov and A. M. Shkrob: *Membrane Active Complexones*, Elsevier, New-York, Amsterdam (1974). (b) B. C. Pressman: *Ann. Rev. Biochem.* **45,** 501 (1976). (c) S. B. Hladky and D. A. Haydon: *Biochim. Biophys. Acta* **274,** 294 (1972).
5. A. Agtarap and J. W. Chamberlain: *Antimicrob. Agents Chemother.* (1961–1970), 359 (1968); R. W. Briggs and J. F. Hinton: *Biochemistry* **17,** 5576 (1978); G. R. Painter and B. C. Pressman: *Metal Ions in Biological Systems*, Ch. 9, p. 229 (1985); M. Pinkerton and L. K. Steinrauf: *J. Mol. Biol.* **49,** 533 (1970); D. L. Ward, K. T. Wei, J. G. Hoogerheide and A. Popov: *Acta Crystallogr. Sect. B*, **B 34,** 110 (1978); F. G. Riddell and M. K. Hayer: *Biochim. Biophys. Acta* **817,** 313 (1985); F. G. Ridell, S. Arumugam and B. G. Cox: *J. Chem. Soc., Chem. Commun.* 1891 (1987); R. Sandeaux, J. Sandeaux, C. Gavach and B. Brun: *Biochim. Biophys. Acta* **684,** 127 (1982); D. C. Shungu and R. W. Briggs: *J. Mag. Reson* **77,** 491 (1988); D. S. Shungu, D. C. Buster and R. W. Briggs: *J. Mag. Reson.* **89,** 102 (1990).
6. (a) C. J. Pedersen: *J. Am. Chem. Soc.* **89,** 7017 (1967); C. J. Pedersen and H. K. Frensdorff: *Angew. Chem. Int. Ed. Engl.* **11,** 16 (1972). (b) D. W. McBride, Jr., R. M. Izatt, J. D. Lamb and J. J. Christensen: *Inclusion Compounds*, J. L. Atwood, J. E. D. Davies and D. D. MacNicol (eds.), Academic Press, London, Vol. 3., p. 571 (1984). (c) W. E. Morf, D. Amman, R. Bissig, E. Pretsch, W. Simon: *Progress in Macrocyclic Chemistry*, Vol. 1, p. 1, Wiley, New York, 1987. (d) W. Simon, W. E. Morf, and P. C. Meyer: *Struct. Bonding (Berlin)* **16,** 113 (1973).
7. S. R. Byrn: *Biochemistry* **13,** 5186 (1974); G. Eisenman, J. Sandblom and E. Neher: in *Metal-Ligand Interaction in Organic Chemistry and Biochemistry*, B. Pullman and B. Goldblum (eds.) Reidel Dordrecht, Part B, p. 1. (1977); P. C. Jordan: *J. Phys. Chem.* **91,** 6582 (1987); D. A. Langs: *Science* **241,** 188 (1988); E. A. Liberman and V. P. Popaly: *Biochim. Biophys. Acta* **163,** 125 (1965); V. P. Meyers and D. A. Haydon: *Biochim. Biophys. Acta* **274,** 313 (1972); P. Mueller and D. O. Rudin: *Biochem. Biophys. Res. Commun.* **26,** 398 (1965); B. C. Pressman: *Proc. Natl. Acad. Sci. U.S.A.* **53,** 1076 (1965); G. N. Ramachandran and R. Chandrasekhar: *Indian J. Biochem. Biophys.* **9,** 1 (1972); R. Sarges and B. Witkop: *J. Am. Chem. Soc.* **87,** 2011 (1965); D. C. Tosteson, T. E. Andreoli, M. Tiefenberg and P. Cook: *J. Gen. Physiol.* **51,** 373 S (1968); D. W. Urry: *Top. Curr. Chem.* **121,** 67 (1985); D. W. Urry, M. C. Goodall, J. D. Glickson and D. F. Mayers: *Proc. Natl. Acad. Sci. U.S.A.* **68,** 672 (1971); W. R. Veatch, E. T. Fossel and E. R. Blout: *Biochemistry* **13,** 5249 (1974).
8. R. Nagaraj and P. Balaram: *Acc. Chem. Res.* **14,** 356 (1981); R. O. Fox Jr. and F. M. Richards: *Nature* **300,** 325 (1982).
9. (a) J. H. Furhop and U. Liman: *J. Am. Chem. Soc.* **106,** 4643 (1984). J. H. Furhop, U. Liman and H. H. David: *Angew. Chem. Int. Ed. Engl.* **24,** 339 (1985); J. H. Furhop, H. H. David, J. Mathieu, U. Liman, H. J. Winter and E. Boekema: *J. Am. Chem. Soc.* **108,** 1785 (1986); J. H. Fuhrop, U. Liman and V. Koesling: *J. Am. Chem. Soc.* **110,** 6840 (1988); C. J. Stankovic, S. H. Heinemann, J. M. Delfino, F. J. Sigworth and S. L. Schreiber: *Science* 813 (1989); C. J. Stankovic, S. H. Heinemann, J. M. Delfino, F. J. Sigworth and S. L. Schreiber: *Science* 813 (1989); C. J. Stankovic, S. H. Heinemann and S. L. Schreiber: *J. Am. Chem. Soc.* **112,** 3702 (1990). (b) U. F. Kragten, F. M. Roks, and R. J. M. Nolte: *J. Chem. Soc., Chem. Commun.* 1275 (1985); O. E. Sielcken, L. A. Van De Kuil, W. Drenth and R. J. M. Nolte: *J. Chem. Soc., Chem. Commun.* 1232 (1988). (c) V. E. Carmichael, P. J. Dutton, T. M. Fyles, T. D. James, J. A. Swan and M. Zojaji: *J. Am. Chem. Soc.* **111,** 767 (1989); A. Nakano, Q. Xie, J. V. Mallen, L. Echegoyen and G. W. Gokel: *J. Am. Chem. Soc.* **112,** 1287 (1990). (d) I. Tabushi, Y. Kuroda and K. Yokota: *Tetrahedron Lett.* **23,** 4601 (1982). (e) J. D. Lear, Z. R. Wasserman and W. F. De Grado: *Science* 1177 (1988); M. Mutter and S. Vuillemier:

Angew. Chem., Int. Ed. Engl. **28,** 535 (1989). (f) T. M. Fyles, K. C. Kaye, T. D. James and D. W. M. Smiley: *Tetrahedron Lett.* **31,** 1233 (1990); F. M. Menger, D. S. Davis, R. A. Persichetti and J. J. Lee: *J. Am. Chem. Soc.* **112,** 2451 (1990).

10. (a) J.-M. Lehn, J. Malthête, A.-M. Levelut: *J. Chem. Soc. Chem. Commun.* 1794 (1985). (b) R. Heng and J.-M. Lehn: unpublished results; R. Heng: Thèse de Doctorat ès Sciences, Université Louis Pasteur, Strasbourg, 1985. (c) J.-P. Behr, J.-M. Lehn, A.-C. Dock and D. Moras: *Nature* **195,** 526 (1982).
11. For a preliminary report see: L. Jullien and J.-M. Lehn: *Tetrahedron Lett.* **29,** 3803 (1988).
12. J. P. Behr, J. M. Girodeau, R. C. Hayward, J.-M. Lehn and J. P. Sauvage: *Helv. Chim. Acta* **63,** 2096 (1980).
13. M. Guggi, M. Oehme, E. Pretsch and W. Simon: *Helv. Chim. Acta* **59,** 2417 (1976). N. N. L. Kirsh, R. J. J. Funck, E. Pretsch and W. Simon: *Helv. Chim. Acta* **60,** 2326 (1977).
14. F. Carrière: *Bull. Soc. Chim. Fr.* 3266 (1967).
15. J. Fugger, J. M. Tien and I. M. Hunsberger: *J. Am. Chem. Soc.* 1843 (1955).
16. T. W. Greene: *Protective Groups in Organic Synthesis*, J. Wiley and Sons, 1981; V. J. Lee, A. R. Branfman, T. R. Herrin and K. L. Rinehart Jr: *J. Am. Chem. Soc.* **100,** 4225 (1978).
17. E. Breitmaier and W. Voetter: *Carbon-13 NMR Spectroscopy*, 3nd Ed., VCH, 163; G. C. Levy, R. L. Lichter and G. L. Nelson: *Carbon-13 Nuclear Magnetic Resonance Spectroscopy*, 2nd Ed., John Wiley and Sons (1980).
18. G. Wipff, P. Weiner and P. Kollman: *J. Am. Chem. Soc.* **104,** 3249 (1982).
19. J. D. Dunitz, M. Dobler, P. Seiler and R. P. Phizackerley: *Acta Crystallogr. Sect. B*, **B 30,** 2733 (1974); J. D. Dunitz and P. Seiler, *ibid.* **B 30,** 2739 (1974).
20. G. Ranghino, S. Romano, J.-M. Lehn and G. Wipff: *J. Am. Chem. Soc.* **107,** 7873 (1985).
21. J. P. Behr and J.-M. Lehn, D. Moras and J.-C. Thierry: *J. Am. Chem. Soc.* **103,** 701 (1981).
22. J. J. Daly, P. Schönholzer, J.-P. Behr and J.-M. Lehn: *Helv. Chim. Acta* **64,** 1444 (1981).
23. For ion binding by macrocycles related to the core unit see: J. P. Behr, J.-M. Lehn and P. Vierling: *Helv. Chim. Acta* **65,** 1853 (1982).

Journal of Inclusion Phenomena and Molecular Recognition in Chemistry **12**: 75–119, 1992.

Consequences of the 'Pedersen Papers' on Crown Type Chemistry at Würzburg and Bonn Universities: From Heteroaromatic Crowns and Podands to Large Molecular and Crystalline Cavities Including Multisite Receptors, Cascade Molecules, Chromoionophores, Siderophores, Surfactant-Type, and Extreme Ligands[#]

FRITZ VÖGTLE and EDWIN WEBER
Institut für Organische Chemie und Biochemie der Universität Bonn, Gerhard-Domagk-Straße 1, D-5300 Bonn-1, Germany

(Received: 30 November 1990; in final form 15 January 1991)

Abstract. We thank the editors of this issue for the opportunity to present the historic development of crown chemistry at the Universities of Würzburg and Bonn in memory of C. J. Pedersen, the originator of the crown ethers. His legacy of science has tremendously stimulated research at both universities.

Key words. Crown-type compounds, podands, multisite receptors, chromoionophores, siderophores, surfactant-type crowns, large molecular cavities, crystalline hosts, complexation chemistry, crystal structures.

1. Introduction

1.1. GENERAL REMARKS

Evolution in chemistry is difficult to predict. The moment you finish a chemical experiment you hardly know whether you have only produced a boring compound or opened a door to a fruitful scientific future. Sometimes a primer from outside is required to bring your chosen course of science to detonation and ring in a new chemical era. We had this experience thanks to the discovery of the crown ethers by Charles J. Pedersen [1]. The memory of this lucky break in our research work unleashed by the revolutionary Pedersen papers on crown ethers in 1967 [2, 3] is still alive.

1.2. PREVIOUS HISTORY

In the period before Pedersen, one of the main interests of the Vögtle group was in cyclophanes [4], a general type of macrocycle incorporating aliphatic and aromatic ring segments (Figure 1). In particular, the efforts were directed to the study of highly strained [5] and multiply bridged [6] cyclophanes and heterophanes (cf. **1**–**5**)

[#] This paper is dedicated to the memory of the late Dr C. J. Pedersen.

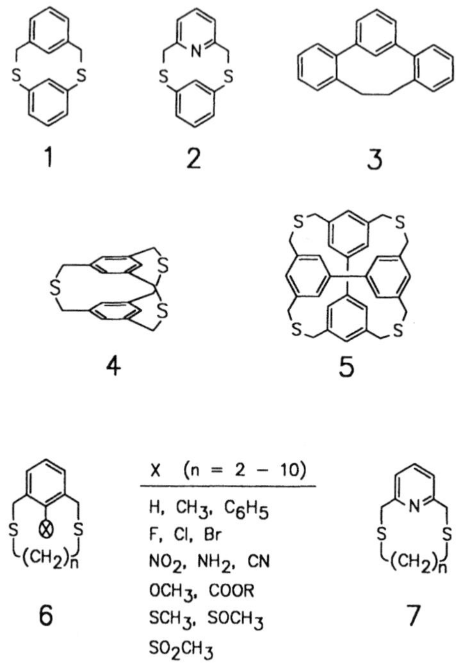

X (n = 2 − 10)
H, CH_3, C_6H_5
F, Cl, Br
NO_2, NH_2, CN
OCH_3, COOR
SCH_3, $SOCH_3$
SO_2CH_3

Fig. 1. Typical cyclophane-type macrorings of our pre-Pedersen era [4–8].

or similar macrocycles which were fitted with intraannular substituents with the aim of measuring their steric demands [7] (cf. **6**). In the case where these compounds contained heteroatoms, such as nitrogen, oxygen, or sulphur, the heteroatoms were only in the minority. On the other hand, some of the typical macrorings studied at that time embody a long (CH_2)-segment bridged in the 2,6-position via sulphur atoms across a pyridine nucleus (**7**) [8].

However, in the late sixties, the first signs of a macrocyclic complexation chemistry emerged from the research work of the Vögtle group. They relate to the compounds specified in Figure 2, i.e. macrocycles **8** [9], **9** [10] and **10** [11], and a macrobicycle **11** [12] with appropriately located sulphur, nitrogen and oxygen (donor) atoms, and a cavity. Correspondingly, **11** yielded a stable complex with Cu^{2+} and Hg^{2+} ions [12]. Thus these macrocycles may be considered a preliminary operation in preparing the ground for the development of crown chemistry to come.

2. Crown Research at Würzbury University (Vögtle, Weber) Stimulated by C. J. Pedersen

2.1. WHAT GAVE US THE IDEA?

In the strict sense, the prime Pedersen crown, dibenzo-18-crown-6 (**12**) [1–3] may be understood as a heteracyclophane [13] (Figure 3), but substitution of individual CH_2-groups in **7** at proper positions leads to a novel type of crown compound with heteroaromatic units incorporated in the macroring to meet a crown

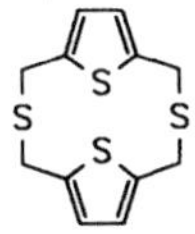

8

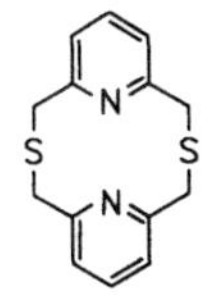

9

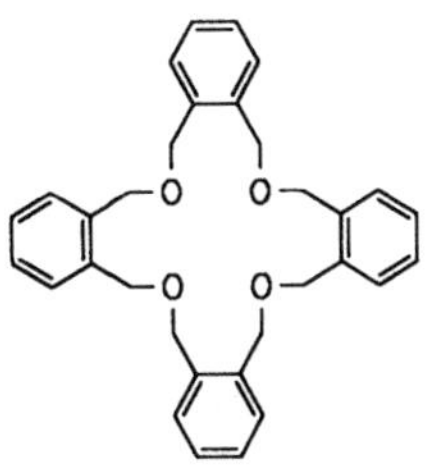

10

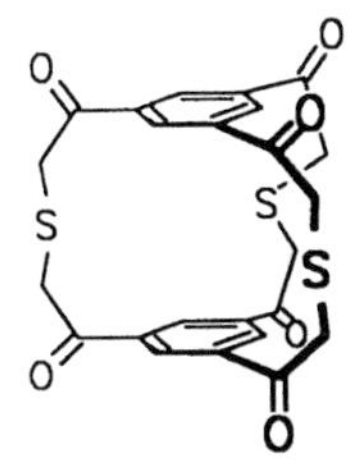

11
(complex with Cu^{2+})

Fig. 2. The first signs of a macrocylic and three-dimensional complexation chemistry [9–12].

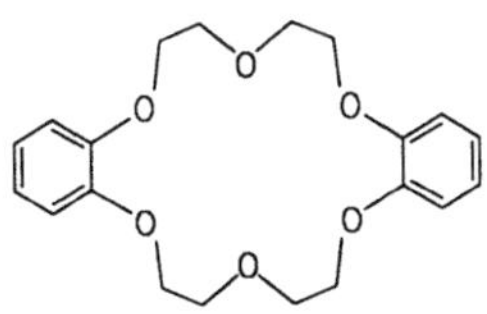

12
Dibenzo–18–crown–6 ≙ "Heterg–Phane"

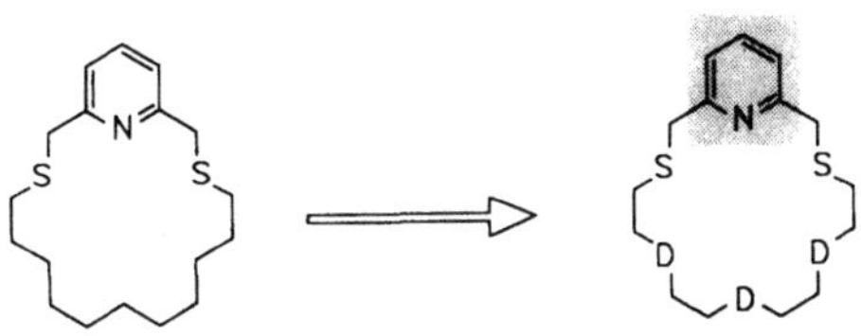

Hetero–Phane | Heteroaromatic–Crown

Fig. 3. The prime idea: from heterophane to heteroaromatic crown (D = donor hetero atom).

analogous arrangement of donor atoms such as that exemplified in Figure 3. By doing that, the cyclophane and the crowns get married, and the union promises much for the future. This is the initial break in our research that owed thanks to Pedersen's ingenious first papers [2, 3]. It was during the time at Würzburg University.

2.2 REALIZATION OF A NEW CROWN TYPE – THE HETEROAROMATIC CROWNS

Edwin Weber (now Professor at Bonn University) had the privilege of performing the marriage rites between cyclophanes and crowns. During the time under consideration (1973) he began his doctoral work [14] with Prof. Vögtle on exactly this

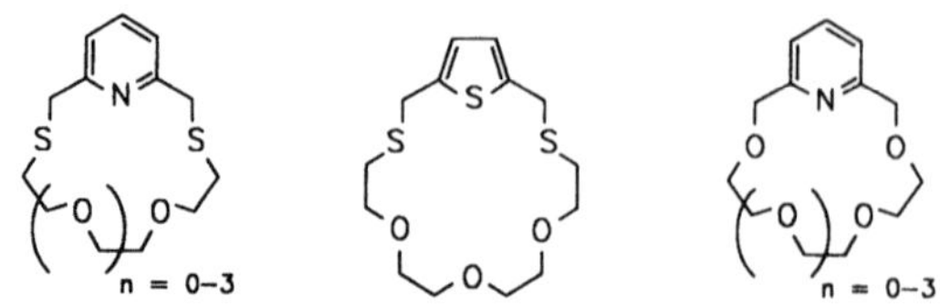

13 14 15

Fig. 4. First generation heteroaromatic crown compounds [15, 16].

heteroaromatic crown type compound (Figure 3). Since we had mastered cyclophane synthesis, the success came quickly. Within some few weeks the first compounds of the heteroaromatic crown class (**13, 14**; Figure 4) had been prepared in the Würzburg laboratory. We realized that the compounds had quite unexpected complexation behaviour with respect to Na^+ ions, i.e. they easily solubilized $NaMnO_4$ into organic solvents, as contrasted with $KMnO_4$. Thus we had discovered a new efficient crown family (actually the first – and foremost – crown compounds synthesized in Germany). A paper for *Angewandte Chemie* was prepared, smoothly accepted, and we looked forward to its appearance [15].

Although we knew about the importance of our discovery, it hardly entered into our heads that at the same time somebody else could have had a similar idea that would lead him to the same matter. But the unbelievable became true. Just a few days before our paper on the pyridinodithia-crowns appeared [15], we became aware of a short note in the 'blue pages' (*Nachr. Chem. Techn. Lab.*) [16] referring to a talk by D. J. Cram indicating that he had worked on similar pyridino (all-oxygen) crowns **15** (Figure 4). Obviously the new heteroaromatic crown type has been developed independently in the two different laboratories. We were overcome by a fateful event of overlap in chemistry relating to the origin of research work.

Nevertheless we took advantage of the fact, having published the first real scientific paper in German and English editions [15] on the new heteroaromatic crown type having versatile complexation properties as illustrated in Figure 5 [17]. Due to the coexistence of hard (oxygen, nitrogen) and soft (sulphur) donor atoms, coordination to alkali/alkaline earth and transition metal ions is possible (cf. Ba^{2+} and Cu^{2+} complexes of Figure 5) giving rise to extended conformational re-arrangements.

2.3. FIRST USE OF INTRAANNULAR SUBSTITUENTS

Fortunately we did not suffer the same fate with the intraannularily substituted or modified crown compounds **16** and **17** (Figure 6) [18, 19], some of which were also described in our first paper [15]. Compounds **16** arise from **6** when the alkyl chain is replaced by a crown analogous oligo-ethyleneoxy segment. This crown type, or more generally speaking, the idea of intraannular substitution of crowns, became very important in chromoionophore design [20] and still lives on in the recent spherands of Cram [21] (another overlapping of interests between both research groups). Other crown constitutions developed at that early time are represented by formulae **18–20** (Figure 6) [19].

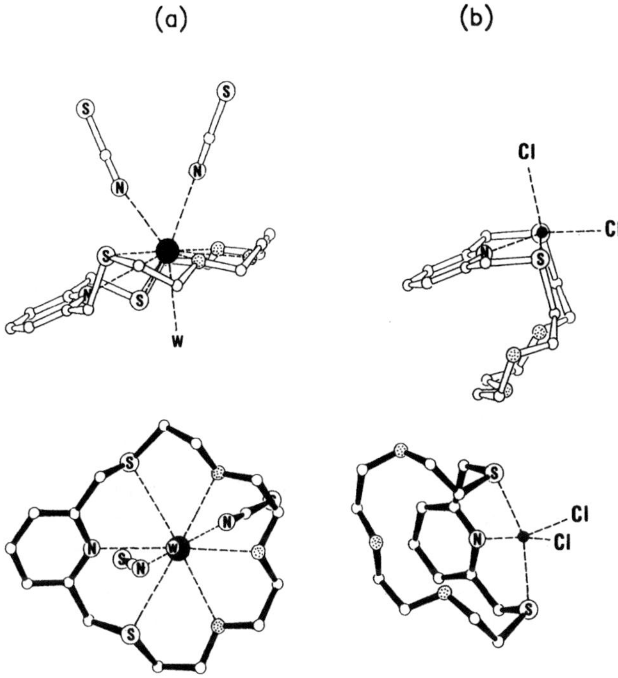

Fig. 5. Perspective views of the hydrated $Ba(SCN)_2$ (a) and the $CuCl_2$ complex (b) of dithiapyridino-18-crown-6 (**13**, $n = 2$; Fig. 3) from crystal structures (H atoms omitted) [17].

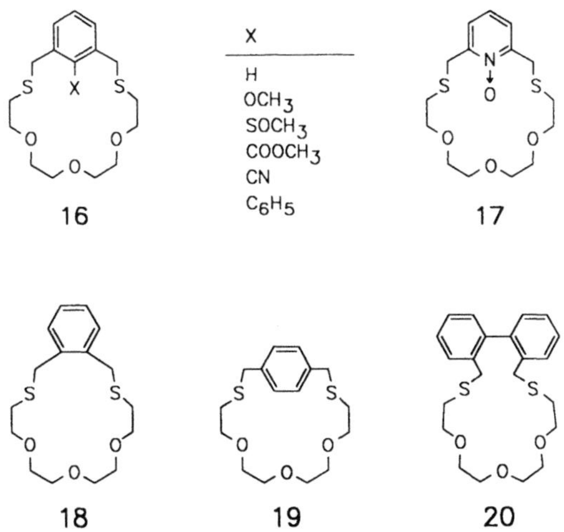

Fig. 6. Intraannularily substituted crown compounds and other early developed crown constitutions [18, 19].

2.4 HANDLING OF DONOR ATOMS AND LINKAGE POSITIONS

It is typical of the early period of our crown research that we used sulphur as a linking atom in a benzylic position (cf. Figures 4 and 6). This changed in subsequent work since nitrogen (donor) atoms such as those in **21** and **22** (Figure 7) were increasingly used to close the ring [19, 22]. On the other hand, sulphur

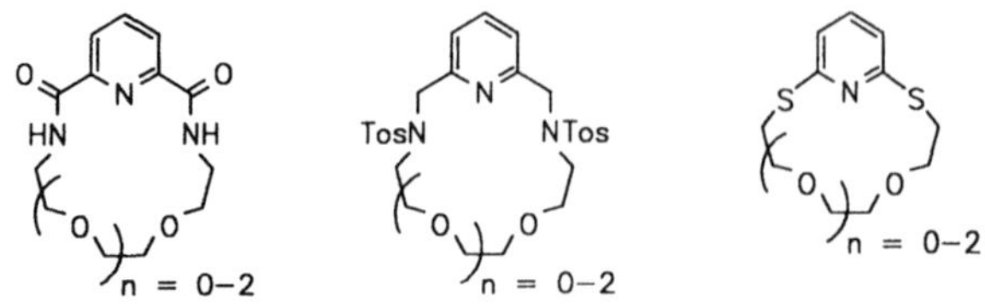

Fig. 7. Heteroaromatic crowns with modified linkage atoms and linkage positions [19, 22].

(donor) atoms were still used for ring closure, but rather in a non-benzylic position to yield heteroaromatic crowns **23** (Figure 7) of quite uncommon ring sizes [19, 22].

It was also Edwin Weber's work at the laboratory bench which replaced the oxygens in the oligoether segment (cf. **13**, **14**, etc.; Figure 4) by sulphur atoms, i.e. he synthesized all-thia and mixed thia-aza aromatic crowns, exemplified by formulae **24**–**26** (Figure 8) [23]. Replacement of the oxygens in the oligoether segment by nitrogen, such as in **27**–**29** (Figure 8), was undertaken by other members of the Vögtle group [22]. These heteroaromatic crowns with only thia and aza donors showed particular complexation properties in respect of transition metal ions [22, 23].

In addition to the mixed thiaoxa, thiaaza, and all-thia or all-aza crowns, we wanted to have crowns with oxygen and nitrogen (cf. **15**; Figure 4) or only oxygen donors. It was expected that they would behave as more efficient complexants for alkali and alkaline earth metal ions [24]. The first step into this direction was again undertaken by Edwin Weber. The key diphenol compound **30** of his synthetic approach (Figure 9), however, came from Pedersen's early work [3]. Thanks to that he succeeded in the synthesis of a large number of oxaaza and all-oxa crowns

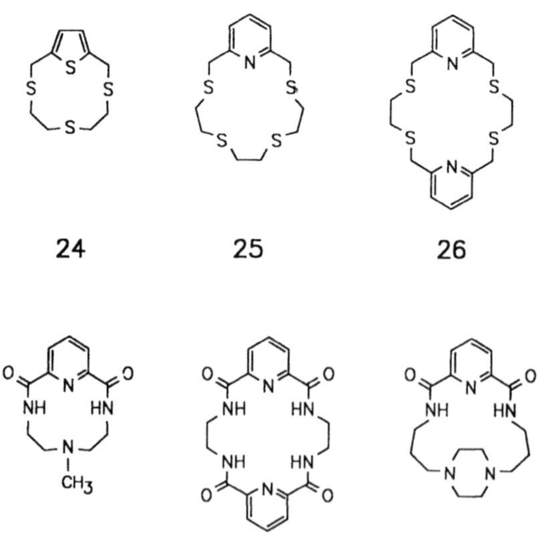

Fig. 8. All-thia, mixed thiaaza, and all-aza heteroaromatic crown modifications [22, 23].

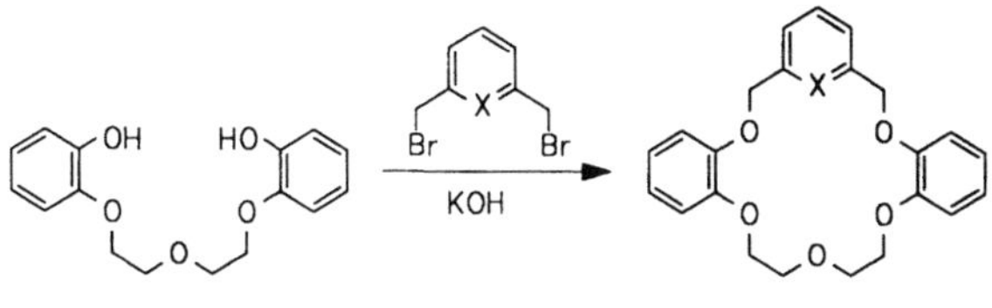

30 31

X = N, N→O, C–H, C–F, C–OCH_3, C–NO_2, C–$SOCH_3$

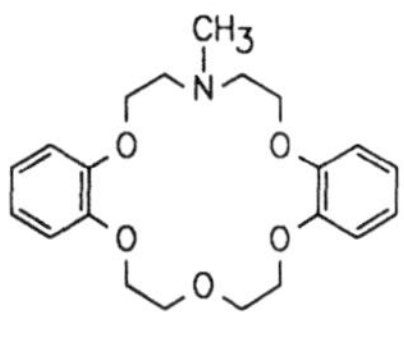

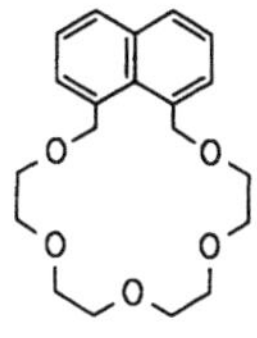

32 33

Fig. 9. Just as we thought! Synthesis of oxaaza and all-oxa (intraannularily-substituted) crowns [19] based on the 'Pedersen diphenol'.

exemplified in Figure 9 (**31**–**33**) [19]. For the most part, they showed strong binding of alkali and alkaline earth metal ions and useful solubility properties of their salts [19, 25].

2.5 THE NONCYCLIC VARIETY OF CROWN COMPOUNDS: PODANDS

The next innovative step of the crown research at Würzburg University stems again from a joint Vögtle/Weber idea. For application of crown compounds, e.g. as phase transfer catalysts, it would be an advantage to make cheap crown analogs available [26–28]. What about using noncyclic analogs of crowns (later called 'podands')? They allow the saving of a ring formation stage and reduce the preparation to a conventional synthetic problem. However, we knew from Pedersen's work [29] that simple glymes were not capable of efficient complexation. If we refined the glyme terminals by so-called strong and rigid donor end-groups, such as quinoline residues, would it give the desired effect? We did the experiment and, in fact, it worked out fine.

The first compounds of this new ligand class were the bis-(quinolino)oligoether **34** and its pyridine analog **35** (Figure 10) [30]. They yielded a great many stoichiometric crystalline complexes with different alkali, alkaline earth, and transition metal salts. Thus the door to the promising field of non-cyclic crown analogs – termed podands [31] – was opened.

It took us only a short time to synthesize a large number of podands with different end groups and chain lengths (cf. **36** in Figure 10) [32], and to study their complex formation properties [33, 34]. The complexes gave very interesting crystal structures [32, 35] ranging from butterfly conformations to helices and spheres (Figure 11a–c) [36–38]. Consequently, a so-called 'end group concept' [32, 39] was

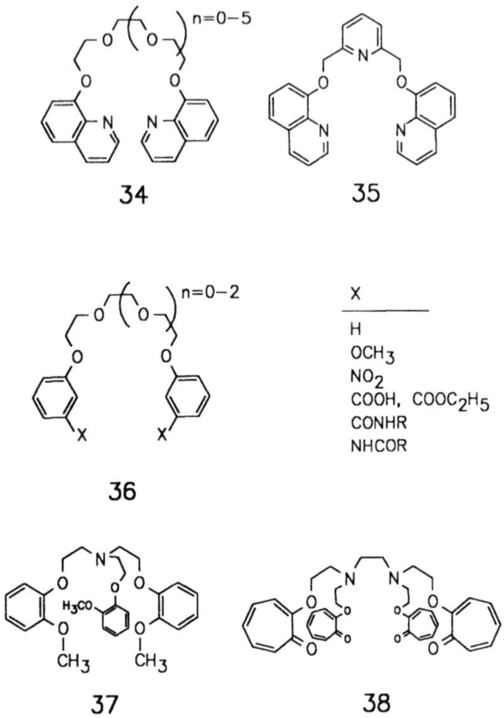

Fig. 10. The noncyclic crown/cryptand version (podand and open-chain cryptands) [30, 39, 46, 47].

developed, which made an efficient podand design possible [40–44]. It is still being used on other places to discover new ion carriers [45].

The same principle was applied a little later to tripod and tetrapod constitutions such as are exemplified by formulae **37** and **38** (Figure 10) [46, 47]. Due to their strong complexation with alkali and alkaline earth metal ions, and for reasons of topology, they were termed ‘open-chain cryptands’ [32]. Indeed, crystal structures of the complexes show efficient wrapping of the cation (Figure 11d) [35].

2.6. THE TENTACLE APPROACH: OCTOPUS MOLECULES

Possibly the most popular compounds developed during the joint Vögtle/Weber era were the ‘octopus molecules’ (Figure 12) [48]. To some extent they relate to the noncyclic cryptands mentioned above but they were synthesized earlier. A characteristic of this compound type is the high number of coordination ‘tentacles’ (e.g. six in **39**) which make them efficient complexants and carriers for alkali and alkaline earth cations. The phenomenal structure, resembling that of an octopus (though not in the number of tentacles), has imparted the nick-name ‘octopus molecule’. The term is apt because the donor atoms act like the suction cups with which a real octopus grasps food (Figure 12). For compound **39**, the meal consists of metallic ions and, as mentioned, its appetite for particular main Group I and II cations is voracious. Later, it was found that ‘octopus’-type molecules are also hosts for uncharged organic molecules [49].

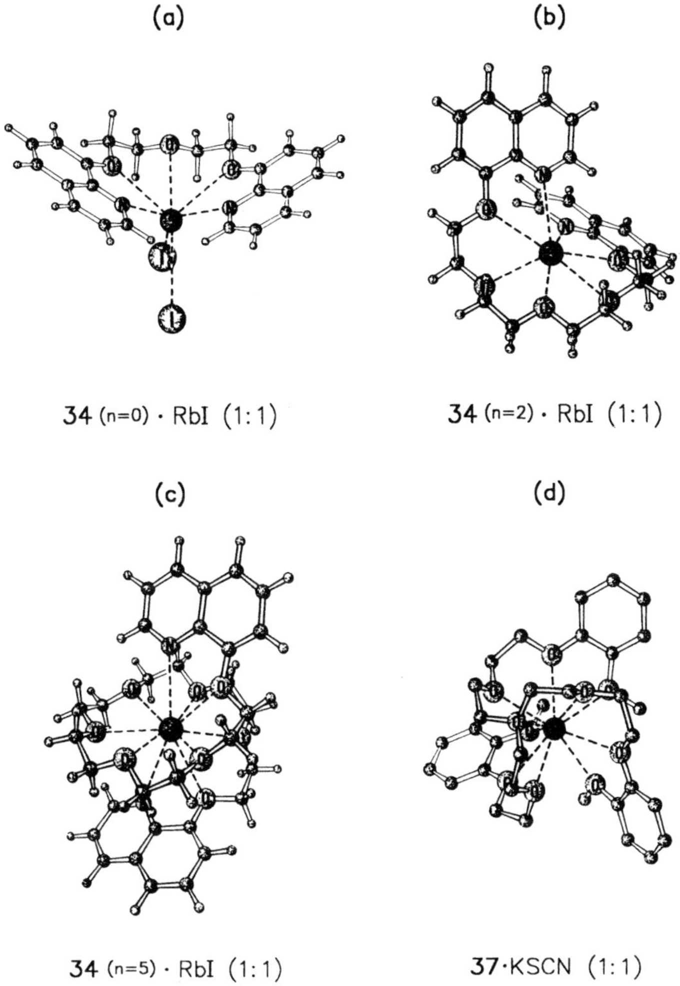

Fig. 11. (a) Butterfly, (b) helical, (c) spherical, and (d) wrapping complexation of metal ions by non-cyclic crown compounds (podands; crystal structures) [35–38].

At the end of 1975, the Vögtle group moved from Würzburg to Bonn University. In 1976, Edwin Weber finished his doctoral thesis [14] on the various topics mentioned above and began his own work. From now, we have two independent research groups working at Bonn. In the following, we will look separately into the highlights these two groups have achieved from about 1976 (Weber 1978) up to the present time, stimulated by C. J. Pedersen's work.

3. Vögtle Group Research at Bonn University

3.1. NEW PLACE, NEW TOPOLOGY, NEW BUILDING BLOCKS

On the one hand, the removal of the Vögtle group from Würzburg to Bonn University became manifest in a topical extension oriented along the three-dimensional cryptands [50–53]. On the other hand, a logical continuation of the previ-

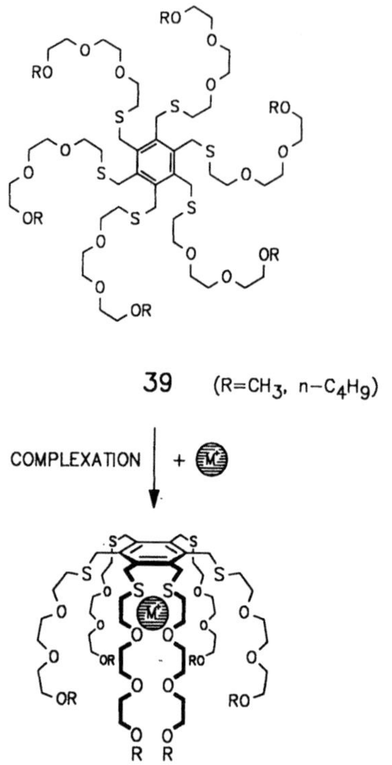

Fig. 12. Imitating the animal: 'octopus molecules' grasp cations [48].

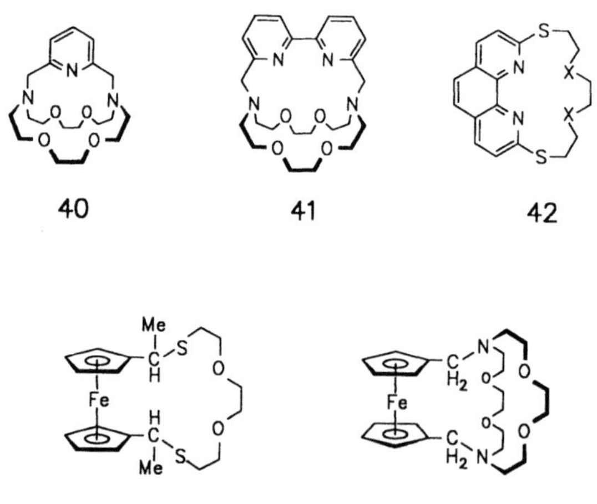

Fig. 13. Crowns (coronands) and cryptands incorporating pyridino, bipyridino, phenanthrolino, and ferroceno subunits [50, 51, 55, 58].

ously pursued course of applying new building blocks and donors is also evident [5, 52, 54–58]. This is expressed in Figure 13 showing examples of the first pyridino and bipyridino cryptands (**40** or **41**) to be synthesized [50, 51, 53] as well as of phenanthrolino crowns (**42**) [55] and of crowns/cryptands (**43**, **44**) for the first time incorporating a ferroceno unit [58]. These particular building blocks are still important structural parameters in recent host design [59, 60].

Another new building block development refers to the use of structural constituents, as in natural bioactive compounds or synthetic drugs. The underlying idea was to use crown modification to equip ordinary pharmacophoric substances with cation selectivity in order to make them affine to particular cation-containing tissues, such as those in bones, or to effect local ion concentrations. Thus, a series of crowns and podands bearing haptophoric and pharmacophoric groups (papaverin, adrenalin, apomorphin, cinnarizin, procain, etc.) have been synthesized and studied [61–65]. Examples are given in Figure 14 (**45**–**49**).

Moreover, the typically amphiphilic lipids, as another class of natural compounds, were also modified in a crown or podand analogous manner to give 'hydrophilic lipids' (**50**, **51**; Figure 14). These podands, based on the glyceryl backbone, show interesting cation carrier properties [66].

3.2. COLOR RESPONSIVE CROWNS: CHROMOIONOPHORES

Pedersen had shown in his initial papers [2, 3] that the UV spectrum of dibenzo-18-crown-6 (**12**, Figure 3) is slightly effected by complexation, depending on the cation

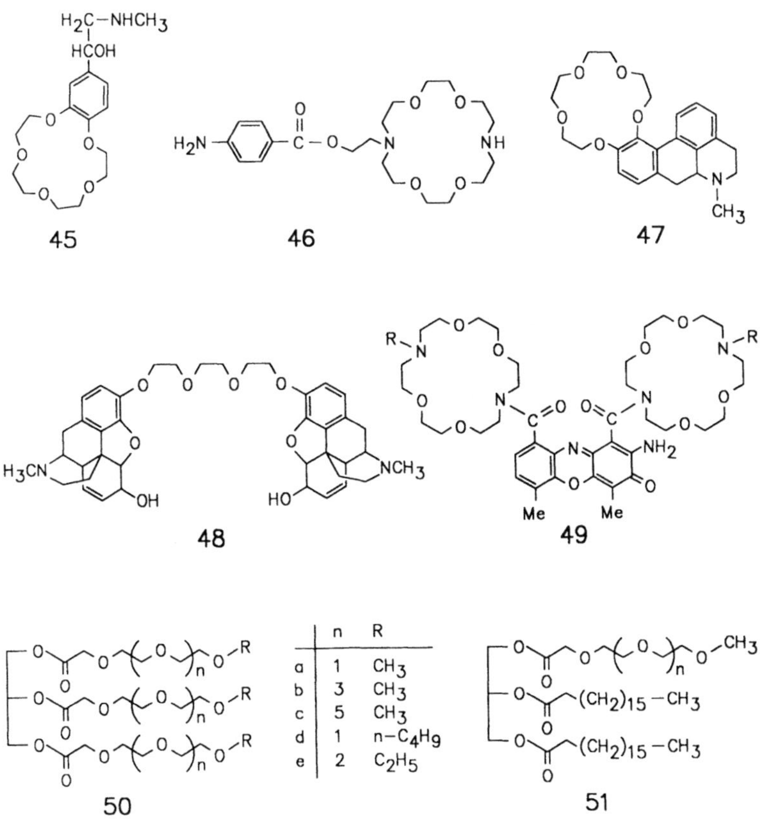

	n	R
a	1	CH_3
b	3	CH_3
c	5	CH_3
d	1	n-C_4H_9
e	2	C_2H_5

Fig. 14. Natural compound and drug related crowns/podands [61–66].

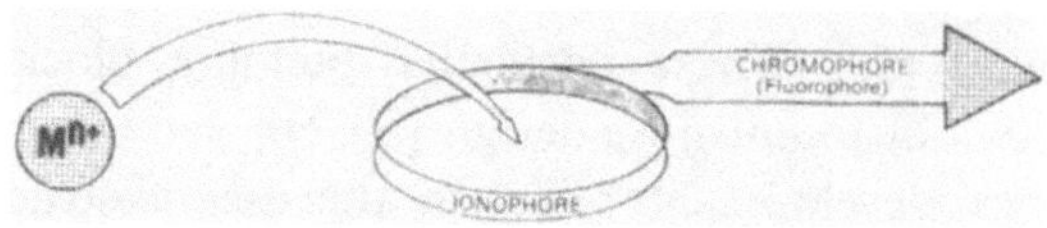

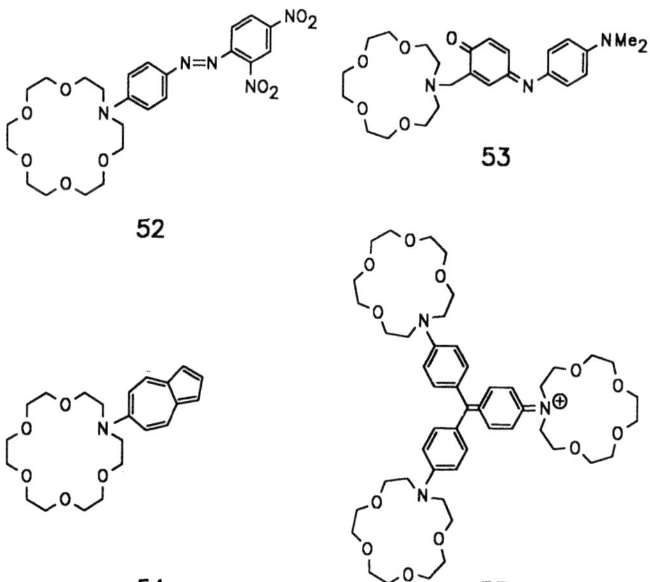

Fig. 15. Chromoionophores: molecular construction and examples of compound [20a, 67–71].

type. He used the property as a method for establishing complex formation [3]. Stimulated by this early work of Pedersen, the chromoionophores were developed [20a]. This particular class of dyestuff-analogous crown compound is based on the general idea of making selective cation complexation of crown compounds visible through a color effect in the same molecule [67]. It requires, besides the ligand-typical (coordination active) part, a chromophore in the molecule, and an electronical coupling between both, as illustrated in Figure 15.

A number of chromoionophores were designed on the basis of this principle. They include chromophore elements such as azo, quinone, quinone imine, azulene or cyanine groups [68–71]. Typical examples of compounds are depicted in Figure 15 (**52**–**55**). Examples of the chromophore effect that such compounds provide on the addition of different cations are shown in Figure 16. The quinone imine type chromoionophore **53** gives significant bathochromic shifts with enhanced extinction [69]. After this we proceeded to chiral dyestuff crowns [72] which were expected to give different color effects with guest enantiomers.

Independently, Japanese scientists developed similar types of chromoionophores [73], but with proton ionizable (acidic) groups [20].

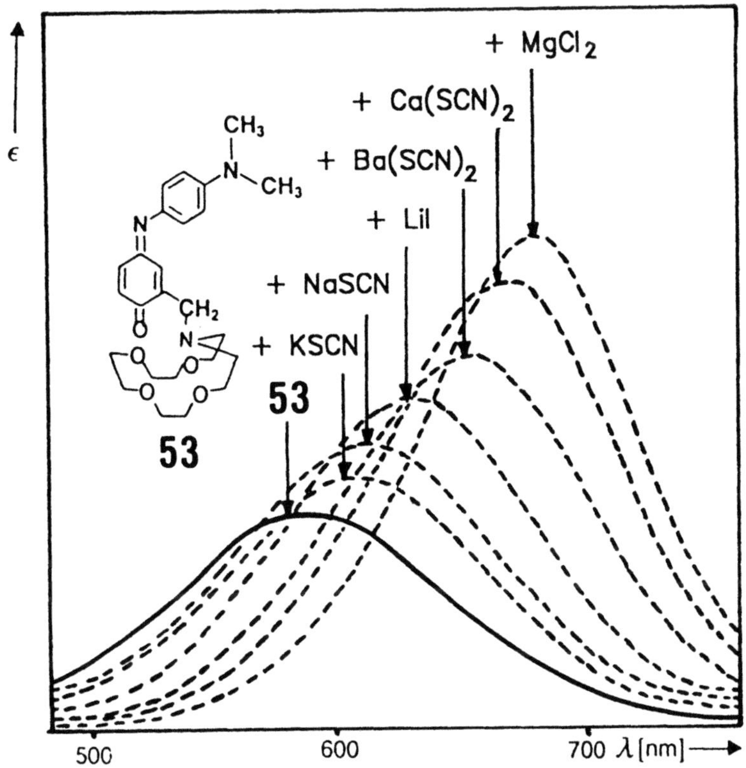

Fig. 16. Selective ion determination via cation selective light absorption: bathocromic shifts upon addition of salts to a solution of the chromoionophore **53** in acetonitrile [69].

3.3. BEGINNING OF THE SUPRAMOLECULAR ERA: COMPLEXATION OF UNCHARGED GUEST MOLECULES BY CROWNS AND PODANDS

It was also Pedersen who stimulated our initial work on crown and podand chemistry with uncharged guests. In one of his early papers [74], he mentioned the isolation of crystalline complexes between dibenzo-18-crown-6 and thiourea or related compounds but their structures remained unclear. More remarkable is that a crystalline urea complex was not included, although Pedersen stated that he had found enhanced solubility of dibenzo-18-crown-6 (**12**, Figure 3) on addition of urea [74].

To succeed with crystalline urea and thiourea complexes became a spur to us because of the importance of these molecules. Fortunately, we were successful in isolating the first crystalline urea complexes of oligoethers which were simple podands [75, 76]. We were even able to solve the crystal structures of the corresponding thiourea complexes (Figure 17a, b) [77, 78]. This was the beginning of our complexation chemistry with uncharged molecules as guests [79].

Following on from this we prepared and studied a series of further complexes of crowns, heterocrowns, and podands with uncharged OH- NH-, CH- and SH-containing guests such as dimethyl sulfate, anilines, hydrazines, phenols, water, and others [80, 81, 82]. Typical examples of complex structures involving 18-crown-6 [83, 84] are illustrated in Figure 17c and d. Some of these complexes were used as

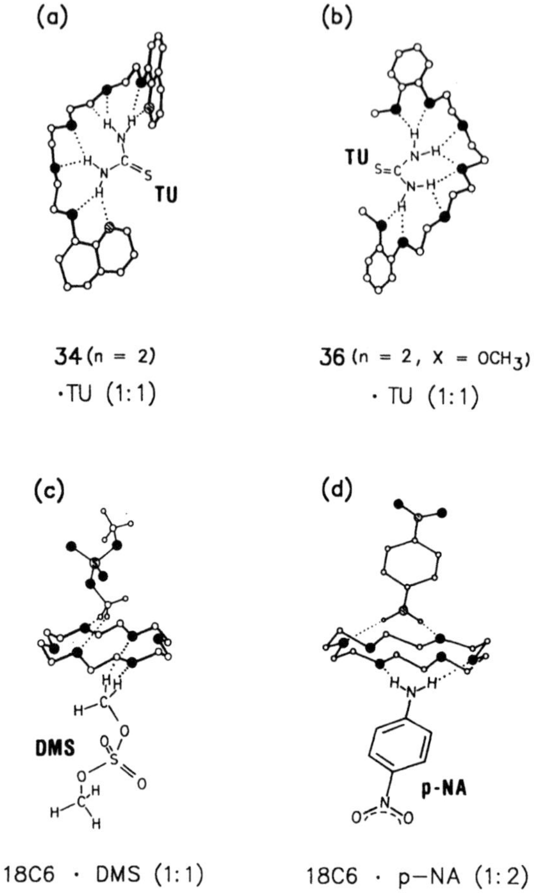

Fig. 17. Crystal structures of typical uncharged molecule complexes of podands (a, b) and 18-crown-6 (c, d) of the early supramolecular period (H-bonded complexes, H-bonds dotted) [74, 78, 82, 83].

reagents with the complexed species exhibiting modified solubility and reactivity properties [85–87].

3.4. DESIGNED HOST TOPOLOGIES FOR UNCHARGED MOLECULE COMPLEXATION

Believe it or not, even the three-dimensional host structures shown here owe their origin to some degree to Pedersen if one considers his 'lanterns' [32] and other compounds mentioned in one of his many patents (see Ref. [1]). It is also foreshadowed in his work [32] that trigonal host symmetry is something fundamental. Despite this support from basic principles, the development of hosts which will specifically complex organic uncharged molecules is difficult to manage [79, 88].

In this respect, one of the successes of the Vögtle group refers to the synthesis of host molecule **56** [89] – a macro-cyclic hexa-amide with inherent three-fold symmetry (Figure 18). Previous host molecules of this type had two-fold symmetry, which makes it possible to achieve a relatively simple synthesis from two molecular halves [88]. Hosts such as the ones we consider here, having a three-fold molecular axis,

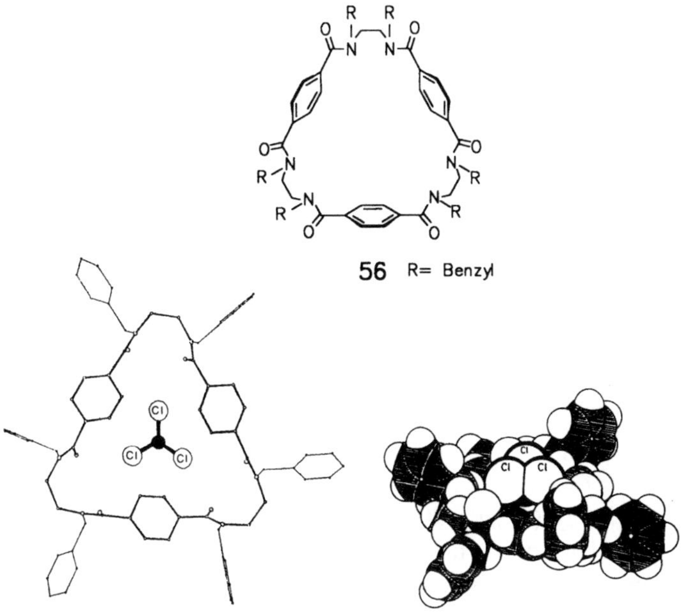

56 R= Benzyl

Fig. 18. A host with threefold symmetry (**56**) recognizes chloroform to give a crystalline inclusion compound selectively [89, 91].

are much more difficult to synthesize. In recompense, the 'endoreceptor' **56** yielded highly selective inclusion complexes with haloform-type guest molecules (CHX_3) [89]; to our knowledge this is one of the first uncharged host–uncharged guest complexes to have been discovered. The complexes are even stable in solution, as measurements using ion selective electrodes demonstrated [90]. The crystal structure of the $CHCl_3$ complex shows a well-fitting supramolecular host–guest aggregate (Figure 18) [91]. Clearly, the correspondence of trigonal symmetry between host and guest is operating here.

Another interesting host topology was materialized in the form of molecules shaped like a basket or a cup [92]. Examples are given in Figure 19 (**57**, **58**). Basket compound **57a** forms a stable inclusion complex with dioxane [92].

In subsequent work it was a challenge to proceed from a molecular basket to a closed cavity [93, 94]. This requries the use of two trivalent building blocks (**B**) that

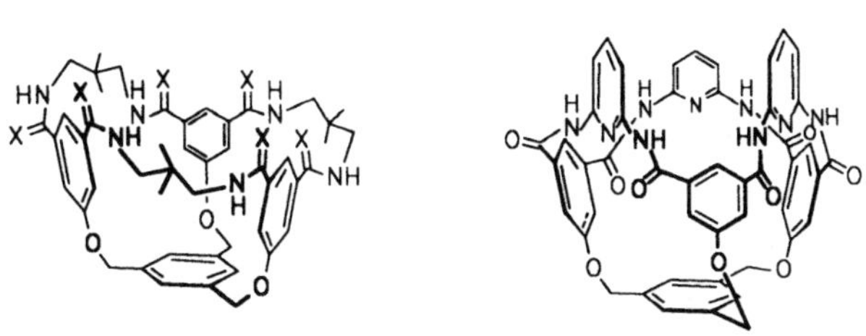

57 a: X= H_2 b: X= O 58

Fig. 19. Basket-type hosts [92].

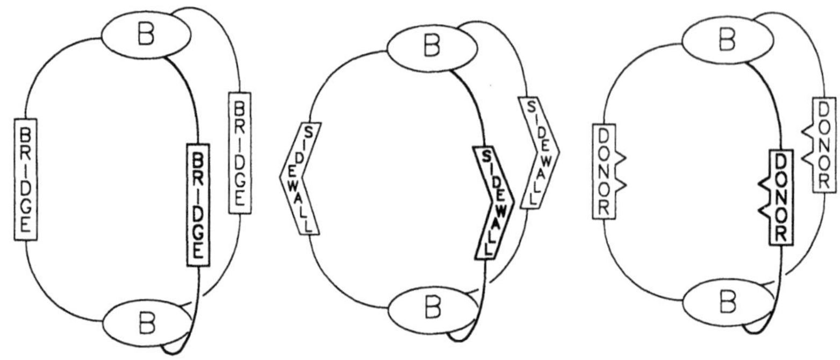

Fig. 20. Scheme of macrobicyclic cavity types [93, 94].

can be bridged appropriately (Figure 20). The first successful approach based on the structural principle of a common cryptand (cf. **40, 41** in Figure 13), i.e. two trivalent nitrogen atoms were brought in bridgehead positions. Bridging of these atoms by rigid spacers (cf. II in Figure 20) such as *o*-terphenylene yielded a cavity molecule **59** with a lenticular shape (Figure 21) [95].

Host molecule **60** shown in Figure 21, contains one triphenylamine and one triphenylmethane bridgehead unit [96]. These are bridged by three flexible chains which makes adaptation of the host molecule possible. Nevertheless, **60** has a permanent cavity suitable for the selective inclusion of disk-shaped naphthalenediols.

Replacement of the triphenylamine bridgehead unit by another triphenylmethane building block yielded host molecules **61** and **62** (Figure 21) [97]. They are topological isomers (*out/out*, *out/in*). A typical behaviour of the *out/out* isomer **61** is the inclusion of the highly lipophilic spherical adamatane molecule, which contrasts with the inclusion behaviour of **60**. The *out/in* isomer **62** also failed to enclose adamantane [97].

Host molecules with different cavity dimensions were designed using a symmetrically trisubstituted benzene nucleus as bridgehead unit [98–100]. Examples of molecules are **63** and **64** in Figure 21. The cavity of **63** is ideally suited for the molecular inclusion – and solubilization (in acidic water) – of large discoid lipophilic guest molecules like triphenylene and pyrene [98, 99].

The macrocyclic host molecule **64** containing three azobenzene units functions in a responsive manner [100]. It alters the cavity size and shape upon irradiation. All four possible *E/Z*-isomers of the molecule have been isolated.

3.5. HOST CAVITIES WITH DONOR OUTFIT AND FUNCTIONAL GROUPS

At a first glance, and if topological considerations are well to the fore, there seems to be no vast difference between a cavity with inside donor atoms or functional groups and one without (Figure 20). However, the fact is that there is really a vast difference. Generally it is difficult to force polar donor sites or functional groups inside the cavity into converging positions [21]. This holds for synthetic and conformational aspects as well.

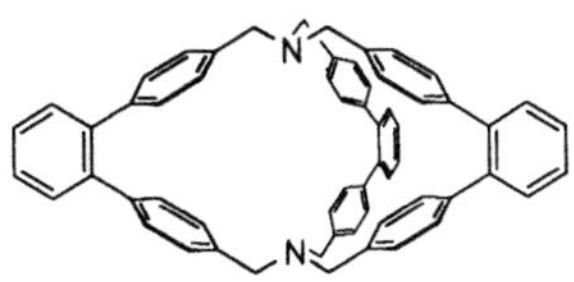

59

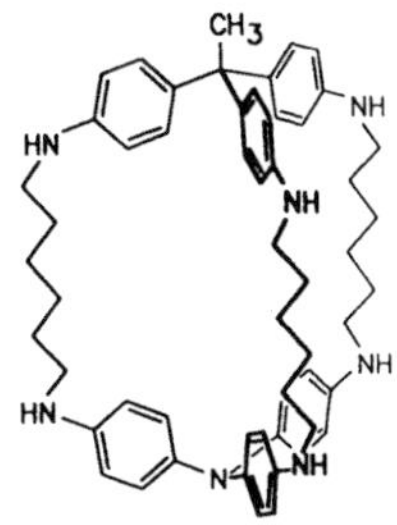

60

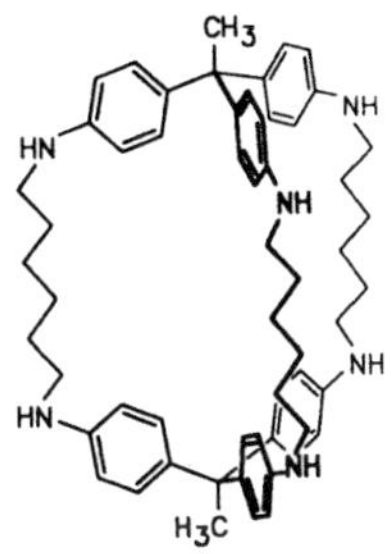

(out/out)
61

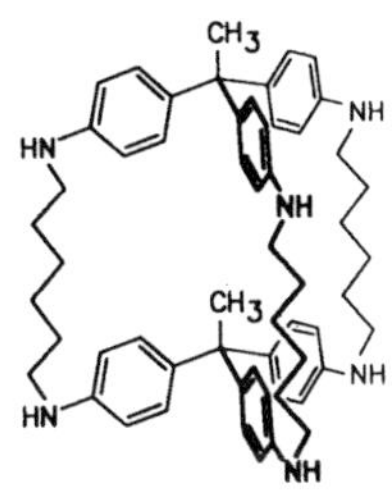

(out/in)
62

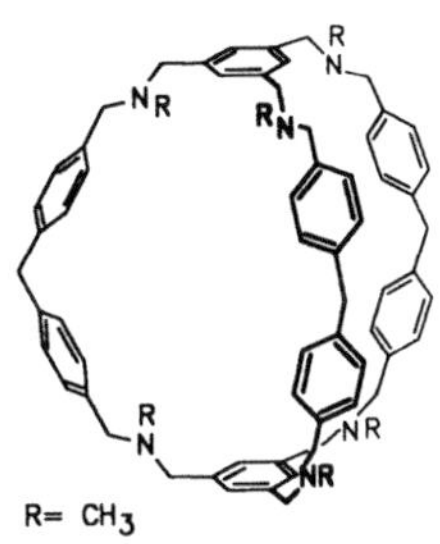

63

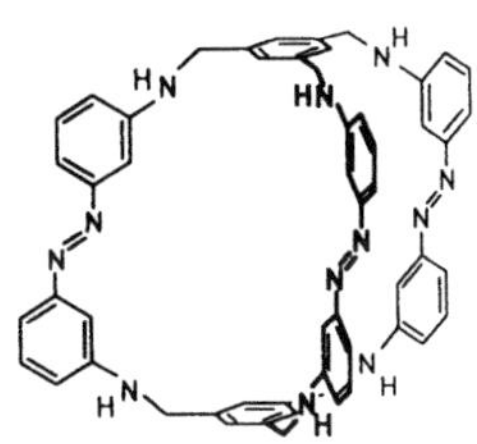

64

Fig. 21. Macrobicyclic hosts to include lipophilic guest molecules [95–100].

Consequently, we chose the easiest route and started rather trivially, i.e. we synthesized the rigid cryptands **65** and **66** (Figure 22) [95]. These cryptand type molecules combine typical oligoether segments of the cation complexing spacers such as in the cavity molecule **59** (cf. Figure 21). Hence, they are special cases of cryptands depending on the number of oligoether and aromatic segments. Due to the presence of a diaza-18-crown-6 moiety, **65** complexes Na^+ ions with high selectivity [95].

Whereas compounds **65** and **66** contain the usual nitrogen bridgeheads, hosts **67** and **68** (Figure 22) [101] with aromatic bridgehead units are close to the cavity molecules **63** and **64** (Figure 21). Nevertheless, because of the oligoether segments in **67** and **68**, they complex cations rather than uncharged organic molecules [101].

Topologically, the cavity molecules **69** and **70** (Figure 22) [102–104] are also cryptands. They have the same nitrogen or trisubstituted benzene bridgeheads as before. However, what makes these structures so complex is that three complete

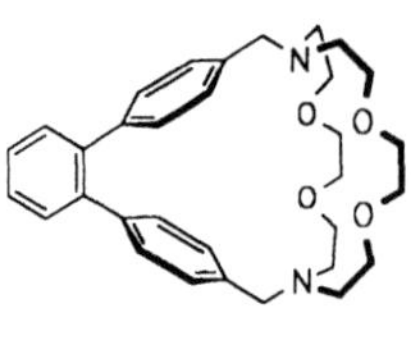

65

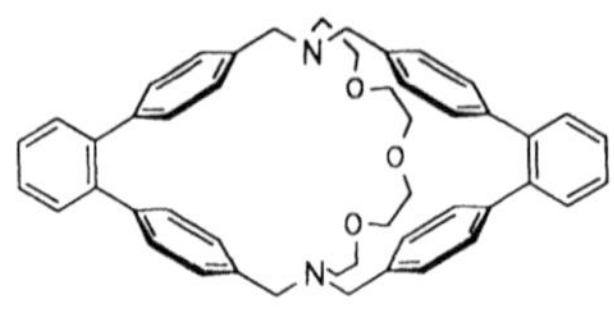

66

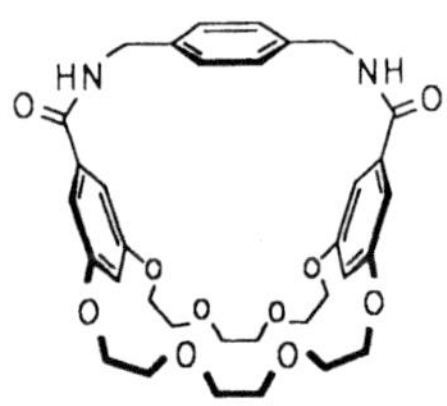

67

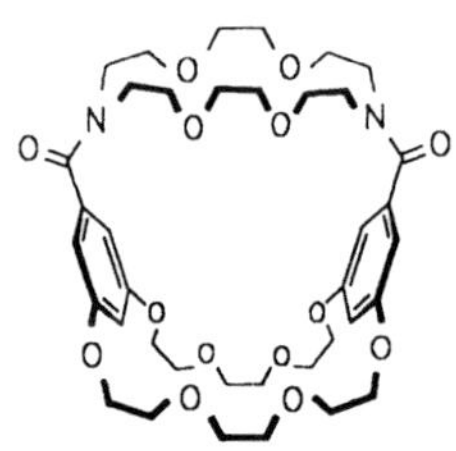

68

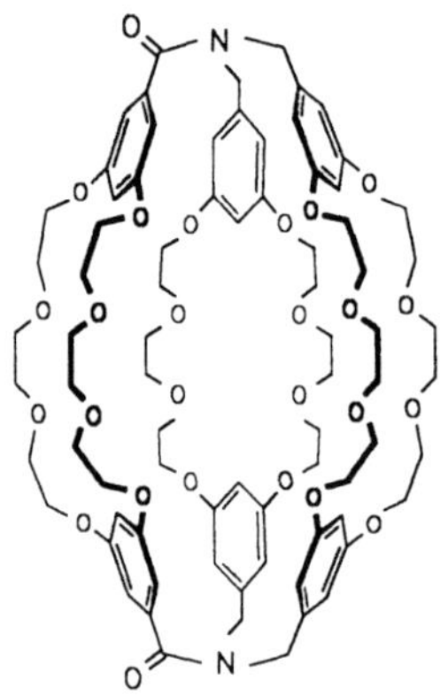

69

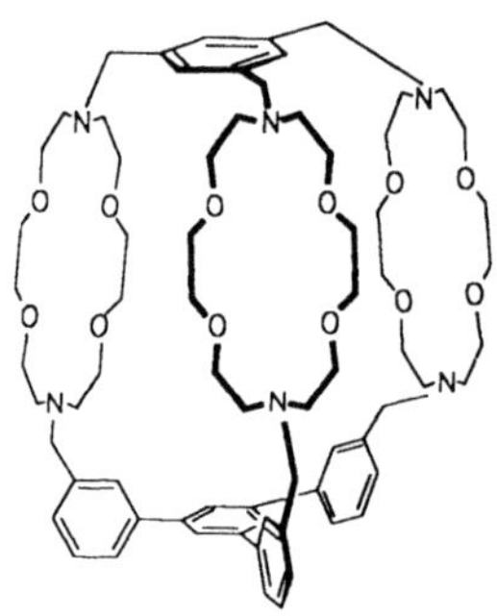

70

Fig. 22. Macropolycyclic hosts incorporating crown units (polyether segments) [95, 101–104].

crown rings form the bridges. As expected, the macropolycyclic tris-crown hosts **69** and **70** complex alkali metal ions [102–104]. Interestingly, host **70** was also found to complex oligophenolic molecules [104].

How about the problem with the functionalized cavities? Could it be solved? Well, we crossed that bridge when we came to it, as is obvious from Figures 23 and 24, which show logical series of functionalized host cavities **71** – **74** and **77** – **80** [93, 94, 105]. They are derived from typical catechol or bipyridine building blocks and different spacer (bridgehead) elements, thus leading to cavities with various functions, sizes, and shapes.

Most of these host cavities efficiently include cations and phenolic hosts, but some provide amazing complexation properties. Examples of such complexes (**75**, **76**, **81**, and **82**) are illustrated in Figures 23 and 24. For instance, the macrobi-

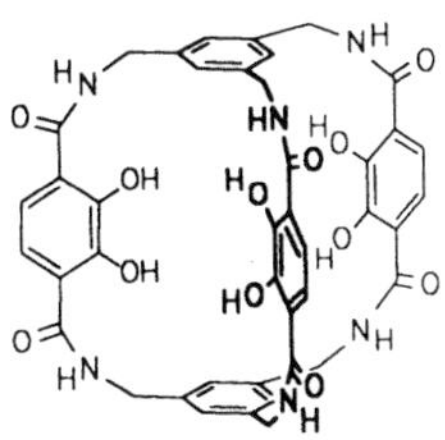

71

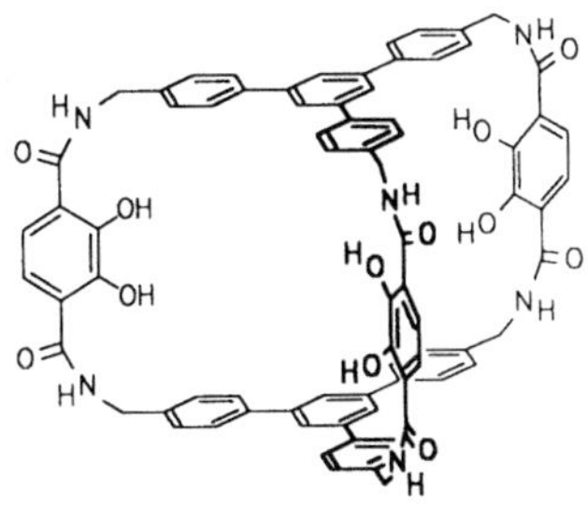

72

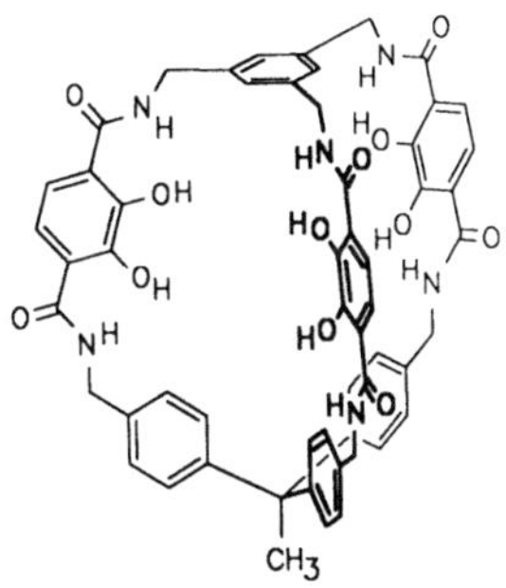

73

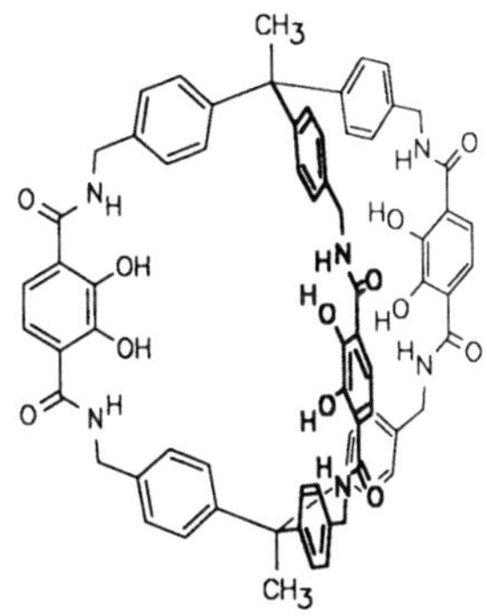

74

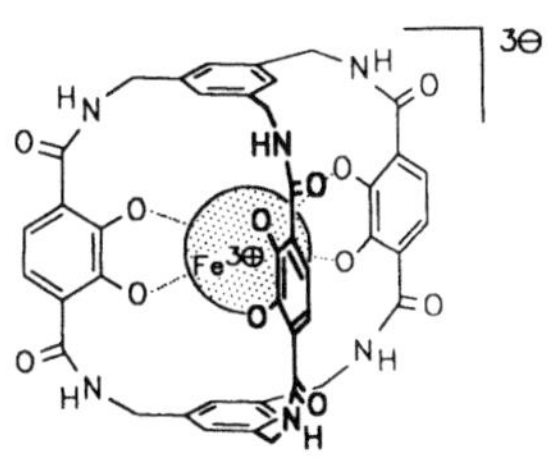

75

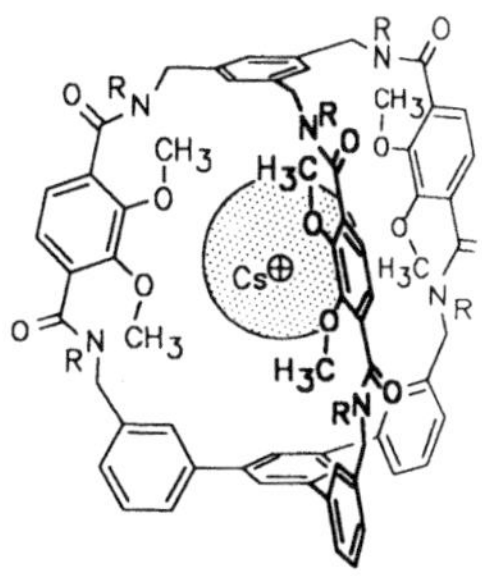

R= Benzyl 76

Fig. 23. Macrobicyclic tris-catechol hosts and typical complexes [93, 94, 105–107,110].

cyclic tris-catechol host **71** complexes Fe^{3+} as in **75** (Figure 23) [106] with $K_s \approx 10^{59}\ l\ mol^{-1}$ [107]! It is thus an even stronger Fe^{3+} binder than the natural siderophore enterobactin ($K_s \approx 10^{52}\ l\ mol^{-1}$) [105].

Another complex of high stability, **81** (Figure 24), is formed between tris-bipyridine host **77** and Ru^{2+} ions [108]. This ruthenium complex exhibits double the lifetime of the excited triplet state in comparison with $Ru(bipy)_3^{2+}$ but at the same time has a higher photostability, by a factor of 10^4 [109]. For that reason, the present Ru^{2+} complex of **77** is very promising in photophysics [94].

The analogous hexamethoxy host molecule, such as in **76** (Figure 23), forms a selective inclusion complex with Cs^+ ions which is very interesting in analytical

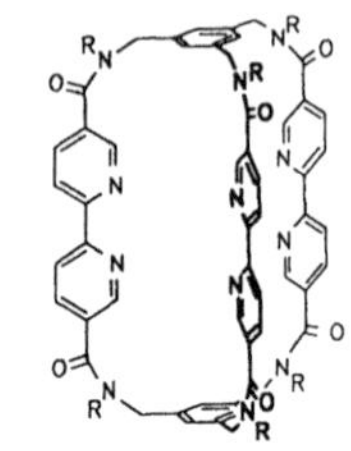

77

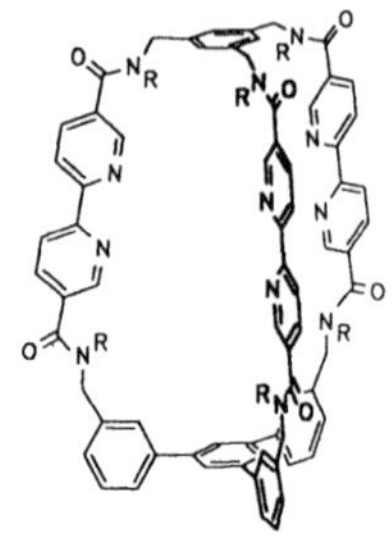

78

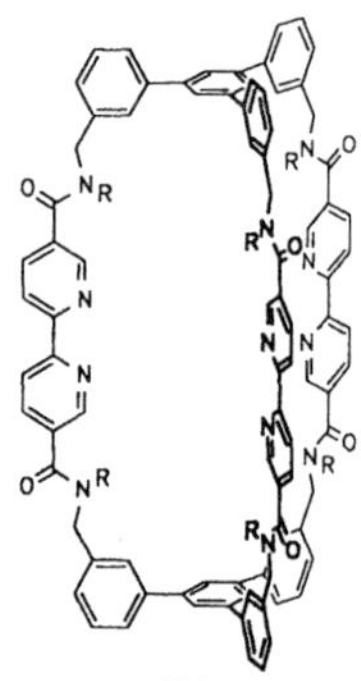

79

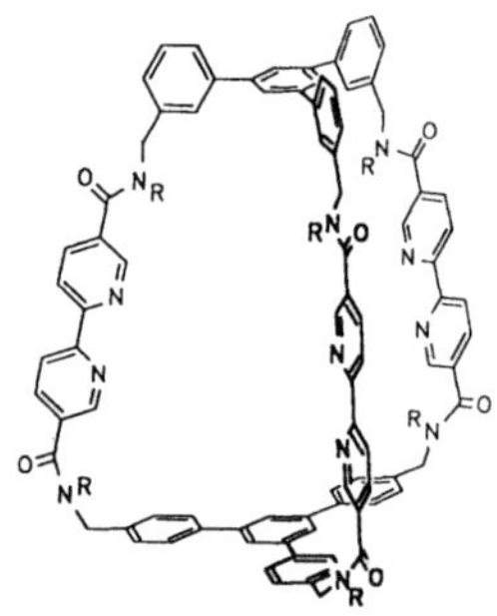

80

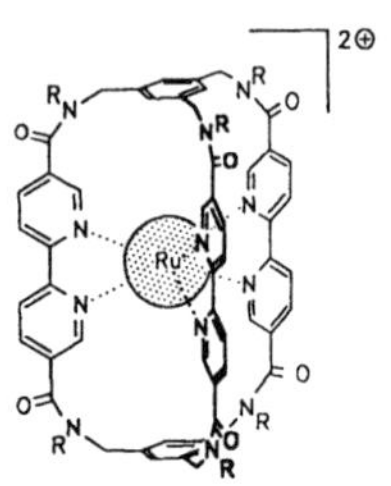

81

R = Benzyl

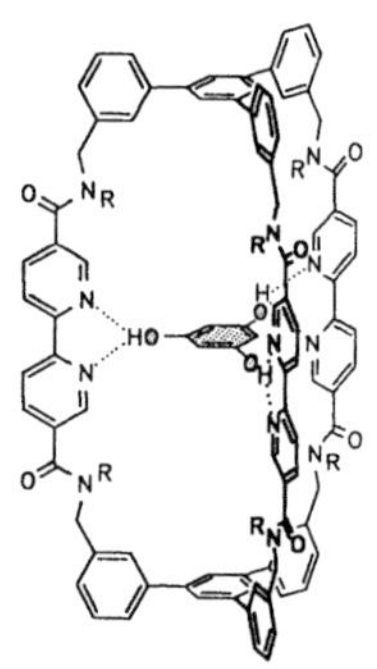

82

Fig. 24. Macrobicyclic tris-bipyridine hosts and typical complexes [93, 94, 108, 109, 111].

and radio medicinal chemistry [110]. Moreover, the tris-bipyridine host **79** (Figure 24) encloses phloroglucinol with an association constant K_{ass} of more than $10^4\,l\,mol^{-1}$ [111]. A proposed structure of this complex (**82**), where hydrogen bonds are involved, is shown in Figure 24.

The modular synthetic strategy (cf. Figure 20) used for the synthesis of these macrobicyclic hosts allows broad structural variation [93, 94]. We are currently following this line in order to make available new, selective and strongly complexing host cavities with various sizes, shapes, polarities, and endo-functions [112] – similar to Pedersen, who had demonstrated it in outline with his original crown compounds.

4. Weber Group Research at Bonn University

4.1. THE MULTI-LOOP APPROACH: COMPLEXATION OF CATION ARRANGEMENTS

At the beginning in 1978, the Weber research group was only two persons, including Weber himself. Accordingly it was difficult to study crown chemistry over a wide range and one had to decide on one or two particular topics. It appeared very promising to develop ligands which enable a close association of more than one alkali or alkaline earth metal ion. Hence we settled on this topic.

The novel host design which was considered first features a spiro-linked assembly of two to four individual macrorings differing in ring size, rigidity, and donor characteristics [113]. Typical compound examples are specified in Figure 25 (**83**–**85**). Due to their particular topology, these new hosts were dubbed 'multi-loop

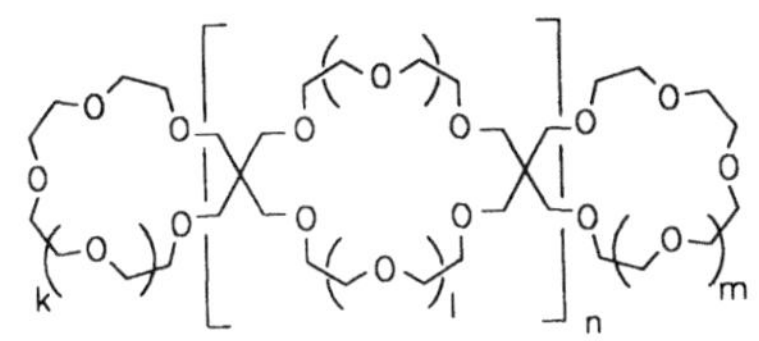

83

k = 0–4; l = 1, 2
m = 0–2; n = 0, 1, 2

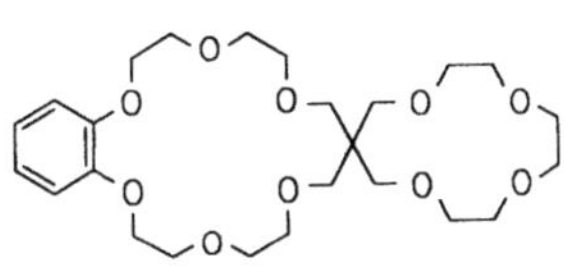

84

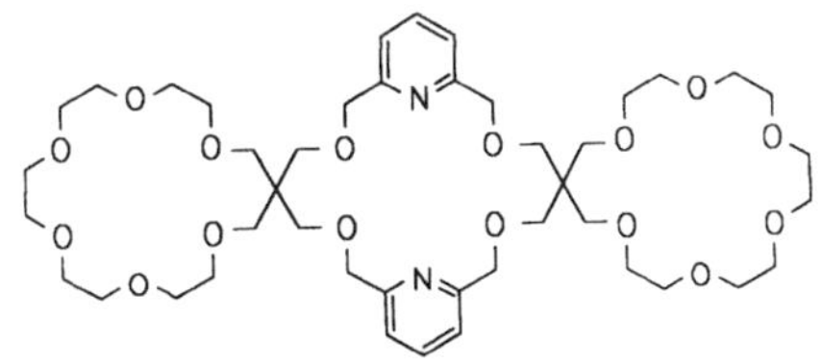

85

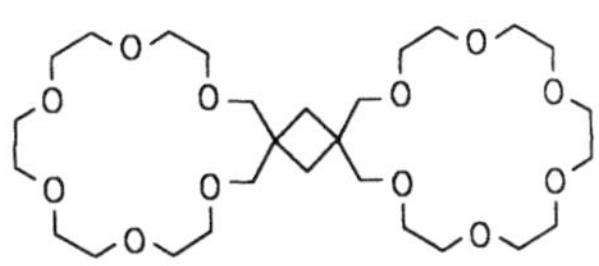

86

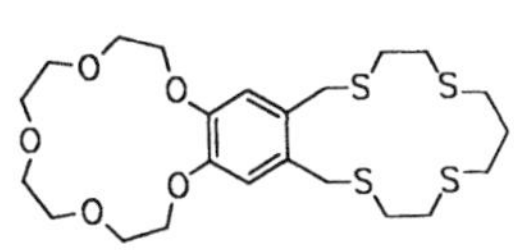

87

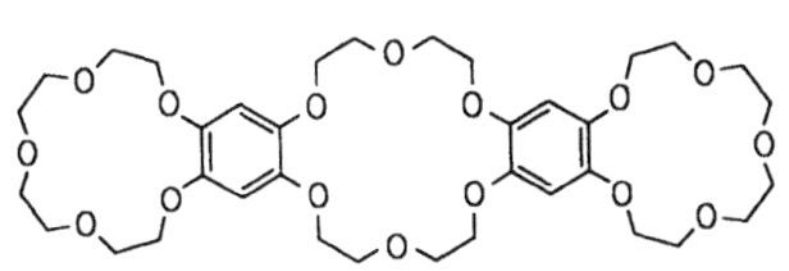

88

Fig. 25. Multi-loop crown compounds for complexing cation arrangements [113–115].

crowns' [114]. The synthesis of these species is based on the pentaerythritol unit providing the spiro linkage and involves skillful use of blocking/deblocking procedures and high-dilution techniques. Later we, extended our methodology to cyclobutane dispiro systems, as in crown compound **86** (Figure 25) [114] and, more recently, the spiro linkage was replaced by an aromatic unit to yield another new multi-loop crown topology, such as in compounds **87** and **88** (Figure 25) [115].

As expected, these new ligands are efficient multisite receptors for cations [113–115]. They were shown to be generally suitable hosts for the common incorporation of identical or different metal ions (to give homo- or heteronuclear complexes, respectively), corresponding to the specific binding characterization of the individual coordination compartments combined in the ligand skeleton. However, cooperativity and allosteric behaviour is also observed in the complexation properties, which originates from an interaction of adjacent binding sites. Thermodynamic and kinetic data of the complexation from ^{23}Na NMR studies are available [116] and the transport property of Pr^{3+} across a phosphatidylcholine vesicle membrane mediated by a diloop crown (**83**; $k = 4, m = 0, n = 0$) is reported in Ref. 117. We were also successful in isolating a great many crystalline multinuclear cation complexes [113–115].

In one case of a metal ion complex, we dispose of a crystal structure (Figure 26) which shows several interesting points [118]. Each loop of the crown compound (diloop) contains one Li^+, thus exhibiting a biloculi receptor behaviour. There is also one water molecule in each loop, the function of which is to fill the host cavity (19-membered crown ring) which is definitely too large for Li^+. The included water molecule is coordinated to the Li^+ and to two of the ether oxygens. A second coordinated water molecule lies outside the ligand cavity.

4.2. RINGS AND CHAINS: THE DONOR SIDE ARM APPROACH

Next we attempted to enhance the complexation strength of crown compounds by using something other than the cryptand strategy which causes unfavourable kinetics [119]. A possible solution considered was to use a crown with secondary

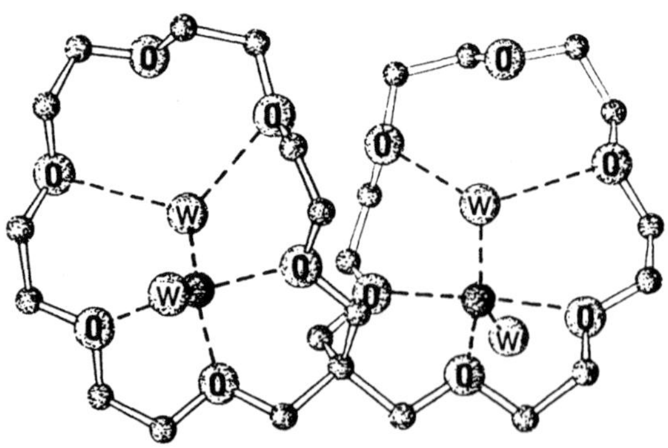

Fig. 26. Perspective view of the dinuclear hydrated LiI complex of diloop crown compound **83** ($k = 2, m = 2, n = 0$): simultaneous complexation of a metal ion and water (W). Neither iodide anions nor hydrogen atoms are included in the figure [118].

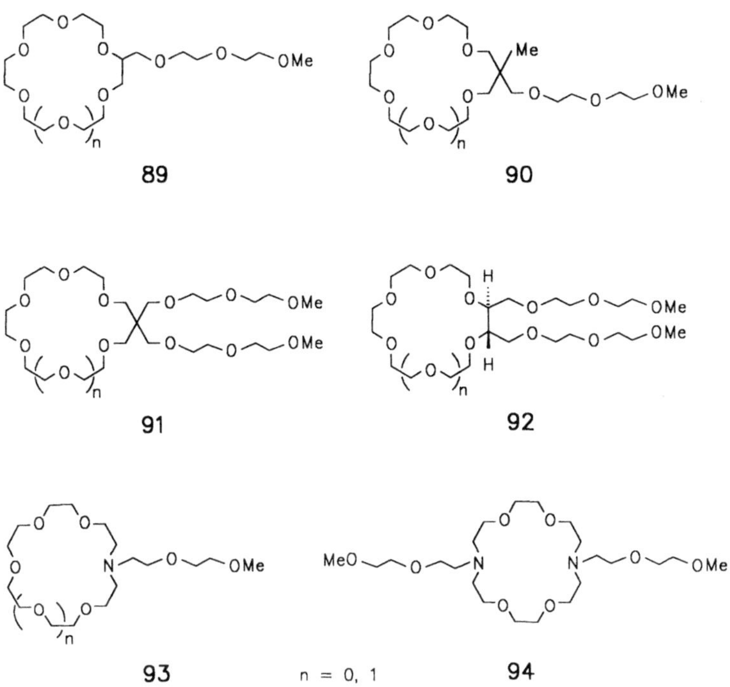

Fig. 27. Crown compounds with donor side arms: lariat ethers [122, 123].

donor groups bound to the macroring by flexible side arms, later termed 'lariat compounds' [120]. It happened: compounds of this type were synthesized in Bonn and almost simultaneously by the group led by G. Gokel [121]. Apparently (as for the pyridino crowns, see Section 2.2) two groups had the same idea at nearly the same time.

Compound examples (**89**–**94**) typical of our lariat design are illustrated in Figure 27 [122, 123]. They involve differently sized crown rings with one or two lariat arms attached to carbon or nitrogen pivot atoms. These compounds are relatively strong binders of cations (e.g. Na^+ as determined by ^{23}Na NMR spectroscopy) [122] and show efficiency in solid-to-liquid and liquid-to-liquid phase transfer of cations [123]. Since that early work, a great number of analogous lariat compounds have been synthesized and studied globally [120, 121], mostly for the purpose of the solvent extraction of cations [124].

4.3. PARTLY CROWN, PARTLY SURFACTANT: CROWN COMPOUNDS WITH LIPOPHILIC SIDE ARMS

In another approach for enhancement of the cation extraction and transport properties of cations, alkyl side arms have been appended to crown rings in order to provide additional lipophilicity. This led to highly lipophilic compounds such as **95**–**100** (Figure 28) [122–125]. They show quite remarkable solid-to-liquid and liquid-to-liquid phase transfer properties of cations as well as cation transport across a liquid model membrane [123, 125]. Crystalline complexes with alkali and alkaline earth cations were also isolated [125, 126].

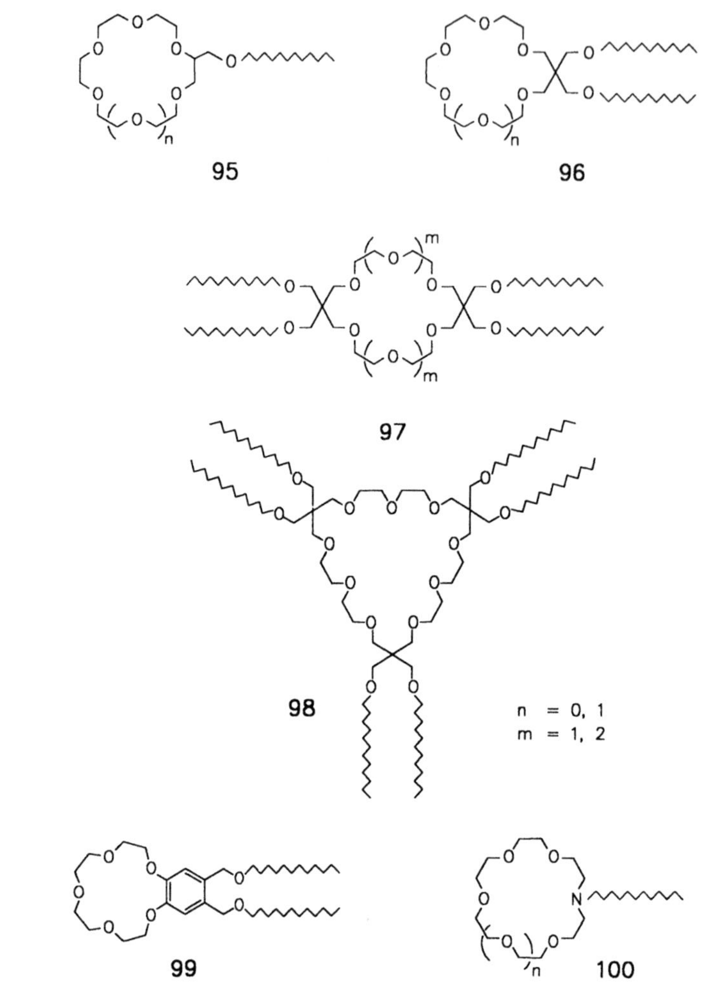

Fig. 28. Crown compounds with lipophilic side arms [125].

However, due to the particular structure (combination of a hydrophilic crown unit with a distinct lipophilic constituent) some of the molecules in Figure 28 behave as a typical surfactant, showing micelle formation and a cloud point [125].

In the solid state, the cation complex between **96** ($n = 0$) and Na^+, which was studied by X-ray crystallography (Figure 29) [126], reveals interesting parallels to certain membrane-forming lipids, including an analogous helical kink in one of the alkyl chains and the double layer structure characteristic of lipid crystals. Thus the surfactant type crown ethers are suitable as model compounds for biogenic lipid molecules and are suggested for use as ion-selective membrane carriers [127]. Aza compound **100** was a model to develop the 'anilides' [128].

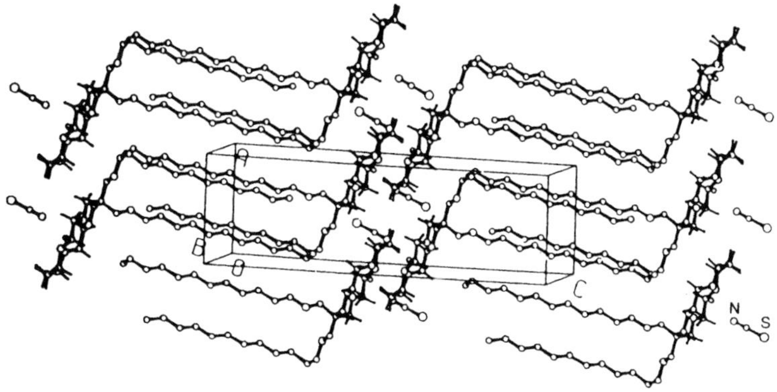

Fig. 29. Molecular packing in the crystal of the NaSCN complex of surfactant-type crown compound **96** ($n = 0$, Fig. 28) showing the double layer formation and the helical kink [126].

4.4. DISCOID AND SPHERICAL OLIGO-CROWNS: NEW TOPOLOGIES OF SURFACTANT MOLECULES

The type of compounds (**101**–**103**) exemplified in Figure 30 involved perhaps the greatest surprise to use. Originally, these molecules were intended as multisite cation receptors (cf. multiloop structures, Section 4.1.) having a new spherical topology. However, interest in this class of substances developed in a different way. Although the type of compounds specified in Figure 30 does not consist of a lipophilic and a hydrophilic part as in a typical surfactant structure, the substances behave in a similar manner [129]. Instead, these molecules are characteristic of a variable number of crown subunits (three to six) attached to a central aromatic or heteroaromatic core using different linkage elements (cf. Figure 30).

The compounds are highly soluble in water, wet a contiguous glass wall, and show foaming ability. The surfactant character is illustrated even more clearly by the occurrence of cloud points in water (between 24 and 85°C, 10^{-3} M solution), which are directly related to the molecular structure (the cloud point decreases with increasing conformational flexibility of the molecular framework and *vice versa*). Critical concentrations for micelle formation are in the range of 10^{-4} with aggregation numbers of about 50–80 [129].

Thus, the oligotopic crown compounds of type **101**–**103** are surfactants having a novel construction principle [130]. The compounds may also be understood as structural forerunners of the now rather important discoid liquid crystal forming substances. [131].

4.5. HIGHLY BENZO-CONDENSED CROWN VERSIONS

Although dibenzo-18-crown-6 (**12**; Figure 3) was the initial crown ether discovered by Pedersen [1, 2], analogous crown compounds with a higher number of aromatic rings and with benzo condensations in different positions remained practically unexplored [3, 132] until we set to work on it. As we saw it, such compounds combine several properties which show promise for selective cation complexation, including steric shielding, molecular rigidity, and reduced donor ability. Thus, to fill

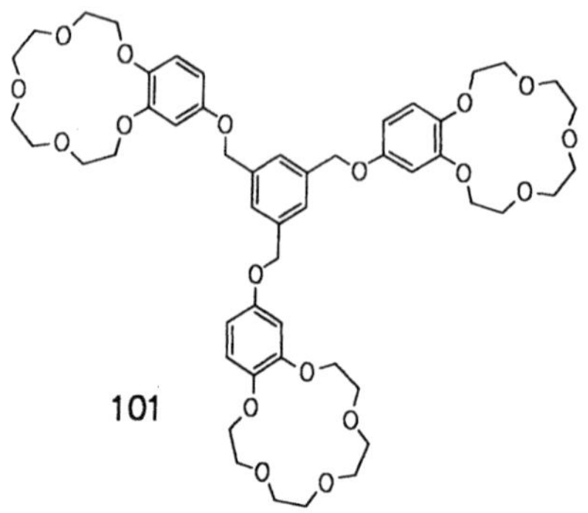

101

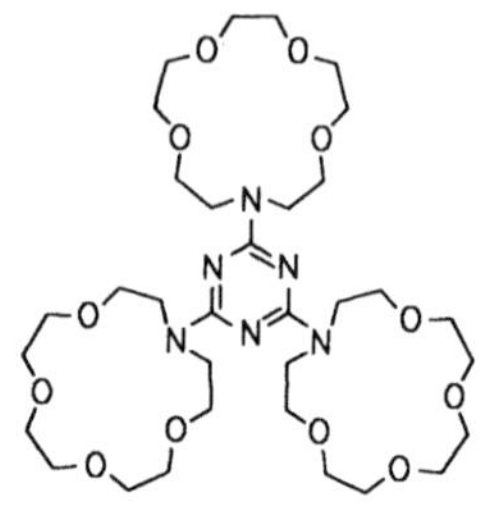

102

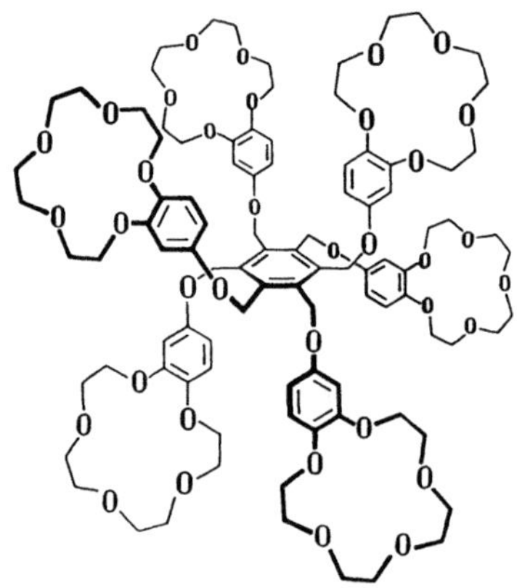

103

Fig. 30. Discoid and spherical multiloop crowns with surfactant behaviour [129].

this gap, we synthesized hitherto absent members of the series of 18-crown-6 rings with oligo-benzo condensation [133]. Characteristic examples of compounds are specified in Figure 31.

In fact, the oligo-benzo crowns responded with remarkable selectivity to cations, when used in liquid membrane electrodes [133]. In calorimetric titration studies, complexation of protons (in acetonitrile) correlates well with the number of benzo units present in these compounds (complexation decreases with increasing number of

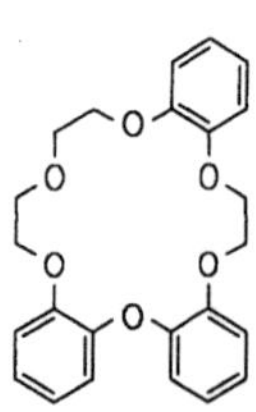

104

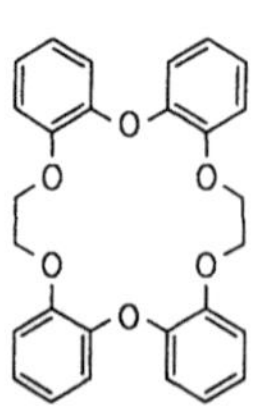

105

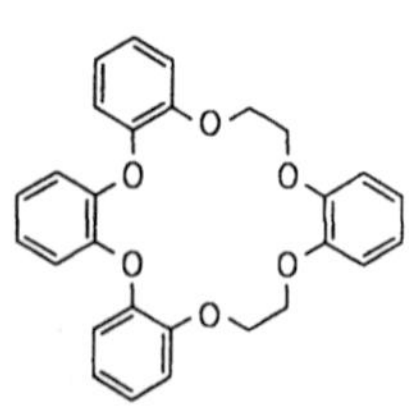

106

Fig. 31. Highly benzo-condensed 18-crown-6 derivatives [133].

benzo units) [134]. Also, some of these macrorings (e.g. **106**) yield crystalline inclusion complexes with water and small organic molecules, as a result of their rigid cavity structure [133].

4.6. DESIGN OF HETEROAROMATIC AND CAVITAND-TYPE CROWN RECEPTORS FOR UNCHARGED ORGANIC MOLECULES

There is no reasonable explanation why the discovery by Charles Pedersen at the beginning of the seventies that crown ethers are capable of complexing uncharged organic molecules [74] (cf. Section 3.4.) remained so long on the shelf before stimulating people. Actually, up to the end of the seventies, complexes between crown compounds and uncharged organic molecules were very rare; mostly such studies related to complexes of 18-crown-6 and they were rather accidental discoveries [80, 81]. Later, crystalline complexes of crown compounds with CH– and NH–acidic uncharged guest molecules became more and more numerous, but the formation of analogous complexes with OH–acidic guests, such as alcohols, remained unsuccessful over a long period [135].

However, the answer followed quickly when we considered ligands with one or two pyridino units and a certain number of aromatic groups gathered in a macroring, of which compounds **107–112** (Figure 32) are typical [136–141]. These macrorings may be seen as descendants of compound **31** (X = N, Figure 9), which also shows crystalline complex formation with uncharged organic molecules (mainly of dipolar aprotic nature) [142], but compounds **107–112** and their analogs are more efficient. They form crystalline complexes with organic guests in a broad range, including alcohols [135–141]. Nevertheless, crystalline complex formation is highly selective for each macroring, depending on the structural parameters of host and guest.

For instance, host compound **107** (R = H) yields a stable 1 : 2 stoichiometric inclusion compound with methanol (hydrated) [136], whereas **107** (R = CMe_3) which is substituted by a voluminous *t*-butyl group, cannot [137]. The latter macrocycle, however, binds 2-propanol highly selectively.

When complexing methanol by **107** (R = H), the conformation of the host molecule changes significantly from a closed to an open structure, as evidenced by crystal structure analysis [143]. A unique host–guest order is found for the methanol complex of **107** (R = H) involving an almost linear O—H···N host–guest hydrogen bond and an intra-cavity oriented methyl group of the guest, as shown in Figure 33a [143]. The high selectivity of **107** (R = H) for small unbranched alcohols is now evident for spatial reasons. This is probably the first example of neutral guest complexation via hydrogen bonding inside a host cavity niche [88, 135].

Replacement of the benzo rings by naphtho units, as in **110**, gives rise to a molecular conformation with two deep clefts. They are potential sites for locating guest molecules. Figure 33b illustrates the crystal structure of the 1 : 1 dioxane complex of **110** [140].

Another unique property occurring in this class of compounds is face differentiation with reference to the host macroring and uncharged guest molecules. This is obvious when comparing Figure 33c with Figure 33d, which show the crystal structures of the complexes of macroring **111** with MeCN (1 : 2) and $MeNO_2$ (1 : 1), respectively [144, 145].

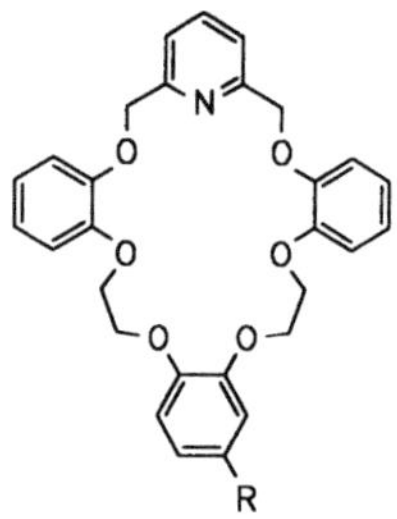

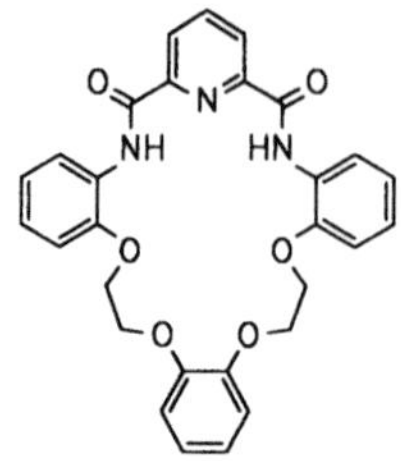

107 R = H, Me, CMe_3 108

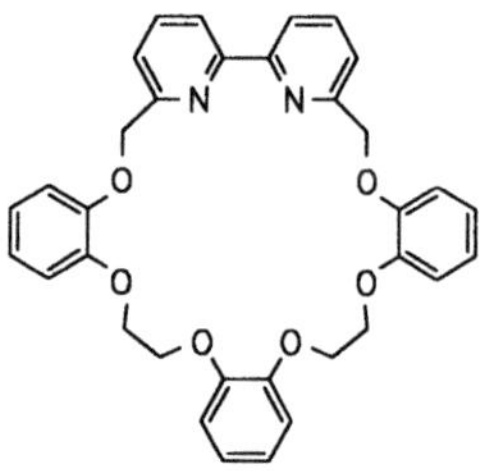

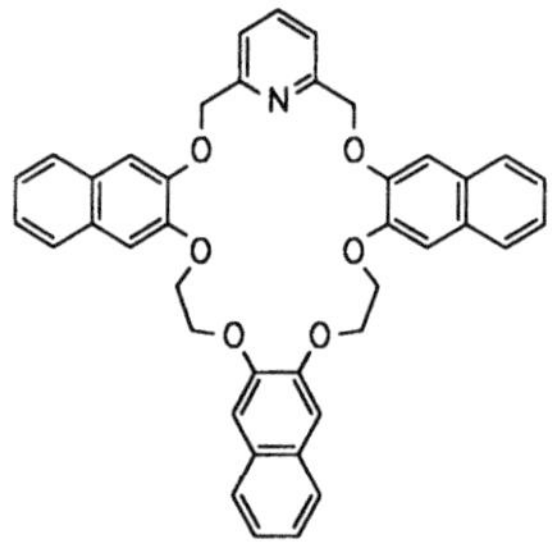

109 110

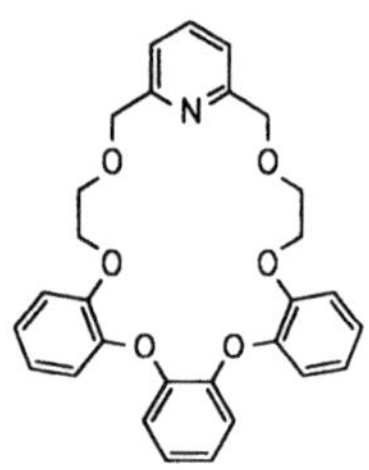

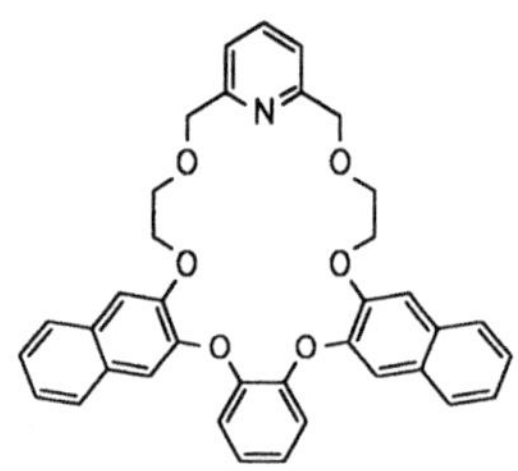

111 112

Fig. 32. Heteroaromatic crown hosts which form uncharged molecule complexes [136–141].

The net conformational geometry of **111** is the same in both complexes, providing the host molecules with two concave faces (A and B in Figure 33a and 33b). One (A) has the rough appearance of an ice-cream cone and offers a relatively deep and narrow cavity, the other (B) is similar to a book-rest with a relatively wide and low shielded hollow. For steric reasons, the MeCN and $MeNO_2$ guests, which are rod-shaped and Y-shaped molecules, respectively, are correspondingly distributed between A and B in the present complexes (Figures 33a and 33b) [144, 145]. Besides the spatial relations, motifs of polarity and hydrogen bond complementarity between the different host faces and the two guests are also evident.

There are many more uncharged guest complexes and free molecules of this host class which have had their structures solved, involving macrocycles with different aromatic and heteroaromatic building blocks, donor sites, ring sizes, and rigidity, as well as guest molecules of different sizes, shapes, and polarity [146–151]. Thus, the complexation chemistry of semi-rigid macrocyclic pyridino-, bipyridino-, and

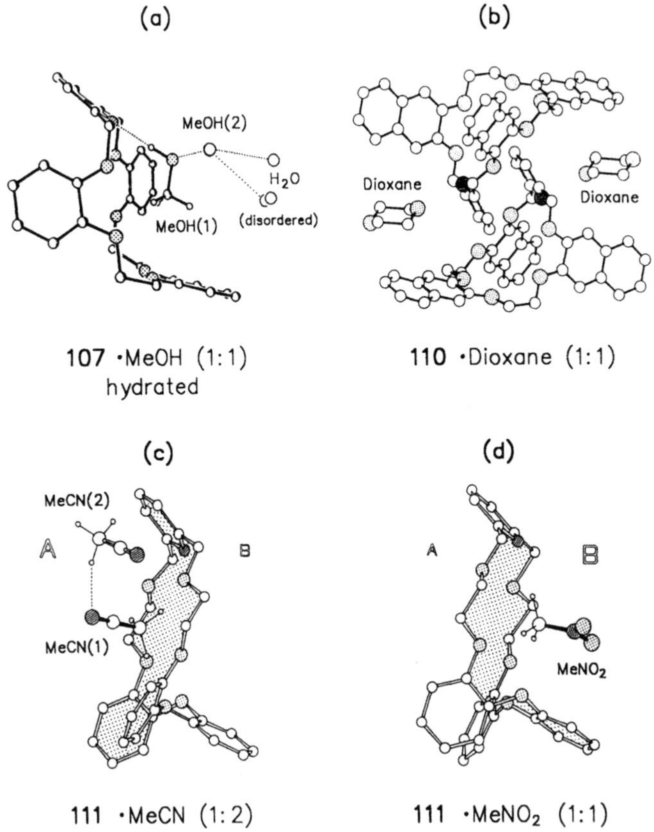

Fig. 33. Crystal structures of typical pyridino crown complexes with uncharged organic molecules (O atoms dotted, N atoms hatched) [140, 143–145].

phenanthrolino-containing hosts with uncharged organic guests is highly developed at the present time [135].

Recent investigations in our laboratory are keyed to the development of macrocyclic receptors which have a defined cavity and specific inward-facing functional groups, e.g. strong proton donor and acceptor sites. Typical examples are the macrocycles **113** and **114** (Figure 34a) [152].

Macroring **113** forms a unique 1 : 1 inclusion compound with DMF, the structure of which is shown in Figure 34b [152]. The guest molecules of DMF (disordered) are in the center of the macroring, aligned with an intra-host channel. Macrocycle **114** is promising as a selective extractant (solvent–solvent) for small alkaline earth metal ions [153].

4.7. MACRORING VERSUS LATTICE CAVITY: HOST–GUEST CHEMISTRY IN THE SOLID-STATE, INCLUDING CLATHRATES, COORDINATOCLATHRATES, AND CRYSTALLINE COMPLEXES

Since the middle of the eighties, research in our group shifted increasingly from the crown cation complexes and molecular inclusion compounds to the lattice-type

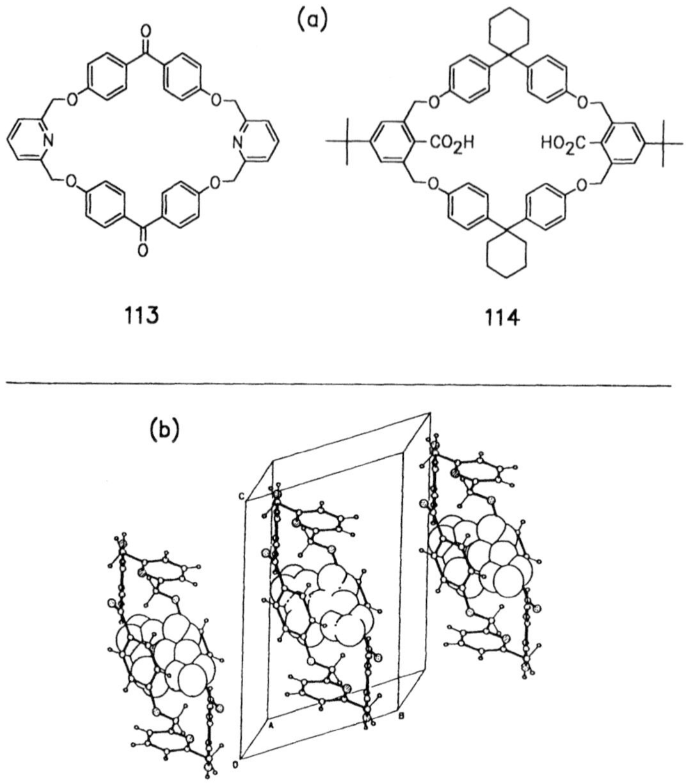

113 · DMF (1:1)

Fig. 34. Cavitand-type hosts (a) and crystal structure of an inclusion compound (b) [152].

inclusions (clathrates) [154, 155] and supramolecular crystals [156]. They are now the main topics of our research. To steer the discussion to the point, Pedersen never worked with host lattices and clathrate compounds. Actually, the existence of such species was recorded long before Pedersen [157]. Nevertheless, by his promotion of the general inclusion idea, he also stimulated, albeit unintentionally, these particular fields of host–guest chemistry, giving them a tremendous impulse in the last few years [45, 154–159].

The attraction of the lattice (clathrate) approach for forming inclusion compounds lies in the fact that you can use small host molecules for cavity formation and thus avoid a troublesome synthesis of macrocyclic host cavities (multimolecular vs. monomolecular cavity formation, cf. Figure 35) [160]. On the other hand, the small clathrate hosts were subject to a high degree of randomness, and only very recently have some helpful design principles for clathrate hosts been developed [161].

Possibly the most general ones which came from our group are obvious from Figure 36 and we may state the following [162]: a clathrate host molecule which will crystallize in such a form as to leave hollow spaces in the crystal lattice for guest inclusion should be: (1) bulky in constitution to cause the lattice holes; (2) it should have a rigid conformation to maintain the cavity structure; and also (3) one

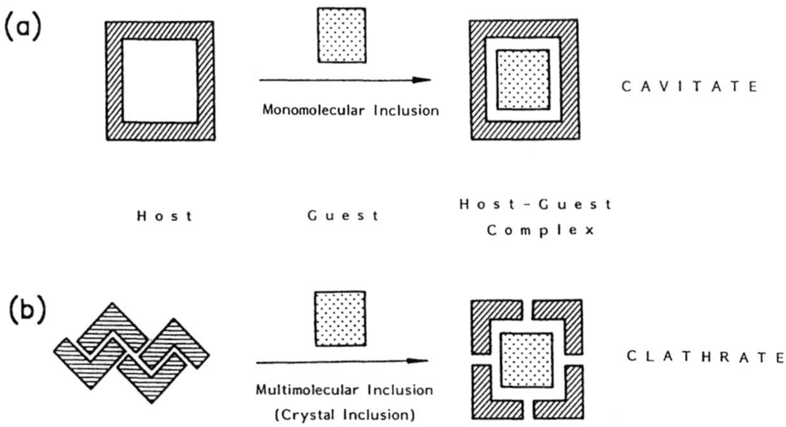

Fig. 35. Strategies of inclusion formation [160].

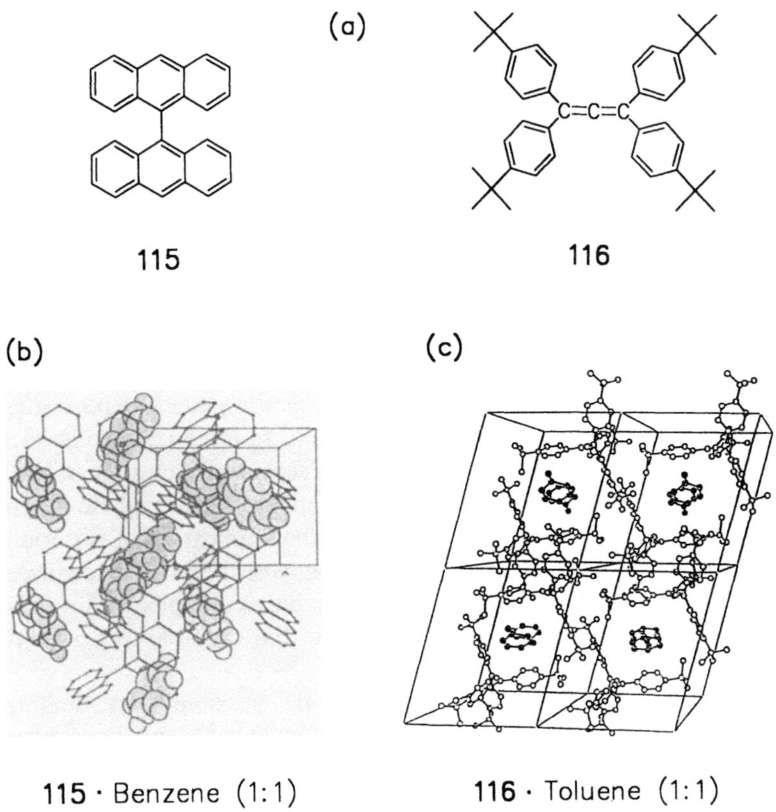

Fig. 36. Apolar lattice hosts (a) and crystal structures of corresponding inclusion compounds (b, c) [163, 164].

should aim for a balanced symmetric overall shape of the molecule, which will help in the stabilization of the crystal lattice.

Compounds **115** and **116** (Figure 36a) are representative examples of host molecules which meet these principles [163–165]. In terms of their shape they belong to the scissor-type constitutions [166, 167]. Other examples relate to a molecular roof [167] or are based on a rigid, small ring [168, 169].

Crystalline inclusion structures involving the host compounds **115** and **116** and benzene or toluene guest molecules, respectively, are illustrated in Figures 36b and 36c [163, 164]. They underline the presence of spacers, arene units and bulky groups arranged in a geometric pattern as being important in creating a porous host lattice with cages (Figure 36b) or channels (Figure 36c) that makes hydrocarbon guest enclosure possible [162]. The inclusion formation is driven by the dense crystal packing [170] which is obtained when guest molecules fill the hollow space of the host lattice rather than by specific molecular interactions [156].

However, inclusion is believed not only to be the result of sterical properties but also of molecular polarity relations where oriented dipole interactions or hydrogen–bonding between host and guest are contributing factors [155, 156]. Clearly, the use of the properties of polar functions or groups which may allow hydrogen bonding would substantially enlarge the scope of crystalline hosts and guests. This prompted us to develop the so-called principle of 'coordination-assisted clathrate formation' [160, 167].

As suggested by the term, a coordinatively assisted clathrate, or 'coordinatoclathrate', involves a hybrid between a complex and a clathrate (Figure 37a) [160, 167]. Thus, coordinatoclathrates combine attributes of coordinative complexes and of lattice dependent clathrates, and this is the reason that they permit a high degree of selectivity in different directions, including chemoselectivity and constitutional selectivity, according to their origin (see Figure 37a). Consequently, a corresponding coordinatoclathrate host (Figure 37b) consists of two components: (1) a bulky basis skeleton that makes available lattice cavities typical of a clathrate (see above); and (2) appended functional groups (sensor groups) that manage the coordination to the included guest substrate [160, 167].

However, the realization of this idea is not an easy task since functional groups are mostly supplied with co-existing proton donor *and* proton acceptor sites, such as in carboxylic acids, amides, or alcohols. They tend to dimerize or oligomerize or influence one another, rather than to complex with a secondary component [160, 162].

We consider two possible ways out of this problem. One is obvious from the scheme given in Figure 38a [160, 166]. It involves the application of a steric barrier for suppression of the functional group contact between hosts in a localized or in a more integral sense. Yet these functions are available to complementary groups of sterically less demanding (smaller) guest molecules that are able to bridge the gap and to fill space.

A typical host molecule that approaches this idea is simple 1,1′-binaphthyl-2,2′-dicarboxylic acid (BNDA) (**117**, Fig 39a) with salient scissor-type shape (Fig. 39b) [171]. Recrystalization of BNDA (**117**) from different alcohols (such as MeOH, EtOH, 1-PrOH, 2-PrOH, *t*-BuOH etc.) results in the formation of well-developed channels of approximately 6–7 Å diameter with a stoichiometric number of solvent molecules inside the channel.

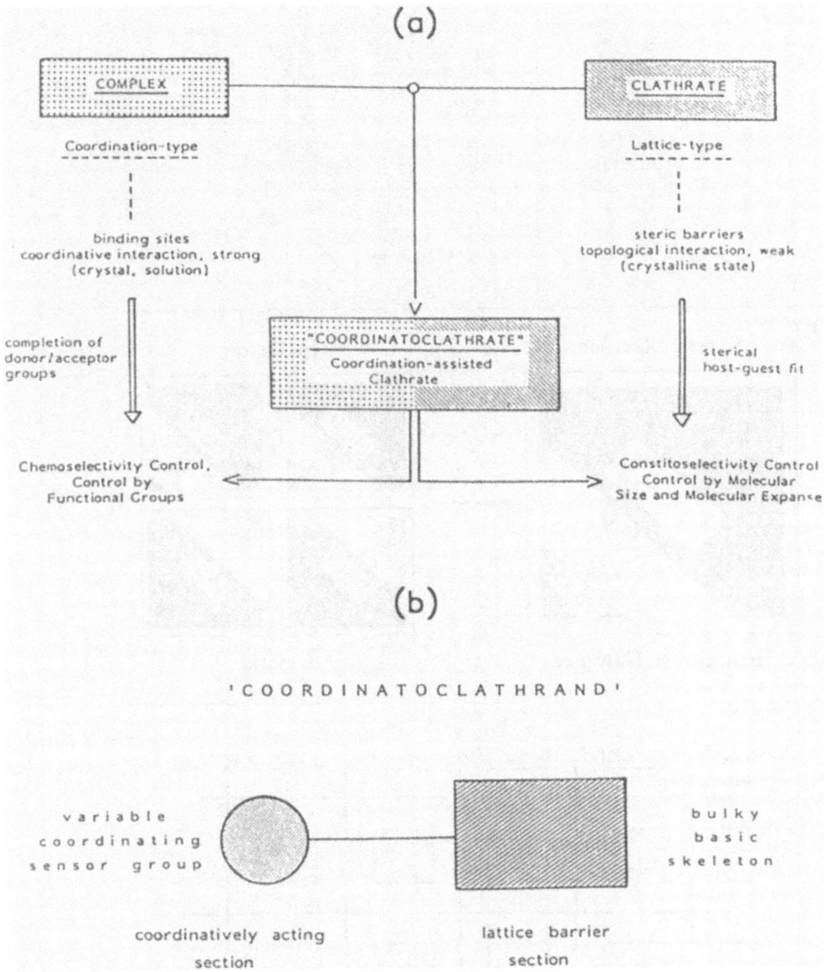

Fig. 37. Coordinatoclathrate concept (a) and abstracted structure of a coordinatoclathrate host (b) [160, 167].

Figure 39 (c–e) illustrates the facts for the inclusion compound of **117** with MeOH (1:2) [160, 167]. The channels (Figure 39c) have mainly an apolar surface that is interrupted periodically by hydrophilic narrowings consisting of the carboxylic groups (Figure 39d). The cross section of electron densities in this channel area (Figure 39e) clearly shows two facing carboxylic groups held at a noninteractive distance that follow closely the basic idea of Figure 38a, namely formation of a specific gap in the crystal lattice. Here are the sites at which the accommodated guest molecules (MeOH) contact the host.

This is more evident in Figure 39f, which shows a detail of the crystal structure of inclusion compound **117** · MeOH (1:2) [171]. Bridging of the gap is *via* two

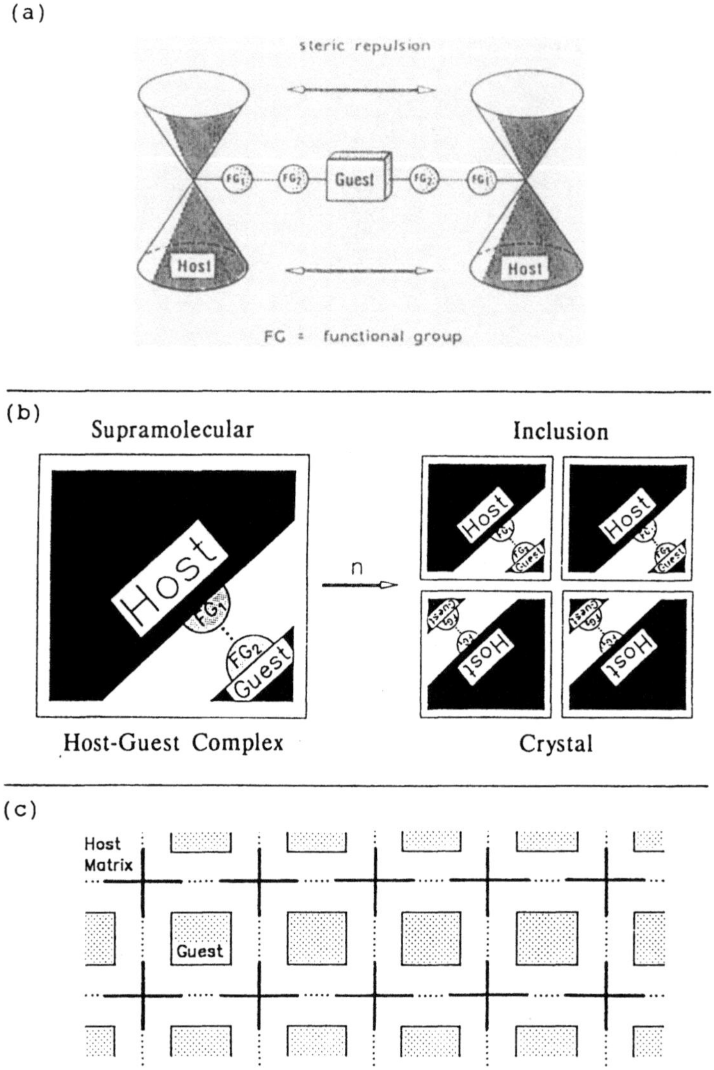

Fig. 38. Principles of crystalline inclusion (clathrate) formation [160, 162, 166].

molecules of methanol providing a set of complementary hydrogen bonds to the carboxylic groups. Thus, a 12-membered ring system of coupled hydrogen bonds is formed.

Alcohols other than MeOH, as demonstrated by crystal structures [171, 172], are involved in the same or in a similar binding pattern of hydrogen bridges with the BNDA host (e.g. a 10-membered ring) depending on spatial properties. Other carboxylic [168, 169, 173–175] or carbonamide hosts [174, 176], such as the roof-shaped compounds **118** [173, 174] and **119** [176] or the small-ring derived molecule **120** [175] behave correspondingly (cf. Figure 38a) and form either cyclic (Figure 40a, b) [174, 176] or helical (Figure 40c) [175] systems of hydrogen bonds between host and guest in their inclusion compounds [160, 162].

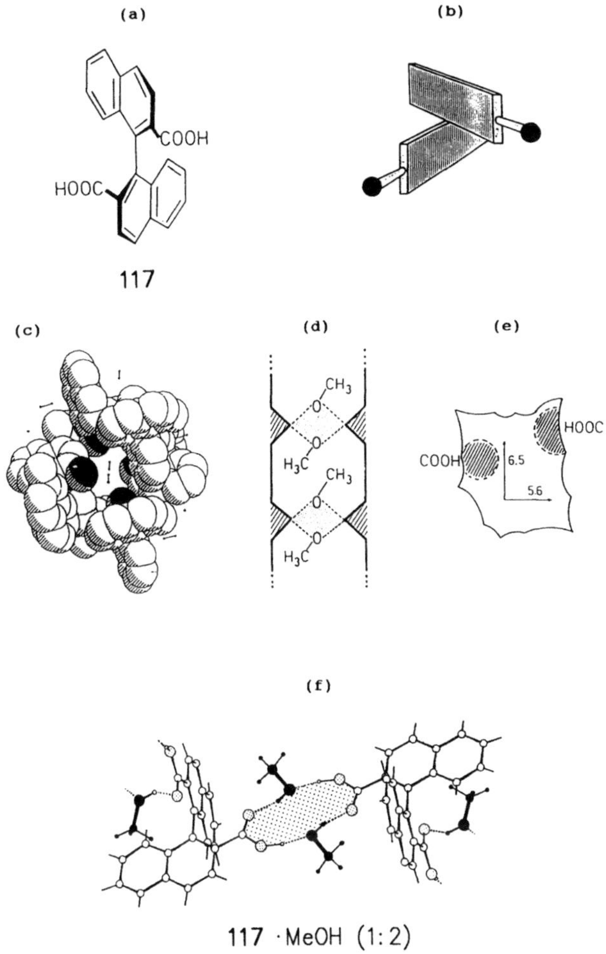

Fig. 39. Prototypical coordinatoclathrate host (a, b) and specification of its inclusion compound with MeOH (c–f) (H bonds shown as dotted lines) [160, 167, 171].

The second way to overcome the problem with complete dimerization of host functional groups interfering with the host–guest binding is to find a more convenient coordinative partner or complementary site than its own sort of functional group, i.e. to form a preferred host–guest complex, a supermolecule, which has an acceptable shape for crystallization (Figure 38b) [162]. An example is given in Figure 41a. The bulky diol host **121** shows high affinity for coordination with amines [177, 178]. When complexing the tertiary amine guest, a smooth surface of the supermolecule is obtained, which is suitable for crystallization.

Apart from using host functional groups for direct binding of polar guests acccording to the principles of Figures 38a and 38b, host functional group dimerization may be helpful for inclusion formation with apolar guests, as schematized in Figure 38c [161, 162] and illustrated in Figure 41b [179]. Dimerization of the carboxylic groups in the case of **117** results in a loose crystal packing. Consequently, a secondary component such as bromobenzene is required for filling the channel space, thus stabilizing the crystal lattice [179].

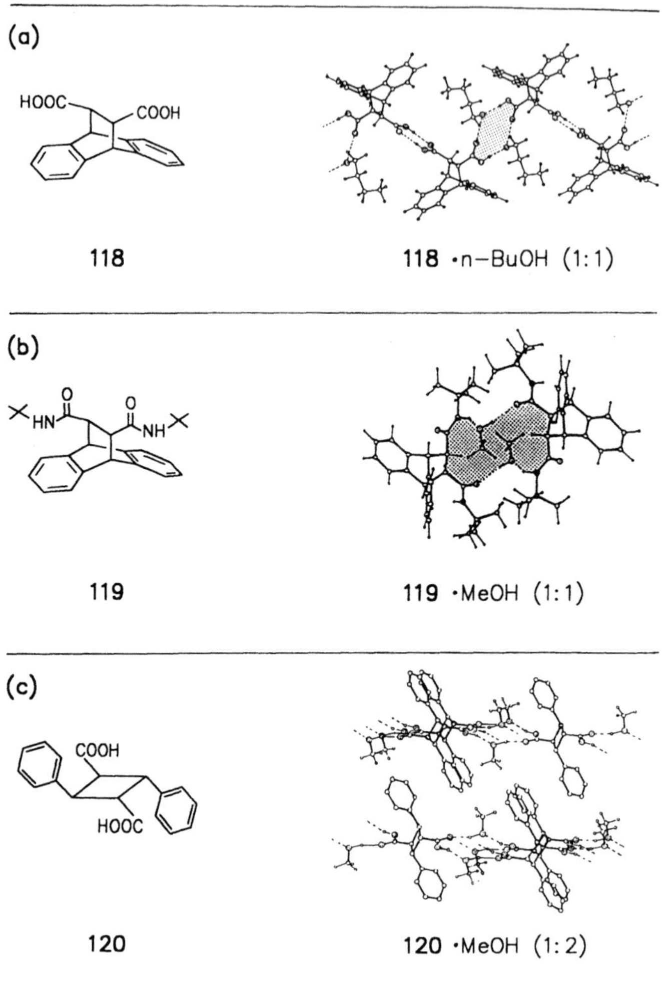

Fig. 40. Host molecules and inclusion structures (packing excerpts) showing the supramolecular binding pattern (H bonds shown as dotted lines; dotted regions indicate H-bonded rings) [174–176].

Similar factors are important in the formation of the inclusion compounds specified in Figure 42 [168, 169, 174, 176–181] and they also hold for the inclusion compounds of a large number of other hosts [182–193] developed by the design principles presented above [160, 162, 166].

This vast number of inclusion compounds were obtained by simple recrystallization of the host substance from the respective guest solvent or of host and guest from an inert solvent.

Guest competition experiments showed that each host compound provides a specific selectivity pattern of inclusion formation, dependent on the molecular parameters. Nevertheless, very simple host constitutions may give rise to unusually high guest selectivities. For example, triphenylmethanol forms a crystalline inclusion compound specifically with methanol from solution mixtures of alcohols [188]; triphenylsilanol does the same, but with ethanol [193]. Other hosts, such as **121** and

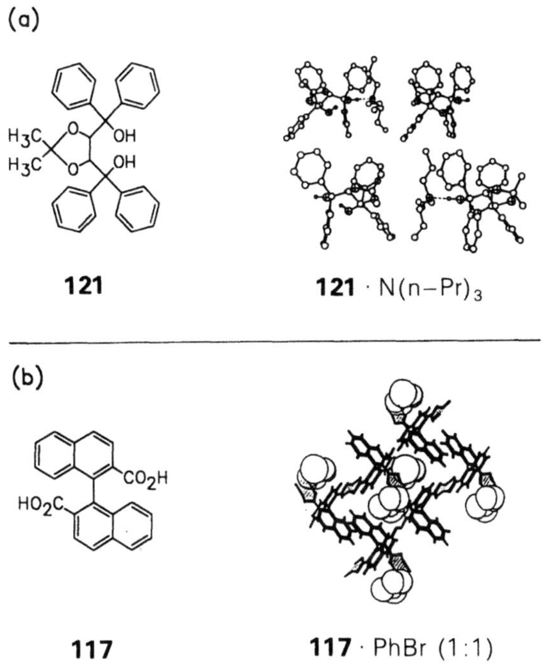

Fig. 41. Host molecules and inclusion structures (packing excerpts) typical of host–guest and host–host H bond interaction. In (a), H bonds are shown as dotted lines; the hatched regions in (b) indicate the H-bonded carboxylic group dimers [177, 179].

125–**127** are selective from solution for a particular type of amine [177, 178, 181, 194].

Moreover, the optically active hosts, which we have recently synthesized, are efficient in enantioselective inclusion of guests; e.g. optically active **125**, which can be derived easily from natural lactic acid, yields highly enantioselective inclusion compounds with chiral ketones [194]. It is also interesting to know that guests in the vapour state are also absorbed highly specifically from a vaporous solvent mixture by certain hosts to form the appropriate inclusion complexes [188, 190, 193, 194].

Thus, the simple compounds mentioned here are useful for sequestering chemical and stereochemical species in the liquid and vapour phase, comparable to the macrocyclic receptors initiated by Charles Pedersen. They are also promising in organic materials design [156, 162] and display some structural similarity, allowing them to mimic proteolytic enzymes [195].

5. Closing Remarks

It is clear that the scientific legacy of Charles J. Pedersen has deeply influenced our work on organic complexes and inclusion compounds from the beginning. We hope that this has been made evident by this paper, which has emphasized structural aspects.

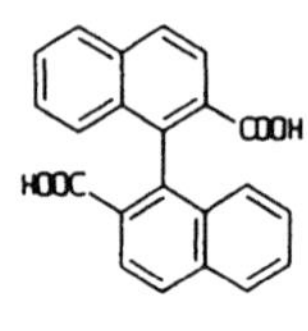	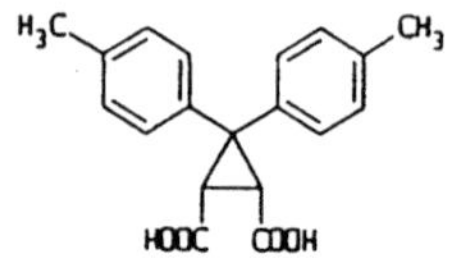	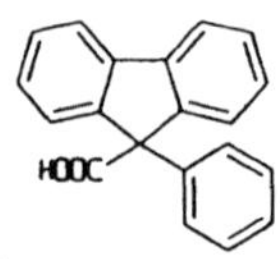
117	122	123
alcohols, acids, amides, MeCN, MeNO$_2$, DMSO, PhBr (30 examples)	alcohols, acids, amides, nitriles, MeNO$_2$, EtNO$_2$, DMSO, ethers (21 examples)	MeOH, EtOH, i-PrOH, t-BuOH, THF, dioxane, benzene
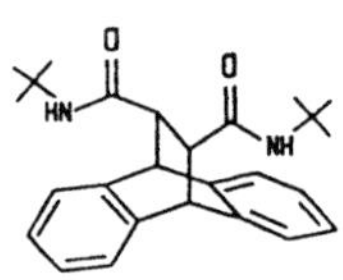	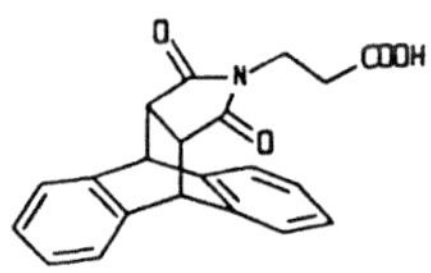	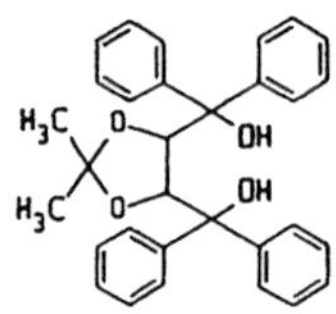
119	124	121
MeOH, HCOOH, MeCOOH, EtCOOH	alcohols, acids, amides, MeCN, MeNO$_2$, EtNO$_2$, DMSO, pyridine, ethers (20 examples)	amines (18 examples)
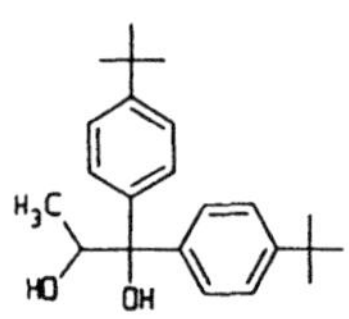	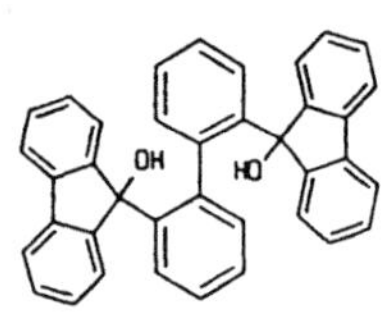	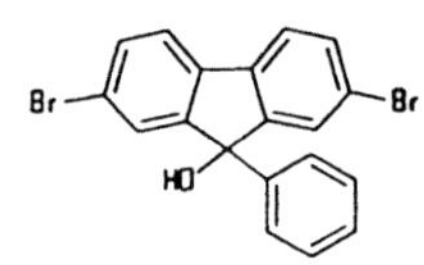
125	126	127
alcohols, amines, ketones, DMF, PrCN, DMSO, dioxane (24 examples)	alcohols, amines, ketones, MeCN, MeNO$_2$, DMF, DMSO, ethers, benzene, xylene (25 examples)	alcohols, amines, MeCN, DMF, DMSO, Me$_2$CO, ethers (12 examples)

Fig. 42. Representative lattice host molecules and inclusion formation [168, 169, 174, 176–181].

Moreover, some general topics related to the host–guest and supramolecular fields with which we are involved, were also greatly stimulated. They include synthetic strategies, which were used before by other scientific groups, such as the 'high dilution reaction principle' [196, 197] which came to fruition, the so-called 'Cs-effect' [198, 199] which was developed for the purpose of strained macrorings, or certain repetitive synthetic approaches which lead to 'cascade' and nonskid-chain-like' molecular topologies, as specified in Figure 43 [79, 200]. Actually the latter synthetic strategy has opened up the recent design of three-dimensional,

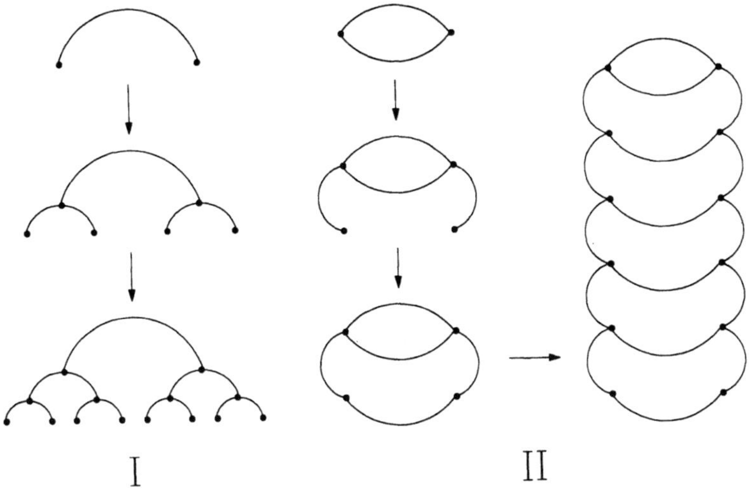

Fig. 43. Repetitive synthetic strategies for 'cascade' (I) and nonskid-chain-like (II) molecular topologies (I equals a second generation, II a fifth generation species) [200].

highly ordered oligomeric and polymeric compounds known as 'arborols' and 'star-burst dendrimers' [201].

Another, not synthetic, but no less important topic is nomenclature. The designation 'crown', coined and introduced by Pedersen [1], originally was only a nickname for a particular type of oligoether macrocycle. Naturally, the immense variety of crown-analogous compounds synthesized over the years called for a concise nomenclature. We have suggested useful nomenclature and notation systems [31, 202].

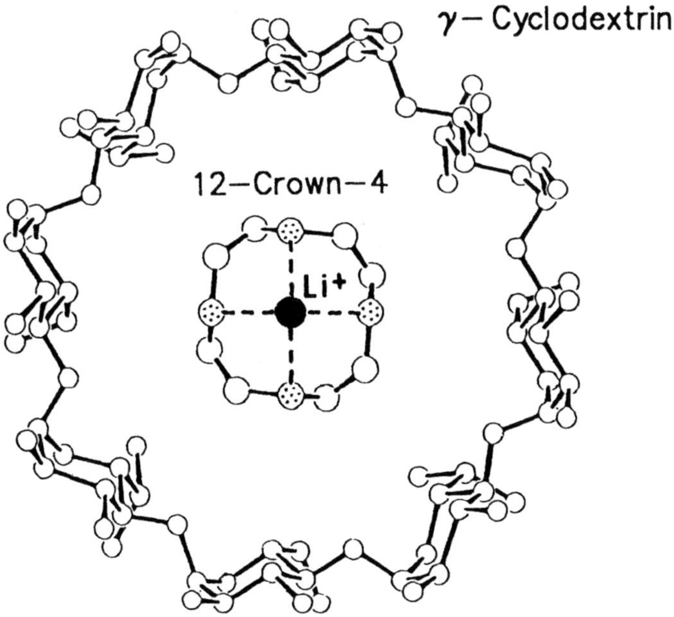

Fig. 44. 'Complexed crown': ternary complex between γ-cyclodextrin, 12-crown-4, and Li^+ [203, 204].

A final plum: perhaps the most appealing structure of crown chemistry is that of the supramolecular complex shown in Figure 44 [203, 204]. It involves Li^+, 12-crown-4 and γ-cyclodextrin nested in three spheres, i.e. a beautiful and symbolic combination of original Pedersen-crown and more recent supramolecular host–guest chemistry.

Unfortunately, it was not granted to both of us to meet Charles J. Pedersen himself in order to gain a lasting remembrance of his greatly esteemed personality. However, his brilliant ideas and scientific genius is still alive in his original papers and will certainly stimulate future topics of our research. Apart from that, the early work of D. J. Cram [205] and J.-M. Lehn [206], who won the 1987 Nobel Prize on chemistry together with C. J. Pedersen [1], was also fruitful in our research, to a degree.

Acknowledgements

We are deeply indebted to our coworkers, whose names appear in the references, for their tireless efforts. Their dedication has made possible the work described in this review. The authors thank the Deutsche Foirschungsgemeinschaft and the Fonds der Chemischen Industrie for financial support of their work. We are also grateful to M. Frank, R. Güther, D. Karbach, A. Lohner, C. Reutel, A. Schröder, and C. Seel for their help with the drawings of the formulae which appear here.

References

1. C. J. Pedersen: *Angew. Chem.* **100**, 1053 (1988); *Angew. Chem. Int., Ed. Engl.* **27**, 1021 (1988).
2. C. J. Pedersen: *J. Am. Chem. Soc.* **89**, 2495 (1967).
3. C. J. Pedersen: *J. Am. Chem. Soc.* **89**, 7017 (1967).
4. (a) F. Vögtle: *Cyclophan-Chemie* (Teubner Studienbücher Chemie), B. G. Teubner, Stuttgart (1990); (b) F. Vögtle and P. Neumann: *Top. Curr. Chem.* **48**, 67 (1974); (c) F. Vögtle and P. Neumann: *Angew. Chem.* **84**, 75 (1972); *Angew. Chem., Int. Ed. Engl.* **11**, 73 (1972); (d) F. Vögtle and P. Neumann: *Chimia* **26**, 64 (1972).
5. F. Vögtle and G. Hohner: *Top. Curr. Chem.* **74**, 1 (1978).
6. W. Kiggen and F. Vögtle: in *Synthesis of Macrocycles – The Design of Selective Complexing Agents*, eds. R. M. Izatt and J. J. Christensen, Progress in Macrocyclic Chemistry, Vol. 3, p. 309, Wiley, New York (1987).
7. F. Vögtle, H. Förster: *Angew. Chem.* **89**, 443 (1977); *Angew. Chem., Int. Ed. Engl.* **16**, 429 (1977).
8. F. Vögtle: *Tetrahedron* **25**, 3231 (1969).
9. F. Vögtle and R. Lichtenthaler: *Chemiker-Ztg.* **94**, 727 (1970).
10. F. Vögtle and L. Schunder: *Chem. Ber.* **102**, 2677 (1969).
11. F. Vögtle and M. Zuber: *Tetrahedron Lett.* 561 (1972).
12. F. Vögtle and R. G. Lichtenthaler: *Angew Chem.* **84**, 588 (1972); *Angew. Chem., Int. Ed. Engl.* **11**, 535 (1972).
13. F. Vögtle and P. Neumann: *Tetrahedron* **26**, 5847 (1970).
14. E. Weber: *Entwurf und Synthese neuartiger Kronenätherligand-Systeme und ihrer Komplexe mit Alkali- und Übergangsmetallionen*, Thesis, University of Würzburg (1976).
15. F. Vögtle and E. Weber: *Angew. Chem.* **86**, 126 (1974); *Angew. Chem., Int. Ed. Engl.* **13**, 149 (1974).
16. See: *Nachr. Chem. Techn.* **22**, pp. 91 (1974).
17. G. Weber: *Inorg. Chim. Acta* **58**, 27 (1982).
18. E. Weber, W. Wieder, and F. Vögtle: *Chem. Ber.* **109**, 1002 (1976).
19. E. Weber and F. Vögtle: *Chem. Ber.* **109**, 1803 (1976).

20. (a) H.-G. Löhr and F. Vögtle: *Acc. Chem. Res.* **18**, 65 (1985); (b) M. Takagi and K. Ueno: *Top. Curr. Chem.* **121**, 39 (1984).
21. (a) D. J. Cram: *Angew. Chem.* **98**, 1041 (1986); *Angew. Chem., Int. Ed. Engl.* **25**, 1039 (1986); (b) K. Paek, C. B. Knobler, E. F. Maverick, and D. J. Cram: *J. Am. Chem. Soc.* **111**, 8662 (1989).
22. F. Vögtle, E. Weber, W. Wehner, R. Nätscher, and J. Grütze: *Chemiker-Ztg.* **98**, 562 (1974).
23. E. Weber and F. Vögtle: *Liebigs Ann. Chem.* 891 (1976).
24. M. Newcomb, G. W. Gokel, and D. J. Cram: *J. Am. Chem. Soc.* **96**, 6810 (1974).
25. G. Weber: *Acta Crystallogr.* **C40**, 592 (1984).
26. E. Weber: *Kontakte (Merck)* 26 (1984/1) and previous papers of this series.
27. E. Weber: in *Phase Transfer Catalysts: Properties and Applications*, p. 33, Merck-Schuchardt, Darmstadt (1987).
28. E. Weber: in *Ullmann's Encyclopaedia of Industrial Chemistry* (5th Int. Ed.), Vol. A8, p. 91, VCH Verlagsgesellschaft, Weinheim (1987).
29. C. J. Pedersen: in *Synthetic Multidentate Macrocyclic Compounds*, eds. R. M. Izatt and J. J. Christensen, p. 1, Academic Press, New York (1978).
30. E. Weber and F. Vögtle: *Tetrahedron Lett.* 2415 (1975).
31. E. Weber and F. Vögtle: *Inorg. Chim. Acta* **45**, L65 (1980).
32. F. Vögtle and E. Weber: *Angew. Chem.* **91**, 813 (1979); *Angew. Chem., Int. Ed. Engl.* **18**, 753 (1979).
33. B. Tümmler, G. Maaß, E. Weber, W. Wehner, and F. Vögtle: *J. Am. Chem. Soc.* **99**, 4683 (1977).
34. B. Tümmler, G. Maaß, F. Vögtle, H. Sieger, U. Heimann, and E. Weber: *J. Am. Chem. Soc.* **101**, 2588 (1979).
35. R. Hilgenfeld and W. Saenger: *Top. Curr. Chem.* **101**, 1 (1982).
36. W. Saenger and B. S. Reddy: *Acta Crystallogr.* **B35**, 56 (1979).
37. W. Saenger, H. Brand, F. Vögtle, and E. Weber: in *Metal–Ligand Interactions in Organic Chemistry and Biochemistry*, Part I, eds. B. Pullman and N. Goldblum, p. 363, Reidel, Dordrecht, Holland (1977).
38. W. Saenger and H. Brand: *Acta Crystallogr.* **B35**, 838 (1979).
39. F. Vögtle and H. Sieger: *Angew. Chem.* **89**, 410 (1977); *Angew. Chem., Int. Ed. Engl.* **16**, 396 (1977).
40. W. Raßhofer, G. Oepen, and F. Vögtle: *Chem. Ber.* **111**, 419 (1978).
41. W. Raßhofer, G. Oepen, W. M. Müller, and F. Vögtle: *Chem. Ber.* **111**, 1108 (1978).
42. F. Vögtle and U. Heimann: *Chem. Ber.* **111**, 2757 (1978).
43. W. Raßhofer, W. M. Müller, and F. Vögtle: *Chem. Ber.* **112**, 2095 (1979).
44. H. Sieger and F. Vögtle: *Liebigs Ann. Chem.* 425, 1980.
45. See: *Book of Abstr.* 'VI. International Symposium on Molecular Recognition and Inclusion', eds. W. Saenger and F. Vögtle, Berlin (1990).
46. F. Vögtle, W. Müller, W. Wehner, and E. Buhleier: *Angew. Chem.* **89**, 564 (1977); *Angew. Chem., Int. Ed. Engl.* **16**, 548 (1977).
47. F. Vögtle, W. M. Müller, E. Buhleier, and W. Wehner: *Chem. Ber.* **112**, 899 (1979).
48. F. Vögtle and E. Weber: *Angew. Chem.* **86**, 896 (1974); *Angew. Chem., Int. Ed. Engl.* **13**, 814 (1974).
49. E. Weber, W. M. Müller, and F. Vögtle: *Tetrahedron Lett.* 2335 (1979).
50. W. Wehner and F. Vögtle: *Tetrahedron Lett.* 2603 (1976).
51. E. Buhleier, W. Wehner, and F. Vögtle: *Chem. Ber.* **111**, 200 (1978).
52. E. Buhleier, K. Frensch, F. Luppertz, and F. Vögtle: *Liebigs Ann. Chem.* 1586 (1978).
53. E. Buhleier, W. Wehner, and F. Vögtle: *Chem. Ber* **112**, 546 (1979).
54. K. Frensch and F. Vögtle: *Tetrahedron Lett.* 2573 (1977).
55. E. Buhleier and F. Vögtle: *Liebigs Ann. Chem.* 1080 (1977).
56. E. Buhleier, W. Wehner, and F. Vögtle: *Liebigs Ann. Chem.* 537 (1978).
57. E. Buhleier and F. Vögtle: *Chem. Ber.* **111**, 2729 (1978).
58. G. Oepen and F. Vögtle: *Liebigs Ann. Chem.* 1094 (1979).
59. R. M. Izatt and J. J. Christensen (eds): *Synthesis of Macrocycles – The Design of Selective Complexing Agents*, Progress in Macrocyclic Chemistry, Vol. 3, Wiley, New York (1987).
60. S. Patai and Z. Rappoport (eds), *Crown Ethers and Analogs*, Wiley, Chichester (1989).
61. F. Vögtle and B. Jansen: *Tetrahedron Lett.* 4895 (1976).
62. F. Vögtle and K. Frensch: *Angew. Chem.* **88**, 722 (1976); *Angew. Chem., Int. Ed. Engl.* **15**, 685 (1976).

63. F. Vögtle and U. Elben: *Chem. Ber.* **111**, 1434 (1978).
64. F. Vögtle and U. Elben: *J. Chem. Res. S* 316 (1978).
65. K. Frensch and F. Vögtle: *Liebigs Ann. Chem.* 2118 (1979).
66. U. Heimann and F. Vögtle: *Liebigs Ann. Chem.* 858 (1980).
67. F. Vögtle: *Pure Appl. Chem.* **52**, 2405 (1980).
68. J. P. Dix and F. Vögtle: *Angew. Chem.* **90**, 983 (1978); *Angew. Chem., Int. Ed. Engl.* **17**, 857 (1978).
69. J. P. Dix and F. Vögtle: *Chem. Ber.* **114**, 638 (1981).
70. H.-G. Löhr, F. Vögtle, W. Schuh, and H. Puff: *Chem. Ber.* **117**, 2839 (1984).
71. H.-G. Löhr and F. Vögtle: *Chem. Ber.* **118**, 905 (1985).
72. G. Hollman and F. Vögtle: *Chem. Ber.* **117**, 1355 (1984).
73. S. Misumi: *Pure Appl. Chem.* **62**, 493 (1990).
74. C. J. Pedersen: *J. Org. Chem.* **36**, 1690 (1971).
75. W. Raßhofer and F. Vögtle: *Tetrahedron Lett.* 309 (1978).
76. U. Heimann and F. Vögtle: *Chem. Ber.* **112**, 3034 (1979).
77. I.-H. Suh and W. Saenger: *Angew. Chem.* **90**, 565 (1978); *Angew. Chem., Int. Ed. Engl.* **17**, 534 (1978).
78. G. Weber and W. Saenger: *Acta Crystallogr.* **B36**, 424 (1980).
79. F. Vögtle: *Chimia* **33**, 239 (1979).
80. F. Vögtle, H. Sieger, and W. M. Müller: *Top. Curr. Chem.* **98**, 107 (1981).
81. F. Vögtle, W. M. Müller, and W. H. Watson: *Top. Curr. Chem.* **125**, 131 (1984).
82. W. H. Watson, F. Vögtle, and W. M. Müller: *J. Incl. Phenom.* **6**, 491 (1988).
83. G. Weber: *J. Mol. Struct.* **98**, 333 (1983).
84. G. Weber: *Z. Naturforsch.* **36b**, 896 (1981).
85. W. Raßhofer, G. Oepen, and F. Vögtle: *Isr. J. Chem.* **18**, 249 (1979).
86. W. Raßhofer and F. Vögtle: *Tetrahedron Lett.* 1217 (1979).
87. F. Vögtle and W. M. Müller: *Chem. Ber.* **113**, 2081 (1980).
88. F. Vögtle and W. M. Müller: *J. Incl. Phenom.* **1**, 369 (1984).
89. F. Vögtle and W. M. Müller: *Angew. Chem.* **94**, 138 (1982); *Angew. Chem., Int. Ed. Engl.* **21**, 147 (1982); *Angew. Chem. Suppl.* 185 (1982).
90. F. Behm, W. Simon, W. M. Müller, and F. Vögtle: *Helv. Chim. Acta* **68**, 940 (1985).
91. F. Vögtle, H. Puff, E. Friedrichs, and W. M. Müller: *J. Chem. Soc., Chem. Commun.* 1398 (1982).
92. L. Wambach and F. Vögtle: *Tetrahedron Lett.* **26**, 1483 (1985).
93. F. Vögtle: *Kemia-Kemi* **16**, 796 (1989).
94. F. Ebmeyer and F. Vögtle: in *Bioorganic Chemistry Frontiers*, ed. H. Dugas, Vol. 1, p. 143, Springer-Verlag, Berlin-Heidelberg (1990).
95. L. Rossa and F. Vögtle: *Liebigs Ann. Chem.* 459 (1981).
96. J. Winkel and F. Vögtle: *Tetrahedron Lett.* 1561 (1979).
97. J. Franke and F. Vögtle: *Angew. Chem.* **97**, 224 (1985); *Angew. Chem., Int. Ed. Engl.* **24**, 219 (1985).
98. F. Vögtle, W. M. Müller, U. Werner, and H.-W. Losensky: *Angew. Chem.* **99**, 930 (1987); *Angew. Chem., Int. Ed. Engl.* **26**, 901 (1987).
99. A. Wallon, J. Peter-Katalinic, U. Werner, W. M. Müller, and F. Vögtle: *Chem. Ber.* **123**, 375 (1990).
100. H. W. Losensky, H. Spelthann, A. Ehlen, F. Vögtle, and J. Bargon: *Angew. Chem.* **100**, 1225 (1988); *Angew. Chem., Int. Ed. Engl.* **27**, 1189 (1988).
101. N. Wester and F. Vögtle: *Chem. Ber.* **112**, 3723 (1979).
102. N. Wester and F. Vögtle: *Chem. Ber.* **113**, 1487 (1980).
103. A. Wallon, U. Werner, W. M. Müller, M. Nieger, and F. Vögtle: *Chem. Ber.* **123**, 859 (1990).
104. F. Vögtle, A. Wallon, W. M. Müller, and U. Werner: *J. Chem. Soc., Chem. Commun.* 158 (1990).
105. P. Stutte, W. Kiggen, and F. Vögtle: *Tetrahedron* **43**, 2065 (1987).
106. W. Kiggen, F. Vögtle, S. Franken, and H. Puff: *Tetrahedron* **42**, 1859 (1986).
107. W. Kiggen and F. Vögtle: *Angew. Chem.* **96**, 712 (1984); *Angew. Chem., Int. Ed. Engl.* **23**, 714 (1984).
108. L. de Cola, F. Barigelletti, V. Balzani, P. Belser, A. v. Zelewsky, F. Vögtle, F. Ebmeyer, and S. Grammenudi: *J. Am. Chem. Soc.* **110**, 7210 (1988).
109. F. Barigelletti, L. de Cola, V. Balzani, P. Belser, A. v. Zelewsky, F. Vögtle, F. Ebmeyer, and S. Grammenudi: *J. Am. Chem. Soc.* **111**, 4662 (1989).

110. J. Peter-Katalinic, F. Ebmeyer, C. Seel, and F. Vögtle: *Chem. Ber.* **122**, 2391 (1989).
111. F. Ebmeyer and F. Vögtle: *Angew. Chem.* **101**, 95 (1989); *Angew. Chem., Int. Ed. Engl.* **28**, 79 (1989).
112. F. Vögtle: *Supramolekulare Chemie*, Teubner, Stuttgart (1989); F. Vögtle: *Supramolecular Chemistry – An Introduction*, Wiley, Chichester (1991).
113. E. Weber: *Angew. Chem.* **91**, 230 (1979); *Angew. Chem., Int. Ed. Engl.* **18**, 219 (1979).
114. E. Weber: *J. Org. Chem.* **47**, 3478 (1982).
115. E. Weber, M. Ouichi, and K. Skobridis: *Bull. Chem. Soc. Jpn.* **63**, 3670 (1990).
116. J. Bouquant, A. Delville, J. Grandjean, and P. Laszlo: *J. Am. Chem. Soc.* **104**, 686 (1982).
117. J. Donis, J. Grandjean, A. Grosjean, and P. Laszlo: *Biochem. Biophys. Res. Commun.* **102**, 690 (1981).
118. M. Czugler and E. Weber: *J. Chem. Soc., Chem. Commun.*, 472 (1981).
119. E. Weber and F. Vögtle: *Top. Curr. Chem.* **98**, 1 (1981).
120. E. Weber: in *Crown Ethers and Analogs*, eds. S. Patai and Z. Rappoport, p. 305, Wiley, Chichester (1989).
121. G. W. Gokel and L. Echegoyen: in *Bioorganic Chemistry Frontiers*, ed. H. Dugas, Vol. 1, p. 115, Springer-Verlag, Berlin-Heidelberg (1990).
122. W. Offermann and E. Weber: *Chem. Ber.* **117**, 234 (1984).
123. M. Ouchi, Y. Inoue, K. Wada, S. Iketani, T. Hakushi, and E. Weber: *J. Org. Chem.* **52**, 2420 (1987).
124. K. Gloe, P. Mühl, and J. Beger: *Z. Chem.* **28**, 1 (1988).
125. E. Weber: *Liebigs Ann. Chem.* 770 (1983).
126. M. Czugler, E. Weber, A. Kálmán, B. Stensland, and L. Párkányi: *Angew. Chem.* **94**, 641 (1982); *Angew. Chem., Int. Ed. Engl.* **21**, 627 (1982); *Angew. Chem. Suppl.* 1273 (1982).
127. P. Laszlo: *Nachr. Chem. Techn. Lab.* **27**, 710 (1979).
128. J. LeMoigne and J. Simon: *J. Phys. Chem.* **84**, 170 (1980).
129. E. Weber: *Angew. Chem.* **95**, 632 (1983); *Angew. Chem., Int. Ed. Engl.* **22**, 616 (1983); *Angew. Chem. Suppl.* 840 (1983).
130. B. W. Werdelmann: *Surfactants in our World – Today and Tomorrow*, Henkel, Düsseldorf (1984).
131. B. Kohne and K. Praefcke. *Chemiker-Ztg.* **109**, 121 (1985).
132. K. Torizuka and T. Sato: *Magn. Res.* **12**, 190 (1979).
133. E. Weber: *Chem. Ber.* **118**, 4439 (1985).
134. H.-J. Buschmann: *Polyhedron* **6**, 1469 (1987).
135. E. Weber: in: *Synthesis of Macrocyles: The Design of Selective Complexing Agents*, eds. R. M. Izatt and J. J. Christensen, Progress in Macrocyclic Chemistry, Vol. 3, p. 337, Wiley, New York (1987).
136. E. Weber and F. Vögtle: *Angew. Chem.* **92**, 1067 (1980); *Angew. Chem., Int. Ed. Engl.* **19**, 1030 (1980).
137. E. Weber, F. Vögtle, H.-P. Josel, G. R. Newkome, and W. E. Puckett: *Chem. Ber.* **116**, 1906 (1983).
138. E. Weber, H.-P. Josel, H. Puff, and S. Franken: *J. Org. Chem.* **50**, 3125 (1985).
139. E. Weber, S. Franken, H. Puff, and J. Ahrendt: *J. Chem. Soc., Chem. Commun.* 467 (1986).
140. E. Weber, H.-J. Köhler, and H. Reuter: *Chem. Ber.* **122**, 959 (1989).
141. E. Weber, H.-J. Köhler, and H. Reuter: *J. Org. Chem.* **56**, 1236 (1990).
142. F. Vögtle, W. M. Müller, and E. Weber: *Chem. Ber.* **113**, 1130 (1980).
143. G. Weber and P. G. Jones: *Acta Crystallogr.* **C39**, 1577 (1983).
144. E. Weber, S. Franken, J. Ahrendt, and H. Puff: *J. Org. Chem.* **52**, 5291 (1987).
145. E. Weber: *Mol. Cryst. Liq. Cryst. Inc. Nonlin. Opt.* **156**, 371 (1988).
146. G. Weber and P. G. Jones: *Acta Crystallogr.* **C39**, 1053 (1983).
147. G. Weber: *Acta Crystallogr.* **C40**, 592 (1984).
148. E. Weber, G. R. Newkome, F. R. Fronczek, and S. Franken: *J. Incl. Phenom.* **6**, 1 (1988).
149. K. K. Chacko, G. A. Ruban, K. Aoki, and E. Weber: *Acta Crystallogr.* **C44**, 352 (1988).
150. E. Weber, H.-J. Köhler, K. Panneerselvam, and K. K. Chacko: *J. Chem. Soc., Perkin Trans. 2*, 1599 (1990).
151. K. Panneerselvam, K. K. Chacko, E. Weber, and H.-J. Köhler: *J. Incl. Phenom.* **9**, 337 (1990).
152. E. Weber, R. Pollex, and M. Czugler: in Ref. 45, p. F24.
153. E. Weber, K. Gloe, and E. Uhlemann: unpublished.

154. *Encyclopaedia Britannica USA, Instant Service Report R-1519*, Britannica Centre, 310 South Michigan Ave, Chicago, IL 60604.
155. E. Weber, (ed.): *Molecular Inclusion and Molecular Recognition – Clathrates I and II*, Topics in Current Chemistry, Vols. 140 and 149, Springer-Verlag, Berlin-Heidelberg (1987 and 1988).
156. G. R. Desiraju (ed.): *Crystal Engineering – The Design of Organic Solids*, Materials Science Monographs, Vol. 54, Elsevier, Amsterdam (1989).
157. J. E. D. Davies, W. Kemula, H. M. Powell, and N. O. Smith: *J. Incl. Phenom.* **1**, 3 (1983).
158. J. L. Atwood, J. E. D. Davies, and D. D. MacNicol (eds): *Inclusion Compounds*, Vols. 1–3, Academic Press, London (1984); Vols. 4 and 5, Oxford University Press (1991).
159. G. Tsoucaris: in *Organic Solid State Chemistry*, ed. G. R. Desiraju, Studies in Organic Chemistry, Vol. 32, p. 207, Elsevier Amsterdam (1987).
160. E. Weber: *J. Mol. Graphics* **7**, 12 (1989).
161. E. Weber: in Ref. 155, Vol. 140, p. 1 (1987).
162. E. Weber: in *Book of Abstr.* 'New Materials 90 Japan', p. 305, Osaka (1990).
163. E. Weber, J. Ahrendt, M. Czugler, and I. Csöregh: *Angew. Chem.* **98**, 719 (1986); *Angew. Chem., Int. Ed. Engl.* **25**, 746 (1986).
164. E. Weber, W. Seichter, and I. Goldberg: *J. Chem. Soc., Chem. Commun.* 1426 (1987).
165. E. Weber, W. Seichter, and I. Goldberg: *Chem. Ber.* **123**, 811 (1990).
166. E. Weber: in Ref. 158, Vol. 4, p. 188 (1991).
167. E. Weber and M. Czugler: in Ref. 155, Vol. 149, p. 45 (1988).
168. E. Weber, M. Hecker, I. Csöregh, and M. Czugler: *J. Am. Chem. Soc.* **111**, 7866 (1989).
169. E. Weber, M. Hecker, I. Csöregh, and M. Czugler: *Mol. Cryst. Liq. Cryst.* **187**, 165 (1990).
170. A. I. Kitaigorodskii: *Molekülkristalle*, Akademie-Verlag, Berlin (1979).
171. E. Weber, I. Csöregh, B. Stensland, and M. Czugler: *J. Am. Chem. Soc.* **106**, 3297 (1984).
172. M. Czugler and E. Weber: *J. Incl. Phenom.* **10**, 355 (1991).
173. I. Csöregh, M. Czugler, and E. Weber: *Z. Kristallogr.* **185**, 201 (1988).
174. E. Weber, I. Csöregh, J. Ahrendt, S. Finge, and M. Czugler: *J. Org. Chem.* **53**, 5831 (1988).
175. I. Csöregh, M. Czugler, A. Kálmán, E. Weber, and M. Hecker: *Bull. Chem. Soc. Jpn.* **64** (1991), in press.
176. I. Csöregh, S. Finge, and E. Weber: *Bull. Chem. Soc. Jpn.*, **64**, 1971 (1991).
177. E. Weber, N. Dörpinghaus, and I. Goldberg: *J. Chem. Soc., Chem. Commun.* 1566 (1988).
178. I. Goldberg, Z. Stein, E. Weber, N. Dörpinghaus, and S. Franken: *J. Chem. Soc., Perkin Trans. 2* 953 (1990).
179. I. Csöregh, M. Czugler, E. Weber, A. Sjögren, and M. Cserzö: *J. Chem. Soc., Perkin Trans. 2* 507 (1986).
180. I. Csöregh, M. Czugler, E. Weber, and J. Ahrendt: *J. Incl. Phenom.* **8**, 309 (1990).
181. E. Weber, N. Dörpinghaus, and I. Csöregh: *J. Chem. Soc., Perkin Trans. 2* 2167 (1990).
182. M. Czugler, J. J. Stezowski, and E. Weber: *J. Chem. Soc., Chem. Commun.* 154 (1983).
183. M. Czugler, E. Weber, and J. Ahrendt: *J. Chem. Soc., Chem. Commun.* 1632 (1984).
184. E. Weber, U. Müller, D. Worsch, F. Vögtle, G. Will, and A. Kirfel. *J. Chem. Soc., Chem. Commun.* 1578 (1985).
185. I. Csöregh, M. Czugler, and E. Weber: in *Molecular Structure: Chemical Reactivity and Biological Activity*, eds. J. J. Stezowski, J.-L. Huang, and M.-C. Shao, p. 390, International Union of Crystallography, Oxford University Press (1988).
186. I. Goldberg and E. Weber: *Z. Kristallogr.* **185**, 215 (1988).
187. I. Csöregh, M. Czugler, K. W. Törnroos, E. Weber, and J. Ahrendt: *J. Chem. Soc., Perkin Trans. 2* 1491 (1989).
188. E. Weber, K. Skobridis, and I. Goldberg: *J. Chem. Soc., Chem. Commun.* 1195 (1989).
189. I. Csöregh, M. Czugler, A. Ertan, E. Weber, and J. Ahrendt: *J. Incl. Phenom,* **8**, 275 (1990).
190. M. R. Caira, L. R. Nassimbeni, M. L. Niven, W.-D. Schubert, E. Weber, and N. Dörpinghaus: *J. Chem. Soc., Perkin Trans. 2* 2129 (1990).
191. E. Weber, W. Seichter, I. Goldberg, G. Will, and H.-J. Dasting: *J. Incl. Phenom.* **10**, 267 (1991).
192. E. Weber, W. Seichter, R.-J. Wang and T. C. W. Mak: *Bull. Chem. Soc. Jpn.* **64**, 659 (1991).
193. S. A. Bourne, L. R. Nassimbeni, K. Skobridis, and E. Weber: *J. Chem. Soc., Chem. Commun.* 282 (1991).
194. E. Weber, C. Wimmer, A. L. Llamas-Saiz, and C. Foces-Foces: *Angew. Chem.*, submitted.

195. M. Czugler, J. Angyan, G. Náray-Szabó, and E. Weber: *J. Am. Chem. Soc.* **108**, 1275 (1986).
196. L. Rossa and F. Vögtle: *Top. Curr. Chem.* **113**, 1 (1983).
197. P. Knops, N. Sendhoff, H.-B. Mekelburger, and F. Vögtle: *Top. Curr. Chem.* **161** (1991), in press.
198. B. Klieser, L. Rossa, and F. Vögtle: *Kontakte (Merck)*, 3 (1984/1).
199. A. Ostrowicki, E. Koepp, and F. Vögtle: *Top. Curr. Chem.* **161** (1991), in press.
200. E. Buhleier, W. Wehner, and F. Vögtle: *Synthesis*, 155 (1978).
201. D. A. Tomalia, A. M. Naylor, and W. A. Goddard III: *Angew. Chem.* **102**, 119 (1990); *Angew. Chem., Int. Ed. Engl.* **29**, 138 (1990).
202. E. Weber and H.-P. Josel: *J. Incl. Phenom.* **1**, 79 (1983).
203. F. Vögtle and W. M. Müller: *Angew. Chem.* **91**, 676 (1979); *Angew. Chem., Int. Ed. Engl.* **18**, 623 (1979).
204. S. Kamitori, K. Hirotsu, and T. Higuchi: *J. Am. Chem. Soc.* **109**, 2409 (1987).
205. D. J. Cram: *Angew. Chem.* **100**, 1041 (1988); *Angew. Chem., Int. Ed. Engl.* **27**, 1009 (1988).
206. J.-M. Lehn: *Angew. Chem.* **100**, 91 (1988); *Angew. Chem., Int. Ed. Engl.* **27**, 89 (1988).

Journal of Inclusion Phenomena and Molecular Recognition in Chemistry **12**: 121–150, 1992.

Conformational Mobility in Chemically-Modified Cyclodextrins [#]

PAUL ELLWOOD, CATRIONA M. SPENCER, NEIL SPENCER,
J. FRASER STODDART* and RYSZARD ZARZYCKI
Department of Chemistry, The University, Sheffield S3 7HF, U.K.

(Received: 17 October 1990; in final form: 28 December 1990)

Abstract. The observation that per-2,6-*O*-methyl-3-*O*-benzoyl-α-cyclodextrin (**1**) displays some unusual conformational behaviour in solution has led to a detailed investigation by (dynamic) NMR spectroscopy of the equilibration process that occurs in solutions of per-2,3-*O*-benzoyl-α-cyclodextrin (**3**) and some related compounds (**7**–**9**) between conformational isomers with averaged C_6 and C_3 molecular symmetries in certain organic solvents such as benzene, dichloromethane, and chloroform. The solvent dependence of the conformational equilibrium is also reflected in a spread of values for the specific optical rotations for **3** from +9° in 1,1,2,2-tetrachloroethane, where there is a degenerate equilibrium between species with C_3 molecular symmetry, to +92° in acetone where a species with averaged C_6 symmetry is present.

Key words. Chemically-modified cyclodextrins, conformational analysis, NMR spectroscopy.

This paper is dedicated to the memory of Charles J. Pedersen whose seminal papers on crown ethers in the late 1960s captured the imagination of the senior author, encouraging him to effect (J. K. N. Jones, J. F. Stoddart, and W. A. Szarek, *Cand. J. Chem.* **47**, 3213 (1969)) the chemical modifications of α-cyclodextrin and β-cyclodextrin to 30- and 35-membered ring compounds containing 12 and 14 oxygen atoms, respectively. This early foray into the synthesis of oxygen-containing macrocycles was to lead directly to the incorporation of carbohydrates into chiral crown ethers (J. F. Stoddart, *Chem. Soc. Rev.* **8**, 85 (1979); *Top. Stereochem.* **17**, 207 (1987)), whilst preserving more than a passing interest (J. F. Stoddart, R. Zarzycki, *Rec. Trav. Chim. Pays-Bas* **107**, 515 (1988)) in one of Nature's most fascinating collection of compounds, the cyclodextrins, which share with Pedersen's wholly synthetic creations, the crown ethers, the ability to form inclusion complexes with other molecules and ions.

1. Introduction

Cyclodextrins (CDs) are a class of cyclic oligosaccharides composed of α-(1→4)-linked D-(+)-glucopyranose residues. They are designated by a Greek letter according to the number of residues. The commonly available CDs have 6, 7 and 8 D-glucopyranose units and are called α-, β- and γ-CD, respectively. In the 100 years [1] that have elapsed since they were first isolated by Villiers [2] they have been the subject of a large amount of research. It is well established that the D-glucopyranose units are in the $^4C_1$ chair conformation, and that the overall shape of the molecules is that of a truncated cone [3]. Their most characteristic feature – their ability to form inclusion complexes – is a direct consequence of this shape, and is being increasingly exploited in technological fields [4].

[#] This paper is dedicated to the memory of the late Dr C. J. Pedersen.
* Author for correspondence. Present address: School of Chemistry, University of Birmingham, Edgbaston, Birmingham B15 2TT, U.K.

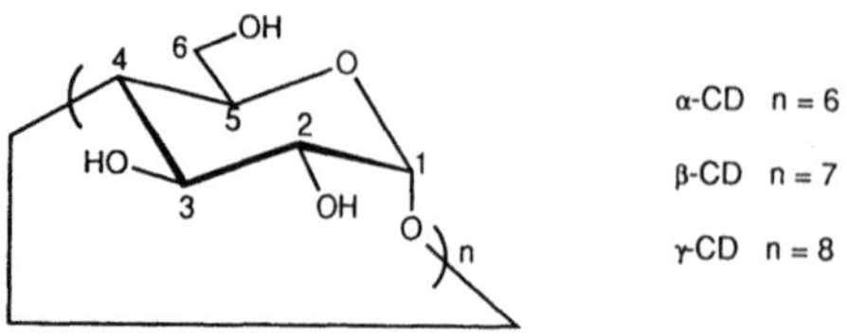

Numerous chemical modifications [5] of cyclodextrins have been performed with the aim of improving the complexing abilities and catalytic properties of the parent molecules [6]. To date, most of the research has concentrated on the modification of the primary face of cyclodextrins because of the relatively high reactivity of the C-6 hydroxyl groups. NMR spectroscopic studies carried out on the parent molecules are facilitated by the fact that rapid inter-residue motion results in NMR spectra which are qualitatively similar to that of one D-glucopyranose residue [7]. A large number of the chemically-modified CDs which have been prepared, maintain the original C_n molecular symmetry (n = number of glucose units) of the parent molecules, and therefore also give relatively simple NMR spectra. However, as the secondary faces of the molecules become [8] more amenable to synthetic elaboration, and the rigidifying effect of the circular network of intramolecular hydrogen bonds is removed, a range of conformationally-mobile CDs is revealed. In a series of elegant X-ray crystallographic studies carried out on per-methylated CDs, Harata [9] has shown that removal of the possibility of intramolecular hydrogen bonding [3] by per-methylation of CD molecules results in derivatives with drastically distorted cavities. In this paper, we wish to discuss the properties of some symmetrically-substituted CD derivatives which, as revealed by their unusually complex NMR spectra, exist in atypical conformations in solution.

2. Experimental

2.1. GENERAL SYNTHETIC PROCEDURES

Unless otherwise stated, chemicals were used as received. Prior to use, α-cyclodextrin was dried at 100°C, under high vacuum, in the presence of P_2O_5 for 24 hours. Literature procedures were employed for drying solvents. Column chromatography was carried out using Silica Gel 60 (Merck 7736 or 9385). Melting points were determined using a Reichert hot-stage apparatus and are uncorrected. Microanalyses were performed by the University of Sheffield Microanalytical Service. Optical rotations were determined on a Perkin Elmer 141 polarimeter. Fast atom bombardment mass spectra (FABMS) were obtained using xenon and were performed on the Kratos MS 80 instrument with m-nitrobenzyl alcohol as the matrix. ^{1}H NMR spectra were recorded on (1) a Bruker WH 400 spectrometer (400.1 MHz) or (2) a Bruker AM 250 spectrometer (250.1 MHz). ^{13}C NMR spectra were recorded on (1) the Bruker WH 400 spectrometer (100.6 MHz) or (2) the Bruker AM 250 spectrometer (62.9 MHz). ^{19}F NMR spectra were recorded on a Bruker WP 80 SY spectrometer (74.9 MHz). All chemical shifts are quoted in ppm on the δ scale. All ^{1}H and ^{13}C NMR spectra were referenced to trimethylsilane (TMS) using either the solvent reference of internal TMS. ^{19}F NMR spectra were referenced to external

$CFCl_3$. In reporting the NMR chemical shift data, the superscripts A, B, C, and D are used to differentiate between different benzoyl and substituted-benzoyl groups.

2^A, 2^B, 2^C, 2^D, 2^E, 2^F, 3^A, 3^B, 3^C, 3^D, 3^E, 3^F-Dodeca-O-benzoyl-α-cyclodextrin (Per-2,3-O-benzoyl-α-CD) (**3**) [12]

Potassium 2-propoxide in 2-propanol was prepared by adding clean potassium metal to distilled 2-propanol. This solution was titrated to determine the concentration of base. Per-2,3,6-*O*-benzoyl-α-cyclodextrin (1.22 g, 0.43 mmol) was added with stirring to distilled benzene (17 mL) under nitrogen and distilled 2-propanol (8.6 mL) was added. The resulting solution was cooled to 0°C and potassium 2-propoxide was added (3.03 mL of 0.847M solution, 2.58 mmol, i.e. 6 molar equiv.). The reaction mixture was allowed to warm up to room temperature and the solution was stirred at room temperature for 0.5 h. It was then neutralised with hydrochloric acid and the organic solvents were removed under vacuum. The resulting residue was shaken with ethanol-free chloroform (50 mL). The chloroform solution was washed with saturated aqueous chloride solution (3 × 30 mL) and dried ($MgSO_4$). Hexane was added to the chloroform solution until the product precipitated. This mixture was allowed to stand overnight and then it was filtered. The resulting cream coloured product was dried *in vacuo*. Column chromatography on silica gel using a gradient elution of chloroform to a chloroform/methanol mixture (9 : 1 v/v) achieved the purification of per-2,3-*O*-benzoyl-α-CD (**3**). (0.48 g, 51%); m.p. 209–210°C; (*Found*: C, 64.1; H, 4.9%; *m*/*z* (positive-ion FABMS), 2244 for $[M+Na]^+$. *Calc.* for $(C_{20}H_{18}O_7)_6$: C, 64.8; H, 4.9%; M, 2221); $[\alpha]_D +71°$ (*c*, 1.2 in CH_3SOCH_3), $[\alpha]_D +11°$ (*c*, 1.0 in $CHCl_3$) (Lit. [12] $[\alpha]_D +96°$(*c*, 1.3 in CH_3SOCH_3), $[\alpha]_D +9°$ (*c*, 1.1 in $CHCl_3$)); δ_H (400 MHz, CD_3COCD_3), 4.29 (12 H, *br m*, 5-H, 6b-H), 4.35 (6 H, *dd*, $J_{3,4}$,$J_{4,5}$ 8 Hz, 4-H), 4.63 (6 H, *dd*, $J_{5,6a} < 2$, $J_{6a,6b}$ 10 Hz, 6a-H), 5.04 (6 H, *dd*, $J_{1,2}$ 4, $J_{2,3}$ 10 Hz, 2-H), 5.52 (6 H, *d*, $J_{1,2}$ 4 Hz, 1-H), 6.14 (6 H, *dd*, $J_{2,3}$ 10, $J_{3,4}$ 8 Hz, 3-H), 6.96 (12 H, *t*, $J_{o,m}$, $J_{m,p}$ 8 Hz, C_m-H), 7.08 (12 H, *t*, $J_{o,m}$, $J_{m,p}$ 8 Hz, C_m-H), 7.29 (6 H, *t*, $J_{m,p}$ 8 Hz, C_p-H), 7.32 (6 H, *t*, $J_{m,p}$ 8 Hz, C_p-H), 7.45 (12 H, *d*, $J_{o,m}$ 8 Hz, C_o-H), and 7.60 (12 H, *d*, $J_{o,m}$ 8 Hz, C_o-H). δ_H (400 MHz, $CDCl_3$) 4.05 (3 H, *m*, $J_{4'5'}$ 9 Hz, 5′-H), 4.3–4.6 (12 H, *m*, 4-H, 6a′-H, 6b-H, 6b′-H), 4.7–4.9 (12 H, *m*, 2′-H, 4′-H, 5-H, 6a-H), 5.42 (3 H, *br s*, 1-H), 5.60 (3 H, *dd*, $J_{1,2}$ 3.5, $J_{2,3}$ 10 Hz, 2-H), 5.70 (3 H, *d*, $J_{1,2}$ 3.5 Hz, 1-H), 6.20 (3 H, *t*, $J_{2,3}$, $J_{3,4}$ 10 Hz, 3-H), 6.28 (3 H, *t*, 3′-H), 6.28 (6 H, *t*, C_m^D-H), 6.62 (6 H, *d*, $J_{o,m}$ 7.5 Hz, C_o^D-H), 6.86 (3 H, *t*, $J_{m,p}$ 7.5 Hz, C_p^D-H), 7.12 (6 H, *t*, $J_{o,m}$,$J_{m,p}$ 7.5 Hz, C_m^C-H), 7.24 (6 H, *t*, $J_{o,m}$, $J_{m,p}$ 8 Hz, C_m^A-H), 7.33 (6 H, *d*, $J_{o,m}$ 7.5 Hz, C_o^C-H), 7.35 (3 H, *t*, $J_{m,p}$ 7.5 Hz, C_p^C-H), 7.43 (3 H, *t*, $J_{m,p}$ 8 Hz, C_p^A-H), 7.47 (6 H, *t*, $J_{o,m}$, $J_{m,p}$ 8 Hz, C_m^B-H), 7.55 (3 H, *t*, $J_{m,p}$ 8 Hz, C_p^B-H), 7.98 (6 H, *d*, $J_{o,m}$ 8 Hz, C_o^B-H), and 8.10 (6 H, *d*, $J_{o,m}$ 8 Hz, C_o^A-H). δ_C (100 MHz, C_5D_5N) 62.4 (C-6), 73.0 (C-2), 73.4 (C-3), 73.6 (C-5), 77.9 (C-4), 98.0 (C-1), 128.2 and 128.3 (C_m), 129.9 and 130.1 (C_o), 132.7 and 133.1 (C_p). δ_C (100 MHz, $CDCl_3$) 52.3 and 64.7 (C-6, C-6′), 68.7 (C-2), 71.8 (C-5), 72.3 and 72.4 (C-3, C-3′), 73.4 and 73.5 (C-2′, C-5), 74.9 (C-4′), 77.9 (C-4), 94.5 (C-1′), 98.2 (C-1), 127.2 (C_m^D), 127.6 (C_i), 127.6 (C_m^A), 127.8 (C_m^C), 128.0 (C_m^B), 128.9 (C_i), 129.0 (C_o^D), 129.2 (C_i), 129.3 (C_o^C), 130.5 (C_o^A), 130.6 (C_i), 130.8 (C_o^B), 131.7 (C_p^D), 131.9 (C_p^A), 132.4 (C_p^C), and 132.6 (C_p^B).

2^A, 2^B, 2^C, 2^D, 2^E, 2^F, 3^A, 3^B, 3^C, 3^D, 3^E, 3^F-Dodeca-O-benzoyl-6^A, 6^B, 6^C, 6^D, 6^E, 6^F-Hexa-O-p-toluenesulphonyl-α-cyclodextrin (Per-2-3-O-benzoyl-6-O-tosyl-α-CD) (**10**) [12]

Dry per-2,3-O-benzoyl-α-CD (**3**) (100 mg, 4.5×10^{-5} mol) was dissolved in pyridine (4.5 mL) under nitrogen. Freshly recrystallised *p*-toluenesulphonyl chloride (0.85 g, 4.5 mmol, 100 equiv.) was added to this solution. The reaction mixture was stirred overnight at room temperature. When the product was poured, with stirring, into ice/water, a fine white precipitate was formed. This precipitate was separated by fast (*ca.* 3000 rpm) centrifugation. The pyridine/water was decanted and the solid was dissolved in dichloromethane. Addition of light petroleum (bp. 60–80°C) to the solution reprecipitated the crude product which was again isolated by fast centrifugation followed by the decantation of the mother liquor. Subjecting the crude material to column chromatography on silica gel, using chloroform as the eluant, yielded a white solid that was characterised as per-2,3-*O*-benzoyl-6-*O*-tosyl-α-CD (120 mg, 85%); m.p. 155–156°C; (*Found*: C, 61.6; H, 4.58; S, 5.92%; *m/z* (positive-ion FABMS), 3168 for $[M + Na]^+$. *Calc.* for $(C_{27}H_{24}O_9S)_6$: C, 61.8; H, 4.61; S, 6.11%, M, 3145). $[\alpha]_D + 109°$ (*c*, 0.9 in $CHCl_3$) (Lit. [12] $[\alpha]_D + 104°$ (*c*, 1.1 in $CHCl_3$));δ_H (250 MHz, $CDCl_3$), 2.51 (18 H, *s*, Me), 3.95 (6 H, *dd*, $J_{3,4}$9, $J_{4,5}$9 Hz, 4-H), 4.40–4.67 (18 H, *m*, 5-H, 6a-H and 6b-H), 4.71 (6 H, *dd*, $J_{1,2}$3.5, $J_{2,3}$10.5 Hz, 2-H), 5.05 (6 H, *d*, $J_{1,2}$3.5 Hz, 1-H), 6.00 (6 H, *dd*, $J_{2,3}$10.5, $J_{3,4}$9 Hz, 3-H), 6.74 (12 H, *t*, $J_{o,m}$, $J_{m,p}$8 Hz, C_m-H), 6.81 (12 H, *t*, $J_{o,m}$, $J_{m,p}$8 Hz, C_m-H), 7.05 (12 H, *tt*, $J_{o,p}$1.5, $J_{m,p}$8 Hz, C_p-H), 7.08 (12 H, *dd*, $J_{o,m}$, $J_{o,p}$1.5 Hz, C_o-H), 7.27 (12 H, *dd*, $J_{o,m}$8, $J_{o,p}$1.5 Hz, C_o-H), 7.44 (12 H, BB′ of an AA′BB′ system, $J_{A,B}$8 Hz, C_m-H of tosylate), and 7.91 (12 H, AA′ of an AA′BB′ system, $J_{A,B}$8 Hz, C_o-H of tosylate). δ_C (63 MHz, $CDCl_3$) 22(Me), 69.0 (C-6), 70.0 (C-2), 71.0 (C-3), 71.8 (C-5), 78.5 (C-4), 98.5 (C-1), 127.4, 128.1, 129.3, 129.6, 130.1, 132.0, 132.4 (aromatic C—H), 127.7 (*C*—Me), 129.1, 132.9 (*C*—CO), 145.3 (C—SO_2), 164.3 and 165.8 (C=O).

2^A, 2^B, 2^C, 2^D, 2^E, 2^F, 3^A, 3^B, 3^C, 3^D, 3^E, 3^F-Dodeca-O-benzoyl-6^A, 6^B, 6^C, 6^D, 6^E, 6^F-Hexa-O-methanesulphonyl-α-cyclodextrin (Per-2,3-O-benzoyl-6-O-mesyl-α-Cd) (**11**)

Dry per-2,3-*O*-benzoyl-α-CD (**3**) (100 mg, 0.045 mmol) was dissolved in dichloromethane (5 mL) containing freshly-distilled triethylamine (1 mL) at 0°C under nitrogen. Freshly distilled methanesulphonyl chloride (1 mL) was added over a period of 5 min. The colour of the solution turned bright yellow and a precipitate was formed. After stirring the solution for about 15 min, the solvents were removed under reduced pressure to give a pale yellow precipitate. An attempted recrystallisation using dichloromethane/hexane led to a precipitate, which was shown by TLC (chloroform/methanol, 9 : 1 v/v) to be mostly a side-product, presumably $Et_3NH^+MeSO_3^-$. It also gave a mother liquor which was enriched in the desired product. The mother liquors were removed under vacuum to give a white residue which was passed through a short chromatography column packed with silica gel using dichloromethane and then chloroform as an eluant. Removal of solvents under vacuum gave a white solid. The recrystallised product was dissolved in dichloromethane (5 mL) and this solution was washed successively with hydrochloric acid (0.05M), dilute sodium bicarbonate, and water (all 3 × 5 mL). The resulting product was again further purified by passing it through a short chromatography column packed with silica gel using dichloromethane followed by chloroform as the eluants. Removal of solvents under vacuum gave a white solid which was characterised as per-2,3-*O*-benzoyl-6-*O*-mesyl-α-CD. (40 mg, 33%); m.p. 215–216°C; (*Found*: C, 56.3; H, 4.5; S, 7.3%; *m*/*z* (positive ion FABMS), 2712 for $[M + Na]^+$. $(C_{21}H_{20}O_9S)_6$ requires C, 56.3; H, 4.5; S, 7.1%; M, 2689). $[\alpha]_D + 97°$ (*c*, 0.7 in $CHCl_3$); δ_H (400 MHz, $CDCl_3$), 3.26 (18 H, *s*, $-SO_2Me$), 4.15 (1 H, *dd*, $J_{3,4}$10, $J_{4,5}$10 Hz, 4-H), 4.65 (6 H, *m*, $J_{4,5}$10 Hz, 5-H), 4.79 (6 H, *d*, $J_{5,6b}$1.5, $J_{6a,6b}$11.5 Hz, 6b-H), 4.89 (6 H, *dd*, $J_{5,6a}$1.5, $J_{6a,6b}$11.5 Hz, 6a-H), 4.96 (6 H, *dd*, $J_{1,2}$3.5, $J_{2,3}$10.5 Hz, 2-H), 5.48 (6 H, *d*, $J_{1,2}$3.5 Hz, 1-H), 6.19 (6 H, *dd*, $J_{2,3}$10.5, $J_{3,4}$10 Hz, 3-H), 6.76 (12 H, *t*, $J_{o,m}$, $J_{m,p}$7.5 Hz, C_m-H), 6.81 (12 H, *t*, $J_{o,m}$, $J_{m,p}$7.5 Hz, C_m-H), 7.05 (6 H, *tt*, $J_{o,p}$1.5, $J_{m,p}$7.5 Hz, C_p-H), 7.07 (6 H, *tt*, $J_{o,p}$1.5, $J_{m,p}$7.5 Hz, C_p-H), 7.25 (12 H, *dd*, $J_{o,m}$7.5, $J_{o,p}$1.5 Hz, C_o-H), and 7.32 (12 H, *dd*, $J_{o,m}$7.5, $J_{o,p}$1.5 Hz, C_o-H).

2^A, 2^B, 2^C, 2^D, 2^E, 2^F, 3^A, 3^B, 3^C, 3^D, 3^E, 3^F, 6^A, 6^B, 6^C, 6^D, 6^E, 6^F-Octadeca-O-p-fluorobenzoyl-α-cyclodextrin (Per-2,3,6-O-p-fluorobenzoyl-α-CD)

Dry α-CD (1.2 g, 1.2 mmol) was added with stirring to a solution of freshly distilled *p*-fluorobenzoyl chloride (12 mL, 14.4 g, 0.096 mol, i.e. 80 molar equiv.) in dry pyridine (50 mL) under nitrogen. When the resulting white suspension was heated to 100°C, a clear solution was obtained. This temperature was maintained for 4 days. The dark red mixture was poured, with stirring, into water (300 mL), to give a beige-coloured precipitate which was filtered. The filtered precipitate was subjected to column chromatography on silica gel with a gradient elution from dichloromethane to 5% methanol in dichloromethane. A second column was required in order to isolate a pure white solid which was characterised as per-2,3,6-*O*-*p*-fluorobenzoyl-α-CD. (2.4 g, 61%); m.p. 171–172°C; (*Found*: C, 61.1; H, 3.62%; *m*/*z* (positive-ion FABMS), 3192 for $[M + Na]^+$. $(C_{27}H_{19}O_8F_3)_6$ requires C, 61.4; H, 3.60%; M, 3169). $[\alpha]_D + 37°$ (*c*, 0.4 in $CHCl_3$); δ_H (250 MHz, $CDCl_3$), 4.15 (6 H, *t*, $J_{3,4}$, $J_{4,5}$9 Hz, 4-H), 4.73 (6 H, *br d*, $J_{6a,6b}$11 Hz, 6b-H), 4.80 (6 H, *m*, 5-H), 4.84 (6 H, *dd*, $J_{1,2}$3, $J_{2,3}$10 Hz, 2-H), 5.01 (6 H, *br d*, $J_{6a,6b}$11 Hz, 6a-H), 5.52 (6 H, *d*, $J_{1,2}$3 Hz, 1-H), 6.15 (6 H, *dd*, $J_{2,3}$10, $J_{3,4}$9 Hz, 3-H), 6.55 (12 H, BB′ part of a AA′BB′X system, $J_{A,B} = J_{o,m}$9 Hz, $J_{B,X} = J_{m,F}$9 Hz, secondary benzoyl C_m-H), 6.58 (12 H, BB′ part of an AA′BB′X system, $J_{A,B} = J_{o,m}$9 Hz, $J_{B,X} = J_{m,F}$9 Hz, secondary benzoyl C_m-H), 7.16 (12 H, BB′ part of an AA′BB′X system, $J_{A,B} = J_{o,m}$9 Hz, $J_{B,X} = J_{m,F}$9 Hz, primary benzoyl C_m-H), 7.29 (12 H, AA′ part of an AA′BB′X system, $J_{A,B} = J_{o,m}$9 Hz, $J_{A,X} = J_{o,F}$6 Hz, secondary benzoyl C_o-H), 7.35 (12 H, AA′ part of an AA′BB′X system, $J_{A,B} = J_{o,m}$9 Hz, $J_{A,X} = J_{o,F}$6 Hz, secondary benzoyl C_o-H), and 8.15 (12 H, AA′ part of an AA′BB′X system, $J_{A,B} = J_{o,m}$9 Hz, $J_{A,X} = J_{o,F}$6 Hz, primary benzoyl C_o-H). δ_C (63 MHz, $CDCl_3$), 63.3 (C-6), 70.4 (C-2), 71.6 (C-3), 72.9 (C-5), 79.0 (C-4), 98.6 (C-1), 114.8 (*d*, $J_{C,F}$22 Hz, 2 × secondary benzoyl C_m), 115.8 ($J_{C,F}$22 Hz, primary benzoyl C_m), 124.1, 125.2, 125.8 (*d*, each $J_{C,F}$2.5 Hz, *C*—CO), 131.8, 132.2, 132.5 (*d*, each $J_{C,F}$9 Hz, C_o), 163.7, 165.0, 165.1 (C=O), 165.4, 165.7, and 166.0 (*d*, each

$J_{C,F}$255 hz, C—F). δ_F (75.4 MHz, $CDCl_3$), −105.4 (6 F, *tt*, $J_{o,F}$9, $J_{m,F}$6 Hz, secondary benzoyl F), −105.5 (6 F, *tt*, $J_{o,F}$9, $J_{m,F}$6 hz, secondary benzoyl F), and −106.0 (6 F, *tt*, $J_{o,F}$9, $J_{m,F}$6 Hz, primary benzoyl F).

2^A, 2^B, 2^C, 2^D, 2^E, 2^F, 3^A, 3^B, 3^C, 3^D, 3^E, 3^F-Dodeca-O-p-fluorobenzoyl-α-cyclodextrin (Per-2,3-O-p-fluorobenzoyl-α-CD) (**7**)

Per-2,3,6-*O*-*p*-fluorobenzoyl-α-CD (50 mg, 1.6×10^{-5} mol) was dissolved in dry benzene (1 mL) and dry 2-propanol (1 mL). This solution was cooled to 0–5°C (ice-bath) under nitrogen. A solution of potassium 2-propoxide in 2-propanol (0.56 ml of a 0.17 M solution, 9.6×10^{-5} mol, i.e. 6 molar equiv.) was added and the reaction mixture was stirred at 0–5°C for 1 h. It was allowed to warm up to room temperature and then it was stirred for 1.5 h. The reaction mixture was neutralised with hydrochloric acid (*ca.* 0.1M) and the solvents were removed under reduced pressure to give a white residue. This crude product was subjected to column chromatography on silica gel using a gradient elution of dichloromethane to 10% methanol in dichloromethane to afford a white solid which was characterised as per-2,3-*O*-*p*-fluorobenzoyl-α-CD (13 mg, 34%); m.p. 304–305°C; (*Found*: C, 58.5; H, 4.59%; *m/z* (positive-ion FABMS), 2460 for $[M+Na]^+$. $(C_{20}H_{16}O_7F_2)_6$ requires C, 59.1; H, 3.97%; M, 2437); $[\alpha]_D + 10°$ (*c*, 0.2 in CH_3COCH_3); δ_H (250 MHz, CD_3COCD_3) 4.20–4.30 (12 H, *m*, 6a-H, 6b-H), 4.32 (6 H, *t*, $J_{3,4}$, $J_{4,5}$9 Hz, 4-H), 4.45 (6 H, *br s*, OH), 4.53 (6 H, *br d*, $J_{4,5}$9 Hz, 5-H), 4.91 (6 H, *dd*, $J_{1,2}$3.5, $J_{2,3}$10 Hz, 2-H), 5.54 (6 H, *d*, $J_{1,2}$3.5 Hz, 1-H), 6.05 (6 H, *dd*, $J_{2,3}$10, $J_{3,4}$9 Hz, 3-H), 6.77 (12 H, BB′ part of an AA′BB′X system, $J_{A,B} = J_{o,m}$9 Hz, $J_{B,X} = J_{m,F}$9 Hz, C_m-H), 6.81 (12 H, BB′ part of an AA′BB′X system, $J_{A,B} = J_{o,m}$9 Hz, $J_{B,X} = J_{m,F}$9 Hz, C_m-H), 7.45 (12 H, AA′ part of an AA′BB′X system, $J_{A,B} = J_{o,m}$9 Hz, $J_{A,X} = J_{o,F}$6 Hz, C_o-H), and 7.63 (12 H, AA′ part of an AA′BB′X system, $J_{A,B} = J_{o,m}$9 Hz, $J_{A,X} = J_{o,F}$6 Hz, C_o-H). δ_F (75.4 MHz, CD_3COCD_3) −106.8 (F, *tt*, $J_{o,F}$9, $J_{m,F}$6 Hz), and −107.5 (6 F, *tt*, $J_{o,F}$9, $J_{m,F}$6 Hz). The compound was only poorly soluble in $CDCl_3$, so the ^1H NMR spectrum recorded in this solvent exhibited a very poor signal/noise ratio. However, the general appearance of the spectrum was broad and complex, with the atypical doubling-up of expected resonances.

2^A, 2^B, 2^C, 2^D, 2^E, 2^F, 3^A, 3^B, 3^C, 3^D, 3^E, 3^F, 6^A, 6^B, 6^C, 6^D, 6^E, 6^F-Octadeca-O-p-methoxybenzoyl-α-cyclodextrin (Per-2,3,6-O-p-methoxybenzoyl-α-CD)

Dry α-CD (1 g, 1.02×10^{-3} mol) was dissolved in distilled pyridine (50 mL). *p*-Methoxybenzoyl chloride (14 g, 0.082 mol, i.e. 80 molar equiv.) was added along with some more pyridine (20 mL). The reaction mixture was heated at 100°C for 4 days under nitrogen. It was then cooled and poured, with stirring, into ice/water (450 mL). The resulting beige coloured precipitate was filtered off and subjected to column chromatography on silica gel using a gradient elution of dichloromethane to chloroform. This chromatography achieved only a partial purification. Fractions corresponding to the major component were combined and the solvents were removed under vacuum to give a white solid that was characterised as per-2,3,6-*O*-*p*-methoxybenzoyl-α-CD. The isolated yield of the pure material after one column was 1.0 g, 29%; m.p. 153–155°C; (*Found*: C, 63.4; H, 4.46%; *m*/*z* (positive-ion FABMS), 3410 for $[M + Na]^{+}$. $(C_{30}H_{28}O_{11})_6$ requires C, 63.8; H, 4.96%; M, 3387). $[\alpha]_D + 37°$ (*c*, 0.45 in $CHCl_3$); δ_H (250 MHz, $CDCl_3$), 3.65 (36 H, *s*, 2 × OMe), 3.84 (18 H, *s*, OMe), 4.20 (6 H, *t*, $J_{3,4}$, $J_{4,5}$ 9 Hz, H-4), 4.77 (6 H, *m*, 5-H), 4.85 (6 H, *m*, 6b-H), 4.87 (6 H, *dd*, $J_{1,2}$ 3.5, $J_{2,3}$ 10 Hz, 2-H), 5.02 (6 H, *m*, $J_{6a,6b}$ 12 Hz, 6a-H), 5.51 (6 H, *d*, $J_{1,2}$ 3.5 Hz, 1-H), 6.19 (6 H, *dd*, $J_{2,3}$ 9, $J_{3,4}$ 10 Hz, 3-H), 6.31 (12 H, $(BB')^A$ of an $(AA'BB')^A$ system, $J_{o,m}$ 9 Hz, secondary benzoyl C_m^A-H), 6.34 (12 H, $(BB')^B$ of an $(AA'BB')^B$ system, $J_{o,m}$ 9 Hz, secondary benzoyl C_m^B-H), 6.96 (12 H, $(BB')^C$ of an $(AA'BB')^C$ system, $J_{o,m}$ 9 Hz, primary benzoyl C_m^C-H), 7.31 (12 H, $(AA')^A$ of an $(AA'BB')^A$ system, $J_{o,m}$ 9 Hz, secondary benzoyl C_o^A-H), 7.35 (12 H, $(AA')^B$ of an $(AA'BB')^B$ system, $J_{o,m}$ 9 Hz, secondary benzoyl C_o^B-H), and 8.11 (12 H, $(AA')^C$ of an $(AA'BB')^C$ system, $J_{o,m}$ 9 Hz, primary benzoyl C_o^C-H). δ_C (63 MHz, $CDCl_3$), 54.9 (2 × OMe), 55.3 (OMe), 63.2 (C-6), 70.5 (C-2), 71.4 (C-3), 72.6 (C-5), 79.0 (C-4), 98.6 (C-1), 112.6, 112.7, 113.8 (C_m), 120.8, 122.3, 122.5 (*C*—OMe), 131.5, 131.9, 132.0 (C_o), 162.3, 162,7, 163.3 (*C*—CO), 164.2, 165.6, and 165.9 (C=O).

2^A, 2^B, 2^C, 2^D, 2^E, 2^F, 3^A, 3^B, 3^C, 3^D, 3^E, 3^F-Dodeca-O-p-methoxybenzoyl-α-cyclodextrin (Per-2,3-O-p-methoxybenzoyl-α-CD) (**8**)

Per-2,3,6-*O*-*p*-methoxybenzoyl-α-CD (100 mg, 0.0295 mmol) was dissolved in distilled 2-propanol (2 mL) and distilled benzene (3 mL) at 0–5°C (ice-bath) under nitrogen. A solution of postassium 2-propoxide in 2-propanol (0.32 mL of a 0.56M solution, 1.77×10^{-4} mol, 6 molar equiv.) was added and the solution was stirred at 0–5°C for 4 h before being allowed to warm to room temperature and then being stirred for a further 3 days. The reaction mixture was neutralised with hydrochloric acid (*ca.* 0.1M) and the solvents were removed under reduced pressure to give a white residue. This crude product was subjected to column chromatography on silica gel using a gradient elution of dichloromethane to 10% methanol in dichloromethane to give pure per-2,3-*O*-*p*-methoxybenzoyl-α-CD. (25 mg, 33%), m.p. 195–196°C; (*Found*: C, 61.9; H, 5.80%; *m*/*z* (positive-ion FABMS), 2604 for $[M + Na]^{+}$. $(C_{22}H_{22}O_9)_6$ requires C, 61.5; H, 5.45%; M, 2581); $[\alpha]_D + 19°$ (*c*, 0.9 in $CHCl_3$), $[\alpha]_D + 105°$ (*c*, 0.4 in CH_3COCH_3); δ_H (250 MHz, CD_3COCD_3), 3.74 36 H, *s*, 2 × OMe), 4.2–4.4 (18 H, *m*, 4-H, 5-H, 6b-H), 4.57 (6 H, *br d*, $J_{6a,6b}$ 9 Hz, 6a-H), 4.87 (6 H, *dd*, $J_{1,2}$ 3.5, $J_{2,3}$ 10 Hz, 2-H), 5.47 (6 H, *d*, $J_{1,2}$ 3.5 Hz, 1-H), 6.06 (6 H, *dd*, $J_{2,3}$ 10, $J_{3,4}$ 9 Hz, 3-H), 6.49 (12 H, *d*, $J_{o,m}$ 9.5 Hz, C_m^A-H), 6.57 (12 H, *d*, $J_{o,m}$ 9.5 Hz, C_m^B-H), 7.42 (12 H, *d*, $J_{o,m}$ 9.5 Hz, C_o^A-H), and 7.54 (12 H, *d*, $J_{o,m}$ 9.5 Hz, C_o^B-H). δ_H (250 MHz, $CDCl_3$), 3.55, 3.72, 3.75, 3.76 (each 9 H, *s*, OMe), 3.97 (3 H, *br m*, 5′-H), 4.2–4.5 (15 H, *m*, 5-H, 6a-H, 6b-H, 6a′-H, 6b′-H), 4.57–4.82 (9 H, *m*, 4-H, 4′-H, 2′-H), 5.30 (3 H, *br s*, 1′-H), 5.50 (3 H, *dd*, $J_{1,2}$ 3.5, $J_{2,3}$ 9 Hz, 2-H), 5.60 (3 H, *d*, $J_{1,2}$ 3.5 Hz, 1-H), 5.78 (6 H, $(BB')^D$ of an $(AA'BB')^D$ system, $J_{A,B} = J_{o,m}$ 9 Hz, C_m^D-H), 6.08 (3 H, *t*, $J_{2,3}$, $J_{3,4}$ 9 Hz, 3-H), 6.15 (3 H, *t*, $J_{2,3}$, $J_{3,4}$ 9 Hz, 3-H), 6.58 (12 H, *m*, $(BB')^C$ of an $(AA'BB')^C$ system, $J_{A,B} = J_{o,m}$ 9 Hz, C_m^C-H and 6 H, (AA′) of an $(AA'BB')^D$ system, $J_{A,B} = J_{o,m}$ 9 Hz, C_o^D-H), 6.68 (6 H, $(BB')^A$ of an $(AA'BB')^A$ system, $J_{A,B} = J_{o,m}$ 9 Hz, C_m^A-H), 6.94 (6 H, $(BB')^B$ of an $(AA'BB')^B$ system, $J_{A,B} = J_{o,m}$ 9 Hz, C_m^B-H), 7.25 (6 H, $(AA')^C$ of an $(AA'BB')^C$ system, $J_{A,B} = J_{o,m}$ 9 Hz, C_m^C-H), 7.94 (6 H, $(AA')^B$ of an $(AA'BB')^B$ system,

$J_{A,B} = J_{o,m}$ 9 Hz, C_m^B-H), and 8.03 (6 H, $(AA')^A$ of an $(AA'BB')^A$ system, $J_{A,B} = J_{o,m}$ 9 Hz, C_m^A-H).

2^A, 2^B, 2^C, 2^D, 2^E, 2^F, 3^A, 3^B, 3^C, 3^D, 3^D, 3^F, 6^A, 6^B, 6^C, 6^D, 6^E, 6^F-Octadeca-O-p-bromobenzoyl-α-cyclodextrin (Per-2,3,6-O-p-bromobenzoyl-α-CD)

Dry α-CD (500 mg, 0.51 mmol) was dissolved in distilled pyridine (20 mL) and *p*-bromobenzoyl chloride (9 g, 0.041 mol, i.e. 80 molar equiv.) added under an atmosphere of nitrogen. At room temperature a thick solid quickly formed which prevented stirring of the reaction mixture. However, this solid dissolved as the mixture was warmed up to 100°C and stirring was continued at this temperature for 3 days. The solution was cooled and poured into water (200 mL). This gave a copious amount of a brown precipitate which was separated by centrifugation. The solid was stirred with chloroform and the insoluble material was filtered. Column chromatography on silica gel, with dichloromethane as the eluent, afforded a pure sample of per-2,3,6-*O*-*p*-bromobenzoyl-α-CD (1.62 g, 74%); m.p. 191–192°C; *Found*: C, 45.5; H, 2.55; Br, 34.0%. $(C_{27}H_{19}O_8Br_3)_6$ requires C, 45.6; H, 2.67; Br, 33.8%. $[\alpha]_D + 49°$ (*c*, 1.0 in $CHCl_3$); δ_H (250 MHz, $CDCl_3$), 4.14 (6 H, *t*, $J_{3,4}$, $J_{4,5}$ 9 Hz, 4-H), 4.71 (6 H, *br d*, $J_{4,5}$ 10 Hz, 5-H), 4.78 (6 H, *br dd*, $J_{5,6b}$ 4, $J_{6a,6b}$ 12 Hz, 6b-H), 4.85 (6 H, *dd*, $J_{1,2}$ 3.5, $J_{2,3}$ 10 Hz, 2-H), 5.01 (6 H, *br d*, $J_{5,6a} < 1$, $J_{6a,6b}$ 12 Hz, 6a-H), 5.54 (6 H, *d*, $J_{1,2}$ 3.5 Hz, 1-H), 6.15 (6 H, *dd*, $J_{2,3}$ 10, $J_{3,4}$ 9 Hz, 3-H), 7.08 (12 H, $(BB')^C$ of an $(AA'BB')^C$ system, $J_{A,B} = J_{o,m}$ 9 Hz, secondary benzoyl C_m^C-H), 7.13 (12 H, $(BB')^B$ of an $(AA'BB')^B$ system, $J_{A,B} = J_{o,m}$ 9 Hz, secondary benzoyl C_m^B-H), 7.15 (12 H, $(AA')^C$ of an $(AA'BB')^C$ system, $J_{A,B} = J_{o,m}$ 9 Hz, secondary benzoyl C_o^C-H), 7.22 (12 H, $(AA')^B$ of an $(AA'BB')^B$ system, $J_{A,B} = J_{o,m}$ 9 Hz, secondary benzoyl C_o^B-H), 7.64 (12 H, $(BB')^A$ of an $(AA'BB')^A$ system, $J_{A,B} = J_{o,m}$ 9 Hz, primary benzoyl C_m^A-H), and 7.99 (12 H, $(AA')^A$ of an $(AA'BB')^A$ system, $J_{A,B} = J_{o,m}$ 9 Hz, primary benzoyl C_o^A-H). δ_C (63 MHz, $CDCl_3$), 63.4 (C-6), 70.4 (C-2), 71.7 (C-3), 72.8 (C-5), 78.8 (C-4), 98.4 (C-1), 126.5, 127.7, 128.31 (C—Br), 128.33, 128.68, 128.70 (*C*—CO), 130.74, 130.98, 131.17 (C_o), 131.20, 131.39, 132.04 (C_m), 164.0 (CO), and 165.3 (2 × CO).

2^A, 2^B, 2^C, 2^D, 2^E, 2^F, 3^A, 3^B, 3^C, 3^D, 3^E, 3^F-Dodeca-O-p-bromobenzoyl-α-cyclodextrin (Per-2,3-O-p-bromobenzoyl-α-CD) (**9**)

Per-2,3,6-*O*-*p*-bromobenzoyl-α-CD (500 mg, 0.117 mmol) was dissolved in benzene (10 mL) and distilled 2-propanol (5 mL). A solution of potassium 2-propoxide in 2-propanol (1.56 mL of 0.45M solution, 0.7 mmol, 6 molar equiv.) was added and the mixture was stirred at room temperature under nitrogen for 1 h. The reaction mixture was neutralised with hydrochloric acid (*ca.* 0.1M) and solvents were removed under reduced pressure to give a white residue. Column chromatography on silica gel with a gradient elution regime employing dichloromethane to 10% methanol in dichloromethane gave only a partial purification of the desired compound. Repeating this procedure achieved a complete separation of by-products and yielded per-2,3-*O*-*p*-bromobenzoyl-α-CD (170 mg, 46%); (*Found*: C, 45.3; H, 3.01%; *m*/*z* (positive-ion FABMS), 3191 for $[M + Na]^+$. $(C_{20}H_{16}O_7Br_2)_6$ requires C, 45.5; H, 3.05%; M, 3168); $[\alpha]_D + 14°$ (*c*, 0.9 in $CHCl_3$), $[\alpha]_D + 69°$ (*c*,

0.8 in THF); δ_H (400 MHz, d_8-THF), 4.11 (6 H, *br s*, 6b-H), 4.27 (6 H, *t*, 4-H), 4.40 (6 H, *m*, 5-H), 4.46 (6 H, *t*, 6b-H), 4.97 (6 H, *dd*, 2-H), 5.48 (6 H, *dd*, 1-H), 6.06 (6 H, *dd*, 3-H), 7.28 (12 H, $(BB')^A$ of an $(AA'BB')^A$ system, C_m^A-H), 7.38 (12 H, $(BB')^B$ of an $(AA'BB')^B$ system, C_m^B-H), 7.42 (12 H, $(AA')^A$ of an $(AA'BB')^A$ system, C_m^A-H), and 7.48 (12 H, $(AA')^B$ of an $(AA'BB')^B$ system, C_m^B-H). δ_H (400 MHz, $CDCl_3$), 3.91 (3 H, *m*, 5′-H), 4.13–4.45 (12 H, *m*, 6a-H, 6a′-H, 6b′-H, 4-H), 4.58–4.72 (12 H, *m*, 2′-H, 4′-H, 5-H, 6b-H), 5.29 (3 H, *br s*, 1-H), 5.52 (3 H, *dd*, 2-H), 5.57 (3 H, *d*, 1-H), 6.08 (3 H, *t*, 3′-H), 6.13 (3 H, *t*, 3-H), 6.37 (6 H, $(BB')^D$ of an $(AA'BB')^D$ system, C_m^D-H), 6.62 (6 H, $(AA')^D$ of an $(AA'BB')^D$ system, C_o^D-H), 7.14 (6 H, $(BB')^C$ of an $(AA'BB')^C$ system, C_m^C-H), 7.34 (6 H, $(AA')^C$ of an $(AA'BB')^C$ system, C_o^C-H), 7.50 (6 H, $(BB')^B$ of an $(AA'BB')^B$ system, C_m^B-H), 7.66 (6 H, $(BB')^A$ of an $(AA'BB')^A$ system, C_m^A-H), 7.78 (6 H, $(AA')^B$ of an $(AA'BB')^B$ system, C_o^B-H), 7.90 (6 H, $(AA')^A$ of an $(AA'BB')^A$ system, C_o^A-H), and 12.6 (*br s*, OH). δ_C (63 MHz, $CDCl_3$), 53.8, 59.0 (C-6 and C-6′), 68.5 (C-2), 71.6 (C-5′), 72.0, 73.0 (C-3, C-3′), 73.4, 74.6 (C-2′, C-5), 77.2 (C-4′), 77.5 (C-4), 94.1 (C-1′), 97.9 (C-1), 125.8, 127.1, 127.3, 128.3, 128.4 (C—Br and *C*—CO), 128.7 (three coincident resonances, C—Br and *C*—CO), 130.0, 130.5, 131.6, 131.9, 132.1 (C_o and C_m), 131.5 (three coincident signals, C_o and C_m), 162.9, 164.2, and 165.5 (C=O).

3. Results and Discussion

During the course of a research programme involving the synthesis of methylated CDs [10, 11], per-2,6-*O*-methyl-3-*O*-benzoyl-α-CD (**1**) and per-2,6-*O*-methyl-3-*O*-benzoyl-β-CD (**2**) were prepared [11]. The ¹H NMR spectrum of **2** shows a slight solvent dependence, whereas the effect of solvent on the ¹H NMR spectrum of the corresponding α-CD derivative (**1**) is greater.

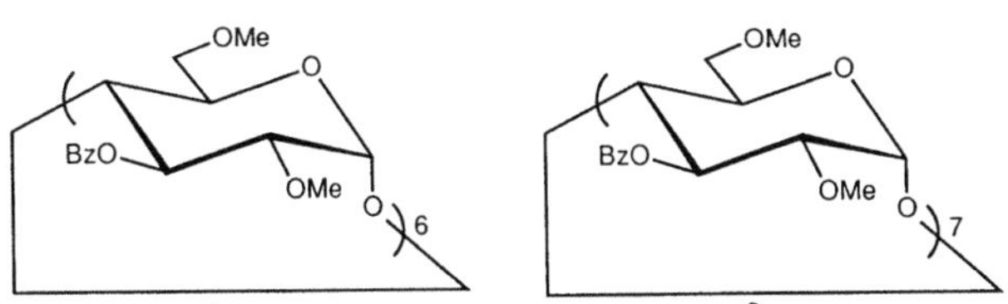

1 **2**

The ¹H NMR spectrum of **1** was recorded (Figure 1) in $CDCl_3$. Extensive line broadening of signals was observed. When the spectra were recorded at elevated temperatures (50°C), the magnitude of the solvent effect was not so pronounced. By contrast, when the ¹H NMR spectrum of **1** was recorded in CD_2Cl_2, the only manifestation of unusual behaviour was observed in the broadening of the singlet for the 2-*O*-Me group. In C_6D_6 (Figure 2), however, the unusual behaviour was so pronounced that several resonances were observed for some protons while other signals were broadened to the extent that they could not be recognised. Although the spectrum recorded in CD_3SOCD_3 was well-resolved, the H-2 resonance and the 2-*O*-Me singlet were obscured by signals for HDO and residual CD_3SOCD_2H, respectively. The best solvent in which to record the ¹H NMR spectrum of **1** was

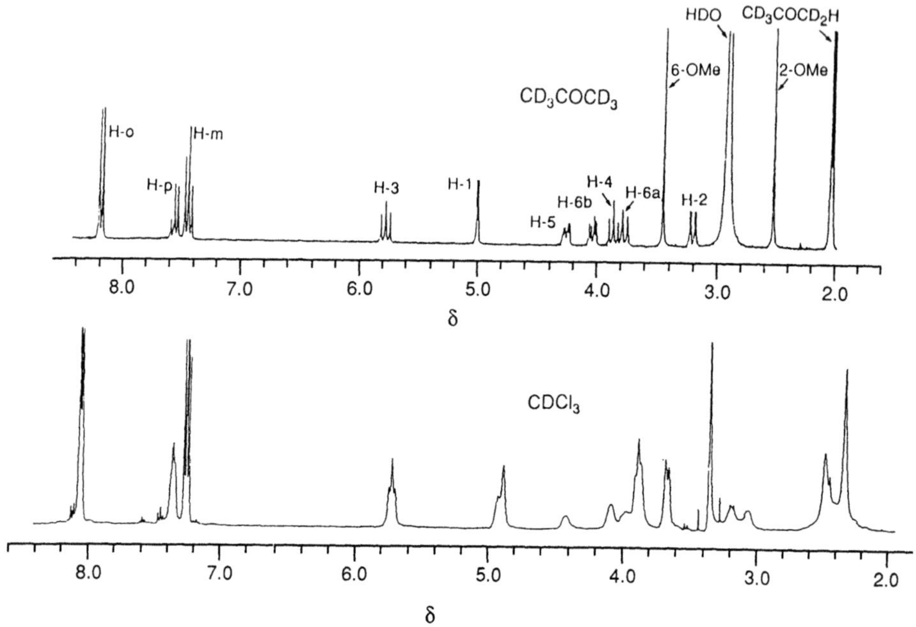

Fig. 1. ^{1}H NMR spectra of per-2,6-*O*-methyl-3-*O*-benzoyl-α-CD (**1**) in CD_3COCD_3 (250 MHz) and in $CDCl_3$ (400 MHz).

found to be CD_3COCD_3 (Figure 1). In this solvent, the resolution of signals was superior to that observed in all the other solvents examined. The resonances were well-dispersed and none of them were obscured by solvent peaks.

It was suspected that the atypical appearance of the spectra recorded in $CDCl_3$ and C_6D_6 might be the result of a slow equilibration process between diastereoisomeric conformations. All attempts to vindicate this hypothesis and identify conformational diastereoisomers by employing dynamic NMR spectroscopy were unsuccessful.

A similar phenomenon was observed [12] by Lehn *et al.* concerning the solvent dependence of the ^1H NMR (Figure 3) and ^{13}C NMR (Figure 4) spectra of

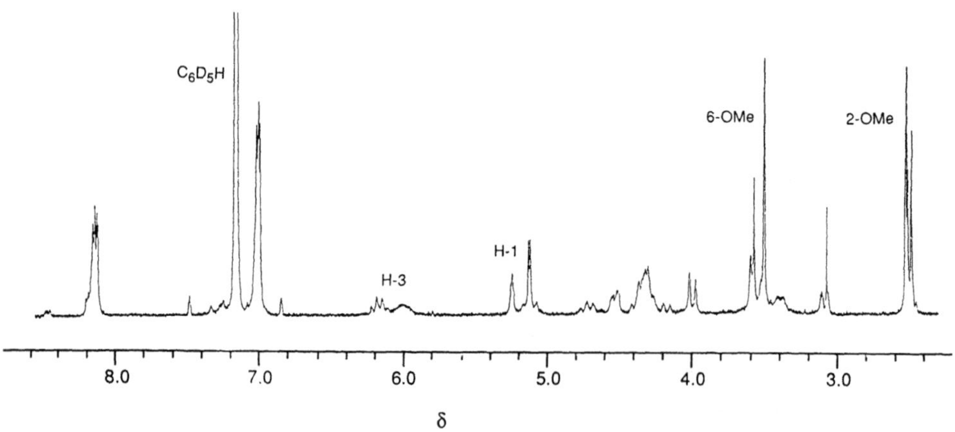

Fig. 2. ^{1}H NMR spectrum (250 MHz) of per-2,6-*O*-methyl-3-*O*-benzoyl-α-CD (**1**) in C_6D_6.

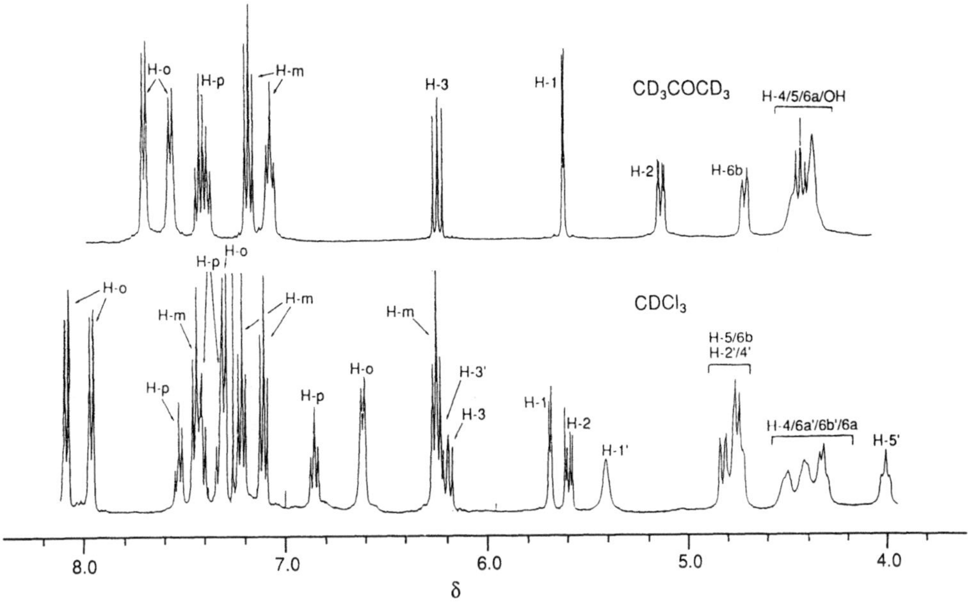

Fig. 3. ^{1}H NMR spectra (400 MHz) of per-2,3-*O*-benzoyl-α-CD (**3**) in CD_3COCD_3 and $CDCl_3$.

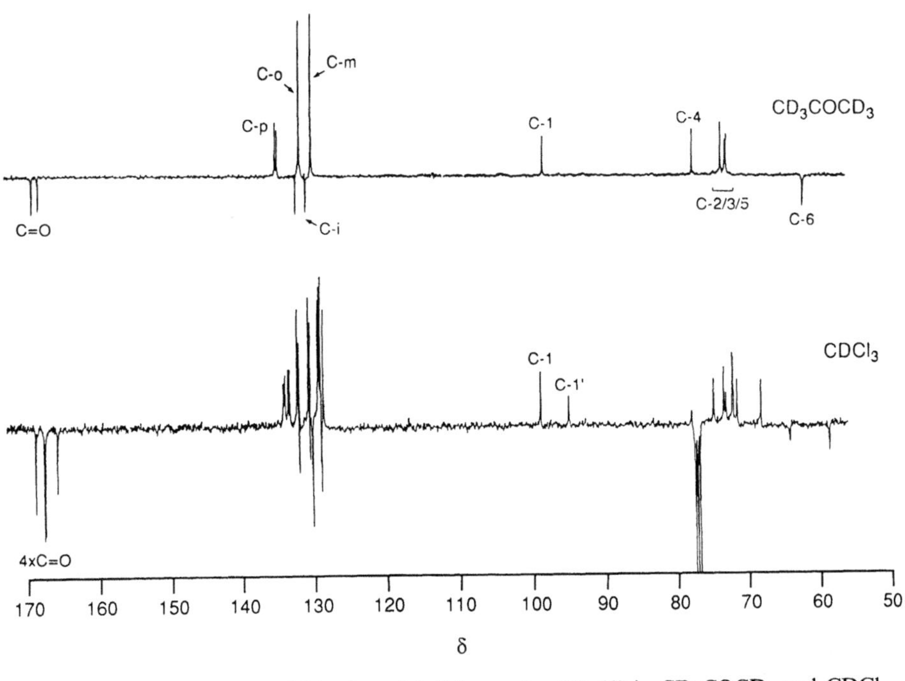

Fig. 4. ^{13}C NMR spectra (100 MHz) of per-2,3-*O*-benzoyl-α-CD (**3**) in CD_3COCD_3 and $CDCl_3$.

per-2,3-*O*-benzoyl-α-CD (**3**). The assignments shown in Figures 3 and 4 were made by us using homonuclear difference decoupling and correlation (both H—H and C—H) spectroscopies.

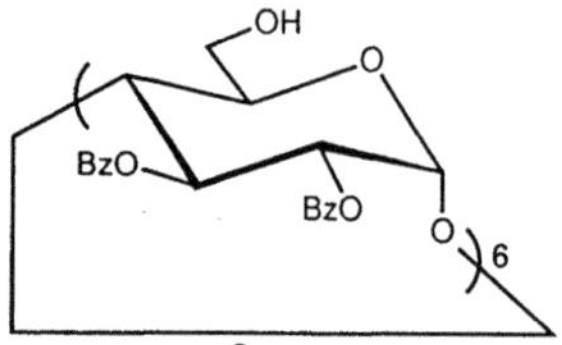

3

The ^{1}H NMR spectrum of **3** in CD_3COCD_3 shows sharp resonances corresponding to an α-CD derivative with C_6 molecular symmetry. In $CDCl_3$ solution, however, the signals, although still quite well-resolved, are doubled-up in number. Furthermore, integration shows that the signals corresponding to any particular proton located in constitutionally identical environments are in an intensity ratio of 1 : 1. The initial, tentative explanation [12] of this unusual phenomena was that **3** undergoes a conformational change in $CDCl_3$ solution, which reduces its C_6 molecular symmetry by a factor of two, i.e. a species with averaged C_3 molecular symmetry (Figure 5) is present in $CDCl_3$ solution. Although the exact nature of this conformation was not discussed in detail by Lehn *et al.*, the importance of intramolecular hydrogen bonding in stabilising an atypical conformation was highlighted by IR spectroscopic data.

Since the conformational change undergone by **3** is more pronounced than that undergone by per-2,6-*O*-methyl-3-*O*-benzoyl-α-CD (**1**), a detailed investigation of **3** was initiated. It was anticipated that not only would a greater understanding of the conformational properties of **3** be achieved, but also that a general appreciation would be gained on how to induce drastic conformational changes into cyclodextrins as a result of simple chemical modifications. At the outset, two issues were addressed. They were the identification of (i) the nature of the conformational species in $CDCl_3$ solution and (ii) the factors which induce the phenomenon to occur. Both spectroscopic and synthetic studies were performed in order to bring a greater understanding to bear on this problem.

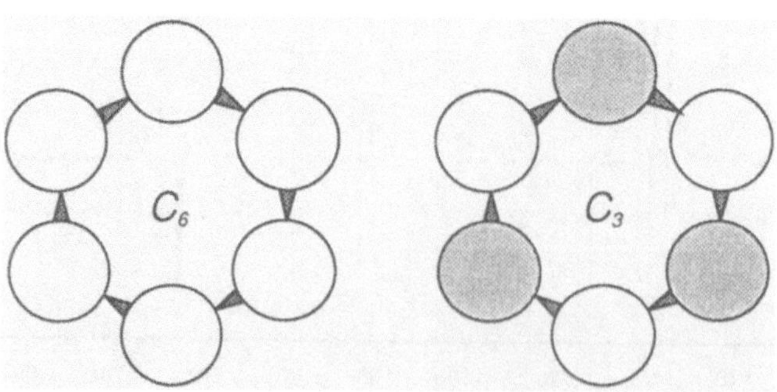

Fig. 5. Schematic representation of cyclodextrin derivatives with C_6 and C_3 symmetries (each circle represents a substituted D-glucopyranose ring and each arrow represents an α-1 → 4 glucosidic linkage).

The ^1H NMR spectrum of **3** was recorded in a wide range of solvents. In CD_3OD, CD_3SOCD_3 and C_5D_5N, well-resolved spectra consistent with α-CD derivatives with averaged C_6 molecular symmetry were observed. As expected, small chemical shift differences were evident in these polar solvents. ^1H NMR spectra were also recorded in CD_2Cl_2, C_6D_6, and $CDCl_2CDCl_2$ solutions. These spectra were all complex and signals were broad relative to those observed in the spectrum of **3** recorded in $CDCl_3$ solution under the same conditions. These observations supported the belief that there is an equilibrium established between conformational diastereoisomers in these solvents.

Two different types of gross conformational change can be identified. (i) The substituted D-glucopyranose residues might exist in $^4C_1$ or $^1C_4$ ring conformations or (ii) the relative orientations of contiguous D-glucopyranose units could be altered. The conformations of D-glucopyranose rings may be established by measuring the vicinal H—H coupling constants and applying the Karplus relationship [13]. Unfortunately, in all the solvents in which the unusual conformational behaviour is observed, the proton resonances are either too broad or they overlap with other resonances. As a result, the vicinal H—H coupling constants could not be determined for both of the resonances corresponding to any one particular ring proton.

An alternative method of determining the D-glucopyranose ring conformations is to measure the $^1J_{CH}$ coupling constants associated with the anomeric centre [14]. It has been shown that the magnitude of the coupling constant depends on the location of the anomeric proton in either an axial or an equatorial position. The two diastereoisomeric chair conformations of cyclodextrin D-glucopyranose rings in **3** – namely the $^4C_1$ and $^1C_4$ conformations – are associated with an equatorial and an axial anomeric proton, respectively. Values of this coupling constant observed for **3** are given in Table I along with appropriate literature values. The data recorded in this table for **3** in $CDCl_3$ solution indicate that the corresponding H-1 protons are in approximately equatorial positions on their respective D-glucopyranose rings. Since the values obtained are not exactly equal, it is possible that the ring conformations of the two D-glucopyranose units are not identical. However, such distortions are considered to be insignificant and could not possibly account for the conformational changes observed for this compound.

Table I. $^1J_{CH}$ Coupling constants of per-2,3-*O*-benzoyl-α-CD (**3**) at the anomeric centre.

Compound	$^1J_{CH}$ for anomeric centre (Hz)
α-D-Glucose[a]	169.5
β-D-Glucose[a]	162.0
α-CD[a]	168.5
Per-acetylated-α-CD[b]	177.5
Per-3,6-anhydro-β-CD[a,c]	164.0
Per-2,3-*O*-benzoyl-α-CD[c]	173.5, 169.0[b]
	172.0[d]

[a] Recorded in D_2O.
[b] Recorded in $CDCl_3$.
[c] Obtained using the INEPT pulse sequence.
[d] Recorded in C_5D_5N.

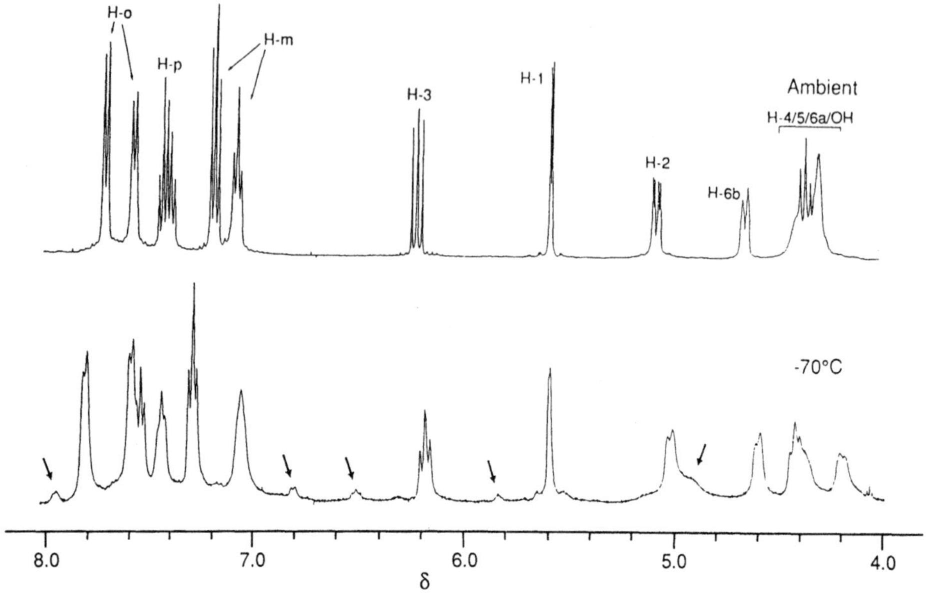

Fig. 6. ^{1}H NMR spectrum (400 MHz) of per-2,3-*O*-benzoyl-α-CD (**3**) in CD_3COCD_3 recorded at ambient temperature and at −70°C.

In order to establish the existence of a conformational equilibrium in solution, a series of dynamic NMR spectroscopic studies were performed on **3**. Considering first of all solutions of **3**, which appear to comprise the averaged C_6 symmetrical species at room temperature, a CD_3COCD_3 solution of **3** was cooled to −70°C. This resulted (Figure 6) in a certain amount of line broadening and a number of additional resonances of low intensity, indicated with arrows in Figure 6, were identified. This line-broadening is unlikely to be a consequence of any increase in the viscosity of the solvent. In fact, it is believed to be a result of a conformational equilibration process involving a minor conformation. The presence of the additional resonances at low temperatures confirm the existence of this minor conformation.

The relatively low boiling point of CD_3COCD_3 limits the scope of the variable temperature NMR studies that can be performed in this solvent. Therefore, a CD_3SOCD_3 solution of **3** was employed for obtaining spectra above room temperature, because this solvent not only boils at a relatively high temperature (at *ca.* 190°C), it also gives similar spectroscopic results to those obtained in CD_3COCD_3. Warming a CD_3SOCD_3 solution of **3** to 60°C led (Figure 7) to a sharpening of the resonances in the spectrum, as a result of the decrease in viscosity of this solvent that occurs as the temperature is raised. The only difference noted in the spectra recorded at the two temperatures concerned the chemical shift of the OH resonance. Under ambient conditions, this proton resonates as a triplet at δ 4.85, a chemical shift characteristic [15] of a non-anomeric hydroxyl group hydrogen bonded to CD_3SOCD_3. On the other hand, the spectrum recorded at 60°C reveals the hydroxyl resonance as a double doublet at δ 4.65. Such a shift of a hydroxyl resonance to lower frequency with an increase in temperature is indicative [16] of hydrogen bond formation between the solute and the solvent. This hydrogen

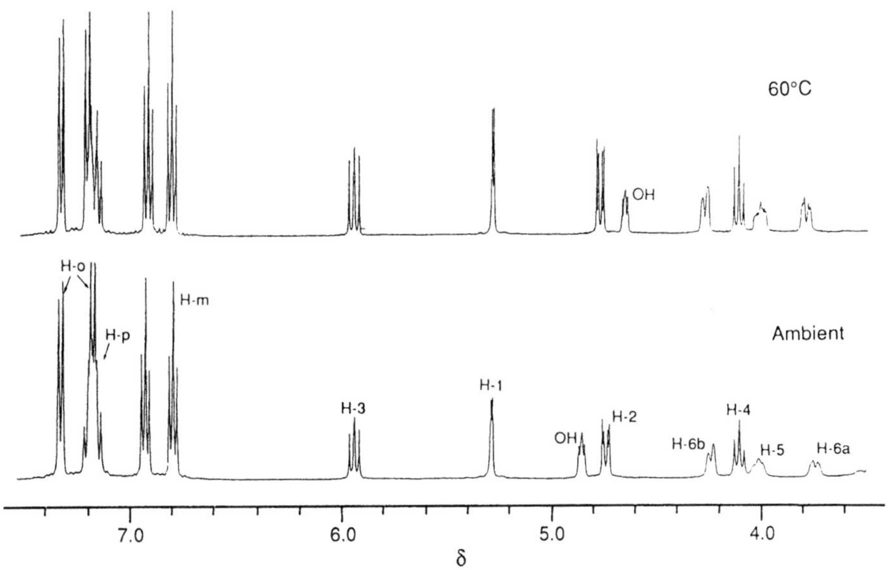

Fig. 7. ^{1}H NMR spectrum (400 MHz) of per-2,3-*O*-benzoyl-α-CD (**3**) in CD_3SOCD_3 recorded at 60°C and at ambient temperature.

bonding interaction between **3** and the solvent is believed to be an important factor in stabilising the C_6 symmetrical structure of **3** in CD_3SOCD_3, and, by implication, in CD_3COCD_3, CD_3OD, and C_5D_5N as well.

When a $CDCl_3$ solution of **3** was warmed to 60°C, broadening of the ^{1}H NMR resonances was observed (Figure 8). However, the complexity of the spectrum, as reflected in the number of resonances, was not altered significantly. Cooling a

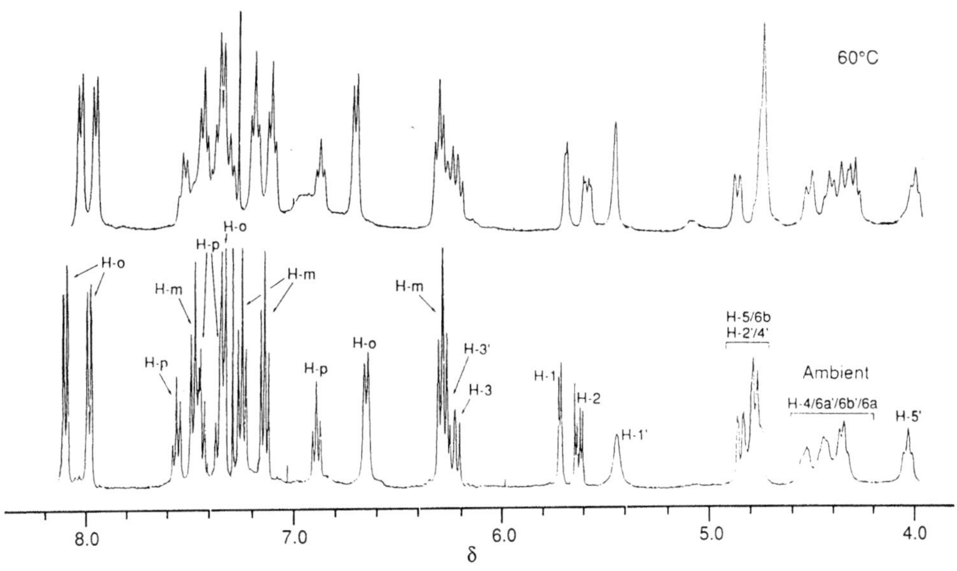

Fig. 8. ^{1}H NMR spectrum (400 MHz) of per-2,3-*O*-benzoyl-α-CD (**3**) in $CDCl_3$ recorded at 60°C and at ambient temperature.

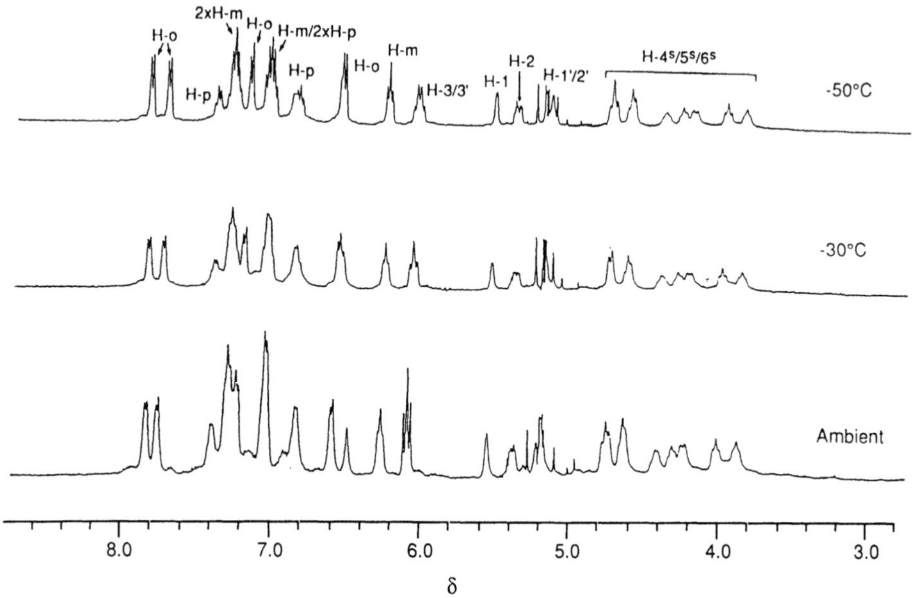

Fig. 9. ^{1}H NMR spectrum (400 MHz) of per-2,3-*O*-benzoyl-α-CD (**3**) in CD_2Cl_2 at −50°C, at −30°C, and at ambient temperature.

$CDCl_3$ solution of **3** to −50°C also resulted in line broadening of all the resonances. However, in this instance, the phenomenon can be attributed to the increased viscosity of $CDCl_3$ at low temperatures.

Although the ^{1}H NMR spectrum of **3** in CD_2Cl_2 under ambient conditions (Figure 9) is both complex and broad – it resembles the ^{1}H NMR spectrum of **3** recorded in $CDCl_3$ in terms of the number of resonances, with the pair of doublets around δ 8 being diagnostic of the presence of the atypical conformational isomer – the resonances are not as well resolved. A tentative explanation for the appearance of this ^{1}H NMR spectrum was that the conformational equilibration process, which occurs in $CDCl_3$, is also occurring in CD_2Cl_2, although the rate of the equilibration is faster in CD_2Cl_2. Therefore it was anticipated that cooling the solution down would result in a better resolved ^{1}H NMR spectrum. As shown in Figure 9, this prediction was supported by our experimental observations. A well-resolved spectrum was obtained at −50°C and integration of the resonances at this temperature again revealed that the areas of peaks corresponding to a proton in a particular constitutional environment were in the ratio of 1 : 1. Warming up the solution above room temperature gave slightly broader spectra from which no further conclusions could be drawn.

The ^{1}H NMR spectrum of **3** in C_6D_6 under ambient conditions (Figure 10) was even broader than that obtained for **3** in CD_2Cl_2. As expected, the signals were dispersed over a much wider chemical shift range, with the characteristic high frequency pair of doublets for the atypical conformation resonating at δ 8.3 and δ 8.6. The broad and complex nature of the spectrum was again rationalised tentatively as being a consequence of a more rapid equilibration (relative to the rate in $CDCl_3$) between the two conformations. Because benzene freezes at 5°C, the

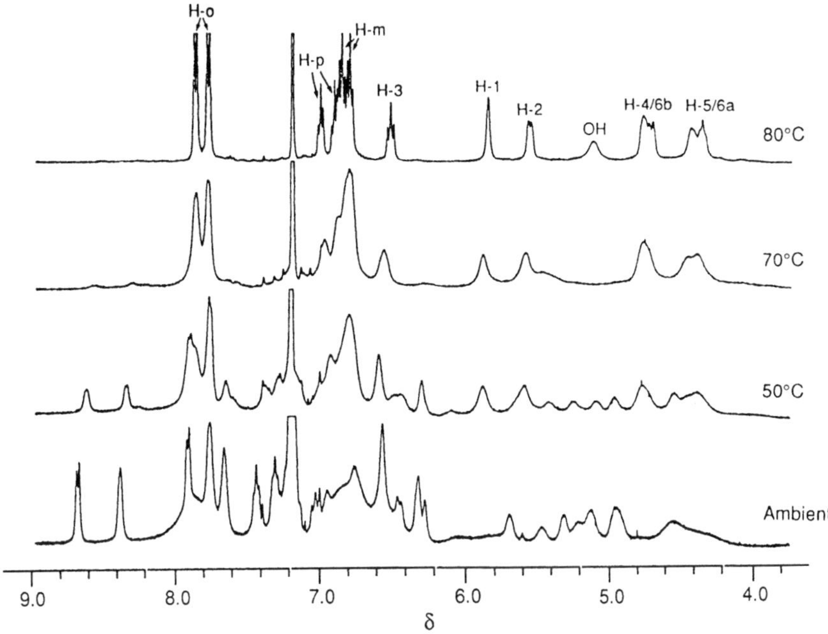

Fig. 10. ^{1}H NMR spectrum (400 MHz) of per-2,3-*O*-benzoyl-α-CD (**3**) in C_6D_6 at 80°C, at 70°C, at 50°C, and at ambient temperature.

solution could not be cooled much below room temperature. Cooling the solution to 10°C did not result in any marked alteration in the appearance of the ^{1}H NMR spectrum. Warming the solution did, however, lead to changes in the spectrum. As the temperature was raised the number of peaks in the spectrum gradually diminished. This simplification in the ^{1}H NMR spectrum is most apparent by considering the two high frequency resonances at δ 8.3 and δ 8.6. Increasing the temperature of the solution leads to a decrease in the intensities of these peaks and eventually to their disappearance. The spectrum obtained at 80°C is characteristic of an α-CD derivative exhibiting the usual averaged C_6 molecular symmetry.

The existence of broad lines in the ^{1}H NMR spectra of **3** in CD_2Cl_2 and C_6D_6 at room temperature clearly reflects the existence of the same conformational process, i.e. an equilibration between two degenerate conformational states. Unfortunately, the low boiling point of CD_2Cl_2 and the high melting-point of C_6D_6 hinder the observation of a change with temperature, from a well-resolved but complex spectrum, to a well-resolved but simple spectrum. However, such a change could be achieved by recording the ^{1}H NMR spectra in $CDCl_2CDCl_2$.

The ^{1}H NMR spectrum of **3** in $CDCl_2CDCl_2$ recorded under ambient conditions showed substantial line-broadening, which probably results from the viscous nature of the solvent. Warming the solution led to a sharpening of the resonances, with the best resolution being achieved around 60°C. The spectrum at this temperature was partially assigned using homonuclear difference decoupling spectroscopy. These experiments succeeded in establishing two distinct resonances corresponding to constitutionally-identical protons on the disubstituted D-glucopyranose ring. These resonances were associated with two heterotopic H-3 protons, appearing as one

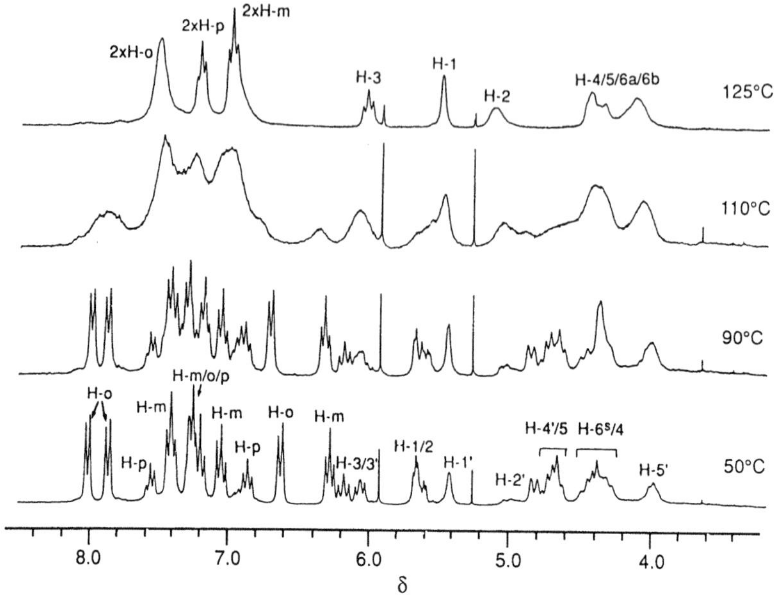

Fig. 11. ^{1}H NMR spectra (250 MHz) of per-2,3-*O*-benzoyl-α-CD (**3**) in $CDCl_2CDCl_2$ at 125°C, at 110°C, at 90°C and at 50°C.

triplet at δ 6.21 ($J_{2,3} = J_{3,4} = 9.5$ Hz) and as another triplet of equal intensity at δ 6.07 ($J_{2,3} = J_{3,4} = 9.0$ Hz). In contrast to all of the ^{1}H NMR spectra of **3** recorded in pure solvents, this is the only spectrum in which resonances associated with constitutionally-identical protons have been resolved completely. Given the limited accuracy with which the vicinal H—H coupling constants can be measured, both heterotopic H-3 protons are associated with D-glucopyranose rings in the same $^4C_1$ chair conformation.

As the CD_2ClCD_2Cl solution was warmed up further, the spectra began to broaden, became featureless at 110°C and finally sharpened at 125°C to give a spectrum of the kind expected for an α-CD derivative with averaged C_6 molecular symmetry. Unfortunately, higher temperatures – and presumably better resolved spectra – could not be obtained because of the constraints imposed by the NMR spectrometer. These results helped to confirm the conclusion, based on the spectra recorded in CD_2Cl_2 (Figure 9) and C_6D_6 (Figure 10), that a degenerate equilibration process (Figure 12) is taking place between two equivalent conformations with C_3 molecular symmetry. In $CDCl_2CDCl_2$ at 50°C, this degenerate equilibration process is slow on the ^{1}H NMR time-scale, and hence, two sets of equal intensity resonances are observed, corresponding to two heterotopic D-glucopyranose residues. As the temperature is raised, the rates of inversion of the two degenerate conformers increases until, at 125°C, the rate is so rapid that averaged resonances are observed for the heterotopic D-glucopyranose residues. The free energy of activation for the process can be determined by considering the two heterotopic H-3 resonances which are separated by 30 Hz at 50°C ($\Delta\nu = 30$ Hz) and which coalesce at 110°C, i.e. $T_c = 110$°C. The free energy of activation for the degenerate equilibration process, calculated using the Eyring equation [17], is 19.4 kcal mol^{-1}.

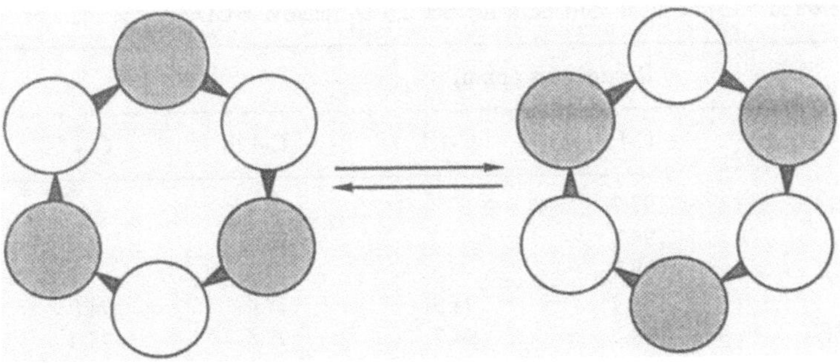

Fig. 12. A schematic representation of the degenerate equilibrium between two cyclodextrin derivatives with averaged C_3 molecular symmetry.

In all of the solvents, and under all of the conditions described above, the magnitudes of coupling constants (both $^3J_{HH}$ and $^1J_{CH}$) associated with the D-glucopyranose rings in **3** indicate that there is no significant distortion of their conformations from the normal $^4C_1$ chair. Therefore, the conformational behaviour of **3** must be associated with a discrete difference in the relative dispositions of the adjacent disubstituted D-glucopyranose residues around the torus of the α-CD derivative.

The relative orientations of two adjacent monosaccharide units in an oligosaccharide are defined [18] in terms of the glycosidic torsion angles, ϕ and ψ. A number of different procedures have been employed to determine these angles. One common method involves measuring vicinal carbon–proton coupling constants across the glycosidic linkage and relating the magnitude of this coupling constant to the relevant torsion angles by applying a 'Karplus-type' relationship [14, 19]. In cyclodextrin derivatives, the anomeric (C-1) and aglyconic (C-4) carbon atoms are also in vicinal relationships with respect to some of the endocylic protons, and so the problem becomes one of the identification and elimination of these endocyclic three-bond carbon–hydrogen couplings from the analysis.

In principle, there are a number of ways in which this can be achieved. Selective deuteration [20] of **3** was not considered to be a feasible method because of the difficulty associated with the synthesis of the deuterated analogue. The recording of a selectively-decoupled ^{13}C NMR spectrum proved to be inconvenient because of the inordinate length of time needed to acquire a spectrum with sufficient resolution. Identification of the small $^3J_{CH}$ coupling constants was achieved by recording the INEPT NMR spectrum of **3**. However, there was no way to establish which coupling constant corresponded to which vicinal carbon–proton pair. Finally, an attempt was made to apply the heteronuclear *J*-resolved technique of Bax and Freeman [21]. Unfortunately, this experiment met with only partial success. However, the vicinal coupling constant in CD_3COCD_3 relating to C-4′—O—C-1—H-1 was found to be 5.2 Hz. The corresponding torsion angle (ϕ) was thus calculated [22] to be 13°. This value is similar to one that was obtained for α-CD from X-ray crystallographic studies [3, 23]. The coupling constant corresponding to ψ (C-1—O—C-4′—H-4′) could not be measured because of the poor resolution of the H-4′ resonance. Overlapping signals and the broad nature of the ^{1}H NMR

Table II. Selected ^{13}C chemical shift data for per-2,3-*O*-benzoyl-α-CD (**3**).

Solvent	Resonances (ppm)			
	C-1	C-1′[a]	C-4	C-4′[a]
CD_3COCD_3	97.7	–	77.4	–
CD_3SOCD_3	96.7	–	76.5	–
C_5D_5N	98.0	–	77.9	–
$CDCl_3$	98.2	94.5	77.9	74.9
$CDCl_2CDCl_2$	98.2	94.7	77.5	75.0

[a] A prime (′) is normally used in cyclodextrin chemistry nomenclature to denote an adjacent glucopyranose residue. In this instance, the glucose units may be divided into two conformationally different types and the prime is used to distinguish between these units.

spectrum in $CDCl_3$ prevented any of the desired coupling constants from being determined for **3** dissolved in this solvent.

The glycosidic torsion angles have been determined in disaccharides and trisaccharides by measuring inter-residue nuclear Overhauser effects (NOEs) [24]. However, overlapping signals made such experiments unsuccessful in the present study.

Finally, recourse was made to the use of the proposed relationships between ^{13}C chemical shifts of anomeric and aglyconic carbon atoms and the glycosidic torsion angles. Such relationships have been discussed with reference to cyclodextrin systems [25], as well as to carbohydrates in general [26]. The relevant ^{13}C chemical shift data collected for **3** are given in Table II.

The magnitude of the torsion angle, ϕ, determined in CD_3COCD_3 suggests that the relative orientation of D-glucopyranose units obtained for this CD derivative in this solvent – and, by implication, in CD_3OD, CD_3SOCD_3, and C_5D_5N – is similar to that of the parent molecules. Therefore, average chemical shift values of *ca.* 97.5 ppm (C-1) and *ca.* 77.3 ppm (C-4) for these three solvents, are believed to be associated with the normal torsion angle values of *ca.* $+10°$. The different frequencies of the anomeric (especially C-1′) and aglyconic (especially C-4′) carbon atom resonances observed in $CDCl_3$ and $CDCl_2CDCl_2$ are believed to be associated with substantially different values for the appropriate torsion angles.

The relationships between ^{13}C chemical shifts and torsion angles are only semiquantitative at best and so must be applied to cyclodextrin derivatives with some caution. Also, the relationships [25] which have been established with particular reference to cyclodextrins were obtained from solid state ^{13}C NMR spectra of the parent molecules. In itself, of course, chemical modification causes alterations in ^{13}C chemical shift values, and further changes in these values may arise, in solution, from, for example, specific solvent effects [27]. With these reservations in mind, a consideration of the various relationships leads to the semiquantitative conclusion [26, 28] that a ± 2 ppm shift of the anomeric carbon corresponds to a change in ψ of $\pm 10°$, and a ± 2 ppm shift of the aglyconic carbon is associated with a change in ϕ of $\pm 5°$. Applying these approximate relationships to the changes in chemical shifts observed in the ^{13}C NMR spectra for **3** in $CDCl_3$ and in $CDCl_2CDCl_2$ resulted in the calculated torsion angle changes listed in Table III.

Table III. Changes in average values of ^{13}C chemical shifts in selected solvents.

Solvent	$\Delta\delta\,^{13}C^a$ (associated change in torsion angle)			
	C-1	C-1′	C-4	C-4′
$CDCl_3$	+0.7 ($\Delta\psi = 3.5°$)	−3.0 ($\Delta\psi' = -15°$)	+0.6 ($\Delta\phi = +1.5°$)	−2.4 ($\Delta\phi' = -6.0°$)
$(CDCl_2)_2$	+1.1 ($\Delta\psi = 5.5°$)	−2.8 ($\Delta\psi' = -14°$)	+0.2 ($\Delta\phi = +0.5°$)	−2.3 ($\Delta\phi' = -5.8°$)

[a] $\Delta\delta\,^{13}C = (\delta_{solvent} - \delta_{average}$ given in Table II) ppm.

Examination of molecular models indicates that such torsion angle changes are consistent with adjacent pairs of D-glucopyranose residues twisting relative to each other about their mutual glycosidic linkages. Such a twist results in the benzoate groups, of the residues with low frequency chemical shift values (those corresponding to C-1′ and C-4′), pointing further in towards the cavity than normal. The benzoate groups of the D-glucopyranose units which exhibit the higher frequency ^{13}C chemical shift values point further away from the cavity than is usual, and consequently, the C-6 hydroxyl groups on the same alternating D-glucopyranose rings are directed towards the centre of the cavity (Figure 13). The calculations suggest that the degree of the twist does not appear to be all that significant. Careful examination of molecular models indicates that changes in ψ' and ϕ' of the order $-30°$ to $-40°$ and a change in ψ and ϕ of the order $+30°$ to $+40°$ are required in order to produce the circular network of hydrogen bonds in **3** described by Lehn *et al.* [12]. The evidence for such a network is very strong. It is based on the O—H stretching frequency observed in the IR spectrum of **3** in $CHCl_3$ (ν O—H stretch, 3260 cm^{-1}), compared with that obtained on recording the IR spectrum in CH_3SOCH_3 (ν O—H stretch, 3430 cm^{-1}). This result was confirmed in the present investigation and extended to include pyridine (ν O—H stretch, 3400 cm^{-1}) as the solvent. By analogy with similar data obtained for calixarenes [29], the relatively

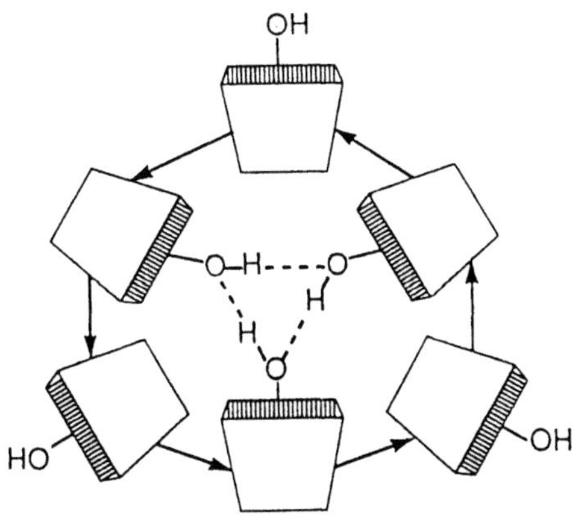

Fig. 13. Schematic representation of the unusual C_3 conformational isomer of **3** showing the relative tilt of the D-glucopyranose residues and circular network of intramolecular hydrogen bonds.

weak O—H stretch noted in the $CHCl_3$ solution probably reflects the presence of a strong intramolecular circular network of cooperative hydrogen bonds. It was also noted that, after shaking a $CDCl_3$ solution of **3** with D_2O and recording the ^1H NMR spectrum, there was no change in the spectrum. Thus, the weakening of the hydrogen bonding network caused by exchanging of H for D is clearly not sufficient to influence the nature of the degenerate conformations in the conformational equilibration process. Clearly, the quantitative uncertainties encountered in applying the chemical shift/torsion angle empirical relationships are such that only a qualitative interpretation can be placed on the data given in Tables II and III, i.e. that contiguous disubstituted D-glucopyranose residues are twisted relative to each other about their mutual glycosidic linkages.

The sensitivity of $[\alpha]_D$ values for cyclodextrins towards changes in D-glucopyranosidic torsional angles is well known [30]. Values of $[\alpha]_D$ for **3**, which were measured in a variety of solvents, are listed in Table IV. Thus, the large difference of $[\alpha]_D$ values recorded in acetone and chloroform must reflect significant changes in disubstituted D-glucopyranosidic torsional angles in **3** in these two solvents. No simple correlation could be achieved between these values and any of the well-known solvent polarity scales [31].

As a result of the spectroscopic and chiroptical studies performed in single solvents, much has been learnt about the conformational behaviour of **3**. In solvents such as methanol, acetone, pyridine, and DMSO, which readily participate in hydrogen bonding with hydroxyl groups, this α-CD derivative exhibits normal averaged C_6 molecular symmetry in which the disubstituted D-glucopyranose residues adopt the expected $^4C_1$ chair conformation. In solution, hydrogen bonds are presumably formed between these solvents and the free C-6 hydroxyl groups on the CD derivative. These intermolecular non-covalent bonds are undoubtedly important in stabilising the gross conformation of the molecule. However, in

Table IV. Chiroptical data for per-2,3-*O*-benzoyl-α-CD (**3**) in selected solvents.

Solvent	$[\alpha]_D^a$
Acetone	+92°
Pyridine	+90°
Ethyl acetate	+86°
Dimethylformamide	+78°
Dimethylsulphoxide[b]	+71°
Acetonitrile	+71°
1,4-Dioxan	+67°
Methanol	+52°
Tetrahydrofuran	+51°
Benzene	+42°
Dichloromethane	+34°
Chloroform[c]	+11°
1,1,2,2-Tetrachloroethane	+9°

[a] *c*, 0.3–1.2 at RT.
[b] Lit. [12] +96° (*c*, 1.3).
[c] Lit. [12] +9° (*c*, 1.1).

dichloromethane, chloroform, tetrachloroethane and benzene, a conformational equilibrium is established (Figure 12), between two degenerate conformations with averaged C_3 molecular symmetry. The reduction in symmetry must be associated with a relative twist in the orientation of adjacent disubstituted D-glucopyranose residues. Furthermore, one of the factors in stabilising the atypical conformation is a circular network of intramolecular cooperative hydrogen bonds (Figure 13) involving primary hydroxyl groups on a trio of alternating disubstituted D-glucopyranose residues.

Following the study of the conformational properties of **3** in pure solvents, an NMR spectroscopic investigation in a mixed solvent system comprised of CD_3COCD_3 and $CDCl_3$ was undertaken. An important point to note at the outset is that, despite being sparingly soluble in acetone, and readily soluble in chloroform, **3** is only just soluble in acetone/chloroform mixtures. Thus, addition of CD_3COCD_3 to a solution of **3** in $CDCl_3$ leads to the precipitation of the compound. Typical concentrations used in the mixed solvent investigations were 4–5 mg in 0.7 mL of solvent, compared with 15–20 mg of compound in the same volume of a pure solvent. In fact, the low solubilities meant that a ^{13}C NMR spectrum of **3** in $CD_3COCD_3/CDCl_3$ (1 : 1 v/v) with a good signal-to-noise ratio could not be obtained under routine conditions.

^1H NMR spectra of **3** in varying proportions of CD_3COCD_3 and $CDCl_3$ were recorded (Figure 14). In $CDCl_3/CD_3COCD_3$ (3 : 1 v/v), a spectrum was observed that was very similar to that obtained in pure $CDCl_3$. However, new peaks appeared, (marked with asterisks in Figure 14) at δ 7.40 (d), δ 7.00 (t), δ5.58 (d) and δ 5.08 (dd). Furthermore, some resonances which were overlapping in the

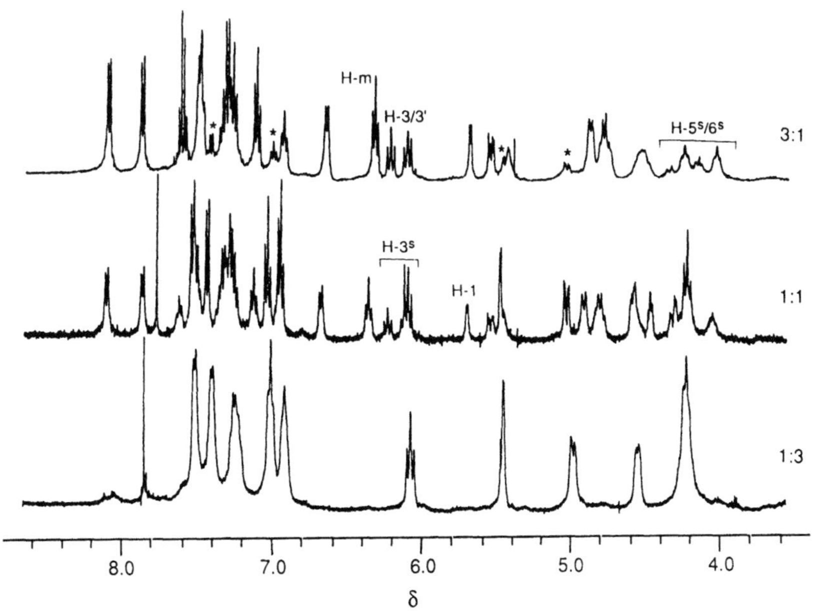

Fig. 14. ^1H NMR spectrum (400 MHz) of per-2,3-*O*-benzoyl-α-CD (**3**) in $CDCl_3/CD_3COCD_3$ with v/v of 3 : 1, 1 : 1, and 1 : 3.

spectrum in pure $CDCl_3$ were well-resolved in the mixed solvent systems. In particular, the resonances in the regions δ 6.05–6.40 (H-3, H-3′, and H-m) and δ 4.0–4.6 (H-5 and H-6) were well separated. Measurements of the coupling constants involving the H-3 and H-3′ resonances showed them to be associated with D-glucopyranose rings in the $^4C_1$ chair conformation.

Employing $CDCl_3/CD_3COCD_3$ (1 : 3 v/v) as the solvent mixture gave a ^{1}H NMR spectrum (Figure 14) similar to that obtained in pure CD_3COCD_3. The only differences are that the spectrum is broader and that the characteristic peaks for the atypical conformation at *ca.* δ 8 are present. The ^{1}H NMR spectrum recorded in $CDCl_3/CD_3COCD_3$ (1 : 1 v/v) is well resolved and also extremely complicated. The relative proton count associated with the integral for the doublet at δ 5.70 (assigned to an anomeric proton) was equated with 1 H. On this basis, the integration of the disubstituted D-glucopyranose ring protons from δ 4.0–6.3 totalled 21 H and the integral corresponding to the benzoyl region (δ 6.30–8.20) equalled 30 H. The relative proton counts indicated by these integrals are consistent with the presence of three different types of disubstituted D-glucopyranose residues present in a ratio of 1 : 1 : 1.

The signals – a triplet and two overlapping triplets – between δ 6.05–6.30 integrate for 3 H and can be assigned to three different H-3 protons in a 1 : 1 : 1 ratio. All vicinal proton–proton coupling constants in this region were consistent with all of the disubstituted D-glucopyranose residues present adopting $^4C_1$ chair conformations.

Values of $[\alpha]_D$ for compound **3** in varying ratios of $CHCl_3$ and CH_3COCH_3 were recorded and an approximately linear relationship (Figure 15) was observed between the specific rotation and the percentage by volume of acetone, which turns out to be approximately equal to the percentages expressed as moles of acetone. Although a number of possible interpretations can be proposed to account for this

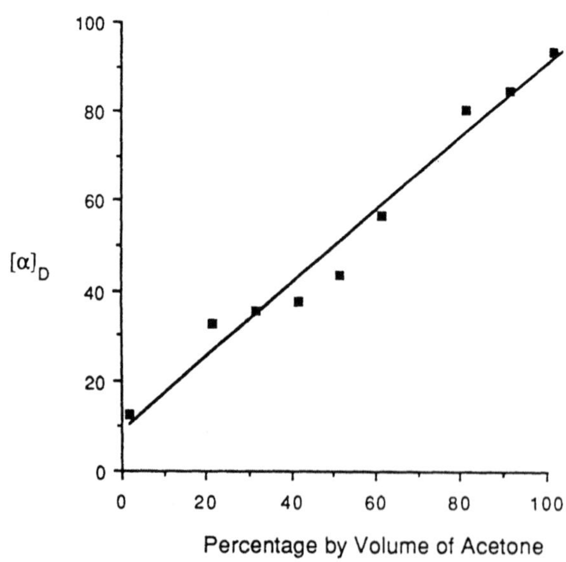

Fig. 15. Plot showing values of $[\alpha]_D$ for solutions of **3** in mixtures of $CHCl_3/CH_3COCH_3$.

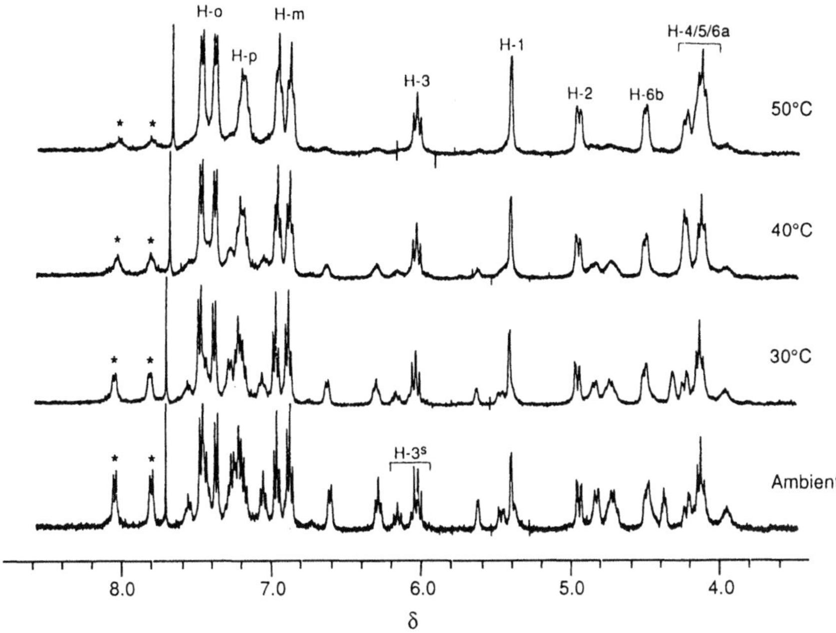

Fig. 16. ^{1}H NMR spectrum (400 MHz) of per-2,3-*O*-benzoyl-α-CD (**3**) in $CDCl_3/CD_3COCD_3$ (1 : 1 v/v) at 50°C, at 40°C, at 30°C and at ambient temperature.

observation, a particularly appealing one is that the linear nature of the relationship is a reflection of the presence in solution of a changing mixture of conformations with averaged C_6 and averaged C_3 molecular symmetries.

The rates of the conformational equilibration processes were investigated by variable temperature ^{1}H NMR spectroscopy in $CD_3COCD_3/CDCl_3$ (1 : 1 v/v). As shown in Figure 16, increasing the temperature of the solution caused many of the signals to 'disappear', e.g. the characteristic resonances for the degenerate conformations with C_3 molecular symmetry at δ 8.10 and δ 7.85 (indicated with asterisks in Figure 16) diminished gradually in intensity and eventually 'vanished'. Also, among the resonances (δ 6.05–6.30) for the H-3 protons, the triplet at δ 6.24 disappeared progressively as the temperature was raised. Furthermore, the quartet centred on δ 6.11 changed in its apparent multiplicity and eventually assumed a triplet character. At 50°C, the ^{1}H NMR spectrum indicates that an α-CD derivative with C_6 molecular symmetry is the predominant conformational isomer present in $CD_3COCD_3/CDCl_3$ (1 : 1 v/v).

The ^{1}H NMR spectra can be interpreted as showing the presence of an equilibrium between conformational isomers with C_3 and C_6 molecular symmetry, as shown schematically in Figure 17. Such a conclusion is consistent with the presence in solution of three different D-glucopyranose residues in a ratio of 1 : 1 : 1.

There is, of course, no fundamental reason why the ratio of $C_3 : C_6$ isomers must be 2 : 1. Changing the solvent ratio would be expected to alter the $C_3 : C_6$ isomer ratio, and it does. Thus, the peaks that were only just visible in the ^{1}H NMR spectrum (Figure 14) recorded in $CDCl_3/CD_3COCD_3$ (3 : 1 v/v) reflect the presence of a small amount of the C_6 conformation present in that solution.

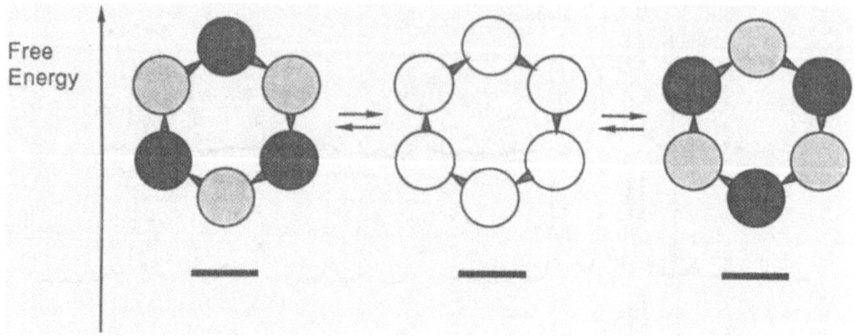

Fig. 17. Schematic energy level diagram representing the equilibration process occurring in $CDCl_3$/ CD_3COCD_3 between different conformational isomers of per-2,3-*O*-benzoyl-α-CD (**3**).

The alteration in the appearance of the ^{1}H NMR spectra occurs over a short temperature range (*ca.* 25°). As a consequence, it can be assumed that there will be negligible changes in the values of the standard free enthalpies and standard free entropies for each of the states. Furthermore, it is apparent, from consideration of the well-known relationship for the standard free energy difference (ΔG^0) between two states (Equation 1), that the influence of ΔS^0 on ΔG^0 increases with a rise in temperature. And so the conformational equilibrium under consideration is probably most profitably analysed in terms of entropy factors.

$$\Delta G^0 = \Delta H^0 - T\Delta S^0 \quad (1)$$

A state containing a conformation with a symmetry number, σ, greater than unity, has its relative entropy reduced by a factor $R \ln \sigma$ [18]. This means that the entropy of the C_6 conformation is reduced relative to that of the C_3 conformation by a factor of $R \ln 2$. This factor is probably less important than the changes in entropy resulting from the different types of hydrogen bonding present in the two conformations. Thus, the C_6 conformation is believed to be stabilised by hydrogen bonds between acetone and the C-6 hydroxyl group. It is a highly solvated but flexible species and can thus be associated with a higher entropy. The C_3 conformation, however, is stabilised by relatively strong intramolecular hydrogen bonds, and, as a consequence, has a much lower entropy. Hence, the change from $C_3 \rightarrow C_6$ is asociated with a positive change in the standard free entropy ($+\Delta S^0$). Consideration of Equation 1 leads to the conclusion that an increase in temperature would mean that the free energy change for $C_3 \rightarrow C_6$ decreases relative to the free energy change for $C_6 \rightarrow C_3$. The outcome is shown schematically in Figure 18. As the temperature is raised, the free energy difference between the C_3 and C_6 conformations will continue to widen, increasing the population of the C_6 conformation even further.

In addition to the NMR spectroscopic experiments which have been performed, many attempts have been made to grow single crystals of **3** suitable for X-ray structural analysis. Clearly it is essential to pay particular attention to the type of solvent molecules present in the crystal lattice. Therefore, crystals were grown by the simple evaporation of the solvent from a solution of **3**. Evaporation of acetone leads to a powder, whereas evaporation of chloroform usually gives single crystals.

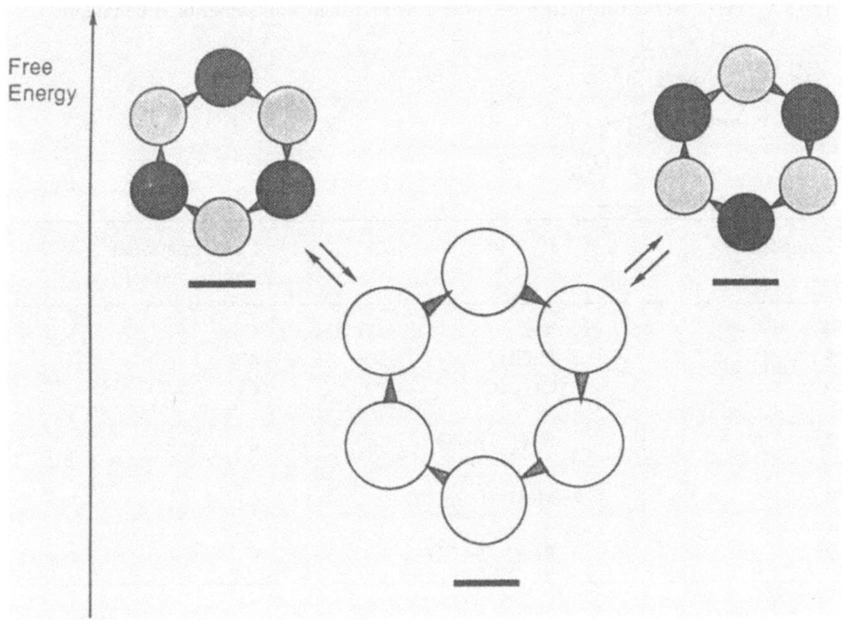

Fig. 18. Schematic representation of the effect of an increase in temperature on the equilibration process involving different conformational isomers (C_6 and C_3) of per-2,3-*O*-benzoyl-α-CD (**3**) (cf. Figure 17).

The practical problem of arresting the evaporation before it was complete meant that the crystals cracked quickly and became unsuitable for solid state structural analysis. Attempts to grow crystals using the vapour diffusion of hexane into a chloroform solution of **3** gave thin, intertwined needles. Good single crystals were eventually obtained by taking advantage of the poor solubility of the compound in chloroform/acetone mixed solvent systems. It was assumed that the precipitate which forms when acetone is added to a chloroform solution of **3** is a consequence of the C_3 conformation present in chloroform being 'insoluble' in acetone. Therefore, vapour diffusion of acetone into a chloroform solution of **3** was attempted and gave excellent single crystals which had a morphology similar to those obtained by the slow evaporation of a neat chloroform solution. Unfortunately, a preliminary crystallographic investigation [32] revealed that the unit cell contained in excess of 1000 non-hydrogen atoms (!) and could not be solved.

For many years, per-2,3-*O*-benzoyl-α-CD (**3**) remained the only well-defined example of a per-2,3-*O*-substituted CD derivative. However, the development of the *tert*-butyldimethylsilyl protecting group to block the primary face of CDs [8], has led to the synthesis of derivatives **4**–**6** (Table V). The experimental data given for these compounds suggest that they do not undergo any unusual conformational changes in chloroform solutions.

An attempt was made to prepare a number of substituted-benzoyl derivatives using the method described by Lehn *et al.* [12]. Examples of benzoyl groups were chosen containing, firstly, an electron withdrawing substituent (—F) in **7**, and secondly, an electron-donating substituent (—OMe) in **8**. The *p*-bromobenzoyl derivative **9** was also synthesised in the hope that the introduction of a heavy atom

Table V. Per-2,3-*O*-substituted α-cyclodextrins and their conformational behaviour.

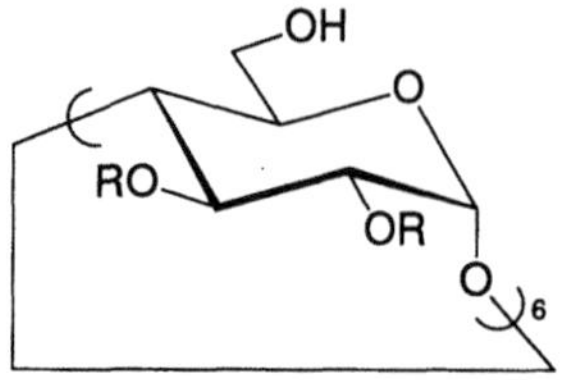

Compound	R	Conformational isomer present
4	Me	C_6
5	$PhCH_2$	C_6
6	CH_3CO	C_6
7	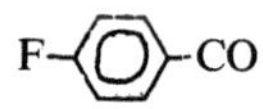	C_3
8	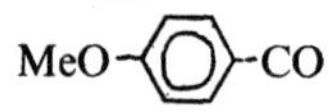	C_3
9	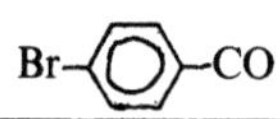	C_3

would aid X-ray crystallographic analysis. Slight modification of the established experimental procedure (see Experimental Section for exact details), making allowance for the different reactivity of the substituted benzoyl groups, enabled compounds **7**–**9** to be prepared in acceptable yields. ^{1}H NMR spectroscopy indicated that all three of these derivatives undergo a conformational change in $CDCl_3$, similar to that observed for per-2,3-*O*-benzoyl-α-CD (**3**).

Lehn *et al.* [12] had shown that completely functionalising the free C-6 hydroxyl groups results in derivatives which adopt conformations with averaged C_6 symmetry. This has been confirmed in the present investigation as a result of the synthesis of per-2,3-*O*-benzoyl-6-*O*-tosyl-α-CD (**10**) and per-2,3-*O*-benzoyl-6-*O*-mesyl-α-CD (**11**).

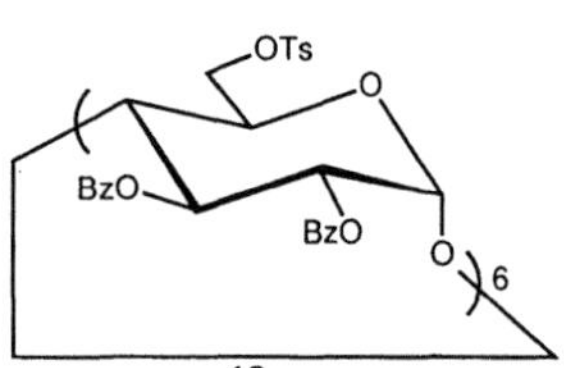

10

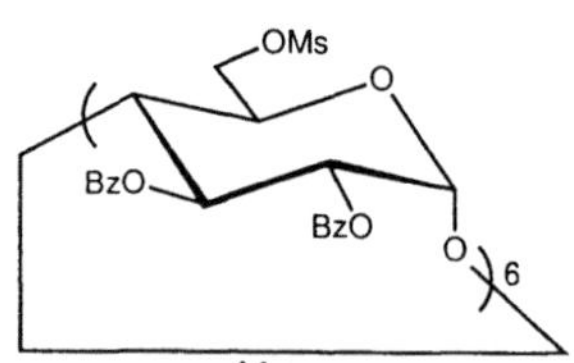

11

Single crystals of per-2,3-*O*-*p*-bromobenzoyl-α-CD (**9**), suitable for X-ray crystallographic analysis, were grown by vapour diffusion of acetone into a chloroform solution of the compound. The crystals are unstable in the absence of solvent and needed to be mounted in the mother liquor. Interestingly, the crystals completely failed to diffract X-rays! The reason for this surprising observation has tentatively been attributed to the glass-like structure of the crystal [32].

The investigations reported in this paper have clearly established that the groups at C-2 and C-3 play an important role in determining the conformational behaviour of compounds like **3**, **7**, **8**, and **9**. It might be expected that the removal of the secondary face intramolecular hydrogen bonding network present in the parent cyclodextrins would render α-CD derivatives conformationally more flexible. However, the reason why benzoyl groups and *para*-substituted benzoyl groups should lead to atypical, but well-defined, conformational states is still unclear. It is likely that, as more chemical modifications of the secondary face of cyclodextrins are carried out, additional examples of unusual conformational behaviour will be uncovered. Indeed, in the fullness of time, it might transpire that the parent CDs and simple CD derivatives with their rigid, well-defined cavities will be the exceptional compounds.

Acknowledgements

We thank the Johnson Matthey Technology Centre (C.A.S.E. awards to R.Z. and P.E.), and the Science and Engineering Research Council in the U.K. for financial support. We gratefully acknowledge the efforts made by Dr David J. Williams (Imperial College) in the collection of X-ray crystallographic data. We thank Drs Barbara Mulloy (National Institute for Biological Standards and Control) and Tom A. Frienkel (MRC Biomedical NMR Centre) for recording the 2-D-*J*-resolved NMR spectra.

References

1. J. F. Stoddart: 'A Century of Cyclodextrins', *Carbohydr. Res.* **192**, xii, (1989).
2. A. Villiers: *C. R. Acad. Sci.* **112**, 536 (1891).
3. W. Saenger: in *Inclusion Compounds*, Vol. 2, eds. J. L. Atwood, J. E. D. Davies, and D. D. MacNicol, Academic Press, London, p. 231 (1984).
4. J. Szejtli: *Cyclodextrins and their Inclusion Complexes*, Akadémiai Kiado, Budapest (1982); D. Duchêne (Ed.), *Cyclodextrins and Their Industrial Uses*, Editions de Santé, Paris, (1987); J. Szejtli, *Cyclodextrin Technology*, Kluwer Academic Publishers, Dordrecht (1988); M. L. Bender and M. Komiyama, *Cyclodextrin Chemistry*, Springer, Berlin (1978).
5. A. P. Croft and R. A. Bartsch: *Tetrahedron* **39**, 1417 (1983).
6. I. Tabushi: in *Inclusion Compounds*, Vol. 3, eds. J. L. Atwood, J. E. D. Davies, and D. D. MacNicol, Academic Press, London, p. 445 (1984). R. Breslow in *Inclusion Compounds*, Vol. 3, eds. J. L. Atwood, J. E. D. Davies, and D. D. MacNicol, Academic Press, London, p. 473 (1984).
7. R. Bergeron: in *Inclusion Compounds*, Vol. 2, eds. J. L. Atwood, J. E. D. Davies, and D. D. MacNicol, Academic Press, London, p. 391 (1984).
8. K. Takeo, K. Uemura, and H. Mitoh: *J. Carbohydr. Chem.* **7**, 293 (1988); K. Takeo, H. Mitoh, and K. Uemura: *Carbohydr. Res.* **187**, 203 (1989).
9. K. Harata, K. Uekama, M. Otagiri, and F. Hirayama: *Bull. Chem. Soc. Jpn.* **60**, 497 (1987); K. Harata: *J. Chem. Soc. Chem. Commun.* 928 (1988); K. Harata, K. Uekama, T. Imai, F. Hirayama, and M. Otagiri: *J. Incl. Phenom.* **6**, 443 (1988).
10. D. R. Alston, P. R. Ashton, T. H. Lilley, J. F. Stoddart, R. Zarzycki, A. M. Z. Slawin, and D. J. Williams: *Carbohydr. Res.* **192**, 259 (1989).
11. C. M. Spencer, J. F. Stoddart, and R. Zarzycki: *J. Chem. Soc. Perkin Trans 2* 1323 (1987).
12. J. Boger, R. J. Concoran, and J.-M. Lehn: *Helv. Chim. Acta* **61**, 2190 (1978).
13. M. Karplus: *J. Chem. Phys.* **30**, 11 (1959).
14. J. A. Schwarz and A. S. Perlin: *Can. J. Chem.* **50**, 3667 (1972); A. Parfondry, N. Cyr, and A. S. Perlin: *Carbohydr. Res.* **59**, 299 (1977).
15. B. Casu, M. Reggiani, G. G. Gallo, and A. Vigevani: *Tetrahedron* **22**, 3061 (1966).

16. M. St Jacques, P. R. Sundarajan, K. J. Taylor, and R. H. Marchessault: *J. Am. Chem. Soc.* **98**, 4386 (1976).
17. I. O. Sutherland: *Ann. Rep. NMR Spectrosc.* **4**, 41 (1971).
18. J. F. Stoddart: *The Stereochemistry of Carbohydrates*, Wiley Interscience (1971).
19. R. U. Lemieux, T. L. Nagabkushan, and B. Paul: *Can. J. Chem.* **50**, 773 (1970).
20. G. K. Hamer, F. Balza, N. Cyr, and A. S. Perlin: *Can. J. Chem.* **56**, 3109 (1978).
21. A. Bax and R. Freeman: *J. Am. Chem. Soc.* **104**, 1099 (1982).
22. B. Mulloy, T. A. Frienkiel, and D. B. Davies: *Carbohydr. Res.* **184**, 39 (1988).
23. B. Klar, B. Hingerty, and W. Saenger: *Acta Crystallogr.* **B36**, 1154 (1980).
24. For example; P. Dais, T. K. M. Shing, and A. S. Perlin: *J. Am. Chem. Soc.* **106**, 3082 (1984); P. Dais and A. S. Perlin: *Magn. Res. Chem.* **26**, 373 (1988); G. M. Lipkind, A. S. Shashkov, S. S. Mamgan, and N. K. Kochetkov: *Carbohydr. Res.* **181**, 1 (1988).
25. R. P. Veregin, C. A. Fyfe, R. H. Marchessault, and M. G. Taylor: *Carbohydr. Res.* **160**, 41 (1987); H. Saito, G. Izumi, T. Mamizuka, S. Suzuki, and R. Tabeta: *J. Chem. Soc. Chem. Commun.* 1386 (1982).
26. K. Bock, A. Brignole, and B. W. Sigurskjold: *J. Chem. Soc. Perkin Trans. 2* 1711 (1986); A. S. Shashkov, G. M. Lipkind, Y. A. Knirel, and N. K. Kochetkov: *Magn. Res. Chem.* **26**, 735 (1988).
27. J. Ronayne and D. H. Williams: *Ann. Rep. NMR Spectrosc.* **2**, 83 (1969).
28. Y. Yamamoto, M. Onda, Y. Takahishi, Y. Inoue, and R. Chûjô: *Carbohydr. Res.* **170**, 229 (1987).
29. D. Gutsche: *The Calixarenes*, Monographs in Supramolecular Chemistry, Series Ed. J. F. Stoddart, RSC, London, Ch. 4 (1990).
30. D. A. Rees: *J. Chem. Soc. (B)* 877 (1970); D. A. Rees and D. Thom: *J. Chem. Soc. Perkin Trans. 2* 191 (1977).
31. E. M. Kosower: in *Introduction to Physical Organic Chemistry*, Wiley, New York, (1968); C. Reichardt: *Angew. Chem. Int. Ed. Engl.* **4**, 29 (1965); E. L. Eliel and O. Hofer: *J. Am. Chem. Soc.* **95**, 8041 (1973).
32. D. J. Williams: unpublished observations.

Journal of Inclusion Phenomena and Molecular Recognition in Chemistry **12**: 151–158, 1992.

Högberg Compounds with a Functionalized Box-Like Cavity#

RICHARD J. M. EGBERINK[a], PETER L. H. M. COBBEN[a], WILLEM VERBOOM[a], SYBOLT HARKEMA[b] and DAVID N. REINHOUDT[a]*
Laboratories of Organic Chemistry[a] and Chemical Physics[b], University of Twente, P.O. Box 217, 7500 AE Enschede, The Netherlands.

(Received: 29 September 1990; in final form: 7 January 1991)

Abstract. Preparations of the Högberg compounds (**Ia-e**) with functionalized box-like cavities, designed for complexation of soft cations like silver and copper(II), are described. The structure of **Ia** was determined by X-ray analysis.

Key words. Högberg compounds, metacyclophane, preorganization.

Supplementary Data. Tables of positional and thermal parameters, bond distances and angles, and observed and calculated structure factors are deposited with the British Library as Supplementary Publication No. SUP 82115 (38 pages).

1. Introduction

After the first reports on the cyclic tetramers, obtained from the condensation of resorcinol and aldehydes were published [1, 2], a number of possible applications for these interesting 'Högberg compounds' has been recognized. Although several geometrical isomers can in principle be formed in this condensation, only two (Figure 1) have been isolated, generally in high yields. The all-*cis* (ccc) isomer of the macrocycle has a bowl-shaped cavity (Figure 1, face view) with eight oxygen atoms, that can be used for the complexation of neutral molecules like sugars and of quaternary ammonium ions [3, 4]. Cram *et al.* have synthesized the so-called cavitands by means of a four-fold ring closure with $BrCH_2Cl$, together with the carcerands, by coupling of two modified cavitands. These compounds are also capable of complexing neutral molecules [5, 6].

In most cases aliphatic aldehydes were preferred over arylaldehydes, probably because of the better solubility properties of the products formed. Although several arylaldehydes have been used in the condensation reaction [7], the only application of a modified arylaldehyde was reported by Beer *et al.* [8] who described the condensation of resorcinol with 4-formylbenzo-15-crown-5. These derivatives are able to bind bipyridinium ions in the crown ether moiety of the molecule.

However, the four aryl groups, which form a concave box-like cavity, in principle offer another possibility to complex cations, as we concluded from CPK model

This paper is dedicated to the memory of the late Dr C. J. Pedersen.
* Author for correspondence.

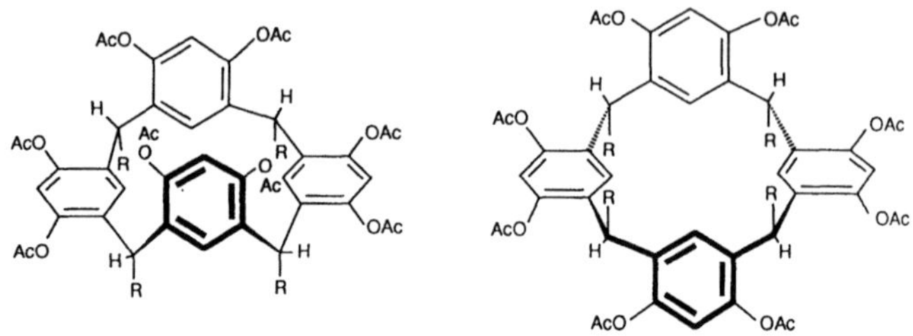

I (CCC) II (CTT)

R =

a 2 - thienyl
b 3 - (methylthio)phenyl
c 3 - methoxyphenyl
d 4 - (benzyloxy)phenyl
e 4 - hydroxyphenyl

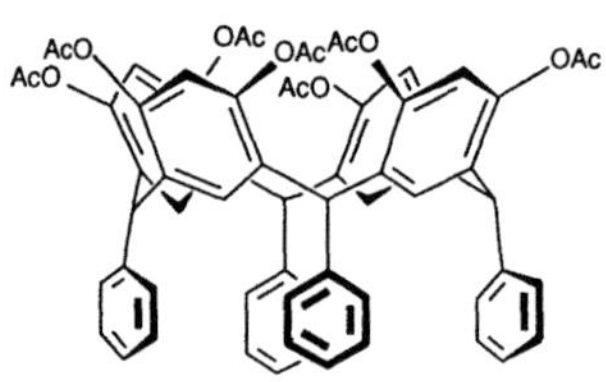

I (face view)

Fig. 1. Structures of the Högberg compounds.

studies. Suitable substituted (benz)aldehydes could, after condensation, yield Högberg compounds with a second set of donor atoms arranged in a more or less square planar fashion. Introduction of sulfur atoms as donor sites might offer a cavity which is able to complex the softer b and ab type cations, such as silver and copper(II). Furthermore, by choosing different substituted aldehydes, and modification of functional groups present in the aldehyde, the size of the box-like cavity can be varied in a simple way.

In this paper we report the synthesis of a number of such 'Högberg compounds' with differently functionalized box-like cavities.

2. Experimental

A small sample of **Ia** was recrystallized from $CHCl_3$ and its crystal structure was determined by X-ray diffraction. Crystal data: $C_{60}H_{48}O_{16}S_4 \cdot 3CHCl_3$, monoclinic, space group $C2/c$; $a = 23.242(2)$Å, $b = 15.864(4)$ Å, $c = 20.435(8)$ Å, $b = 110.86(1)°$; $V = 7041(6)$ Å^3; $T = 193(5)$K; $Z = 4$; $d = 1.43$ g cm^{-3}, $\mu = 5.35$ cm^{-1}. Reflections were measured in the $\omega/2\theta$ scan mode, using graphite monochromated Mo$K\alpha$ radiation [scan width (ω) $1.50 + 0.34 \tan\theta$]. The structure was solved by direct methods (MULTAN [9]) and refined with full-matrix least-squares methods. A total of 2591 reflections with $F_0^2 > 3\sigma(F_0^2)$ was used in the refinement. The molecule has a crystallographic two-fold axis. The C-atom of one of the chloroform molecules is on a two-fold axis implying disorder in the positions of the chlorine atoms. The five-membered rings occupy two positions related by a 180° rotation around the C—C bond, which connects them to the macrocycle. Hydrogen atoms were included in the refinement, except for the hydrogens of the disordered S/C atoms and the disordered chloroform molecule. The number of parameters refined was 522 (scale factor, positional parameters and thermal parameters (anisotropic for the non-hydrogen atoms, isotropic for H-atoms)). The final R factors were $R = 5.8\%$, $R_w = 7.0\%$. All calculations were performed with SDP [10].

All reactions were carried out under an argon atmosphere. FAB mass spectra were recorded using 3-nitrobenzyl alcohol as matrix. Silica gel (230–400 mesh ASTM) was obtained from Merck. Resorcinol was purchased from Merck-Schuchardt, and 2-thiophenecarboxaldehyde, 3-anisaldehyde and 4-(benzyloxy)benzaldehyde were obtained from Janssen Chimica. 3-(Methylthio)benzaldehyde was synthesized starting from 3-bromobenzaldehyde (Janssen Chimica), which was converted into its ethylene acetal. Lithiation and quenching with dimethyl disulfide, followed by deprotection with 2N HCl and distillation under reduced pressure, afforded the pure product. The method is similar to the one described by Euerby [11], who used the Grignard route to obtain the same compound. All newly synthesized compounds showed satisfactory elemental analyses.

2,8,14,20-*Tetra-2-thienyl-pentacyclo*$[19.3.1.1^{3,7}.1^{9,13}.1^{15,19}]$*octacosa*-1(25),3,5,7(28),9,11,13(27),15,17,19(26),21,23-*dodecaene*-4,6,10,12,16,18,22,24-*octol octaacetate* (*Stereoisomers*, **Ia** *and* **IIa**).

To a solution of resorcinol (11.01 g, 0.10 mol) in a mixture of ethyl alcohol (100 mL) and concentrated HCl (25 mL) was added 2-thiophenecarboxaldehyde (11.22 g, 0.10 mol) at room temperature over a period of 15 min. The color changed from orange to purple and after a few minutes a purple precipitate was formed. After stirring at room temperature for 3 h, water (200 mL) was added and the resulting suspension was centrifuged. The liquid was decanted and the residue was washed with water (100 mL) and centrifuged again. The residue was mixed with diethyl ether (100 mL), stirred and filtered. The crude product was dried *in vacuo*, and dissolved in acetic anhydride (75 mL). Pyridine (5 mL) was added and the mixture was stirred at 100°C (oil bath temperature) for 1 h. After cooling to room temperature the solvent was removed under reduced pressure. The residue was stirred with methanol (75 mL) and filtered to remove the last traces of acetic anhydride. Column chromatography (SiO_2, CH_2Cl_2:EtOAc 9:1) afforded the pure isomers **Ia** and **IIa** in yields of 12 and 22%, respectively.

Ia. mp > 300°C(CH_2Cl_2–MeOH). Mass spectrum (EI): (M^+) 1152 (*calc.* 1152). ^1H NMR: δ 7.11 and 6.96 (*s*, 4H, H_c), 7.1–7.0 (*m*, 4H, S—CH), 6.8–6.7 (*m*, 4H, S—CH=C*H*), 6.37 (*d*, 4H, S—C=CH), 6.31 and 6.17 (*s*, 4H, H_b), 5.68 (*s*, 4H, H_a), 2.12 and 2.02 (*s*, 24H, C[O]CH_3). ^{13}C NMR: δ 168.2 and 167.9 (*s*, C=O), 147.0, 146.9, 143.5, 132.6, 130.1 (*s*, aryl-C and thiophene-C), 130.7, 127.1, 127.0, 126.5, 126.4, 124.5, 117.4, 116.3 (*d*, aryl-CH, thiophene-CH and CH_a), 20.6 (*q*, C[O]CH_3).

IIa. mp > 300°C (CH_2Cl_2–MeOH). Mass spectrum (EI): (M^+) 1152 (*calc.* 1152). ^{1}H NMR: δ 7.14 and 6.98 (*s*, 4H, H_c), 7.00 (*dd*, 4H, $J = 5.2$ and 1.2 Hz, S—CH), 6.71 (*dd*, 4H, $J = 3.5$ and 5.2 Hz, S—CH=C*H*), 6.31 (*d*, 4H, $J = 2.7$ Hz, S—C=CH), 6.27 (*s*, 4H, H_b), 5.73 (*s*, 4H, H_a), 2.12 and 2.03 (*s*, 24H, CH_3). ^{13}C NMR: δ 168.3 and 168.1 (*s*, C=O), 146.9, 146.7, 142.2, 131.1, 130.9 (*s*, aryl-C and thiophene-C), 132.4, 128.0, 127.2, 126.8, 124.3, 117.2, 116.6 (*d*, aryl-CH, thiophene-CH, and CH_a), 20.6 (*q*, C[O]*C*H_3).

Synthesis of the Högberg compounds with 3-(methylthio)phenyl substituents (**Ib** *and* **IIb**)

To a solution of resorcinol (4.07 g, 0.037 mol) in a mixture of ethanol (30 mL), water (30 mL) and concentrated HCl (15 mL) was added 3-(methylthio)benzaldehyde (5.62 g, 0.037 mol) at room temperature over a period of 15 min. The reaction mixture was refluxed for 20 h. After cooling down to room temperature, the precipitate was collected by centrifugation and decanting of the liquid. The solid was washed with water and centrifuged again. The almost dry solid was mixed with diethyl ether (100 mL) and stirred. Filtration and successive drying *in vacuo* afforded the crude octol mixture which was acylated as described in the synthesis of **Ia**/**IIa**. After purification by column chromatography (SiO_2, CH_2Cl_2:EtOAc 4:1) **Ib** and **IIb** were isolated in yields of 26 and 51%, respectively.

Ib. mp 277–280°C (CH_2Cl_2–MeOH). Mass spectrum (FAB): *m*/*z* ($[M + 2H]^+$) 1314 (*Calc.* 1314). ^{1}H NMR: δ 7.15 and 6.87 (*s*, 4H, H_c), 7.05–6.9 (*m*, 8H, aryl-H), 6.7–6.3 (*m*, 8H, aryl-H), 6.06 and 5.80 (*s*, 4H, H_b), 5.36 (*s*, 4H, H_a), 2.32 (*s*, 12H, SCH_3), 2.06 and 1.99 (*s*, 24H, C[O]CH_3). ^{13}C NMR: δ 168.1 (*s*, C=O), 44.9 (*d*, CH_a), 20.7 (*q*, C[O]*C*H_3), 15.7 (*q*, SCH_3).

IIb. mp 289–292°C (dec) (CH_2Cl_2–MeOH). Mass spectrum (FAB): *m*/*z* ($[M + 2H]^+$) 1314 (*Calc.* 1314). ^{1}H NMR: δ 7.08 and 6.93 (*s*, 4H, H_c), 7.05–6.9 (*m*, 8H, aryl-H), 6.7–6.3 (*m*, 8H, aryl-H), 6.18 and 5.78 (*s*, 4H, H_b), 5.43 (*s*, 4H, H_a), 2.33 (*s*, 12H, SCH_3), 2.06 (*s*, 24H, C[O]CH_3). ^{13}C NMR: δ 168.5 and 168.1 (*s*, C=O), 44.6 (*d*, CH_a), 20.6 (*q*, C[O]*C*H_3), 15.7 (*q*, SCH_3).

Synthesis of the Högberg compounds with 3-methoxyphenyl substituents (**Ic** *and* **IIc**)

Starting from resorcinol (5.51 g, 0.05 mol) and 3-anisaldehyde (6.81 g, 0.05 mol) compounds **Ic** and **IIc** were prepared analogously to the synthesis of **Ib** and **IIb**. The crude mixture was purified by chromatography (SiO_2, CH_2Cl_2:EtOAc 9:1) to give **Ic** (25%) and **IIc** (49%).

Ic. mp > 300°C (dec) (CH_2Cl_2–MeOH). Mass spectrum (FAB): *m*/*z* ($[M + 2H]^+$) 1250 (*Calc.* 1250). ^{1}H NMR: δ 7.13 and 6.87 (*s*, 4H, H_c), 7.0–6.9 (*m*, 4H, aryl-H), 6.58 (*dd*, 4H, $J = 8.1$ and 2.2 Hz, aryl-H), 6.4–6.05 (*m*, 8H, aryl-H), 6.08 and 5.85 (*s*, 4H, H_b), 5.37 (*s*, 4H, H_a), 3.62 (*s*, 12H, OCH_3), 2.05 and 2.00 (*s*, 24H, C[O]CH_3). ^{13}C NMR: δ 168.2 (*s*, C=O), 55.7 (*q*, OCH_3), 44.9 (*d*, CH_a), 20.7 and 20.5 (*q*, C[O]*C*H_3).

IIc. mp 286–289°C (dec) (CH_2Cl_2–MeOH). Mass spectrum (FAB): *m*/*z* ($[M + 2H]^+$) 1250 (*Calc.* 1250). ^{1}H NMR: δ 7.0–6.9 (*m*, 8H, H_c and aryl-H), 6.65–6.55 (*m*, 4H, aryl-H), 6.35–6.15 (*m*, 4H, aryl-H), 6.22 and 5.87 (*s*, 4H, H_b),

5.44 (*s*, 4H, H_a), 3.62 (*s*, 12H, OCH_3), 2.05 (*s*, 24H, $C[O]CH_3$). ^{13}C NMR: δ 168.5 and 168.1 (*s*, C=O), 54.8 (*q*, OCH_3), 44.6 (*d*, CH_a), 20.6 (*q*, $C[O]CH_3$).

Synthesis of the Högberg compounds with 4-(*benzyloxy*)*phenyl substituents* (**Id** *and* **IId**)

Starting from resorcinol (5.51 g, 0.05 mol) and 4-(benzyloxy)benzaldehyde (10.62 g, 0.05 mol) compounds **Id** and **IId** were prepared analogously to the preparation of **Ib** and **IIb**. Chromatography (SiO_2, CH_2Cl_2:EtOAc 4:1) afforded **Id** (26%) and **IId** (50%).

Id. mp 232–235°C (dec) (CH_2Cl_2–MeOH). Mass spectrum (EI): (M^+) 1552 (*Calc.* 1552). ^1H NMR: δ 7.4–7.2 (*m*, 20H, aryl-H), 7.10 and 6.85 (*s*, 4H, H_c), 6.8–6.6 (*m*, 8H, aryl-H), 6.11 and 6.04 (*s*, 4H, H_b), 5.34 (*s*, 4H, H_a), 4.97 (*s*, 8H, OCH_2Ar), 2.05 and 2.00 (*s*, 24H, $C[O]CH_3$). ^{13}C NMR: δ 168.2 (*s*, C=O), 70.0 (*t*, OCH_2Ar), 44.2 (*d*, CH_a), 20.6 and 20.5 (*q*, $C[O]CH_3$).

IId. mp > 300°C (CH_2Cl_2–MeOH). Mass spectrum (EI): (M^+) 1552 (*Calc.* 1552). ^1H NMR: δ 7.35–7.2 (*m*, 20H, aryl-H), 7.14 and 6.90 (*s*, 4H, H_c), 6.8–6.6 (*m*, 8H, aryl-H), 6.29 and 6.14 (*s*, 4H, H_b), 5.45 (*s*, 4H, H_a), 4.94 (*s*, 8H, OCH_2Ar), 2.06 and 2.01 (*s*, 24H, $C[O]CH_3$). ^{13}C NMR: δ 168.5 and 168.2 (*s*, C=O), 69.9 (*t*, OCH_2Ar), 43.9 (*d*, CH_a), 20.6 and 20.5 (*q*, $C[O]CH_3$).

Synthesis of the Högberg compound with 4-*hydroxyphenyl substituents* (**Ie**)

To a solution of **Id** (0.90 g, 0.58 mmol) in a mixture of benzene (25 mL, thiophene-free) and methanol (10 mL) was added 10% Pd on C (20 mg) and the mixture was stirred for 18 h at room temperature under a hydrogen atmosphere. Filtration and subsequent evaporation of the solvent afforded a nearly quantitative yield of **Ie**. mp 271–74°C (CH_2Cl_2). Mass spectrum (EI): (M^+) 1192 (*Calc.* 1192). ^1H NMR (DMSO-d_6): δ 6.86 and 6.70 (*s*, 4H, H_c), 6.56 (*d*, 8H, $J = 8.8$ Hz, aryl-H), 6.47 (*d*, 8H, $J = 8.8$ Hz, aryl-H), 6.15 and 6.03 (*s*, 4H, H_b), 5.17 (*s*, 4H, H_a), 1.94 (*s*, 24H, $C[O]CH_3$). ^{13}C NMR: δ 168.0 (*s*, C=O), 44.1 (*d*, CH_a), 20.5 (*q*, $C[O]CH_3$).

3. Results and Discussion

Thiophene sulfur atoms are not commonly used as donor sites in macrocyclic chemistry. However, Lucas *et al.* [12] showed that the thiophene sulfur atom can be used as a binding site for copper(II) ions. CPK model studies indicated that the ccc isomer of a Högberg compound with four 2-thienyl substituents (**Ia**) could form an almost square plane of sulfur atoms (diametrical S—S distance 5.5–6.0 Å) for complexation of cations like silver and copper(II). Condensation of 2-thiophenecarboxaldehyde and resorcinol yielded a mixture of ccc and ctt isomers (**Ia** and **IIa**) which were isolated as the corresponding octaacetates in a total yield of 34%. The ratio between ccc and ctt was approximately 1:2, as followed from the integrals of the H_a absorptions (5.68 and 5.73 ppm, respectively) in the ^1H NMR spectrum (Figure 2). After separation of the isomers by chromatography, the structure of the ccc isomer was unambiguously established by X-ray diffraction (Figure 3). The X-ray diffraction experiment showed that in the solid state the thiophene rings

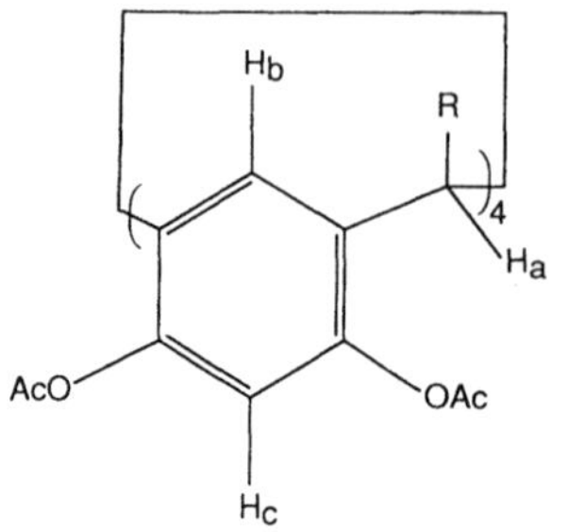

Fig. 2. Schematic drawing of a Högberg compound.

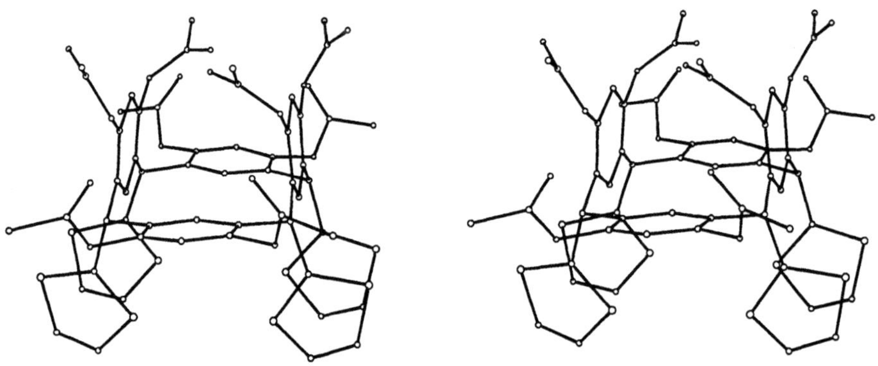

Fig. 3. Stereo view of **1a**.

occupy two positions related by a rotation of 180° around the 'exocyclic' bond. The occupancy of both positions is not the same. For one of the two crystallographically independent sulfur atoms the probability of being inside the cavity is approximately 50%, for the other this number is 30%. The distances between the partially occupied adjacent S-positions within the cavity are 3.949(7) and 4.963(5) Å.

From CPK models we concluded that the ccc isomer **Ib** of a Högberg compound derived from 3-(methylthio)benzaldehyde would have a plane of sulfur atoms with a diametrical S—S distance of about 4.0–4.5 Å. The condensation of 3-(methylthio)benzaldehyde and resorcinol gave (after acylation and purification by chromatography) **Ib** and **IIb** in a yield of 77% with a ccc:ctt ratio of 1:2. The structures of the isomers were proven by dynamic NMR spectroscopy. Previously, Högberg has shown that at elevated temperatures, in general a ccc isomer **I** can equilibrate with the two C_{2v} structures, which results on the NMR time scale in the formation of a C_{4v} isomer [1]. This C_{4v} isomer shows one singlet for the H_b protons (Figure 2), whereas the C_{2v} isomer exhibits two singlets of equal intensity for these protons. The ctt isomer **II** cannot equilibrate because it is much more rigid. Therefore compounds **II** exhibit, even at elevated temperatures, two singlets of equal intensity for H_b. Compound **Ib** shows two singlets (6.05 and 5.90 ppm) for H_b at room temperature in DMSO-d_6; at approximately 100°C these change into one broad singlet. The ctt isomer **IIb** exhibits two singlets at room temperature for H_b (6.20 and 5.85 ppm) which showed no coalescence upon heating to 105°C.

The corresponding oxygen derivative, synthesized from 3-methoxybenzaldehyde, also gave a 1:2 mixture of ccc and ctt isomers **Ic** and **IIc**, respectively, isolated as their acetates in a total yield of 74%. Characterization of the isomers was performed in the same way as for the 3-methylthio derivatives (**Ib** and **IIb**). Compound **Ic** at room temperature in DMSO-d_6 exhibits two singlets for H_b (6.06 and 5.95 ppm) that change into one broad singlet at 100°C. Compound **IIc** even at 100°C in DMSO-d_6, shows two singlets of equal intensity for H_b at 6.23 and 5.95 ppm. Condensation of 3-bromobenzaldehyde or 3-(benzyloxy)benzaldehyde with resorcinol, which would give macrocycles with cavities which could more easily be modified by standard methods, in our hands gave only polymeric material, from which no Högberg compound could be isolated. This may be caused by the fact that the bromo- and the benzyloxy-substituents are too large, thus favoring linear polymerization.

Condensation of 4-(benzyloxy)benzaldehyde and resorcinol followed by acylation with acetic anhydride yielded 76% of a 1:2 mixture of the ccc and ctt acetates **Id** and **IId**, respectively. After isolation of the ccc isomer **Id** via column chromatography, this compound was debenzylated with hydrogen on Pd/C (10%) in a nearly quantitative yield. The product still had the ccc configuration as followed from the ^{1}H NMR spectrum which, at room temperature in DMSO-d_6, exhibits two singlets for H_b at 6.15 and 6.03 ppm, which show coalescense upon heating up to 100°C. Compound **Ie** can be further modified by choosing the appropriate alkylating or acylating agents to give macrocycles with different cavities.

In the present paper we have discussed that Högberg compounds with a functionalized box-like cavity are easily accessible. In the near future, we will investigate the further modification of this cavity and its complexation properties.

Acknowledgement

This investigation was financially supported by the Netherlands Technology Foundation (STW), Future Technical Science Branch of the Netherlands Organization for Advanced and Pure Research (NWO). We also acknowledge J. M. Visser and J. L. M. Vrielink for recording the NMR spectra, T. W. Stevens for recording the mass spectra, A. Montanaro-Christenhusz for performing the elemental analyses, and J. G. Vrieze for his help in the crystal structure determination.

References

1. A. G. S. Högberg: *J. Am. Chem. Soc.* **102**, 6046 (1980).
2. A. G. S. Högberg: *J. Org. Chem.* **45**, 4498 (1980).
3. Y. Aoyama, Y. Tanaka, and S. Sugahara: *J. Am. Chem. Soc.* **111**, 5397 (1989).
4. H.-J. Schneider, D. Güttes, and U. Schneider: *J. Am. Chem. Soc.* **110**, 6449 (1988).
5. D. J. Cram, S. Karbach, H.-E. Kim, C. B. Knobler, E. F. Maverick, J. L. Ericson, and R. C. Helgeson: *J. Am. Chem. Soc.* **110**, 2229 (1988).
6. D. J. Cram, S. Karbach, Y. H. Kim, L. Baczynskyj, K. Marti, R. M. Sampson, and G. W. Kalleymeyn: *J. Am. Chem. Soc.* **110**, 2554 (1988).
7. J. A. Tucker, C. B. Knobler, K. N. Trueblood, and D. J. Cram: *J. Am. Chem. Soc.* **111**, 3688 (1989).
8. P. D. Beer, E. L. Tite, and A. Ibbotson: *J. Chem. Soc., Chem. Commun.* 1874 (1989).
9. G. Germain, P. Main, and M. M. Woolfson: *Acta Crystallogr.* **A27**, 368 (1971).

10. *Structure Determination Package*: B. A. Frenz and Associates, Inc., College Station, TX, and Enraf-Nonius, Delft (1983).
11. M. Euerby and R. D. Waigh: *Synth. Commun.* **11**, 849 (1981).
12. C. R. Lucas, S. Liu, M. J. Newlands, J.-P. Charland, and E. J. Gabe: *Can. J. Chem.* **67**, 639 (1989), Another example of thiophene used in complexation was published by: M. G. B. Drew, P. C. Yates, J. Trocha-Grinshaw, K. P. McKillop, and S. M. Nelson: *J. Chem. Soc., Chem. Commun.* 262 (1985). For more general information about thiophenes used in macrocyclic chemistry, see: G. R. Newkome, J. D. Sauer, J. M. Roper, and D. C. Hager: *Chem. Rev.* **77**, 513 (1977).

Journal of Inclusion Phenomena and Molecular Recognition in Chemistry **12**: 159–173, 1992.

Influence of Quaternary Onium Salts, Crown Ethers and Cryptands on Olefin Epoxidations Promoted by $HOCl/ClO^-$ in the Presence of Mn(III)-tetrakis(2,6-dichlorophenyl)porphyrin Chloride[#]

STEFANO BANFI, FERNANDO MONTANARI*, and SILVIO QUICI
Centro CNR and Dipartimento di Chimica Organica e Industriale dell'Università, Via Golgi 19, 20133 Milano, Italy.

GAGUIK TOROSSIAN
Chemical Department, Polytechnical Institute, 375009 Erevan, Armenia, USSR.

(Received: 5 November 1990; in final form: 11 December 1990)

Abstract. Reaction rates of alkene epoxidations, promoted by aqueous NaOCl and catalyzed by Mn(III)-tetrakis(2,6-dichlorophenyl)porphyrin chloride **1** (P) in the presence of a lipophilic axial ligand (L) (e.g. *N*-hexylimidazole) and carried out under H_2O/CH_2Cl_2 two phase conditions at 0°C, are strongly enhanced by lowering the pH of the aqueous phase from 12.7 to 9.5. Under these conditions, a further relevant increase in the reaction rates can be achieved by adding phase-transfer catalysts (PT), e.g. quaternary ammonium salt **3**, lipophilic crown ether **4** or cryptand **5**, provided that the amount of L is very small (L/P = 1 for very reactive alkenes, e.g. cyclooctene, and 10 for poorly reactive ones, e.g. 1-dodecene). In the case of cyclooctene epoxidation, the use of 0.006–0.03 mol. equiv. of PT completes the reaction in 1–10 min., the initial rates being up to 600 turnovers/min. with $(2.2.2,C_{10})$ cryptand. In the absence of the axial ligand, the quaternary ammonium salt **3** and cryptand **5** show an inhibitory effect. Such an effect is due to the formation of the poorly reactive Mn(P)Cl species, by Cl^- extraction to the organic phase. However, dibenzo crown ether **4** does not show this effect. In the presence of **4**, and with L/P = 1, the 1-dodecene epoxidation reaches 94% in 1 min. The 'unique' behavior of crown ethers can be explained by their ineffectiveness in extracting alkali chlorides, providing a very low concentration of Cl^- in the organic phase and thus avoiding the Mn(III)-porphyrin deactivation.

Key words. Mn(III)-tetraarylporphyrin, olefin epoxidations, $HOCl/ClO^-$, two-phase conditions, influence of phase-transfer catalysts, effect of Cl^-, competitive extraction of Cl^- and ClO^-, quaternary onium salts, crown ethers, cryptands.

1. Introduction

The use of synthetic metalloporphyrins as biomimetic catalysts in organic syntheses has aroused growing interest in recent years [1]. In hydrocarbon oxygenations, a great effort has been made to combine the catalytic efficiency and the chemical stability of the porphyrin. However, in only a few cases have satisfactory turnover numbers been achieved, even when particularly robust porphyrins were used [2, 3].

Aqueous sodium hypochlorite is one of the most versatile of a great variety of oxidants used as single oxygen atom donors. It was first used by Tabushi for alkene

[#] This paper is dedicated to the memory of the late Dr C. J. Pedersen.

* Author for correspondence

epoxidations under aqueous/organic phase-transfer conditions [4]. Later on, Meunier [5] increased the reaction rates by adding huge amounts of pyridines or imidazoles as axial ligands (Equation 1).

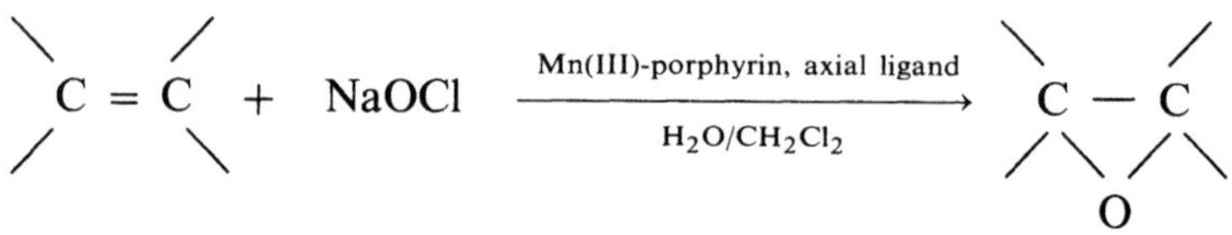

(1)

We have found [2a, b, 6] that a further strong enhancement of the reaction rates is obtained by lowering the pH of the NaOCl aqueous phase from 12.7 to 9.5. At pH lower than 9.5 the concentration of Cl_2 increases [7] so that alkene chlorination becomes unacceptably high [2a, b, 6].

At a pH buffered in the range 9.5–10.5 a significant amount of HOCl ($pK_a = 7.54$) is extracted from the aqueous phase into CH_2Cl_2 where the reaction occurs. HOCl becomes the effective oxygen donor, hence the presence of a phase-transfer catalyst is not strictly necessary [2a, b, 6]. The use of lipophilic axial ligands, such as *N*-hexylimidazole or 4-*tert*-butylpyridine, allowed us to drastically reduce the ligand/porphyrin ratio, and this is very important from the practical point of view [2a, b]. With unhindered electron-rich alkenes, e.g. *cis*-cyclooctene, the reaction proceeds even in the absence of an axial ligand; under such conditions 100 000 overall turnovers were reached without any appreciable degradation of the metalloporphyrin [8]. The results obtained using chemically robust porphyrins with reactive substrates indicate that this catalytic system could be employed on a large scale.

However, in spite of these improvements several aspects still remained unclear:

(i) Other authors [9] found only modest increases of reaction rates by lowering the pH of the NaOCl aqueous phase when working with an excess of imidazole or pyridine with respect to the porphyrin in the presence of quaternary ammonium salts.

(ii) Our preliminary data [2a] indicated that, in the absence of axial ligands, quaternary ammonium salts had an inhibiting effect on the reactions, even at low pH and with reactive substrates.

(iii) Using dibenzo-18-crown-6 as the phase-transfer catalyst, we had found [10], at pH 9.5–10.5 and with very small amounts of axial ligand, that poorly reactive alkenes could be epoxidized in a few minutes and with good selectivity. Under the same conditions quaternary ammonium salts once more showed an inhibiting effect.

These aspects prompted us to further investigate alkene epoxidations promoted by $HOCl/ClO^-$ and catalyzed by chemically robust Mn(III)-porphyrins. Particular attention was given to the influence of quaternary ammonium salts, crown ethers and cryptands [11] on reaction rates, both in the presence and absence of the axial ligand.

2. Experimental

2.1. GENERAL

UV–vis spectra were obtained with a Perkin-Elmer Lambda 6 spectrophotometer. Potentiometric titrations were performed with a Metrohm 670 Titroprocessor equipped with a Metrohm 665 Dosimat. GC analyses were performed on a Varian model 3700 gas chromatograph flame ionization instrument (20 × 0.125 in OV-101-5% on CHP 100–125 mesh column), with VISTA CDS 401 Varian chromatography data system. Oxidations were carried out in a 10 mL flask equipped with a Teflon-lined screw cap and magnetic stirrer, thermostatted at 0 ± 0.2°C with circulating ethanol by a Colora Misstechnick GMBH Lorch/Württ cryostat. Stirring speed was maintained at 1300 ± 50 rpm. The molarity of the NaOCl aqueous solution was determined by iodometric titration and its pH was measured with a Orion pH meter model SH 250 with pH electrode Model 91–03 (semimicro glass body). The alkenes and the phase transfer catalysts were of the highest purity commercially available and were used without further purification. Mn(III)-tetrakis(2,6-dichlorophenyl)porphyrin chloride **1** was synthesized according to a published procedure [12]. Pyrrole and boron trifluoride etherate were distilled before use. CH_2Cl_2 was distilled from $CaCl_2$ when used as solvent in the synthesis of the porphyrin.

2.2. GENERAL PROCEDURE OF ALKENE EPOXIDATION

The flask was charged with: (a) 1 mL of CH_2Cl_2 solution containing 0.5 mmol (S/P = 200) or 2.5 mmol (S/P = 1000) of substrate and 0.25 mmol (or 1.25 mmol) of decane as internal standard; (b) 1 mL of 0.0025M CH_2Cl_2 solution of Mn(III)-porphyrin **1**; (c) the required amount of *N*-hexylimidazole was added with a microsyringe; 10 μL (0.25M) and 50 μL (1.25M) of CH_2Cl_2 solutions for L/P = 1 and 25, respectively; (d) 50 μL of a 0.3M CH_2Cl_2 solution of phase-transfer catalyst (PT/P = 6). Aqueous 0.40M NaOCl solution (4.4 mL), buffered at pH 10.5 or 9.5 by adding 130 mg or 265 mg of solid $NaHCO_3$ to 20 mL of NaOCl (initial pH = 12.7), was then layered over the organic phase. The mixture was stirred and samples were taken at different times and analyzed by GC.

2.3. STABILITY OF Mn-PORPHYRINS 1

In the epoxidation experiments, a 50 μL sample of the organic phase was withdrawn before the addition of the ligand and diluted in 5 mL of CH_2Cl_2 (zero time). Samples (50 μL) were withdrawn at different times and diluted in 5 mL of CH_2Cl_2. Mn-porphyrin decomposition was followed by UV-vis spectroscopy in the 350–700 nm range, measuring the percentage decrease of the absorbance at the λ_{max} referred to the sample taken at zero time.

2.4. POTENTIOMETRIC ClO^- AND Cl^- TITRATION OF AQUEOUS NaOCl SOLUTION

(a) Two mL of 0.04 NaOCl was diluted with 20 mL of distilled water, acidified with 10% HNO_3 and titrated with 13.42 mL of 10^{-2}M $AgNO_3$ solution. This corresponds to 0.067M Cl^- initially present.

(b) Two mL of 0.04M NaOCl were diluted with 20 mL of distilled water, acidified with 10% HNO_3, treated with 0.14 g (1.2 mmol) of Na_2SO_3 and titrated with 21.37 mL of 10^{-2}M $AgNO_3$ solution. This corresponds to 0.107M Cl^- derived from the reduction of ClO^- plus the Cl^- initially present. The ClO^- concentration

(0.04M) was given by the difference between these two values. The molarity of ClO^- found by this method corresponds exactly to that obtained by iodometric titration.

2.5. POTENTIOMETRIC TITRATION OF ClO^- AND Cl^- EXTRACTED IN THE CH_2Cl_2 SOLUTION BY PHASE-TRANSFER CATALYSTS **3**–**5** AT 0°C

Ten mL of 0.4M aqueous NaOCl at pH 12.7 or 9.5 was stirred together with 25 mL of 8.0×10^{-2}M CH_2Cl_2 solution of PT and after 15 min the phases were separated. Two aliquots of 10 mL of the organic layer were poured into about 30 mL of methanol and titrated with 10^{-2}M solution of $AgNO_3$ in the presence and in the absence of the reducing agent, as described above. Results are reported in Table III.

3. Results

Alkene epoxidations were carried out at 0°C under CH_2Cl_2/H_2O two-phase conditions with 0.35–0.7M aqueous NaOCl (OX) at pH = 9.5–12.7 in the presence of Mn(III)-tetrakis(2,6-dichlorophenyl)porphyrin chloride **1** (P) [3a, 12] and with *N*-hexylimidazole **2** as the axial ligand (L). Three commercially available phase-transfer catalysts (PT) were used: tetrabutylammonium hydrogensulphate **3**, dibenzo-18-crown-6 **4**, and (2.2.2,C_{10}) cryptand **5**. *cis*-Cyclooctene and 1-dodecene were used as model substrates (S) of very reactive and poorly reactive alkenes, respectively.

3.1. CYCLOOCTENE EPOXIDATION (TABLE I)

At pH 12.7, 0°C and in the absence of both phase transfer catalyst and axial ligand, only 30% conversion occurs within 90 min. using 5×10^{-3} mol. equiv. of Mn(III)porphyrin **1** (entry 1). The addition of 3×10^{-2} mol. equiv. of quaternary ammonium salt **3** produces only 7% conversion in 1 h (entry 6). In both cases the presence of *N*-hexylimidazole (L/P = 1 and 25, respectively) brings about a modest increase of the reaction rate (51% and 11% conversion, respectively, in 45 min. [13], entries 2, 6, 7).

Lowering the pH of the NaOCl aqueous phase (9.5–10.5) leads to much faster reaction rates [2a, b, 6] and makes the comparison of the different phase transfer catalysts more reliable.

At pH 10.5 and molar ratios P : L : PT : S : OX = 1 : 1 : 6–10 : 200 : 700 reactions are complete in 2–7 min. at 0°C with very high selectivity (entries 9, 10, 13). Cryptand **5** is the most active PT catalyst and, with S/P = 1000, the epoxidation is over in 3 min. (350 turnovers in the first min., entry 17).

The highest rates are observed at pH 9.5 (entries 10, 11) although generally a lower selectivity is obtained. At this pH the difference in efficiency of PT catalysts **3**–**5** becomes more evident with S/P = 1000, the order being **5** $\gg$ **4**–**3** $\gg$ no PT (entries 14–16, 18, and Figure 1). In particular, with cryptand **5**, 600 turnovers are obtained in the first min. A comparison of the 18-crown-6 and its dibenzo-derivative **4** (entries 10, 12) highlights the importance of the liphophilicity of PT catalysts.

Table I. Epoxidation od cyclooctene.[a,b]

No.	pH	L/P	PT	PT/P	S/P	OX/P	React. time (min.)	Conv. (%)	Selec. (%)	Turnovers[c] (at 1 min.)	Residual P (%)[d]
1	12.7	–	–	–	200	700	90	30	100	–	100
2	12.7	1	–	–	200	700	45	51	100	–	100
3	10.5	–	–	–	200	700	50	100	90	4	100
4	10.5	1	–	–	200	700	13	97	95	52	100
5	9.5	1	–	–	200	700	7	100	85	66	95
6	12.7	–	**3**	12	200	700	60	7			
7	12.7	25	**3**	6	200	700	45	11			
8	10.5	–	**3**	6	200	700	5[e]	17			
9	10.5	1	**3**	6	200	700	7	100	96	60	100
10	10.5	1	**4**	10	200	700	7	100	100	60	85
11	9.5	1	**4**	6	200	700	2	96	95	94	79
12	10.5	1	18-C-6[f]	10	200	700	20	95	100	48	77
13	10.5	1	**5**	6	200	700	2	100	100	170	70
14	9.5	1	–	–	1000	2000	50	97	92	170	86
15	9.5	1	**3**	6	1000	2000	15	96	90	220	73
16	9.5	1	**4**	6	1000	2000	10	95	90	210	70
17	10.5	1	**5**	6	1000	2000	3	98	93	350	75
18	9.5	1	**5**	6	1000	2000	2	100	85	600	50

[a]In CH_2Cl_2–H_2O, 2 : 5 v/v, 0°C.

[b]S = cyclooctene, P = Mn(III)-porphyrin chloride **1**, L = *N*-hexylimidazole, PT = phase-transfer catalyst, OX = ClO^-/HOCl.

[c]Based on the converted substrate.

[d]At the end of the reaction.

[e]The reaction stops at this time.

[f]18-crown-6.

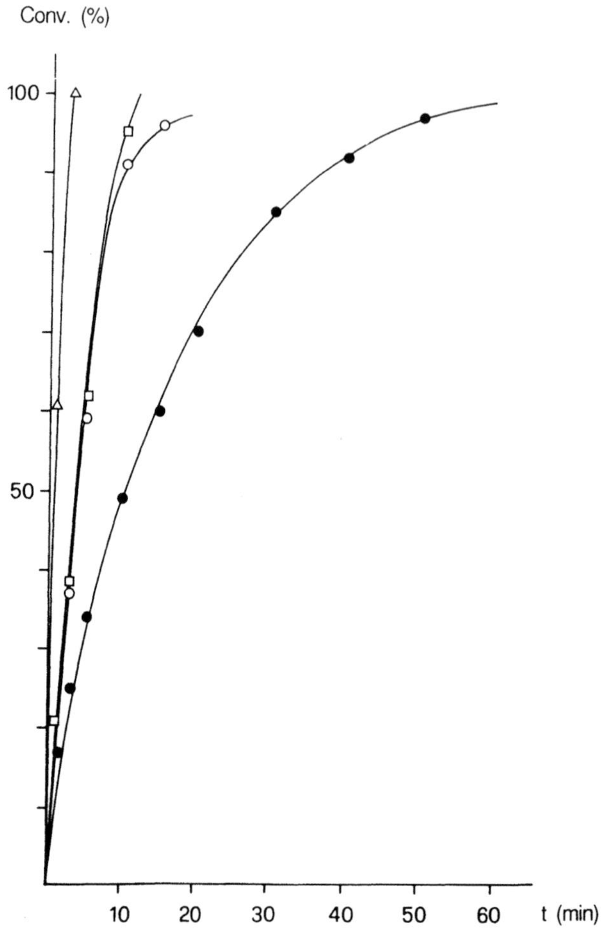

Fig. 1. Influence of phase-transfer catalyst (PT) on the epoxidation of cyclooctene (S) with $HOCl/ClO^-$ (OX), catalyzed by Mn(III)-porphyrin **1** (P) and *N*-hexylimidazole (L) at pH 9.5 and 0°C: absence of PT catalyst (●); $Bu_4N^+ HSO_4^-$ **3** (○); dibenzo-18-crown-6 **4** (□); [2.2.2,C_{10}]cryptand **5** (△). Reagent molar ratio P : L : PT : S : OX = 1 : 1 : 6 : 1000 : 2000.

As previously reported [2a], at pH 9.5 and 10.5 the reactions also proceed in the absence of PT, complete conversion being obtained in 13 and 21 min., respectively (entries 4, 5). In the absence of both axial ligand and PT, the cyclooctene epoxidation is much slower, the reaction being over in 55 min. at pH 10.5 (entry 3). Surprisingly, the addition of the quaternary ammonium salt and cryptand inhibits the reaction (entry 8 and Figure 2). This inhibition is less evident with the crown-ether and complete conversion of cyclooctene is achieved in 70 min. at 0°C, thus the time taken is only slightly longer than that required in the absence of PT.

3.2. 1-DODECENE EPOXIDATION (TABLE II)

At pH = 9.5 in the absence of PT and with molar ratios P : L : S : OX = 1 : 1 : 200 : 700 the reaction is initially very fast, but it suddenly slows down after

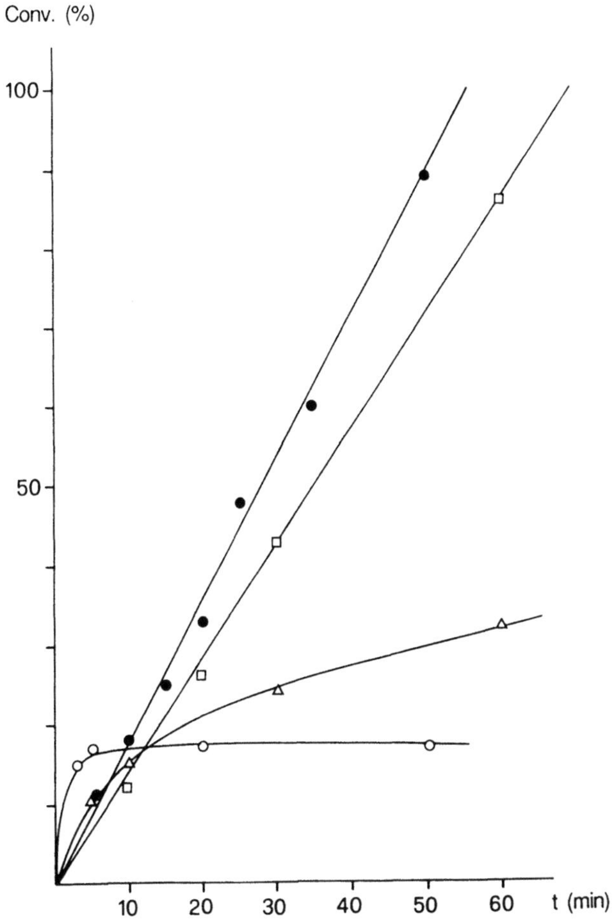

Fig. 2. Influence of phase-transfer catalyst (PT) on the epoxidation of cyclooctene (S) with HOCl/ClO^-(OX), catalyzed by Mn(III)-porphyrin **1** (P) in the absence of the axial ligand at pH 10.5 and 0°C: absence of PT catalyst (●); $Bu_4N^+HSO_4^-$ **3** (○); dibenzo-18-crown-6 **4** (□); [2.2.2,C_{10}] cryptand **5** (△). Reagent molar ratio P : PT : S : OX = 1 : 6 : 200 : 700.

5 min. at about 50% conversion, due to the oxidative destruction of the axial ligand [2a]. As already shown, only an excess of ligand (L/P = 25) allows complete conversion in 5 min. (148 turnovers in the first min.) and good selectivity (entries 1, 2). The catalytic activity decreases on adding the ammonium salt, but with 0.03 mol. equiv. of crown ether **4** 94 and 92% conversion is obtained in 1 and 3 min. at pH 9.5 and 10.5, respectively (188 and 106 turnovers in the first min., entries 3–7). Under these same conditions, cryptand **5** was found less efficient than **4**; however, when L/P = 10, the catalytic efficiency of **4** and **5** is the same and the reaction is nearly complete in 1 min. (entries 8–10 and Figure 3). At pH = 9.5, with **4** and S/P = 1000, 53% conversion is reached in 10 min., with initial rate of 440 turnovers/min. Without PT catalyst, conversion is 33% in 50 min. (entries 11, 12). As previously noted for cyclooctene epoxidation, the absence of the heterocyclic

Table II. Epoxidation of 1-dodecene.[a,b]

No.	pH	L/P	PT	PT/P	S/P	OX/P	React. time (min.)	Conv. (%)	Selec. (%)	Turnovers[c] (at 1 min.)	Residual P (%)[d]
1	9.5	1	–	–	200	700	30	67[e]	90	66	93 (30)
2	9.5	25	–	–	200	700	5	100	89	148	70 (30)
3	10.5	1	**3**	6	200	700	20[f]	49	85	40	75 (30)
4	9.5	1	**3**	6	200	700	1[f]	30	80	60	
5	9.5	10	**3**	6	200	700	1[f]	84	89	168	
6	10.5	1	**4**	6	200	700	3	92	98	106	68 (5)
7	9.5	1	**4**	6	200	700	1	94	100	188	78 (1)
8	9.5	10	**4**	6	200	700	3	96	93	182	
9	10.5	1	**5**	6	200	700	10	81	95	60	75 (10)
10	10.5	10	**5**	6	200	700	1	90	97	180	
11	9.5	1	–	–	1000	2000	50[f]	33	73	140	45 (50)
12	9.5	1	**4**	10	1000	2000	10[f]	53	86	440	0 (10)

[a]In CH_2Cl_2–H_2O, 2 : 5 v/v, 0°C.
[b]S = 1-dodecene, P = Mn(III)-porphyrin chloride **1**, L = *N*-hexylimidazole, PT = phase-transfer catalyst, OX = ClO^-/HOCl.
[c]Based on the converted substrate.
[d]Time (min.) in parentheses.
[e]50% conversion after 5 min.
[f]The reaction stops at this time.

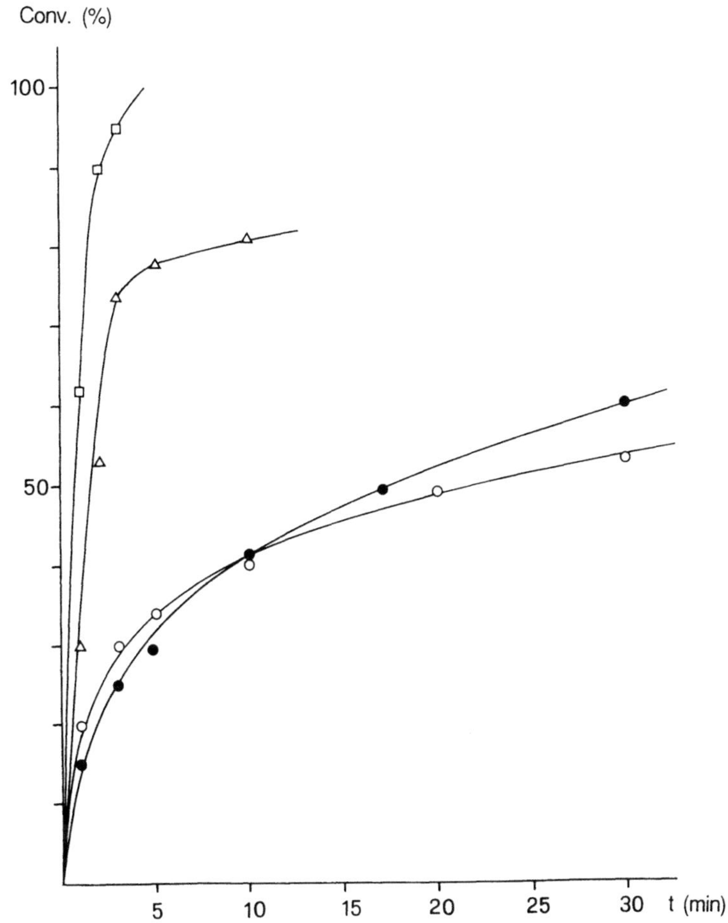

Fig. 3. Influence of phase-transfer catalyst (PT) on the epoxidation of 1-dodecene (S) with $HOCl/ClO^-$ (OX), catalyzed by Mn(III)-porphyrin **1** (P) and *N*-hexylimidazole (L) at pH 10.5 and 0°C: absence of PT catalyst (●); $Bu_4N^+HSO_4^-$ **3** (○); dibenzo-18-crown-6 **4** (□); [2.2.2,C_{10}]cryptand **5**(△). Reagent molar ratio P : L : PT : S : OX = 1 : 1 : 6 : 200 : 700.

nitrogen base causes the reactivity of Mn(III)-porphyrin to be inhibited by all the PT catalysts.

3.3. COMPETITIVE EXTRACTION OF $HOCl/ClO^-$ IN THE ORGANIC PHASE BY PT CATALYSTS 3–5

Extraction experiments were performed by stirring the aqueous NaOCl solution at pH 12.7 and 9.5 at 0°C, with the CH_2Cl_2 solution of PT catalyst **3**–**5**. After 10 min., two aliquots of the organic phase were titrated potentiometrically, one without and one with Na_2SO_3 as the reducing agent. Titration in the absence of Na_2SO_3 allows the direct determination of the (Cl^-PT^+) complex; titration of the second aliquot after its reduction with Na_2SO_3 gives $HOCl/ClO^-$ and Cl^-. The amount of $HOCl/ClO^-$ is obtained by the difference between the two values. Results are reported in Table III.

Table III. Competitive extraction of $HOCl/ClO^-$ and Cl^- in the organic phase by PT-catalysts **3**–**5**.[a,b]

PT[c]	pH	$HOCl/ClO^-$	Cl
3	12.7	70.3	11.6
3	9.5	106.3	20.6
4	9.5	2.5	1.5
5	9.5	117.0	31.5

[a]Ions % with respect to $(PT)^+$.
[b]Values measured after 10 min. stirring at 0°C of the aqueous solution of sodium hypochlorite (0.40M in NaOCl and 0.67M in NaCl) with CH_2Cl_2 (10 : 25 v/v).
[c]8×10^{-3}M in CH_2Cl_2.

4. Discussion

The proposed mechanism for the epoxidation catalyzed by Mn(III)-porphyrins and promoted by NaOCl either at pH = 12.7 or pH = 9.5–10.5 is depicted in Scheme 1.

The formation of the metal-oxene **8**, which is the oxidising species of the catalytic cycle [1], is favored by electron donation of the coordinated nitrogen base. When HOCl is the oxygen donor, the positive charge on the oxygen is the driving force for

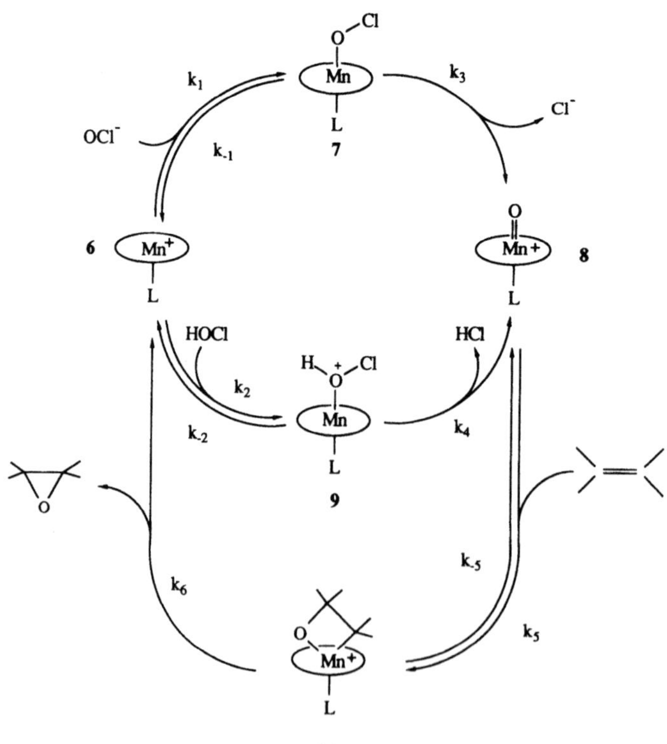

Scheme 1

the conversion of **9** into **8**. Also in this case, the presence of the axial ligand improves the reaction rate but it is not really necessary for the reaction to take place [2a].

The role of the axial ligand on the epoxidation rate can be rationalized through equilibria (2) and (3).

$$\mathrm{Mn(P)} + \mathrm{L} \overset{K_1}{\rightleftarrows} \mathrm{Mn(P)L} \quad (2)$$

$$\mathrm{Mn(P)L} + \mathrm{L} \overset{K_2}{\rightleftarrows} \mathrm{Mn(P)L_2} \quad (3)$$

$$\beta_2 = K_1 \cdot K_2 \quad (4)$$

The values of K_1 and β_2 are obtained by spectrophotometric measurements [2a, 6d, 14].

The concentration of Mn(P)L, which is by far more active with respect to the non-coordinated and bis-coordinated species Mn(P) and $\mathrm{Mn(P)L_2}$, can be calculated by Equation (5).

$$\mathrm{Mn(P)L} = \frac{K_1[\mathrm{L}][\mathrm{Mn(P)}]_0}{1 + K_1[\mathrm{L}] + \beta_2[\mathrm{L}]^2} \quad (5)$$

When the first derivative of Equation (5) with respect to [L] is equal to zero, $[\mathrm{L}]_{max}$ is obtained from Equation (6). $[\mathrm{L}]_{max}$ is the concentration of the free ligand at the equilibrium at the maximum Mn(P)L concentration.

$$[\mathrm{L}]_{max} = \sqrt{\frac{1}{\beta_2}} \quad (6)$$

Hence the optimum L/P initial ratio can easily be calculated in order to avoid an excess of axial ligand in the reaction medium. In fact, the non-coordinated nitrogen base behaves as a competitive substrate and its oxidation by the metal-oxene **8** is well documented [2a].

In this context phase-transfer catalysts play an intriguing role in the olefin epoxidations catalyzed by Mn(III)-porphyrin **1** and promoted by NaOCl; in fact, depending on the reaction conditions, quaternary ammonium salt and cryptand can either promote or inhibit the reaction, while the crown ether displays a particular behavior.

The epoxidations of cyclooctene and 1-dodecene with the aqueous NaOCl solution buffered at pH 9.5–10.5 indicate that quaternary salt **3** and the lipophilic cryptand **5** increase the reaction rate only if a heterocyclic nitrogen base is also present, the amount required depending on the reactivity of the substrate (L/P = 1 and 25 for cyclooctene and 1-dodecene, respectively; Tables 1 and II and Figure 1). However, in the absence of the axial ligand, the phase-transfer catalysts **3** and **5** show an evident inhibiting effect (Fig. 2).

On the other hand, in the presence of the axial ligand (L/P = 1), the crown ether **4** favors both the cyclooctene and 1-dodecene epoxidations, the effect with the former being similar to that of the quaternary salt **3**. With 1-dodecene, the crown-ether **4** is noticeably more effective than cryptand **5**, whereas with the same L/P ratio the quaternary salt **3** again displays an inhibiting effect (Figure 3). Only with L/P = 10 do cryptand **5** and crown ether **4** behave in a similar way (Table II).

The possible reason underlying the intriguing effect of catalysts **3–5** is to be found in the extraction values of Cl^- and ClO^- from NaOCl aqueous phase into CH_2Cl_2 at 0°C (Table III).

At pH 12.7 quaternary salt **3**, in the presence of a high molar excess (about 20 mol. equiv.) of aqueous NaOCl containing NaCl (about 30 mol. equiv.) transfers into CH_2Cl_2 70% of ClO^- and 11.3% of Cl^- with respect to the ammonium cation. At pH 9.5 the overall molar equivalents of the anionic species transported into the organic phase greatly exceed those of PT. At this pH a similar behavior is shown by the lipophilic cryptand **5**. This clearly indicates the presence of the associated species $(PT)^+X^- \cdot (HOCl)_n$ ($X^- = Cl^-, ClO^-$). Similar complexes have already been identified for a relevant number of acids such as HF [15], HCl [16], and CH_3COOH [16]. As far as HOCl is concerned, it was demonstrated by Sasson [17] that the oxidation of aldehydes by NaOCl under PT conditions (quaternary ammonium salt as PT catalyst) proceeds at the maximum rate at pH 9.5–11.0, the range where there is the highest extraction of ClO^- and HOCl into the organic phase [18]. Unlike **3** and **5**, crown ether **4** transfers only a very small amount of Cl^- and ClO^-. This behavior is in agreement with the values already reported for the extraction of halides and pseudohalides by quaternary onium salts, liphophilic crown ethers and cryptands [11, 15].

Based on the anion extraction values, the results of the epoxidations can be tentatively rationalized according to Scheme 2 [26].

In the presence of **3** or **5**, Mn-porphyrin **1**, the amount of which in the organic phase is 6–10 times lower than that of PT catalysts, is complexed by the highly coordinating anion Cl^- [23], and affords species **11**. The equilibria of this species with both **12a** and the corresponding protonated species **12b**, should be noticeably shifted towards **11**, in agreement with the observed very low catalytic activity of the porphyrin. On the contrary, if a lipophilic nitrogen base is present, **11** is easily transformed into **13** due to the electron donation of L on the metal [27].

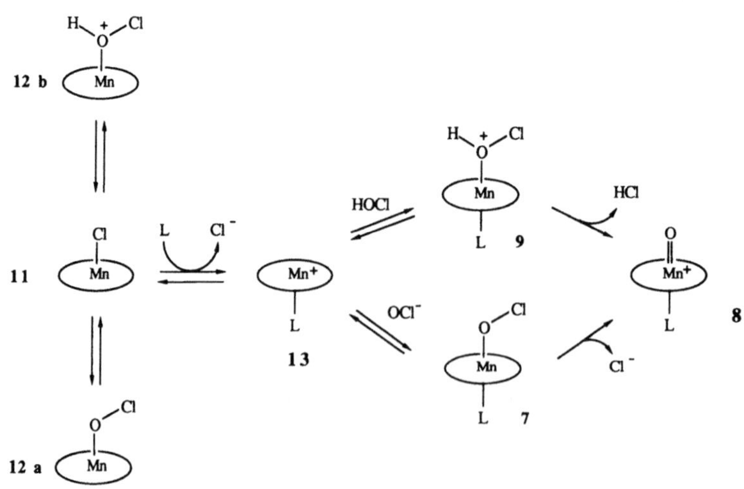

Scheme 2

The intermediate **13** affords **9** (or **7**), this process being favored by both the increased concentration of HOCl (or ClO^-), due to the presence of PT, and by the irreversible formation of the metal-oxene **8**. As the transformation of **7** into **8** is much slower than that of **9** to **8** ($k_3 \ll k_4$, Scheme 1), the catalytic activity at pH 12.7 is expected to be low, as is experimentally observed (Table 1).

As already observed, the extraction capability of crown ether **4** for ClO^- and Cl^- is very low, but with **4** at pH 9.5–10.5 and L/P = 1, the reactivity of Mn(III)-porphyrin **1** is always very high. Furthermore, at this pH range, **4** does not inhibit cyclooctene epoxidation in the absence of L, the reaction rate being only slightly lower than that observed without PT (Figure 2). These results can be explained by the ineffectiveness of **4** in extracting Cl^- anions into CH_2Cl_2; their concentration in the organic phase remains very low, thus limiting the deactivation of Mn(III)-porphyrin **1**.

Under the epoxidation conditions reported above, Mn(III)-porphyrin **1** can undergo partial demolition, its stability decreasing with the increasing efficiency of the catalytic system (Tables I and II).

A last comment concerns a recent observation [9] that there was only a small increase in reaction rate when the pH of the aqueous NaOCl is decreased. The explanation can be related to the very large excess of axial ligand used (4-picoline, L/P = 520, which is twice the amount of the substrate). This excess causes equilibrium (3) to shift to the right, producing almost exclusively the nonreactive bis-coordinated species $Mn(P)L_2$. In fact, we have measured the binding constants between Mn(III)-porphyrin **1** and 4-*tert*-butylpyridine ($k_1 = 420\ mol^{-1}$, $\beta_2 = 3.0 \times 10^5\ mol^{-2}$), following a previously reported procedure [2a, 6d]. At the porphyrin concentration ($2.5 \times 10^{-3}\ mol\ L^{-1}$) and with the amount of axial ligand (L/P = 520) used [9], the calculated $Mn(P)L_2$ percentage is 99.9. The binding constants between **1** and 4-picoline are most probably similar to those found for 4-*tert*-butylpyridine; even taking into account the partial solubility of 4-picoline in the NaOCl aqueous phases (20% with $CH_2Cl_2/H_2O = 25/75$ v/v, our measurements), the amount of $Mn(P)L_2$ at the equilibrium should still be higher than 99%.

Thus, with such an amount of ligand, the reactivity of Mn(III)-prophyrins is minimized [28], and a lowering of pH will have only a negligible effect on the reaction rate.

5. Conclusions

Conditions have been set up for the NaOCl alkene epoxidations catalyzed by the chemically robust Mn(III)-porphyrin **1** in the presence of catalytic amounts of phase transfer agents. Lipophilic crown ethers and cryptands are particularly efficient.

Reaction rates thus obtained are probably the highest reported to date, especially those found for poorly reactive substrates such as 1-dodecene.

Acknowledgement

This work was supported in part by the Ministero Pubblica Istruzione, Roma.

References

1. (a) J. E. Baldwin and P. Perlmutter: *Top. Curr. Chem.* **121**, 181 (1984); (b) P. R. Ortiz de Montellano: *Cytochrome P-450, Structure, Mechanism and Biochemistry*, Plenum Press, New York and London (1986); (c) B. Meunier: *Bull. Soc. Chim. France* 578 (1986); (d) D. Mansuy: *Pure Appl. Chem.* **59**, 759 (1987); (e) *ibid.* **62**, 741 (1990); (f) B. Morgan and D. Dolphin: *Struct. Bond.* **64**, 115 (1987); (g) B. Kräutler: *Chimia* **41**, 277 (1987); (h) I. Tabushi: *Coord. Chem. Rev.* **86**, 1 (1988); (i) T. C. Bruice: *Mol. Struct. Energ.* **9**, 227 (1988); (l) C. L. Hill (Ed.): *Activation and Functionalization of Alkanes*, Wiley, New York (1989); (m) D. Mansuy, P. Battioni and J. P. Battioni: *Eur. J. Biochem.* **184**, 267 (1989); (n) T. Okamoto, K. Sasaki and M. Tachibana: *Bull. Inst. Chem. Res., Kyoto Univ.* **67**, 169 (1989); (m) S. Banfi, A. Maiocchi, F. Montanari and S. Quici: *Chim. Ind.* **72**, 304 (1990).
2. (a) S. Banfi, F. Montanari and S. Quici: *J. Org. Chem.* **54**, 1850 (1989); (b) F. Montanari, S. Banfi and S. Quici: *Pure Appl. Chem.* **61**, 1631 (1989); (c) P. L. Anelli, S. Banfi, F. Montanari and S. Quici: *J. Chem. Soc., Chem. Commun.* 779 (1989); (d) S. Banfi, A. Maiocchi, A. Moggi, F. Montanari and S. Quici: *ibid.*, 1794 (1990).
3. (a) P. S. Traylor, D. Dolphin, and T. G. Traylor: *J. Chem. Soc., Chem. Commun.* 279 (1984); (b) T. G. Traylor, J. C. Marsters, T. Nakano and B. E. Dunlap: *J. Am. Chem. Soc.* **107**, 5537 (1985); (c) A. Robert and B. Meunier: *New J. Chem.* **12**, 885 (1988); (d) C. Querci and M. Ricci: *J. Chem. Soc., Chem. Commun.* 889 (1989); (e) P. Hoffman, G. Labat, A. Robert and B. Meunier: *Tetrahedron Lett.* **31**, 1991 (1990).
4. I. Tabushi and N. Koga: *Tetrahedron Lett.* 3681 (1979).
5. (a) E. Guilmet and B. Meunier: *Tetrahedron Lett.* **23**, 2449 (1982); (b) B. Meunier, E. Guilmet, M. E. De Carvalho and R. Poilblanc: *J. Am. Chem. Soc.* **106**, 6668 (1984); (c) B. Meunier, M. E. De Carvalho and A. Roberts: *J. Mol. Cat.* **41**, 185 (1987); (d) B. Meunier: *Gazz. Chim. Ital.* **118**, 485 (1988); (e) J. P. Collman, J. I. Brauman, B. Meunier, T. Hayashi, T. Kodadek and S. A. Raybuck: *J. Am. Chem. Soc.* **107**, 2000 (1987).
6. (a) F. Montanari, M. Penso, S. Quici and P. Viganò: *J. Org. Chem.* **50**, 4888 (1985); (b) S. Banfi, F. Montanari, M. Penso, V. Sosnovskikh, and P. Viganò: *Gazz. Chim. Ital.* **117**, 689 (1987); (c) S. Banfi, F. Montanari and S. Quici: *J. Org. Chem.* **53**, 2863 (1988); (d) S. Banfi, F. Montanari and S. Quici: *Recl. Trav. Chim. Pays-Bas* **109**, 117 (1990).
7. (a) Kirk-Othmer: *Encyclopedia of Chemical Technology*, 3rd ed.; J. Wiley and Sons: New York, 1979; Vol. 5, pp. 585–611; (b) *Ullmann's Encyclopedia of Industrial Chemistry*; VCH: Weinheim, 1986; Vol. A6, pp. 486–487.
8. Unpublished results from our laboratory.
9. A. W. Van der Made, J. M. C. Bax, R. J. M. Nolte, and W. Drenth: *Recl. Trav. Chim. Pays-Bas* **108**, 185 (1989).
10. G. O. Torossian, N. G. Oganyan, S. Banfi, and F. Montanari: *Metallorg. Khimya* **3**, 549 (1990).
11. F. Montanari, D. Landini, and F. Rolla: *Top. Curr. Chem.* **101**, 147 (1982).
12. A. W. van der Made and E. J. H. Hoppenbrouwer, W. Drenth, R. J. M. Nolte: *Recl. Trav. Chim. Pays-Bays* **107**, 15 (1988).
13. Under similar conditons (P : L : PT : S : OX = 1 : 25: 2 : 160 : 280) but at 25°C, Meunier reported 95% conversion of cyclooctene in 12 h. Taking into account the different temperatures and the S/P ratios, our results are comparable with those of Meunier.
14. (a) F. A. Walker, M.-W. Lo, and M. T. Ree: *J. Am. Chem. Soc.* **98**, 552 (1976); (b) L.-C. Yuan and, T. C. Bruice: *J. Am. Chem. Soc.* **108**, 1643 (1986).
15. D. Landini, A. Maia, and A. Rampoldi: *J. Org. Chem.* **54**, 328 (1989).
16. N. L. Richer, E. F. Pittman, and C. J. King: *J. Separ. Proc. Technol.* **1**, 23 (1980).
17. S. Abramovici, R. Neumann, and Y. Sasson: *J. Mol. Cat.* **28**, 291 (1985).
18. We observed a similar behavior in the oxidation of alcohols by NaOCl catalyzed by oxammonium salts [19], and the same was observed by others in a wide series of oxidations promoted by NaOCl, e.g. oxidations catalyzed by ruthenium salts [20], conversion of hydroquinones into quinones [19], conversion of arenes into arene epoxides [22], epoxidations catalyzed by Ni-salen [23], oxidation of aromatic alcohols [24].

19. (a) P. L. Anelli, C. Biffi, F. Montanari, and S. Quici: *J. Org. Chem.* **52**, 2559 (1987); (b) P. L. Anelli, S. Banfi, F. Montanari, and S. Quici: *ibid.* **54**, 2970 (1989); (c) P. L. Anelli, F. Montanari, and S. Quici: *Organic Synthesis* **69**, 212 (1990).
20. Y. Sasson, G. D. Zappi, and R. Neumann: *J. Org. Chem.* **51**, 2880 (1986).
21. F. Ishii and K. Kishi: *Synthesis* 706 (1980).
22. H. E. Fonouni, S. Krishnan, D. G. Kuhn and G. A. Hamilton: *J. Am. Chem. Soc.* **105**, 7672 (1983).
23. H. Yoon, T. R. Wagler, K. J. O'Connor, and C. J. Burrows: *J. Am. Chem. Soc.* **112**, 4586 (1990).
24. S. Abramovici, R. Neumann, and Y. Sasson: *J. Mol. Cat.* **29**, 299 (1985).
25. D. Landini, A. Maia, F. Montanari, and F. M. Pirisi: *J. Chem. Soc., Perkin Trans 2* 46 (1980).
26. Kinetic measures are in progress.
27. (a) K. M. Kadish, and S. L. Kelly: *Inorg. Chem.* **18**, 2968 (1979); (b) S. L. Kelly and K. M. Kadish: *ibid.* **21**, 3631 (1982).
28. From the published data [9], rates of 0.1 and 0.15 turnovers/min. at 5°C can be calculated for the epoxidation of cyclohexene with Mn(III)-porphirin **1** at pH 13.5 and 10.0, respectively.

Journal of Inclusion Phenomena and Molecular Recognition in Chemistry **12**: 175–186, 1992.

Macrobicyclic Aminals[#]

ANNALISE ATKINSON, NEIL A. BAILEY, DAVID E. FENTON*, PAUL D. HEMPSTEAD and GARY P. WESTWOOD
Department of Chemistry, The University, Sheffield, S3 7 HF, U.K.

(Received 31 July 1990; in final form: 26 December 1990)

Abstract. The synthesis of Schiff base macrocycles in the absence of templating cations and in non-protic and non-coordinating solvent media is reported. The reaction of pyridine-2,6-dicarboxaldehyde with the diamines 1,2-bis(2-aminophenoxy)ethane and 1,4-bis(2-aminophenoxy)butane has been found to give the corresponding macrobicyclic aminals. The chemical properties of the aminals are reported together with a preliminary study of their metal complexation reactions. The X-ray crystal structure of the aminal derived from pyridine-2,6-dicarboxaldehyde and 1,2-bis(2-aminophenoxy)ethane has been determined and verifies the macrobicyclic nature of the product.

Key words. Macrobicyclic, aminals, X-ray crystal structure, tetraimine macrocycles, Schiff base.

Supplementary Data relating to this article are deposited with the British Library as Supplementary Publication No. sup 82116 (16 pages).

1. Retrospect

In 1969 I* was appointed as a Senior Scientific Officer at the Agricultural Research Council's Unit of Structural Chemistry in London. This unit was led by the late Professor Sir Ronald Nyholm (as Honorary Director) and Professor Mary R. Truter and was studying the coordination chemistry of the alkali and alkaline earth metals. A few months later we were joined by a quiet and unassuming American chemist who had recently retired from a long career with du Pont and who had published, in 1967, a seminal paper on the coordination of alkali metals by cyclic polyethers. The unit was involved in solving the X-ray crystal structures of these complexes and part of my task, as a resident chemist, was to prepare complexes and grow crystals suitable for examination. Charlie Pedersen's visit to London was, of course, very influential. Not only was it possible to learn the techniques of polyether synthesis directly from their discoverer but it was possible to benefit from Charlie's wide-ranging chemical experience. It was quite exciting, some time later, to share in the discovery of the dinucleating ability of the cyclic polyethers, perhaps the one combination that had eluded Charlie in his investigations. Aside from the chemistry Charlie had a love for England and one abiding memory is of a day spent exploring the Essex and Cambridgeshire countryside and watching the morris dancing on the green at Finchingfield, reputed to be one of the prettiest villages in England. A visit to du Pont and to New Salem in 1971 served to reinforce impressions gained in London that Charlie was a chemist extraordinary and it is an honor to present this paper in tribute to Charlie and his achievements.

[#] This paper is dedicated to the memory of the late Dr C. J. Pedersen.
* Author for correspondence

2. Introduction

The susceptibility of the imine bond to nucleophilic attack is well known [1]. Addition of water will regenerate the precursor carbonyl and primary amine via the generally unstable carbinolamine intermediate. The addition of alcohols, alkoxides, thiols and thiolates to the imine bond to give isolable products is also known and often occurs when the imine nitrogen is coordinated to a metal [2]. When the nucleophile is a primary amine the intermediate *gem*-diamine (aminal) addition product is generally not stable, although it is known to be stabilised by coordination to a metal centre. The addition of a range of nucleophiles across the imine bonds present in cyclic tetrameric complexes derived from *o*-aminobenzaldehyde serves as a representative example [3]. The nucleophiles either add singly to *trans*-imine linkages (**1**) or, if difunctional as are the bis(2-hydroxyethyl)sulphide and bis(2-hydroxyethyl)methylamine anions, they can add across the molecule to produce 'basket-like' structures (**2**).

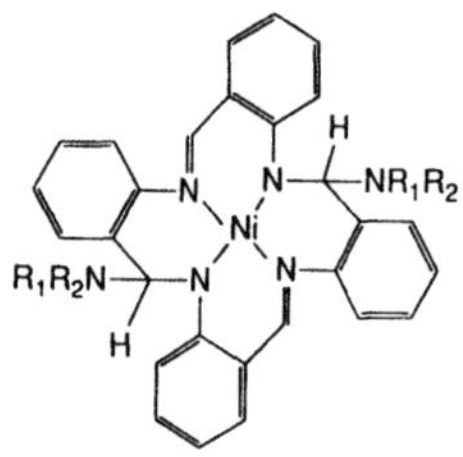

1

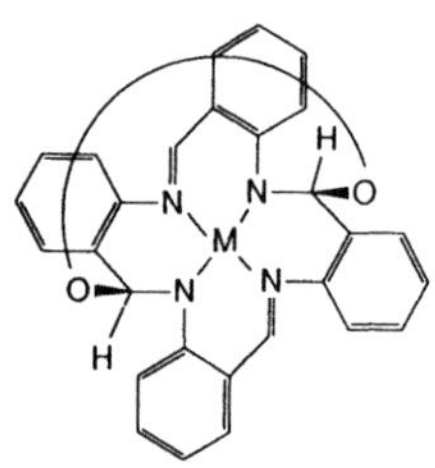

2
The link is $CH_2CH_2XCH_2CH_2$
(X = S or NMe ; M = Cu, Ni)

In a study of the formation of Schiff base macrocycles via the cyclocondensation of pyridine-2,6-dicarboxaldehyde with α,ω-diamines in alcoholic media in the presence of alkaline earth templating devices we noted that it was possible to recover the metal-free macrocycles (**3**) when magnesium salts were used as potential templates [4]. The facile formation of the macrocycles was interpreted as suggesting that the macrocyclic di-imine precursors could be formed in a non-template process and that subsequent addition of an alcohol across an imine occurs in order to relieve angular strain in the Schiff base macrocycle. It was then established that (**3**) could be prepared in the absence of the metal template.

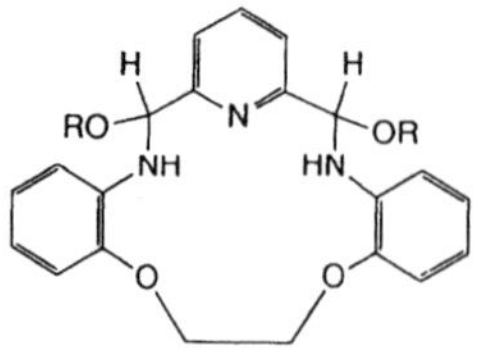

3
R = Me, Et, *n*-Bu

In this paper we present results concerning the synthesis of Schiff base macrocycles in the absence of templating cations and in non-protic and non-coordinating solvent media. The X-ray crystal structure of the novel macrobicyclic aminal derived from pyridine-2,6-dicarboxaldehyde and 1,2-bis(2-amino-phenoxy)ethane is also presented.

3. Experimental

Pyridine-2,6-dicarboxaldehyde [5], and the diamines 1,2-bis(2-aminophenoxy) ethane, 1,3-bis(2-aminophenoxy)propane and 1,4-bis(2-aminophenoxy)butane [6] were prepared according to literature procedures.

3.1. PREPARATION OF THE '2 + 2' TETRAIMINE MACROCYCLE (4)

A solution of pyridine-2,6-dicarboxaldehyde (0.27 g, 2 mmol) in dry (sodium) benzene (30 mL) was added to a stirred solution of 1,2-bis(2-aminophenoxy)ethane (0.48 g, 2 mmol), in dry benzene (30 mL). The mixture was refluxed for 4 h and the product filtered off. Yield 68%; m.p. 136–138°C.; IR (KBr disc): $\nu_{C=N}$ 1640 cm^{-1}, $\nu_{CH-Benzene}$ 690 cm^{-1}; m.s. (e.i.): 686 amu.; Microanalysis (%) *found* (required) for $C_{42}H_{34}N_6O_4 \cdot C_6H_6$: C 74.99 (75.38), H 5.08 (5.27), N 10.84 (10.99).

3.2. DIGESTION OF (4) IN ALCOHOLS

A suspension of the macrocycle (0.69 g, 1 mmol) in absolute ethanol (70 mL) was brought to reflux. After 1 h all the solid had dissolved, and after heating for a further hour the hot solution was filtered and left to cool. The product crystallised out as long white filaments and was confirmed as the bis-ethoxy addition product of the di-imine macrocycle (**3**,R = Et) by comparison with an authentic sample [4] Yield 65%; m.p. 174–176°C.; IR (KBr disc): ν_{NH} 3400 cm^{-1}, ν_{C-O} 1050 cm^{-1}; m.s. (e.i.): 435 amu.; Microanalysis (%) *found* (required) for $C_{25}H_{29}N_3O_4$: C 68.95 (68.95), H 6.57 (6.71), N 9.63 (9.65).

The corresponding bis-methoxy addition product was also prepared by the same method, in methanol solvent, and isolated as fine white needles. Yield 59%; m.p. 130–132°C.; IR (KBr disc): ν_{NH} 3380 cm^{-1}, ν_{C-O} 1045 cm^{-1}; m.s. (e.i.): 407 amu.; Microanalysis (%) *found* (required) for $C_{23}H_{25}N_3O_4$: C 68.02 (67.80), H 6.06 (6.18), N 10.31 (10.31).

3.3. AMINAL (5a)

A solution of pyridine-2,6-dicarboxaldehyde (0.68 g, 5 mmol) in acetonitrile (250 mL) was added dropwise to a stirred solution of 1,2-bis(2-aminophenoxy)ethane (1.22 g, 5 mmol) in acetonitrile (250 mL) over a period of 2 h. A white precipitate formed over a period of 12 h. The crude product was recrystallised from a chloroform/acetonitrile mixture as colourless crystals. The same product can be obtained by using bench benzene or toluene as the solvent and the yield of the reaction can be increased using a diamine : dicarbonyl ratio of 2 : 1. Yield 77%; m.p. 156–158°C.; IR (KBr disc): ν_{NH}3370 cm^{-1}; m.s. (e.i.): 587 amu.; Microanalysis (%) *found* (required) for $C_{35}H_{33}N_5O_4$: C 71.39 (71.53), H 5.60 (5.66), N 11.86 (11.92); ^{1}H NMR (CD_2Cl_2) δ: 7.55–6.60 (19H, *m*, ArH), 5.98 (*t*, CH), 5.53 (*d*, NH), 4.30 (8H, *s*, CH_2).

3.4. AMINAL (**5b**)

The compound was synthesised by the above method using 1,4-bis(2-aminophenoxy)butane (1.36 g, 5 mmol). Yield 74%; m.p. 106–108°C.; IR (KBr disc): ν_{NH}3415 cm^{-1}; m.s. (pos. f.a.b.): 643 amu.; Microanalysis (%) *found* (required) for $C_{39}H_{41}N_5O_4$: C 72.25 (72.76), H 6.25 (6.42), N 10.66 (10.88); ^{1}H NMR ($CDCl_3$) δ: 7.60–6.60 (19H, *m*, ArH), 6.00 (*t*, CH), 5.45 (*d*, NH), 4.15–3.80 (8H, *m*, CH_2), 2.10–1.80 (8H, *m*, CH_2).

3.5. PRODUCT (**5d**)

The preparation was exactly as described previously but using 1,3-bis(2-aminophenoxy) propane (1.29 g, 5 mmol). The brown product was insoluble in common organic solvents preventing recrystallisation and further analysis. Yield 68%; m.p. 300+°C.; IR (KBr disc): ν_{NH}3380 cm^{-1}.

3.6. PREPARATION OF Cu(**4**)(ClO_4)$_2$ AND Cu(**4**)(NO_3)$_2$

A solution of the appropriate copper(II) salt (perchlorate or nitrate) (1 mmol) in absolute ethanol (10 mL) was added to a suspension of the macrocycle (0.69 g, 1 mmol) in ethanol (15 mL). The mixture was refluxed for 30 min and the resultant green solution was filtered. On cooling the product precipitated out as a golden brown solid and was washed with a little ethanol and dried.

Cu(**4**)(NO_3)$_2$ · 2 H_2O: Yield 57%; m.p. 230°C. (dec.); IR (KBr disc): $\nu_{C=N}$1595 cm^{-1}, ν_{N-O}1380 cm^{-1}; m.s. (pos. f.a.b.): 812 [Cu(**4**)(NO_3)]$^+$, 750 amu. [Cu(**4**)]$^+$; Microanalysis (%) *found* (required) for $CuC_{42}H_{38}N_8O_{12}$: C 54.89 (55.41), H 3.84 (4.21), N 12.32 (12.31).

Cu(**4**)(ClO_4)$_2$ · 2 H_2O: Yield 59%; IR (KBr disc): $\nu_{C=N}$1595 cm^{-1}, ν_{ClO}1095, 625 cm^{-1}; m.s. (pos. f.a.b.): 848 [Cu(**4**)(ClO_4)]$^+$, 750 amu. [Cu(**4**)]$^+$; Microanalysis (%) *found* (required) for $CuC_{45}H_{38}N_6O_{14}Cl_2$: C 51.20 (51.20), H 3.67 (3.89), N 8.43 (8.53), Cl 7.22 (7.20).

3.7. REACTION OF IRON(III) CHLORIDE WITH LIGAND (**4**)

The method was exactly as described for the previous complexation, but using iron (III) chloride hexahydrate (0.27 g, 1 mmol). The product, Fe(**6a**)Cl_3 was isolated as a dark green solid. Yield 61%; m.p. 300+°C.; IR (KBr disc): $\nu_{C=N}$1595 cm^{-1}; m.s. (pos. f.a.b.): 469 [Fe(**6a**)Cl_2]$^+$, 434 [Fe(**6a**)Cl]$^+$, 399 amu. [Fe(**6a**)]$^+$; Microanaylsis (%) *found* (required) for $FeC_{21}H_{17}N_3O_2Cl_3$ · 1.5 H_2O: C 50.85 (50.74), H 3.98 (4.05), N 8.37 (8.45), Cl 14.19 (14.26).

3.8. REACTION OF MANGANESE(II) PERCHLORATE WITH LIGAND (4)

The preparation was exactly as before, using manganese(II) perchlorate hexahydrate. The product was obtained as a yellow solid and all analyses correspond to those of an authentic sample of Mn(**6a**)$(ClO_4)_2$ [10]. Yield 27%; IR (KBr disc): $\nu_{C=N}$ 1590 cm^{-1}, ν_{Cl-O} 1120, 620 cm^{-1}; m.s. (pos. f.a.b.): 497 [Mn(**6a**)$(ClO_4)]^+$, 398 amu. [Mn(**6a**)]$^+$; Microanalysis (%) *found* (required) for $MnC_{21}H_{17}N_3O_{10}Cl_2$: C 42.29 (42.23), H 3.26 (2.87), N 6.91 (7.04), Cl 11.48 (11.87).

3.9. REACTION OF α,ω-DIAMINES WITH LIGAND (4)

A solution of the appropriate α,ω-diamine (1 mmol) in acetonitrile (25 mL) was added to a suspension of the tetraimine macrocycle (**4**) (0.34 g, 0.5 mmol) in acetonitrile (25 mL). The mixture was refluxed for 1 h. The solution was filtered and left to cool. Crystals of the product deposited and were filtered off, washed with a little acetonitrile and dried.

3.9. *a. Preparation of* (**5a**) *from Ligand* (**4**)

Using 1,2-bis(2-aminophenoxy)ethane as the diamine in the above procedure leads to isolation of ligand (**5a**) (yield 67%). The product was characterised as the macrobicyclic aminal after comparison with an authentic sample of (**5a**) which had been prepared and characterised as described above.

3.9. *b. Preparation of* (**5c**) *from Ligand* (**4**)

The reaction was carried out, using the above method, with 1,4-bis(2-aminophenoxy)butane (0.27 g, 1 mmol). The product (**5c**) was isolated as white needles. Yield 69%; m.p. 114–116°C.; IR (KBr disc): ν_{NH} 3400 cm^{-1}; m.s. (pos. f.a.b.): 615 amu.; Microanalysis (%) *found* (required) for $C_{37}H_{37}N_5O_4$: C 75.15 (72.18), H 6.16 (6.06), N 11.12 (11.37); ^{1}H NMR ($CDCl_3$) δ: 7.55–6.60 (19 H, *m*, ArH), 5.98 (*t*, CH), 5.65 (*d*, NH), 5.40 (*d*, NH), 4.35–3.90 (8 H, *m*, CH_2—O), 2.05 (4 H, *m*, CH_2).

3.10. REACTION OF MANGANESE(II) PERCHLORATE WITH LIGAND (5a)

A solution of manganese(II) perchlorate hexahydrate (0.36 g, 1 mmol) in acetonitrile (15 mL) was added to a suspension of (**5a**) (0.59 g, 1 mmol) in acetonitrile (30 mL). The mixture was stirred at room temperature for 5 min. The starting materials quickly dissolved and the resultant orange solution was filtered. The product was precipitated out as a yellow powder by addition of ether to the filtrate. The solid was washed with a little ether and dried (yield 24%). Comparison with an authentic sample [10] showed it to be Mn(**6a**)$(ClO_4)_2$.

3.11. REACTION OF MANGANESE(II) PERCHLORATE WITH LIGAND (5b)

Carrying out the same complexation with ligand (**5b**) leads to the isolation of an orange solid, characterised as the complex Mn(**6b**)$(ClO_4)_2$ [6]. (Yield 27%). IR

(KBr disc): $\nu_{C=N}$ 1595 cm^{-1}, ν_{Cl-O} 1110, 630 cm^{-1}; m.s. (pos. f.a.b.): 525 $[Mn(\mathbf{6b})(ClO_4)]^+$, 426 $[Mn(\mathbf{6b})]^+$, 371 amu. $[(\mathbf{6b})]^+$; Microanalysis (%) *found* (required) for $MnC_{23}H_{21}N_3O_{10}Cl_2 \cdot H_2O$: C 42.71 (42.94), H 3.85 (3.60), N 6.41 (6.53), Cl 11.16 (11.02).

3.12. DIGESTION OF LIGAND (**5a**) IN ALCOHOL

A suspension of ligand (**5a**) (0.59 g, 1 mmol) in ethanol (50 mL) was refluxed for 4 h. During this time the solid dissolved and the resultant solution was filtered and left to cool. The product precipitated out as white needles (yield 57%) and was characterised as the bis-ethoxy addition product of the di-imine macrocycle (**3**), R = Et [4].

The bis-methoxy product was prepared in the same way by digestion in methanol (yield 51%).

4. Crystal Structure Data and Determination

Crystal data for the macrobicyclic aminal (**5a**); $C_{35}H_{33}N_5O_4$; $M = 587.68$; crystallises from chloroform/acetonitrile as colourless rhombic prisms; crystal dimensions $0.95 \times 0.50 \times 0.30$ mm. Triclinic, $a = 11.334(12)$, $b = 12.155(18)$, $c = 12.155(13)$ Å, $\alpha = 85.58(10)$, $\beta = 71.15(8)$, $\gamma = 70.12(10)°$, $U = 1496.8(32)$ Å^3; $D_c = 1.304$ g cm^{-3}, $Z = 2$. Space group $P\bar{1}$. MoK_α radiation ($\lambda = 0.71069$ Å), $\mu(\text{Mo}K\alpha) = 0.81$ cm^{-1}.

Three-dimensional, room temperature X-ray data were collected in the range $3.5 < 2\theta < 40°$ on a Nicolet R3 four-circle diffractometer by the omega scan method. The 1965 independent reflections (of 2813 measured) for which $|F|/\sigma(|F|) > 3.0$ were corrected for Lorentz and polarisation effects. The structure was solved by direct methods and Fourier techniques and refined by blocked cascade least squares methods. Hydrogen atoms bound to carbon atoms were included in calculated positions, with isotropic thermal parameters related to those of the supporting atom, and refined in riding mode. Hydrogen atoms bound to nitrogen atoms were refined with isotropic thermal parameters related to those of the supporting atom. Refinement converged at a final R 0.060, $R_w = 0.055$ (409 parameters, maximum shift/e.s.d. 0.006), with allowance for the thermal anisotropy of all non-hydrogen atoms. A final difference electron density synthesis showed peaks of -0.25 and $+0.20$ e Å^{-3}. Complex scattering factors were taken from Ref. [7] and from the program package SHELXTL [8] as implemented on the Data General Nova 3 computer, which was used for structure solution and refinement. A weighting scheme $w^{-1} = [\sigma^2(F) + 0.00066(F)^2]$ was used in the latter stages of the refinement. Table I lists atomic positional parameters with estimated standard deviations. Tables II and III list the bond lengths and angles.

5. Results and Discussion

The reaction of pyridine-2,6-dicarboxaldehyde and 1,2-bis(2-aminophenoxy)-ethane in 1 : 1 ratio in dry benzene gave an off-white powdery material. This product gave a band at 1641 cm^{-1} in the IR spectrum, which could be attributed to an imine

Table I. Atom coordinates (× 10^4) and temperature factors (Å × 10^3)

atom	x/a	y/b	z/c	U
N(1)	4420(4)	3063(3)	−1843(3)	49(2)*
Hn(1)	4341(46)	2817(40)	−2471(38)	60
Hn(2)	1540(39)	2417(34)	−3673(34)	66
Hn(4)	4390(39)	734(35)	−1471(34)	66
Hn(5a)	2616(38)	407(34)	−2514(33)	66
N(2)	837(4)	2261(3)	−3424(3)	55(2)*
N(3)	2233(4)	1973(3)	−1484(3)	48(2)*
N(4)	4642(4)	1136(3)	−1121(4)	57(2)*
N(5)	1955(4)	412(4)	−2713(4)	58(2)*
O(1)	3843(3)	4316(3)	−3627(3)	51(2)*
O(2)	1402(3)	4267(3)	−3926(3)	56(2)*
O(3)	6130(3)	−962(3)	−2342(3)	66(2)*
O(4)	4458(3)	−763(3)	−3703(3)	61(2)*
C(1)	3974(4)	4295(4)	−1733(4)	41(2)*
C(2)	3861(4)	4901(4)	−763(4)	46(3)*
C(3)	3568(5)	6115(5)	−722(4)	57(3)*
C(4)	3359(5)	6746(4)	−1678(4)	61(3)*
C(5)	3421(5)	6183(4)	−2651(4)	52(3)*
C(6)	3716(5)	4976(4)	−2679(4)	45(3)*
C(7)	3284(5)	4901(4)	−4513(4)	54(3)*
C(8)	1786(5)	5303(4)	−4126(4)	55(3)*
C(9)	65(5)	4404(4)	−3483(4)	52(3)*
C(10)	−955(6)	5471(5)	−3307(4)	63(3)*
C(11)	−2277(6)	5504(6)	−2837(5)	68(3)*
C(12)	−2583(5)	4488(6)	−2588(6)	68(3)*
C(13)	−1555(5)	3399(5)	−2797(4)	63(3)*
C(14)	−219(5)	3340(4)	−3238(4)	48(3)*
C(15)	849(5)	1492(4)	−2406(4)	51(2)*
C(16)	983(5)	2126(4)	−1431(4)	50(3)*
C(17)	−108(5)	2842(4)	−590(4)	56(3)*
C(18)	116(6)	3453(4)	192(5)	61(3)*
C(19)	1404(6)	3308(4)	166(4)	58(3)*
C(20)	2436(5)	2552(4)	−690(4)	48(3)*
C(21)	3889(5)	2357(4)	−865(4)	47(3)*
C(22)	5975(5)	632(4)	−1168(4)	49(3)*
C(23)	6556(5)	1117(4)	−550(5)	64(3)*
C(24)	7877(5)	556(5)	−598(5)	65(3)*
C(25)	8632(6)	−464(5)	−1252(5)	65(3)*
C(26)	8055(6)	−951(5)	−1859(5)	66(3)*
C(27)	6741(6)	−409(4)	−1803(4)	54(3)*
C(28)	6578(5)	−960(5)	−3551(4)	70(3)*
C(29)	5810(5)	−1484(4)	−4067(4)	62(3)*
C(30)	3593(5)	−1037(4)	−4148(4)	47(2)*
C(31)	3947(6)	−1842(4)	−5023(5)	60(3)*
C(32)	2986(6)	−2009(4)	−5422(5)	61(3)*
C(33)	1645(6)	−1355(5)	−4920(5)	62(3)*
C(34)	1287(5)	−538(4)	−4035(4)	60(3)*
C(35)	2249(5)	−362(4)	−3624(4)	44(2)*

* Equivalent isotropic U defined as one third of the trace of the orthogonalised U_{ij} tensor.

Table II. Bond lengths (Å).

N(1)—Hn(1)	0.885(56)	N(1)—C(1)	1.411(6)
N(1)—C(21)	1.501(6)	Hn(2)—N(2)	0.837(46)
Hn(4)—N(4)	0.842(53)	Hn(5a)—N(5)	0.857(49)
N(2)—C(14)	1.417(6)	N(2)—C(15)	1.498(6)
N(3)—C(16)	1.344(7)	N(3)—C(20)	1.360(8)
N(4)—C(21)	1.437(6)	N(4)—C(22)	1.408(7)
N(5)—C(15)	1.445(6)	N(5)—C(35)	1.393(7)
O(1)—C(6)	1.401(7)	O(1)—C(7)	1.448(7)
O(2)—C(8)	1.447(7)	O(2)—C(9)	1.389(7)
O(3)—C(27)	1.430(9)	O(3)—C(28)	1.399(6)
O(4)—C(29)	1.421(6)	O(4)—C(30)	1.402(8)
C(1)—C(2)	1.390(8)	C(1)—C(6)	1.418(7)
C(2)—C(3)	1.399(8)	C(3)—C(4)	1.388(8)
C(4)—C(5)	1.387(9)	C(5)—C(6)	1.391(8)
C(7)—C(8)	1.515(7)	C(9)—C(10)	1.388(7)
C(9)—C(14)	1.425(8)	C(10)—C(11)	1.408(10)
C(11)—C(12)	1.380(10)	C(12)—C(13)	1.409(7)
C(13)—C(14)	1.412(8)	C(15)—C(16)	1.540(9)
C(16)—C(17)	1.389(6)	C(17)—C(18)	1.387(10)
C(18)—C(19)	1.399(10)	C(19)—C(20)	1.390(6)
C(20)—C(21)	1.526(8)	C(22)—C(23)	1.416(10)
C(22)—C(27)	1.387(6)	C(23)—C(24)	1.401(8)
C(24)—C(25)	1.377(8)	C(25)—C(26)	1.405(11)
C(26)—C(27)	1.389(9)	C(28)—C(29)	1.534(10)
C(30)—C(31)	1.370(8)	C(30)—C(35)	1.416(6)
C(31)—C(32)	1.406(11)	C(32)—C(33)	1.403(8)
C(33)—C(34)	1.388(8)	C(34)—C(35)	1.420(9)

stretch; no amine or carbonyl bands were observed and so a macrocyclisation was likely to have occurred. The ^1H NMR spectrum, in $CDCl_3$, showed a signal at 8.70 ppm corresponding to an imine proton and no signal assignable to an aldehydic proton, thus reinforcing this suggestion. The spectrum, however, is rapidly complicated by the ingrowth of signals due to (**5a**) during the data collection. The m.s. (E.I. mode) gave a very intense peak at 686 a.m.u. which corresponds to a '2 + 2' cyclocondensation leading to the tetraimine Schiff base (**4**) – a weak intensity peak was noted at 343 a.m.u. The IR spectrum showed a band at 690 cm^{-1} and ^1H NMR spectrum gave a singlet at 7.37 ppm and these, together with the microanalytical data, indicate the presence of one molecule of benzene for every molecule of '2 + 2' macrocycle. This suggests that an inclusion compound is likely, probably as a sandwich species with the benzene providing the filling. The benzene could be interacting with the lateral oxygen or nitrogen donor atoms of the macrocycle or with the aromatic π-cloud of a pyridinyl head-unit.

The macrocycle (**4**) can be converted into (**3**, R = Et) by recrystallisation from ethanol. Also the f.a.b. m.s. of (**4**) run in a 3-nitrobenzyl alcohol (NOBA) matrix shows peaks corresponding to $[(\mathbf{4}) + H]^+$, $[(\mathbf{4}) + NOBA + H]^+$ and $[(\mathbf{4}) + 2\,NOBA + H]^+$, indicating the facile addition of an alcohol across the imine bond. It appears that in the absence of a group capable of adding to the imine and so relieving molecular strain in the '1 + 1' macrocycle, formation of the '2 + 2'

Table III. Bond angles (deg.).

Hn(1)—N(1)—C(1)	112.1(28)	Hn(1)—N(1)—C(21)	106.7(32)
C(1)—N(1)—C(21)	120.4(3)	Hn(2)—N(2)—C(14)	107.0(26)
Hn(2)—N(2)—C(15)	110.4(30)	C(14)—N(2)—C(15)	116.4(3)
C(16)—N(3)—C(20)	118.7(4)	Hn(4)—N(4)—C(21)	119.2(25)
Hn(4)—N(4)—C(22)	113.8(23)	C(21)—N(4)—C(22)	123.9(5)
Hn(5a)—N(5)—C(15)	113.8(25)	Hn(5a)—N(5)—C(35)	115.6(24)
C(15)—N(5)—C(35)	125.5(5)	C(6)—O(1)—C(7)	119.2(3)
C(8)—O(2)—C(9)	118.6(3)	C(27)—O(3)—C(28)	113.7(4)
C(29)—O(4)—C(30)	117.8(4)	N(1)—C(1)—C(2)	123.4(4)
N(1)—C(1)—C(6)	119.7(5)	C(2)—C(1)—C(6)	116.7(4)
C(1)—C(2)—C(3)	122.6(5)	C(2)—C(3)—C(4)	118.9(5)
C(3)—C(4)—C(5)	120.5(5)	C(4)—C(5)—C(6)	119.9(5)
O(1)—C(6)—C(1)	114.0(4)	O(1)—C(6)—C(5)	124.6(5)
C(1)—C(6)—C(5)	121.3(5)	O(1)—C(7)—C(8)	113.6(4)
O(2)—C(8)—C(7)	107.4(4)	O(2)—C(9)—C(10)	124.8(5)
O(2)—C(9)—C(14)	114.8(4)	C(10)—C(9)—C(14)	120.4(5)
C(9)—C(10)—C(11)	119.7(6)	C(10)—C(11)—C(12)	121.2(5)
C(11)—C(12)—C(13)	119.5(6)	C(12)—C(13)—C(14)	120.7(6)
N(2)—C(14)—C(9)	119.5(5)	N(2)—C(14)—C(13)	122.0(5)
C(9)—C(14)—C(13)	118.5(4)	N(2)—C(15)—N(5)	111.0(3)
N(2)—C(15)—C(16)	109.6(4)	N(5)—C(15)—C(16)	108.0(4)
N(3)—C(16)—C(15)	115.1(4)	N(3)—C(16)—C(17)	122.6(6)
C(15)—C(16)—C(17)	122.3(5)	C(16)—C(17)—C(18)	117.9(6)
C(17)—C(18)—C(19)	120.9(4)	C(18)—C(19)—C(20)	117.1(6)
N(3)—C(20)—C(19)	122.7(5)	N(3)—C(20)—C(21)	114.2(4)
C(19)—C(20)—C(21)	123.0(6)	N(1)—C(21)—N(4)	109.3(3)
N(1)—C(21)—C(20)	110.6(4)	N(4)—C(21)—C(20)	109.4(5)
N(4)—C(22)—C(23)	122.6(4)	N(4)—C(22)—C(27)	119.1(6)
C(23)—C(22)—C(27)	118.3(5)	C(22)—C(23)—C(24)	120.2(5)
C(23)—C(24)—C(25)	120.8(7)	C(24)—C(25)—C(26)	119.1(6)
C(25)—C(26)—C(27)	120.4(5)	O(3)—C(27)—C(22)	118.7(5)
O(3)—C(27)—C(26)	120.0(4)	C(22)—C(27)—C(26)	121.2(6)
O(3)—C(28)—C(29)	110.8(5)	O(4)—C(29)—C(28)	108.4(4)
O(4)—C(30)—C(31)	126.1(4)	O(4)—C(30)—C(35)	113.3(4)
C(31)—C(30)—C(35)	120.6(6)	C(30)—C(31)—C(32)	120.7(5)
C(31)—C(32)—C(33)	120.1(5)	C(32)—C(33)—C(34)	119.2(7)
C(33)—C(34)—C(35)	121.3(5)	N(5)—C(35)—C(30)	117.6(5)
N(5)—C(35)—C(34)	124.3(4)	C(30)—C(35)—C(34)	118.1(5)

macrocycle is preferred. The latter is presumably less strained than the '1 + 1' macrocycle and in any given situation the least strained system is the one produced.

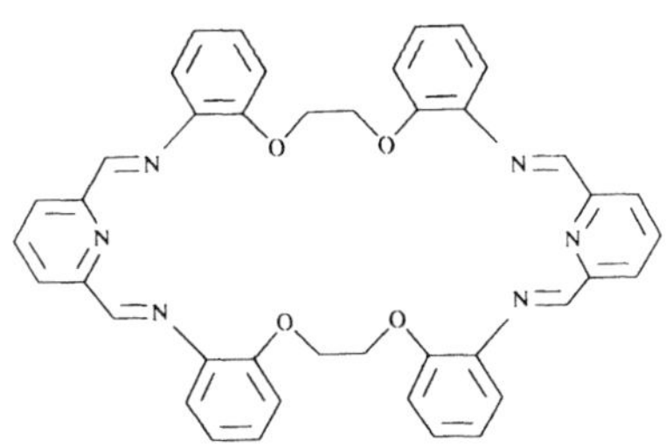

4

The macrocycle (**4**) is also obtainable from dry toluene, but if bench toluene, benzene, or acetonitrile are used as solvents then a different product is obtained. The white powder has no imine or carbonyl stretches in the IR spectrum but shows a band at 3370 cm^{-1} indicative of a secondary amine; the ^1H NMR spectrum is more complex and without a signal due to a aldehydic proton. The f.a.b. m.s. in NOBA gave a strong peak at 334 a.m.u., corresponding to the '1 + 1' Schiff base macrocycle, and that run in sulfolane gave a highest mass peak at 587 a.m.u. Crystals suitable for X-ray structural analysis were grown from acetonitrile–chloroform and the structure was solved revealing that addition of 1,2-bis(2-aminophenoxy)ethane across the imines of a '1 + 1' Schiff base macrocycle had occurred to give a macrobicyclic aminal (**5a**).

The molecular structure of (**5a**) is shown in Figure 1. Bond lengths and angles are listed in Tables II and III and are unexceptional. The conformation of the molecule is best described in terms of the relative orientations of the four planar 2-aminophenoxy units (i.e. C(1) to C(6), N(1) and O(1); C(9) to C(14), O(2) and N(2); C(22) to C(27), N(4) and O(3); C(30) to C(35), N(5) and O(4) to the pyridyl fragment (C(15) to C(21) and N(3)). The first two aminophenoxy units lie in two mutually perpendicular planes, which are inclined at 78° to the pyridyl unit. The third aminophenoxy unit is inclined at 20° to the pyridyl unit. The fourth aminophenoxy

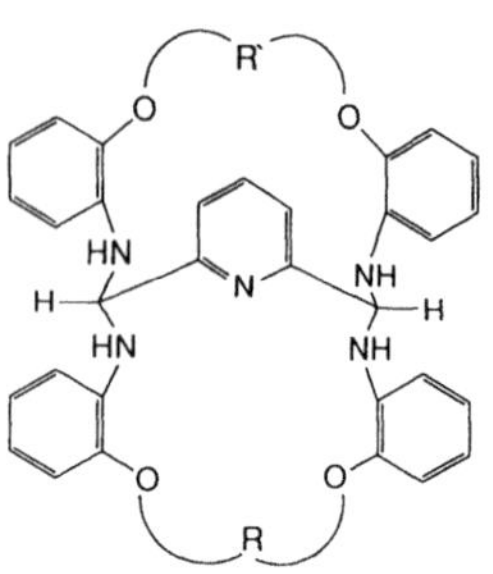

5

5a R = CH_2CH_2 R' = CH_2CH_2

5b R = $(CH_2CH_2)_2$ R' = $(CH_2CH_2)_2$

5c R = CH_2CH_2 R' = $(CH_2CH_2)_2$

5d R = $(CH_2)_3$ R' = $(CH_2)_3$

unit is inclined at 6° to the pyridyl unit and 13° to the third aminophenoxy unit. The overall result is a macrobicycle in which the first two aminophenoxy units and the pyridyl unit define three approximately perpendicular planes (c.f. Mo(CO_3 [(**7**)] · CH_2Cl_2 [9] whereas the pyridyl unit and the other two aminophenoxy units, lie in an approximately planar conformation (c.f. Mn[(**6a**)] ($(ClO_4)_2$) [10].

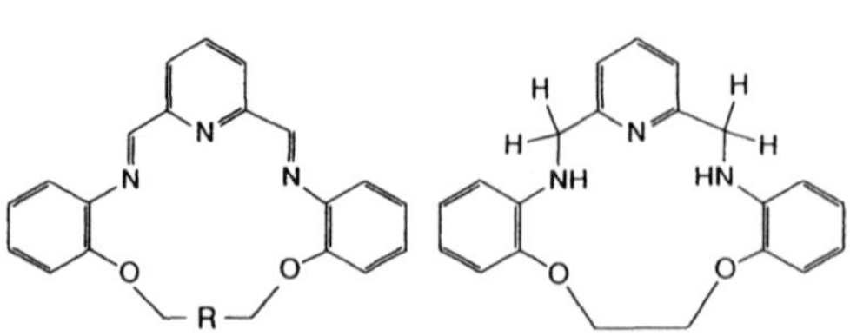

6a, R= CH_2CH_2

6b, R= $(CH_2)_4$

7

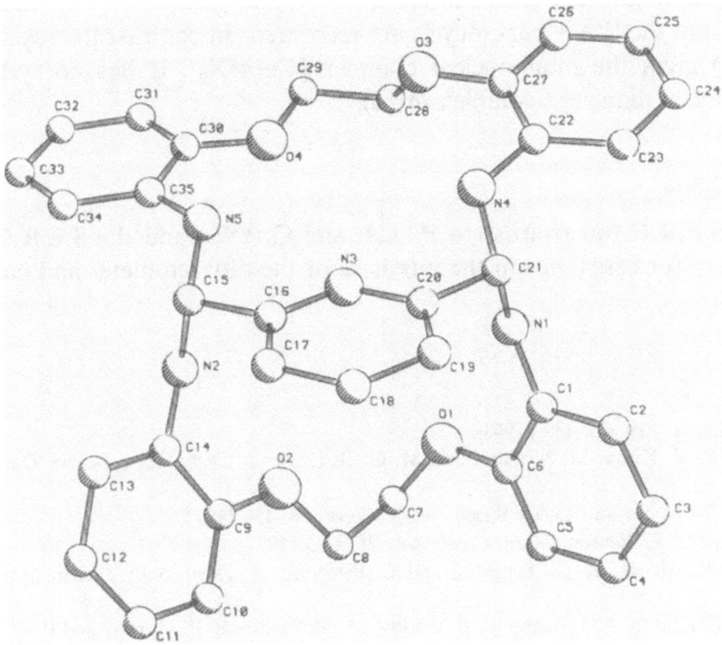

Fig. 1. The molecular structure of the macrobicyclic aminal (**5a**).

The macrobicyclic aminal (**5b**) is readily prepared by reaction of pyridine-2,6-dicarboxaldehyde and 1,4-bis(2-aminophenoxy)butane in 1 : 1 ratio in dry benzene, toluene or acetonitrile. Compound (**5b**) has been characterised by IR, NMR, and m.s. and crystals were grown from a chloroform–acetonitrile solution. The X-ray structure of (**5b**) is similar to that of (**5a**) confirming aminal formation; owing to the presence of disordered solvent molecules it has not, to date, been possible to refine the structure beyond an *R*-factor of 0.175. The reaction of pyridine-2,6-dicarboxaldehyde and 1,3-bis(2-aminophenoxy)propane under the conditions described above gave a product (**5d**) having an IR spectrum similar to those of (**5a**) and (**5b**).The high melting point (> 300°C) and insolubility in common organic solvents suggest that this product may be polymeric in nature.

By increasing the 1,2-bis(2-aminophenoxy)ethane content in the original sythesis to give a 2 : 1 reactant ratio the yield of (**5a**) is greatly enhanced. This suggested that mixed aminals could be made available and addition of the corresponding diamine having a tetramethylene bridge to an acetonitrile solution of (**4**) led to the isolation of the mixed aminal (**5c**) which has been characterised by IR, NMR and m.s. The reaction of (**4**) with the more flexible aliphatic diamine 1,8-diaza-3,6-dioxaoctane did not give a mixed aminal. The macrobicyclic aminal (**5a**) can also be recovered when either (**3**), R = Et, or (**4**) is recrystallised from an acetonitrile–chloroform mixture. Recrystallisation of (**5a**) from ethanol or methanol gave [(**3**), R-Et or Me], further illustrating the fragility of (**5a**) in solution.

Preliminary metal complexation studies show that (**5a**) and (**5b**) react with Mn(II) to give the complexes [Mn(**6a**)X_2] and [Mn(**6b**)X_2] and that (**4**) reacts with

Fe(II) and Mn(II) to give the complexes [Fe(**6a**)X_2] and [Mn(**6a**)X_2]. In each case only products from the '1 + 1' macrocycle are recovered. In contrast the reaction of (**4**) with Cu(II) gives the mononuclear complex [Cu(**4**)X_2]. It has not yet been possible to isolate a dinuclear complex of (**4**).

Acknowledgements

We thank the S.E.R.C for awards (to P.D.H. and G.P.W.) and the S.E.R.C. and the Royal Society for assistance in the purchase of the diffractometer and computing facilities.

References

1. R. W. Layer: *Chem. Rev.* **63**, 489 (1963).
2. S. M. Nelson, C. V. Knox, M. McCann and M. G. B. Drew: *J. Chem. Soc., Dalton Trans.* 1669 (1981).
3. V. Katovic, L. T. Taylor and D. H. Busch: *Inorg. Chem.* **10**, 458 (1971).
4. D. H. Cook and D. E. Fenton: *Inorg. Chim. Acta* **25**, L95 (1977).
5. P. Moore, N. W. Alcock, R. G. Kingston and C. Pierpoint: *J. Chem. Soc., Dalton Trans.* 1941 (1989).
6. A. Bashall, D. E. Fenton, A. J. Leong, L. F. Lindoy, M. McPartlin, B. P. Murphy and P. A. Tasker: *J. Chem. Soc., Dalton Trans.* 2543 (1987).
7. *International Tables for X-ray Crystallography*, Vol. 4, Kynoch Press, Birmingham (1974), distr. Kluwer Academic Publishers, Dordrecht, the Netherlands.
8. G. M. Sheldrick: *SHELXTL*, An integrated system for solving, refining and displaying crystal structures from diffraction data (Revision 4), University of Göttingen, F.R. Germany 1983).
9. H. Adams, N. A. Bailey, D. E. Fenton, P. D. Hempstead, B. A. Hewes, S. J. Kitchen, and M. J. Winter: *Polyhedron* **8**, 1857 (1989).
10. P. Tasker, N. W. Alcock, D. C. Liles and M. McPartlin: *J. Chem. Soc., Chem Commun.* 727 (1974).

Journal of Inclusion Phenomena and Molecular Recognition in Chemistry **12**: 187–197, 1992.

Tribracchial Lariat Ethers: Syntheses, Binding, and Formation of an Intramolecular Macroring-Sidearm Complex in the Absence of Any Cation#

MARA TSESARSKAJA, THOMAS P. CLEARY, STEVEN R. MILLER and JOHN E. TRAFTON
Department of Chemistry, University of Miami, Coral Gables, FL 33124 U.S.A.

SIMON BOTT and JERRY L. ATWOOD
Department of Chemistry, University of Alabama, Tuscaloosa, AL 34587 U.S.A.

GEORGE W. GOKEL
Department of Chemistry, University of Miami, Coral Gables, FL 33124 U.S.A.

(Received: 8 October 1990; in final form: 22 January 1991)

Abstract. Details of a new synthetic approach to 4,10,16-triaza-18-crown-6 ('triaza-18-crown-6') are reported, along with the preparation and binding properties of the following derivatives having the indicated sidearms: $CH_2C{\equiv}CH$, CH_2CH_2OH, $CH_2CH_2OCH_3$, $COCH_2OCH_3$,$CH_2COOCH_2CH_3$, $CH_2C_6H_5$, and $CH_2C_6H_4$-2-NO_2. A key intermediate in the synthesis of triaza-18-crown-6 is 4-*N*-toluenesulfonyl-1,7,13-trioxa-4,10,16-triazacyclooctadecane-9,17-dione. This compound is found by solid state structure analysis to fold to form an intramolecular, doubly-hydrogen bonded complex in which the two N—H protons interact with the two tosyl group oxygens. Details of the structure are reported.

Key words. Cation binding, crown ether, crystal structure, lariat ether, synthesis, triaza-18-crown-6.

Ave Charles Pedersen

In this issue dedicated to the memory of Charles Pedersen, it is fitting to recall a personal interaction with this fine scientist. One of the authors (GWG) was an employee at the du Pont Experimental Station during the Summer of 1974. At that time, the crown ethers were becoming very important subjects of research throughout the world and were of great interest at du Pont because of a project involving their antiviral activity then underway. It was my impression, however, that Pedersen was not very well known (except by reputation) among my contemporaries at the Central Research Department. This was for two reasons. Pedersen made the Elastomer Chemical Department his home during his career and, by 1974 when I arrived, he had already retired. It was only several years later when a Professor visiting my group from the Soviet Union asked to meet Pedersen that I made personal contact. It was a very pleasant meeting, both socially and scientifically, but the strongest recollection I have of it is Pedersen's comment to me that he felt like a father who was forced to give up a child for adoption. He meant, I believe, that he had many ideas that he could not try because of his retirement.

This paper is dedicated to the memory of the late Dr C. J. Pedersen.

One wonders what direction crown chemistry might have taken had Pedersen been involved longer in orchestrating its development.

1. Introduction

As our lariat ether [1] program has evolved over the last decade [2], we gradually increased the number of sidearms present on the macroring in the hope of enhancing the binding strength and selectivity as well as the versatility of this class of compounds. The two-armed or bibracchial lariat ethers (BiBLEs) [3] utilize both sidearms in complexing cations [4] and these sidearms may interact either from the same or opposite sides of the macroring. It seemed a natural question to ask what role a third arm would play since two arms already provided octa-coordination [4]. We set as our target the parent macrocycle 4,10,6-triaza-18-crown-6, previously prepared by Lehn and coworkers [5]. We report here the details [6] of synthesis, homogeneous binding properties, and an unusual intramolecular 'complex' of a key intermediate in the synthesis.

2. Experimental

^{1}H-NMR spectra were recorded on a Varian VXR-400 NMR spectrometer or on a Hitachi Perkin-Elmer R-600 high resolution NMR spectrometer in $CDCl_3$ solvents and are reported in ppm (δ) downfield from internal Me_4Si. ^{13}C-NMR were recorded on a JEOL EX90Q or Varian VXR-400 NMR spectrometer. Infrared spectra were recorded on a Perkin-Elmer 1310 Infrared Spectrophotometer. Melting points were determined on a Thomas Hoover apparatus in open capillaries and are uncorrected. Thin layer chromatographic (TLC) analyses were performed on aluminum oxide 60 F-254 neutral (Type E) with a 0.2 mm layer thickness or on silica gel 60 F-254 0.2 mm layer thickness. Preparative chromatography columns were packed with activated aluminum oxide (MCB 80–325 mesh, chromatographic grade, AX 611) or with Kieselgel 60 (70–230 mesh).

All reactions were conducted under dry nitrogen unless otherwise noted. All reagents were the best grade commerically available and were distilled, recrystallized, or used without further purification, as appropriate. Molecular distillation temperatures refer to the oven temperature of a Kugelrohr apparatus. Combustion analyses were performed by Atlantic Microlab, Inc., Atlanta, GA, and are reported as percentages.

Cation binding constants were measured in absolute MeOH at 25.0 ± 0.1°C using a Corning 476210 electrode and an Orion model 701A 'Ionalyzer' meter according to the method of Frensdorff [7] as described recently in detail [8]. Values for the equilibrium constants are reported as $\log_{10} K_S$.

N,N,-bis(2-Hydroxyethyl)-p-toluenesulfonamide, **1**

Prepared by the method of Lehn and coworkers [5] in 85% yield, mp 100–101°C, and had spectral properties identical to those reported.

N-p-Toluenesulfonyl-2,8-*dioxa*-5-*aza*-1,9-*nonanedicarboxylic acid*, **2**

The disodium salt of **1** (90.0 g, 0.347 mol) was prepared from 10.2 g of NaH (60% in oil) in 2500 mL of THF by stirring the reagents together for 24 h at room temperature with 500 g of 3 mm glass beads in the flask. Chloroacetic acid, sodium salt (101.1 g, 0.867 mol) was added to the reaction flask. The reaction was stirred at reflux temperature for 2 d. The reaction was allowed to cool and 20 mL of water was added. CH_2Cl_2 (400 mL) was added to the reaction mixture, which was then filtered. The filtered solid was dissolved in 600 mL of CH_2Cl_2 and 1200 mL of 1M NaOH. The resulting two-phase system was poured through a Buchner funnel (no filter paper, to remove the glass beads) into a separatory funnel, shaken, and the phases were separated. The organic layer was extracted with 1M NaOH (2 × 100 mL). The combined aqueous phases were extracted with CH_2Cl_2 (100 mL). $BaCl_2 \cdot 2H_2O$ (180 g) was added to the aqueous portion with vigorous stirring. The resulting precipitate was filtered. The precipitate was dissolved in 1200 mL of CH_2Cl_2 and 600 mL of 3M HCl, shaken, and the phases were separated. The organic portion was extracted with 3M HCl (2 × 100 mL). The aqueous portions were extracted with CH_2Cl_2 (3 × 100 mL). The combined organic portions were dried over Na_2SO_4 and evaporated *in vacuo*. Recrystalization (300 mL CH_2Cl_2) afforded 98.0 g (75%) of a white solid, mp 85–87°C (lit. [5c] 87–88°C). The product had spectral properties identical to those reported [5].

N-p-Toluenesulfonyl-2,8,*dioxa*-5-*aza*-1,9-*nonanedicarboxylic acid chloride*, **3**

Oxalyl chloride (24.0 g, 0.189 mol) was added to a suspension of **2** (20.0 g, 0.053 mol) in benzene (200 mL) containing 4 drops of pyridine. The reaction mixture was stirred for 2 d at room temperature, concentrated *in vacuo* and any remaining HCl and oxalyl chloride were removed at high vacuum during 24 h. The pale yellow oil (100% crude yield) had spectral properties identical to those reported [5] and was used in the next step without further purification.

N-p-Toluenesulfonyl-4,10,16-*triaza*-1,7,13-*trioxacyclooctadecane*-9,17-*dione*, **4**

A solution of **3** (21.0 g, 0.050 mol) in benzene (170 mL) was prepared. A second solution of bis(2-aminoethyl)ether (10.48 g, 0.100 mol) in benzene (170 mL) was prepared. Both solutions were pumped into a 3L flask, containing 1L of benzene, during 8 h using a Sage Model 352 syringe pump. The reaction mixture was warmed to 30°C, filtered through celite, and the filtrate was concentrated *in vacuo*. The product was used without further purification and had spectral properties identical to those reported [5]. One sample was purified and the yield was 65%.

Crystal data for compound **4** [mp (after recrystallization from $CHCl_3 : C_6H_6$ 1 : 5 v/v) 115–116°C, lit. [5] mp 117–118°C]: $C_{19}H_{33}O_7N_3S$, FW = 447.55, space group $P\bar{1}$, $a = 7.1357(5)$, $b = 10.1681(4)$, $c = 15.5037(7)$ Å, $z = 2$, $d_c = 1.349$ g cm^{-1}, CuK_α ($\mu = 16.65$ cm^{-1}) $R = 0.021$ for 2119 unique reflections with $I > 2\sigma(I)$ (of 2130 unique data) measured by an Enraf-Nonius CAD4 X-ray diffractometer by $\omega - 2\theta$ scans, $2° < \theta < 50°$.

4,10,16-*Triaza*-1,7,13-*trioxacyclooctadecane*, '*triaza*-18-*crown*-6', **5**

Compound **4** (16.0 g, 0.036 mol) was slowly added as a solid to a stirred suspension of lithium aluminum hydride (27 g, 0.7 mol) in THF (600 mL). The reaction was stirred for 36 h at reflux temperature and was then allowed to cool. Water (50 mL) was added during 1 h and a LiOH solution (15%, 30 mL) was added during 30 min. Anhydrous $MgSO_4$ (35 g) was added followed by 20 g of celite. The reaction was filtered through filter paper and the filtrate was saved. The filtered solid was placed in 500 mL of CH_2Cl_2 in a 1 L beaker. The suspension was stirred with a 20-cm long, 3-cm diameter, 4-blade stirrer at 1200 rpm for 1 min. The suspension was filtered through filter paper. This procedure was repeated a second time. The combined filtrates were concentrated *in vacuo*. The solid residue was recrystallized from hexanes (400 mL) to give a light yellow-green crystalline solid which was then sublimed (0.02 torr, 130–135°C) using an Aldrich Kugelrohr apparatus. The sublimed solid was recrystallized from 300 mL of hexanes by refluxing hexanes through the Kugelrohr receiver bulb until the solid material dissolved in the refluxing hexanes. The title compound was obtained as a white crystalline solid, 6.1 g (65%), m.p. 132.5–135.5°C (lit. [5] 134–135°C), when the hexane solution was cooled to −5°C. The compound had spectral properties identical to those reported [5].

N,N′N″-tris(*Propargyl*)-4,10,16-*triaza*-18-*crown*-6, **6**

To a solution of **5** (1.95 g, 7.5 mmol) and Na_2CO_3 (3.18 g, 30 mmol) in MeCN (35 mL) was added propargyl bromide (2.85 g, 24 mmol). The reaction was stirred at reflux temperature for 16 h. The reaction was cooled, concentrated *in vacuo*, dissolved in CH_2Cl_2 (200 mL), and filtered through celite. The organic solution was extracted with 1M HCl (100 mL). The aqueous layer was then basified with Li_2CO_3 to pH 10 and extracted with CH_2Cl_2 (3 × 50 mL). The organic portions were dried ($MgSO_4$) and concentrated *in vacuo* to give a pale yellow oil. The oil was distilled (Kugelrohr apparatus, bp 160–165°C, 0.1 torr) to give 1.0 g (36%) of a pale yellow oil. ^{1}H-NMR: 2.17 (*t*, 3H, C≡CH), 2.78 (*t*, 12H,CH_2N), 3.48 (*s*, 6H, NCH_2C≡C), 3.58 (*t*,12H, CH_2O); IR (neat): 3320, 2950, 2880, 2100, 1460, 1370, 1330, 1130(s), 1060, 1000, 900, 640 cm^{-1}; High res. mass. spec. *calcd.* for $C_{21}H_{33}N_3O_3$: 375.2522. *Found*: 375.2530.

N,N′N″-tris(2-*Hydroxyethyl*)-4,10,16-*triaza*-18-*crown*-6, **7**

Alkylated using the method of Kulstad and Malmsten [9]. Ethylene oxide (1.02 g, 23 mmol) was added to a solution of **5** (1.02 g, 3.9 mmol) in MeOH (10 mL) at 0°C. The reaction was heated at reflux temperature for 2 h while the condenser was maintained at 0°C. The condenser was then operated at 20°C and the reaction was heated at reflux temperature for an additional hour. The reaction mixture was cooled and concentrated *in vacuo*. The resulting oil was distilled (bp 210–220°C, 0.1 torr) to give 1.2 g (78%) of a colorless oil. ^{1}H-NMR: 270 (*t*, 18H, CH_2N), 3.55 (*t*, 18H, CH_2O), 3.75 (*s*, broad, 3H, OH); IR (neat): 3370(b), 2960, 2900, 2840, 1450, 1360, 1120(s), 1080, 940, 740 cm^{-1}; High res. mass. spec. *Calcd.* for $C_{18}H_{39}N_3O_6$: 393.2838. *Found*: 393.2823.

N,N′,N″-tris(Methoxymethylcarbonyl)-4,10,16-*triaza*-18-*crown*-6, **8**

To a vigorously stirred solution containing **5** (1.50 g, 5.74 mmol) and triethylamine (0.59 g, 5.8 mmol) in benzene (30 mL) was added a solution containing methoxyacetyl chloride (0.53 g, 5.8 mmol) in benzene (25 mL) during 30 min. After the addition, the reaction mixture was stirred at room tepmerature for 1 h. The reaction mixture was concentrated *in vacuo* and the residue was distilled in a Kugelrohr apparatus (b.p. 210–220°C, 0.02 torr) to give 1.2 g (44%) of a pale yellow oil. ^{1}H-NMR: 3.40 (*s*, 9H, OCH_3), 3.60 (*m*, 24H, crown ring), 4.10 (*s*, 6H, $COCH_2OR$); IR (neat): 2960, 2910, 1670(s), 1490, 1440, 1370, 1210, 1120(s), 1020, 940 cm^{-1}; High res. mass. spec. *Calcd.* for $C_{21}H_{39}N_3O_9$; 477.26863. *Found*: 477.26538.

N,′N″-tris(2-Methoxyethyl)-4,10,16-*triaza*-18-*crown*-6, **9**

Compound **8** (1.0 g, 2.1 mmol) was dissolved in a solution of $BH_3 \cdot THF$ (50 mL, 1.0 M). The reaction was heated at reflux temperature for 24 h, cooled, and then concentrated *in vacuo.* The residue was dissolved in 3M HCl (40 mL) and heated at reflux temperature for 24 h. Water (50 mL) was added and the reaction was extracted with CH_2Cl_2 (3 × 50 mL). The aqueous layer was basified with solid LiOH to pH 10 and was extracted with CH_2Cl_2 (300 mL) in a continuous extraction apparatus for 3 d. The organic phase was concentrated *in vacuo* and the residue was distilled (b.p. 160–165°C, 0.02 torr) in a Kugelrohr apparatus. Column chromatography on neutral alumina (40 g, hexanes/CH_2Cl_2/EtOH, increasingly polar solvent gradient) followed by another distillation (b.p. 160–165°C, 0.02 mm) resulted in 250 mg (31%) of a non-viscous pale yellow oil. ^{1}H-NMR: 2.75 (*t*, 18H, NCH_2R), 3.25 (*s*, 9H,OCH_3), 3.45 (*t*, 18 H, OCH_2R); IR (neat): 2960, 2900, 1470, 1370, 1310, 1210, 1130, 1080, 960, 830 cm^{-1}; High res. mass. spec. *Calcd.* for $C_{21}H_{45}N_3O_6$: 435.3308. *Found*: 435.3328.

N,N′,N″-tris(Carbethoxymethyl)-4,10,16-*triaza*-18-*crown*-6, **10**

A solution of **5** (2.0 g, 7.65 mmol), ethyl chloroacetate (2.91 g, 23.7 mmol) and Na_2CO_3 (2.52 g, 23.8 mmol) in MeCN (50 mL) was heated at reflux temperature for 100 h. The reaction was then cooled, filtered, and concentrated *in vacuo.* The brown residue was dissolved in CH_2Cl_2 (50 mL) and washed with water (50 mL). The compound was then extracted into 1M HCl (50 mL) and the aqueous layer was extracted with CH_2Cl_2 (50 mL). The aqueous layer was basified to pH 10 with solid LiOH, extracted with CH_2Cl_2 (3 × 50 mL), and the combined organic portions were concentrated *in vacuo.* The resulting yellow oil was filtered through alumina (30 g) and then distilled (b.p. 195–200°C, 0.01 torr) to give 1.2 g (30%) of a colorless oil. ^{1}H-NMR: 1.28 (*t*, 9H, CH_3), 2.92 (*t*, 12H, CH_2N), 2.5 (*s*, 6H, NCH_2CO), 3.55 (*t*, 12H, CH_2O), 4.15 (*q*, 6H, OCH_2CH_3); IR (neat): 2950, 2900, 1750, 1460, 1370, 1200, 1120, 930 cm^{-1}; *Anal. Calcd.* for $C_{24}H_{45}N_3O_9$; C, 55.47: H, 8.73. *Found*: C, 55.55; H, 8.76%.

N,N'N''-tris(Benzyl)-4,10,16-triaza-18-crown-6, **11**

A solution of triaza-18-crown-6 (200 mg, 0.77 mmol) and Na_2CO_3 (250 mg, 2.38 mmol) in CH_3CN (10 mL) was prepared. A solution of benzyl chloride (280 mg, 2.2 mmol) in CH_3CN (5 mL) was added during 20 min. The reaction was heated at reflux temperature for 12 h, cooled, filtered, and concentrated *in vacuo*. Chromatography over alumina (10% *i*-PrOH in hexane) gave 370 mg (91%) of a pale yellow oil. ^{1}H-NMR: 2.70 (*t*, 12h, $J = 5$ Hz, CH_2N), 3.45 (*s* + *t*, 18H, $J = 5$ Hz, CH_2O and NCH_2Ar), 7.10 (*s*, 15H, Ar); IR (neat): 3080, 3060, 3030, 2900, 1500, 1455, 1120, 1060, 730, 700 cm^{-1}; *Anal. Calcd.* for $C_{33}H_{45}N_3O_3$: C, 74.58; H, 8.47; N, 7.91. *Found*: C, 74.26; H, 8.70; N, 7.78.

N,N',N''-tris(2-Nitrobenzyl)-4,10,16-triaza-18-crown-6, **12**

Compound **5** (0.50 g, 1.9 mmol), 2-nitrobenzyl chloride (1.00 g, 5.8 mmol), and Na_2CO_3 (1.52 g, 14.3 mmol) were combined in MeCN (13 mL) and stirred at reflux for 20 h. The reaction was cooled, filtered, and concentrated *in vacuo*. The residue was dissolved in CH_2Cl_2 (20 mL) and was then washed with water (20 mL). The organic portion was extracted with 1M HCl (30 mL). The aqueous portion was basified to pH 10 with LiOH and extracted with CH_2Cl_2 (2 × 20 mL). The organic portions were dried ($MgSO_4$), and concentrated *in vacuo* to give 0.60 g (50%) of an orange oil. ^{1}H-NMR ($CDCl_3$) 2.77 (*t*, 12H, CH_2N), 3.50 (*t*, 12H, CH_2O), 3.96 (*s*, 6H, $ArCH_2N$), 7.10–8.00 (*m*, 12H, Ar); ^{13}C-NMR: 54.3, 56.58, 69.48, 124.03, 127.48, 130.88, 132.32, 135.33, 149.41; *Anal. Calcd.* for $C_{33}H_{42}N_6O_9$; C, 59.45; H, 6.35%.

N,N',N''-tris(n-Hexyl)-4,10,16-triaza-18-crown-6, **13**

To a refluxing solution of triaza-18-crown-6 (200 mg, 0.77 mmol) and Na_2CO_3 (250 mg) in CH_3CN (10 mL) was added a solution of 1-bromohexane (370 mg, 2.2 mmole) in CH_3CN (5 mL) dropwise over 20 min. The mixture was then left to reflux for 24 h. The mixture was cooled, filtered, the solvent removed *in vacuo* and the residue chromatographed over alumina (5% 2-propanol/hexanes) to yield 200 mg (51%) of *N,N',N''-tris-(n-hexyl)*-4,10,16-*triaza*-18-*crown*-6. ^{1}H-NMR: 5.5 (*t*, 12H, $J = 5$), 2.4 (*m*, 18H), 1.4–1.7 (*m*, 33H). *Anal. Calcd.* for $C_{30}H_{63}N_3O_3$: C, 70.18; H, 12.28; N, 8.19%. *Found*: C, 70.08; H, 12.50; N, 8.05%.

3. Results and Discussion

The Lehn procedure deserves great credit because it was the first to afford triaza-18-crown-6. The procedure is serviceable on a small scale but, in our hands at least, proved cumbersome on a multi-gram scale. Moreover, Lehn's approach to the *C–C–X* unit in $TsN(CH_2CH_2O-C-C-X)_2$ involved chloromethylation of $TsN(CH_2CH_2OH)$ to give $TsN(CH_2CH_2OCH_2Cl)_2$ followed by reaction with cyanide to give $TsN(CH_2CH_2OCH_2C{\equiv}N)_2$. This, in turn, had to be hydrolyzed to the corresponding diacid. We attempted a more direct route, although the influence of the Lehn approach will be obvious. It is summarized schematically below.

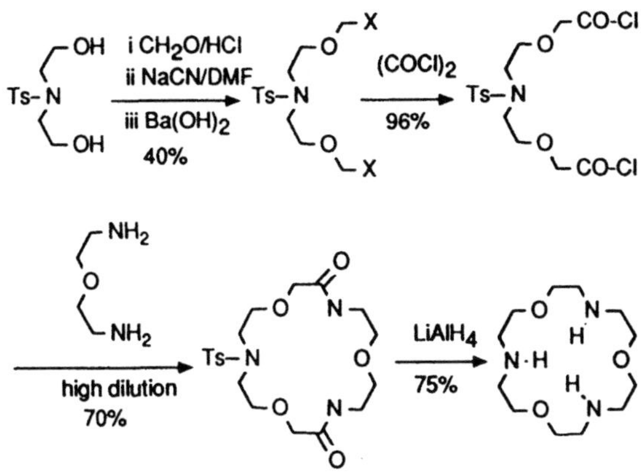

The chloromethylation reaction followed by cyanide treatment and then barium hydroxide catalyzed hydrolysis is obviously an indirect approach to the diacid. Even so, the reaction is successful and the overall yield for Lehn's sequence is 20%.

The previously unreported but obvious reaction sequence:

$$TsN(CH_2CH_2OH)_2 + ClCH_2C{\equiv}N \rightarrow TsN(CH_2CH_2OCH_2C{\equiv}N)_2$$

proved impractical in our hands. We therefore attempted to circumvent all three steps, i.e. chloromethylation, reaction with cyanide, and hydrolysis by using the following sequence:

$$TsN(CH_2CH_2OH)_2 + ClCH_2COONa \rightarrow TsN(CH_2OCH_2COOH)_2.$$

This sequence succeeded, although vigorous agitation (glass beads were present in the reaction mixture) was required for best yields. Once the diacid was in hand, the strategy followed that described by Lehn with some modifications to the experimental procedure. The latter are detailed in the experimental section. The overall procedure reported here is as shown below. This approach has permitted the accumulation of tens of grams of this versatile material [3].

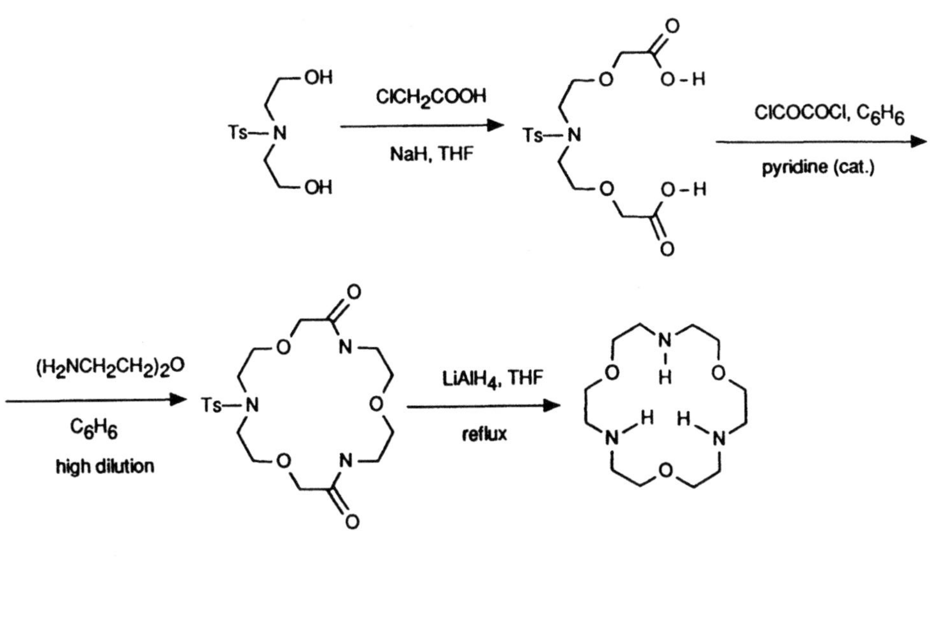

The yields for the formation of each product are as follows: **2**, 75–85% (many replicates); **3**, ~100%; **4**, 65% and **5**, 65%. The overall yield for this sequence is thus 32%, compared to the yield Lehn reported of 20%. Note that we have not been able to achieve quite such high yields as Lehn reported for the penultimate and final steps. Were we able to do so, our overall yield would improve to nearly 40%. All of our yield estimates are based on 75% in the first step although we have sometimes realized 85% in this step. As indicated above, the yield is sensitive to agitation.

3.1. TRIBRACCHIAL LARIAT ETHERS

The three-armed lariat ether (TriBLE) derivatives of triaza-18-crown-6 were prepared by alkylation of the parent compound in acetonitrile. The incipient sidearm was introduced in electrophilic form, usually as its chloride. Sodium carbonate was used as base. Compound **8**, which has three $CH_2CH_2OCH_3$ sidearms, was prepared by acylation of **7** followed by reduction. Using this approach, the yields shown in Table I were obtained.

3.2. CATION BINDING PROPERTIES

The cation binding constants recorded in Table I were determined by ion selective electrode techniques [8] in anhydrous methanol solution at 25.0 ± 0.1°C. The parent crown (**5**, sidearm = H) and the tris(methyl) derivative of **5** exhibit generally low binding. This is expected for **5** since the solid state structure of **5** [6] shows that all three N—H bonds may be focussed into the cavity's center even in solution. The data for the tris(methyl) derivative of **5** were taken from the literature and while

Table I. Cation binding properties of TriBLEs.

Cpd. No.	Sidearm on 5	% yield	log K_S^a Na$^+$	K$^+$	Ca^{2+}
5	H	20–30[b]	1.8	<1.5	<1.5
–	CH_3	—	3.11[c]	2.78[c]	—
6	$CH_2C{\equiv}CH$	36	4.03	5.10	4.12
7	CH_2CH_2OH	78	4.15	4.45	5.58
8	$CO—CH_2—O—CH_3$	44	—	—	—
9	$CH_2CH_2—O—CH_3$	31[c]	4.19	4.93	4.07
10	$CH_2CO—O—CH_2CH_3$	30	5.13	5.87	6.70
10	$CH_2CO—O—CH_2CH_3$	30	1.9[d]	—	4.62[d]
11	$CH_2—C_6H_5$	91	—	—	—
12	$CH_2—C_6H_4—2—NO_2$	50	—	—	—
13	*n*-hexyl	51%	—	—	—

[a]All binding data were determined at 25.0 ± 1.0°C in anhydrous methanol.
[b]Yield range given for the present procedure and that reported in Ref. [5], see text.
[c]Binding constants determined in 90% MeOH–10% H_2O and are from Ref. [11].
[d]Binding constants determined in water.

they are presumed to have been determined under comparable conditions, this is not certain. We did not prepare this derivative nor did we check these values as the sidearms cannot contribute to binding in the anticipated way.

Cation binding strengths for the remaining derivatives, namely **6–13**, are fairly high. It is interesting that Na^+ binding differs little whether the sidearm is $CH_2{-}C{\equiv}CH$, CH_2CH_2OH or $CH_2CH_2OCH_3$. One is tempted to account for this in terms of a π-contribution from the sidearm. Previous studies of π-sidearm BiBLEs have shown that a combination of differences in the entropic and enthalpic contributions can account for these similarities [10].

Two features of the cation binding deserve special note. First, as the sidearm donor groups increase in polarity, cation binding generally rises. This is true for the polarity sequence **8** < **7** < **10**. For this series, Ca^{2+} binding (log K_S) increases in the same order: 4.07, 5.58, and 6.70. Second, cation binding strength for these derivatives is considerable, even in water (log K_S 4.62). This must be due to participation by one or more sidearms.

Although we have no solid state structure of any TriBLE complex, we believe that only two of the three sidearms participate in binding. We draw this conclusion from a comparison of cation binding strengths for two- and three-armed systems in which the sidearms are the same. For the series of compounds noted above, i.e. CH_2CH_2OH (**7**), $CH_2CH_2OCH_3$ (**8**), and $CH_2COOCH_2CH_3$ (**10**), log K_S for Na^+ is, respectively, 4.83, 4.75, and 5.51 for the two-armed compounds compared to 4.15, 4.19, and 5.13 for the TriBLEs. Two possible explanations for the lower binding strength of TriBLEs compared to BiBLEs are obvious. It may be that the TriBLEs use the macroring relatively little and most of the binding strength derives from sidearm donors. Alternately, binding by TriBLEs may occur just as in the BiBLE case (i.e. ring and two sidearms) but the third arm 'gets in the way'. Such interference by the third arm could lower the effectiveness of the remaining two binders. Both of these explanations are essentially enthalpic and entropy must play a greater role when three arms are present in otherwise analogous systems.

The parallel in binding strengths for the Na^+, K^+, and Ca^{2+} ions by the BiBLE and TriBLE (**10**) having $CH_2COOCH_2CH_3$ sidearms is striking. The respective binding constants are: Na^+, 5.51 vs. 5.13; K^+, 5.78 vs. 5.87; and 6.78 vs. 6.70. Again, it is presumptuous to interpret these values using pure enthalpic arguments but one might easily surmise that binding in both cases relies on two sidearms. The data for these systems are summarized in Table I.

3.3. SOLID STATE STRUCTURE OF **4**

A key intermediate in our approach is 4-*N*-tosyloxy-9,17-dioxo-4,10,16-triaza-18-crown-6, **4**, which is, in turn, prepared from the *N*-tosyldiacid chloride, Ts-$N(CH_2CH_2OCH_2COCl)_2$ by high dilution reaction with 2,2,′-diaminodiethyl ether. Compound **1** is critical to our plans since we wish to attach two different kinds of sidearms to this tribracchial lariat ether. We now report X-ray crystal structure data showing that this important intermediate is the first example of a cation-free lariat ether compound that exhibits specific, if weak, sidearm-macroring interaction.

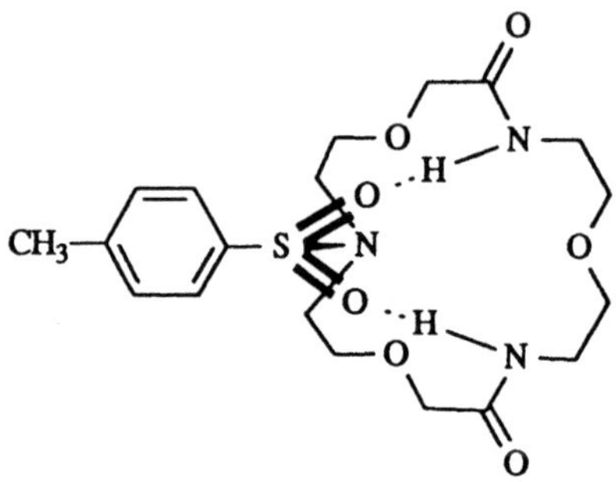

Two views of the solid state structure are shown below. Rather than being planar or nearly so, the macroring is puckered so that the sidearm can reach over the ring. The sidearm is tilted but the ring and amide oxygens approximate a plane and the aromatic ring is almost parallel to it. The two S → O bonds are oriented directly toward the amide nitrogens which are intrannular and at a distance of 3.28 Å and 3.41 Å respectively. The O—H—N angle in the former case is 161° and in the latter it is 154°. The more acute angle is consistent with the longer O···N distance and the clearly weaker interaction. It is, of course, well known that the vast majority of N—H···O hydrogen bonds in amino acids fall within the range 155–175° and relatively few are actually 180° [12].

On the other hand, these hydrogen bonds are long by any standard. The N—H distance in the shorter case is 0.94 Å and in the longer case the corresponding distance is 1.01 Å. The corresponding H···O distances are 2.37 Å and 2.45 Å. There is no obvious packing force that would make this crown pucker so severely were there not some stabilizing interaction present between ring and sidearm. It should be noted that a similar phenomenon is observed in valinomycin which folds into a 'pre-binding' conformation by virtue of hydrogen bond stabilization. Even so, we believe this neutral ring-sidearm interaction to be unique in crown ether chemistry.

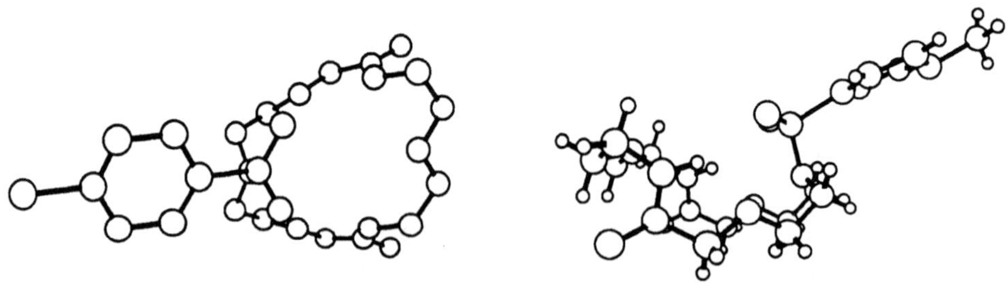

The present structure compares with three other reports of similar compounds or similar interactions. Buchanan, Ripmeester and coworkers [13] have reported a 1 : 1

complex of 18-crown-6 which has a melting point similar to a previously reported [14] complex but differs in stoichiometry from it. The 1 : 1 complex (**14**) is stabilized by hydrogen bond formation between sulfonamide amino nitrogen and two oxygen atoms of 18-crown-6. Thus the O(1)—N and O(7)—N distances are reported to be 3.003 Å and 3.002 Å, respectively [13]. Stoddart, Williams, and coworkers [15] have reported the solid state structure of a diazadibenzo-30-crown-10 disulfonamide which failed to form complexes with 'diquat^{2+}'. In this case, edge-to-face aromatic interactions apparently dominate the structure. No hydrogen bond formation is possible in any event since there are no >NH bonds either in the host or guest. Malinovsky and coworkers [16] have recently reported the structure of **15** which forms a dimer in which the aromatic amine group of one molecule complexes more or less as shown for **14**. Intermolecular distances for the amino nitrogen of one molecule to the macroring of a second molecule are, for three of the five oxygen atoms in the latter, 2.95–3.15 Å (six distances reported) [16].

Despite the somewhat long hydrogen bond distances in the present complex, the sidearm-macroring interaction is clear and, in the most general sense, precedented. This is the first such interaction observed but may presage other observations of this type. The triaza-18-crown-6 system should now prove more accessible and holds forth the promise of considerable interesting complexation chemistry.

Acknowledgment

We warmly thank the NIH for support of this work by research grant GM 36262.

References

1. G. W. Gokel, D. M. Dishong and C. J. Diamond: *J. Chem. Soc., Chem. Commun.* 1053 (1980).
2. (a) D. M. Dishong, C. J. Diamond, M. I. Cinoman and G. W. Gokel: *J. Am. Chem. Soc.* **105**, 583 (1983); (b) R. A. Schultz, B. D. White, D. M. Dishong, K. A. Arnold and G. W. Gokel: *J. Am. Chem. Soc.* **107**, 6659 (1985).
3. V. J. Gatto and G. W. Gokel: *J. Am. Chem. Soc.* **196**, 8240 (1984).
4. K. A. Arnold, L. Echegoyen, F. Fronczek, R. D. Gandour, V. J. Gatto, B. D. White and G. W. Gokel: *J. Am. Chem. Soc.* **109**, 3716 (1987).
5. (a) E. Graf and J.-M. Lehn: *J. Am. Chem. Soc.* **97**, 5022 (1975); (b) J.-M. Lehn: *U.S. Patent* 3,888,887, June 10, 1975; (c) E. Graf and J.-M. Lehn: *Helv. Chim. Acta* **64**, 1040 (1981).
6. Portions of this work have appeared in preliminary form: S. R. Miller, T. P. Cleary, J. E. Trafton, C. Smeraglia, R. D. Gandour, F. R. Fronczek and G. W. Gokel: *J. Chem. Soc., Chem. Commun* 608 (1989).
7. H. K. Frensdorff: *J. Am. Chem. Soc.* **93**, 600 (1971).
8. K. A. Arnold and G. W. Gokel: *J. Org. Chem.* **51**, 5015 (1986).
9. S. Kulstad and L. A. Malmsten: *Acta Chem. Scand.* **B33**, 469 (1979).
10. K. A. Arnold, A. M. Viscariello, M. S. Kim, R. D. Gandour, F. R. Fronczek and G. W. Gokel: *Tetrahedron Lett.* 3025 (1988).
11. P. Vierling and J.-M. Lehn: *Tetrahedron Lett.* **21**, 1323 (1980).
12. T. F. Koetzle and M. S. Lehmann: in *The Hydrogen Bond*, P. Schuster, G. Zundel and C. Sandorfy (Eds.) p. 466, North Holland Publishing Company, Amsterdam (1989).
13. G. W. Buchanan, C. Morat, J. P. Charland, C. I. Ratcliff, and J. A. Ripmeester: *Can. J. Chem.* **67**, 1212 (1989).
14. A. Knoechel, J. Kopf, J. Oehler and G. Rudolph: *J. Chem. Soc., Chem. Commun.* 595 (1978).
15. P. L. Anelli, A. M. Z. Slawin, J. F. Stoddart and D. J. Williams: *Tetrahedron Lett.* 1575 (1988).
16. S. T. Malinovsky, Y. A. Simonov, E. V. Ganin, V. F. Makarov and S. A. Kotlyar: *J. Struct. Chem., U.S.S.R.* **30**, 129 (1989).

Journal of Inclusion Phenomena and Molecular Recognition in Chemistry **12**: 199–211, 1992.

Bis(crown ether)s as Na^+-K^+ ATPase Model in a Liquid Membrane

MITSUO OKAHARA,* YOHJI NAKATSUJI, MASAYUKI SAKAMOTO, and MASAKI WATANABE
Department of Applied Chemistry, Faculty of Engineering, Osaka University, Yamadaoka 2-1, Suita, Osaka 565, Japan.

(Received: 7 September 1990; in final form: 12 November 1990)

Abstract. A new type of bis(crown ether)s containing 15-crown-5 and monoaza-18-crown-6 actively transported Na^+ and K^+ in opposite directions across a dichloromethane membrane by pH control. The effect of the structure of the ion carriers on their transport abilities was examined. A key point for the molecular design of the carriers is how to give them complexing ability toward Na^+ under acidic conditions. A proper choice of the transport conditions was found to be essential for the success of the double uphill transport.

Key words. Synthetic ionophore, active transport, alkali metal cation transport, solvent extraction.

1. Introduction

Our early attention in the field of crown chemistry created by Pedersen [1] was paid to developing facile synthetic methods for crown ethers [2]. The intramolecular cyclization method [3] developed by us for simple crown ethers [4] was successfully applied to the preparation of *N*-substituted azacrown ethers [5], thiacrown ethers [6], and polyether esters [7]. Synthetic procedures for crown ethers with a functional group such as a hydroxyl [8], aminomethyl [9], and bromomethyl [10] enabled the molecular design of highly functionalized derivatives [11–16], which recognize a specific molecule.

One of their important characteristics is the complexing ability toward hard cations such as alkali metal and alkaline earth metal cations, as first shown by Pedersen. This property is especially noteworthy from the viewpoint of the resemblance to the complexation property of naturally occurring antibiotics. In other words, crown ethers are simple model compounds for such natural ionophores [17]. The function of valinomycin and monensin has been realized in an artificial transport system using synthetic ionophores. Although a variety of proton-ionizable crowns has been documented as suitable synthetic ionophores, we succeeded in developing another active transport system using lipophilic azacrown ethers [18]. These compounds are able to selectively transport alkali metal cations against their concentration gradients based on the recognition of the ion sizes with their cavity sizes [19, 20]. The enzyme Na^+-K^+ ATPase actively transports Na^+ and K^+ in opposite directions through a biomembrane [21]. Simple lipophilic azacrown ethers,

This paper is dedicated to the memory of the late Dr C. J. Pedersen.

* Author for correspondence.

however, have no ability to transport a metal ion from an acidic to a basic phase. In order to mimic the function of an enzyme in an artificial transport system, another structural device is needed for the design of the ionophore. Lipophilic monoaza-18-crown-6 can selectively transport K^+ from a basic to an acidic phase against its concentration gradient [18]. If a derivative containing monoaza-18-crown-6 possesses complexing ability toward Na^+ under acidic conditions, it has the possibility of transporting K^+ and Na^+ in opposite directions.

From this point of view, we will describe the synthesis of new bis(crown ether)s and the characterization of their transport ability as the model for the Na^+–K^+ ATPase [22].

2. Experimental

^{1}H NMR spectra were taken at 400 MHz on a JEOL JNM-GSX-400 spectrometer, using tetramethylsilane as the internal standard. IR and UV spectra were obtained with a Hitachi 260-10 spectrometer and a Shimadzu UV-200 spectrophotometer, respectively. Mass spectra were measured with a JEOL JMS-DX303 mass spectrometer. 2-(Bromomethyl)-2-methyl-15-crown-5 (**6**) [10], monoaza-15-crown- 5 (**7a**) [23], monoaza-18-crown-6 (**7b**) [23], *N*-(2-hydroxyethyl)-monoaza-15-crown-5 (**8a**) [13], *N*-(2-hydroxyethyl)-monoaza-18-crown-6 (**8b**) [13], *N*-[2-(2-hydroxyethoxy)ethyl]-monoaza-15-crown-5 (**9a**) [13], *N*-[2-(2-hydroxyethoxy)ethyl]-monoaza-18-crown-6 (**9b**) [13] were prepared according to the literature. *N*-(3-Hydroxypropyl)-monoaza-15-crown-5 (**10a**) was prepared by the reaction of **7a** with 3-chloropropanol in the presence of sodium carbonate at 120°C for 45 h and purified by distillation in a Kugelrohr apparatus (130°C/0.04 torr) to give a slightly yellow oil in a 60% yield. *N*-(3-Hydroxypropyl)-monoaza-18-crown-6 (**10b**) was prepared as above for **10a** from **7b** and purified by distillation in a Kugelrohr apparatus (150°C/0.05 torr) to give a slightly yellow oil in an 84% yield. *N*-Octyl-diethanolamine (**11**) was prepared by the reaction of diethanolamine with *n*-octyl bromide in the presence of sodium carbonate at 120°C for 72 h and purified by distillation in a Kugelrohr apparatus (110°C/0.04 torr) to give a slightly yellow oil in a 65% yield.

2.1. PREPARATION OF **1a**

13-[(2-Methyl-1,4,7,10,13-pentaoxadec-2-yl)methyl]-1,4,7,10-tetraoxa-13-azacyclopentadecane (**1a**) was prepared by the following sequence. A mixture of 2-(bromomethyl)-2-methyl-15-crown-5 (**6**) (3.27 g, 10 mmol), monoaza-15-crown-5 (**7a**) (2.18 g, 10 mmol), and potassium carbonate (1.66 g, 12 mmol) was stirred at 140°C for 62 h under a nitrogen atmosphere. Water (100 mL) was added to the cooled reaction mixture and extracted with dichloromethane (100 mL × 3). The combined organic layer was dried over $MgSO_4$ and concentrated. Low boiling compounds were removed under reduced pressure in a Kugelrohr apparatus (110°C/0.04 torr). The crude product was purified by chromatography over alumina (dichloromethane/dioxane = 10/1 ~ 2/1 (v/v)) to give **1a** (1.87 g, 40%) as a slightly yellow oil. The analytical data for **1a** are as follows: IR (neat) 2880, 1450, 1300, 1250, 1130 cm^{-1}; NMR: δ 1.16 (*s*, 3H), 2.75–3.01(*m*, 6H), 3.40–3.98(*m*, 34H); MS(m/z) 465(M$^+$, 0.8), 233(16), 232(100), 101(2), 57(3), 45(4), 43(12).

Anal. Calcd. for $C_{22}H_{43}NO_9$: C, 56.75; H, 9.31; N, 3.01. *Found*: C, 56.75; H, 9.19; N, 2.88.

2.2. PREPARATION OF **1b**

16-[(2-Methyl-1,4,7,10,13-pentaoxadec-2-yl)methyl]-1,4,7,10,13-pentaoxa-16-azacyclooctadecane (**1b**) was prepared from **6** (4.04 g, 12 mmol), **7b** (3.16 g, 12 mmol), and potassium carbonate (2.16 g, 14 mmol) at 140°C for 45 h under a nitrogen atmosphere as above for **1a**. The crude product was purified by chromatography over alumina (dichloromethane/isopropyl alcohol = 100/1.5 (v/v)) to give **1b** (1.40 g, 23%) as a slightly yellow oil. The analytical data for **1b** are as follows: IR (neat) 2870, 1460, 1300, 1250, 1120 cm^{-1}; NMR: δ 1.15(*s*, 3H), 2.80–2.84(*m*, 6H), 3.40–3.85(*m*, 38H); MS(*m*/*z*) 509(M^+, 0.8), 277(17), 276(90), 232(100), 101(4), 57(4), 45(8), 43(3).

Anal. Calcd. for $C_{24}H_{47}NO_{10}$: C, 56.56; H, 9.30; N, 2.75. *Found*: C, 56.24; H, 9.20; N, 2.94.

2.3. PREPARATION OF **2a**

13-[2-[(2-Methyl-1,4,7,10,13-pentaoxadec-2-yl)methoxy]ethyl]-1,4,7,10-tetraoxa-13-azacyclopentadecane (**2a**) was prepared by the following sequence. After sodium metal (0.15 g, 6.6 mmol) was dissolved in *t*-butyl alcohol (25 mL), **8a** (1.32 g, 5 mmol) was added to the mixture, and the resultant mixture was refluxed for 1 h. **6** (1.08 g, 3.3 mmol) was added to the mixture and then the *t*-butyl alcohol was evaporated off. The residue was stirred at 130°C for 6 h. Water (100 mL) was added to the cooled reaction mixture and extracted with dichloromethane (100 mL × 3). The combined organic layer was dried over $MgSO_4$ and concentrated. Low boiling compounds were removed under reduced pressure in a Kugelrohr apparatus (150°C/0.05 Torr). The crude product was purified by chromatography over alumina (dichloromethane/isopropyl alcohol = 100/3 ~ 20/1 (v/v)) to give **2a** (0.53 g, 32%) as a slightly yellow oil. The analytical data for **2a** are as follows: IR (neat) 2850, 1450, 1300, 1250, 1120 cm^{-1}; NMR: δ 1.18 (*s*, 3H), 2.76–2.85(*m*, 6H), 3.33–3.83(*m*, 34H); fab MS(*m*/*z*) 510(M^+ + 1, 43), 233(18), 232(100), 101(12), 57(45).

Anal. Calcd. for $C_{24}H_{47}NO_{10}$: C, 56.56; H, 9.30; N, 2.75. *Found*: C, 56.32; H, 9.21; N, 2.68.

2.4. PREPARATION OF **2b**

16-[2-[(2-Methyl-1,4,7,10,13-pentaoxadec-2-yl) methoxy] ethyl] -1,4,7,10,13-pentaoxa-16-azacyclooctadecane (**2b**) was prepared from **6** (2.19 g, 6.7 mmol), **8b** (3.07 g, 10 mmol), and sodium metal (0.31 g, 14 mmol) at 130°C for 15 h as above for **2a**. The crude product was purified by chromatography over alumina (dichloromethane/dioxane = 10/1 ~ 3/2 (v/v)) to give **2b** (0.89 g, 24%) as a slightly yellow oil. The analytical data for **2b** are as follows: IR (neat) 2860, 1450, 1290, 1240, 1100 cm^{-1}; NMR: δ 1.20(*s*, 3H), 2.70–2.90(*m*, 6H), 3.41–3.92(*m*, 42H); fab MS(*m*/*z*) 554(M^+ + 1, 82), 277(17), 276(100), 101(13), 100(21), 57(18).

Anal. Calcd. for $C_{26}H_{51}NO_{11}$: C, 56.40; H, 9.28; N, 2.53. *Found*: C, 56.58; H, 9.26; N, 2.42.

2.5. PREPARATION OF **3a**

13-[2-[2-[(2-Methyl-1,4,7,10,13-pentaoxadec-2-yl) methoxy]ethoxy]ethyl]-1,4,7,10-tetraoxa-13-azacyclooctadecane (**3a**) was prepared from **6** (0.82 g, 2.5 mmol), **9a** (1.15 g, 3.75 mmol), and sodium metal (0.11 g, 5 mmol) at 130°C for 6 h as above for **2a**. The crude product was purified by chromatography over alumina (dichloromethane/dioxane = 4/1 (v/v)) to give **3a** (0.61 g, 46%) as a slightly yellow oil. The analytical data for **3a** are as follows: IR (neat) 2870, 1450, 1300, 1250, 1120 cm^{-1}; NMR: δ 1.18(*s*, 3H), 2.68–2.92(*m*, 6H), 3.40–3.88(*m*, 42H); MS(m/z) 553(M^+ + 1, 1), 233(13), 232(100), 101(10), 45(13).

Anal. Calcd. for $C_{26}H_{51}NO_{11}$: C, 56.40; H, 9.28; N, 2.53. *Found*: C, 56.38; H, 9.38; N, 2.46.

2.6. PREPARATION OF **3b**

16-[2-[2-[(2-Methyl-1,4,7,10,13-pentaoxadec-2-yl) methoxy] ethoxy] ethyl]-1,4,7,10,13-pentaoxa-16-azacyclooctadecane (**3b**) was prepared from **6** (1.64 g, 5 mmol), **9b** (2.64 g, 7.5 mmol), and sodium metal (0.23 g, 10 mmol) at 130°C for 6 h as above for **2a**. The crude product was purified by gel permeation chromatography using chloroform as the eluent to give **3b** (1.97 g, 66%) as a slightly yellow oil. The analytical data for **3b** are as follows: IR (neat) 2870, 1460, 1300, 1250, 1120 cm^{-1}; NMR: δ 1.20(*s*, 3H), 2.62–3.00(*m*, 6H), 3.40–3.92(*m*, 46H); MS(m/z) 597(M^+, 0.8), 277(16), 276(100), 232(9), 101(15), 99(8), 57(4), 45(18), 43(7).

Anal. Calcd. for $C_{28}H_{55}NO_{12} \cdot H_2O$: C, 54.62; H, 9.33; N, 2.27. *Found*: C, 54.82; H, 9.20; N, 2.34.

2.7. PREPARATION OF **4a**

13-[3-[(2-Methyl-1,4,7,10,13-pentaoxadec-2-yl)methoxy]propyl]-1,4,7,10-tetraoxa-13-azacyclooctadecane (**4a**) was prepared from **6** (2.62 g, 8 mmol), **10a** (3.33 g, 12 mmol), and sodium metal (0.55 g, 24 mmol) at 130°C for 18 h as above for **2a**. The crude product was purified by chromatography over alumina (dichloromethane/dioxane = 10/1 (v/v)) to give **4a** (0.75 g, 18%) as a slightly yellow oil. The analytical data for **4a** are as follows: IR (neat) 2880, 1460, 1300, 1260, 1100 cm^{-1}; NMR: δ 1.16(*s*, 3H), 1.52–1.92(*m*, 2H), 2.44–2.92 (*m*, 6H), 3.32–3.92 (*m*, 38H); MS(m/z) 523(M^+, 4), 233(30), 232(100), 101(34), 100(15), 57(16), 45(30).

Anal. Calcd. for $C_{25}H_{49}NO_{10}$: C, 57.34; H, 9.43; N, 2.67. *Found*: C, 57.35; H, 9.34; N, 2.36.

2.8. PREPARATION OF **4b**

16-[3-[(2-Methyl-1,4,7,10,13-pentaoxadec-2-yl)methoxy]propyl]-1,4,7,10,13-pentaoxa-16-azacyclooctadecane (**4b**) was prepared from **6** (0.82 g, 2.5 mmol), **10b** (1.21 g, 3.75 mmol), and sodium metal (0.11 g, 5 mmol) at 130°C for 6 h as above for **2a**. The crude product was purified by chromatography over alumina (dichloromethane/dioxane = 4/1 (v/v)) to give **4b** (0.48 g, 34%) as a slightly yellow oil. The analytical data for **4b** are as follows: IR (neat) 2870, 1460, 1300, 1260, 1115 cm^{-1}; NMR: δ 1.16(*s*, 3H), 1.52–1.92(*m*, 2H), 2.44–2.92(*m*, 6H), 3.32–3.92(*m*, 42H); fab MS(*m*/*z*) 568(M^+ + 1, 100), 277(10), 276(60), 101(22), 57(20).

Anal. Calcd. for $C_{27}H_{53}NO_{11}$: C, 57.12; H, 9.41; N, 2.47. *Found*: C, 56.82; H, 9.31; N, 2.16.

2.9. PREPARATION OF 5

N,*N*-[Bis[2-(2-Methyl-1,4,7,10,13-pentaoxadec-2-yl)methoxy]ethyl]octylamine (**5**) was prepared from **6** (3.27 g, 10 mmol), **11** (1.09 g, 5 mmol), and sodium metal (0.29 g, 12.5 mmol) at 130°C for 6 h as above for **2a**. The crude product was purified by chromatography over alumina (dichloromethane/dioxane = 10/1 (v/v)) to give **5** (1.73 g, 49%) as a slightly yellow oil. The analytical data for **5** are as follows: IR (neat) 2930, 2860, 1460, 1290, 1260, 1120 cm^{-1}; NMR: δ 0.88(*t*, 3H), 1.18–1.45(*m*, 18H), 2.50–2.70(*m*, 6H), 3.35–3.83(*m*,44 H); fab MS(*m*/*z*) 710(M^+ + 1, 4), 596(5), 464(100), 462(44), 200(100), 154(23), 57(32).

Anal. Calcd. for $C_{36}H_{71}NO_{12} \cdot 0.5H_2O$: C, 60.14; H, 10.09; N, 1.95. *Found*: C, 59.88; H, 9.94; N, 1.81.

2.10. EXTRACTION PROCEDURE [24]

A mixture of an aqueous solution (10 mL) of alkali metal hydroxide (5×10^{-2} M) and picric acid (5×10^{-4} M) and a dichloromethane solution (10 mL) of an appropriate extractant (5×10^{-4} M) was shaken at 22°C for 9 h. The extractability was obtained from the calculation based on the absorption of picrate anion in the aqueous phase at 354 nm in the UV spectrum.

2.11. CATION TRANSPORT STUDIES

Membrane transport experiments were carried out in a U-type cell (tube diameter: 16 mm; arm height: 150 mm) at 25°C. Both surface areas were 2.0 cm^2. The details for transport conditions are summarized in the footnotes of Tables I–III and the caption of Figure 3. Each cell consisted of a 20 mL membrane phase (CH_2Cl_2, 1.25×10^{-4} M in a carrier, stirred at 500 rpm by a magnetic stirrer) interfaced to two 10 mL of aqueous phases. Both aqueous phases were sampled after appropriate times and analyzed for cation concentration using a Nippon Jarrell-Ash AA-8500 atomic absorption spectrometer. Each experiment was repeated at least three times and the results are reported as the average of the three values. The deviations from the mean were less than ±15% of the value given. In the case of the active transport system (Table III), the data of transported cations (%) denote the mean of the increment of cations in one phase and the decrement of cations in the other phase. The concentration of picrate anion was obtained from the calculation based on the absorption at 354 nm in the UV spectrum. The pH values were measured using a Hitachi-Horiba F7 pH Meter. The concentration of tetramethylammonium ion was determined using ^1H NMR.

3. Results and Discussion

3.1. SYNTHESIS OF BIS(CROWN ETHER)S

The bis(crown ether)s (**1**–**4**) were prepared by the reactions of 2-(bromomethyl)-2-methyl-15-crown-5 (**6**) [10] and the appropriate azacrown ether derivatives. Compound **5** was also prepared by the reaction of **6** and the corresponding diol. The proposed structures were consistent with the data obtained from their IR, NMR, and mass spectra and combustion analyses.

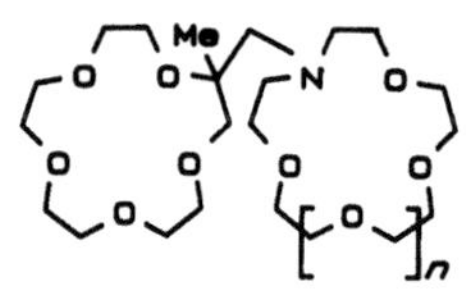

1a: n = 1
1b: n = 2

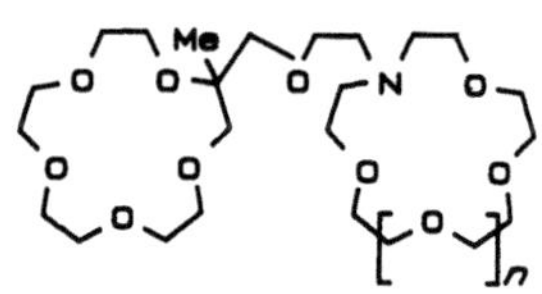

2a: n = 1
2b: n = 2

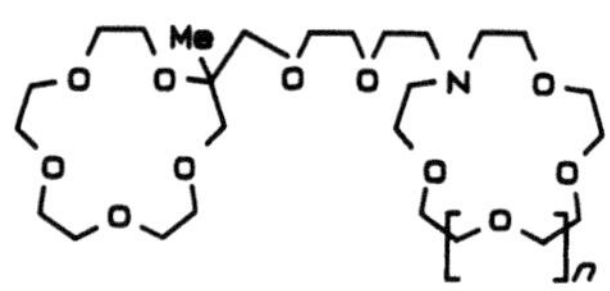

3a: n = 1
3b: n = 2

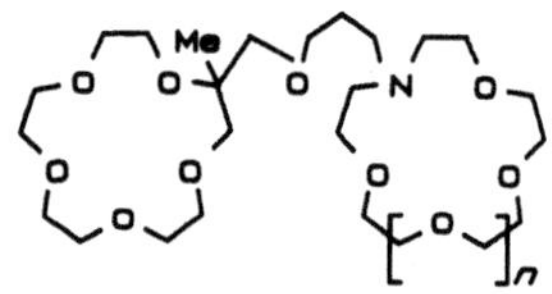

4a: n = 1
4b: n = 2

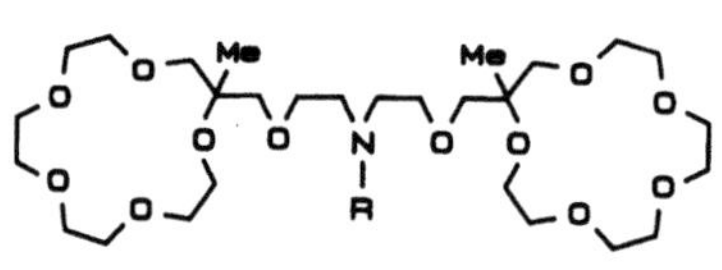

5: R = C_8H_{17}

3.2. SOLVENT EXTRACTION

The complexing ability of **1**–**5** toward alkali metal cations was assessed by the picrate extraction method with dichloromethane as the organic solvent. Figures 1 and 2 show extraction profiles of the derivatives containing two 15-crown rings (**1a**–**4a**, **5**) and the derivatives containing both 15- and 18-crown rings (**1b**–**4b**), respectively. The data are useful for estimating the cation selectivity under basic conditions. Generally speaking, compounds **1b**–**4b** possess higher K^+/Na^+ selectivity than the corresponding analogues (**1a**–**4a**). It is interesting that **4a** and **5** did not show K^+/Na^+ selectivity. This result indicates that both 15-crown rings of **4a** and **5** do not cooperatively work for the uptake of K^+. In the case of **4b**, the cooperative coordination of both crown rings toward K^+ is expected to be insignificant but the monoaza-18-crown-6 ring possesses a strong complexing ability toward K^+.

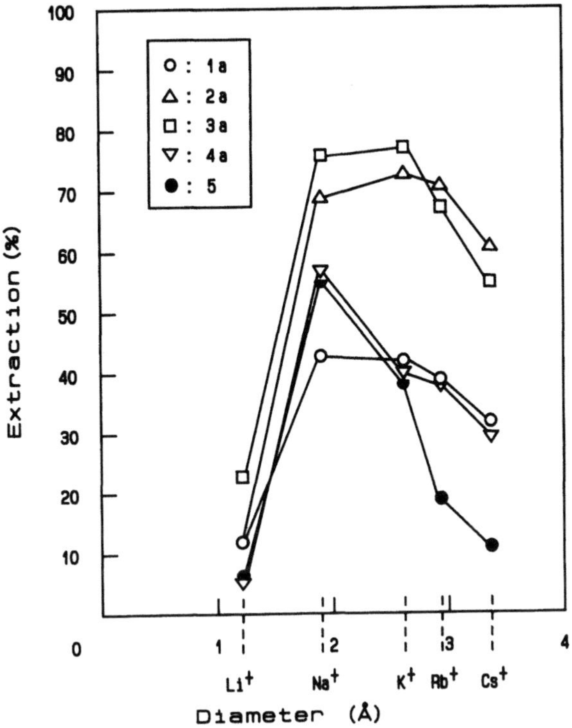

Fig. 1. Alkali-metal picrate extraction with the bis(crown ether)s containing two 15-crown rings.

3.3. PASSIVE TRANSPORT SYSTEM

Transport experiments were carried out at 25°C in a U-type cell. Dichloromethane was used as the liquid membrane. Both aqueous phases were adjusted to be basic and acidic by tetramethylammonium hydroxide and hydrochloric acid. Equimolar amounts of K^+ and Na^+ were added to either aqueous phase according to the transport conditions. The concentrations of cations, pricrate ions, and protons were determined by atomic absorption analysis, UV spectroscopy, and pH titration, respectively. The passive transport data from the basic to the acidic phase and those of the opposite direction are shown in Tables I and II. In the case of passive transport from the basic to the acidic phase, Table I clearly shows that the trend observed in transport selectivity coincides well with that expected from the extraction data. Ionophores **4a** and **5** selectively transported Na^+. The transport ability of **1a** and **1b** was rather lower than that of the corresponding analogues having the oxyethylene connecting group (**2** and **3**). On the other hand, in the passive transport from the acidic to the basic phase, all ionophores showed selectivity toward Na^+ as expected from the structure. In this case, **3** and **4** transported more cations than **1** and **2**. This finding suggests that the distance between the protonation site and the 15-crown-5 ring in the molecule remarkably affects the transport ability. Ionophore **5** carried a lot of cations possibly because each 15-crown-5 ring is able to independently uptake the cation.

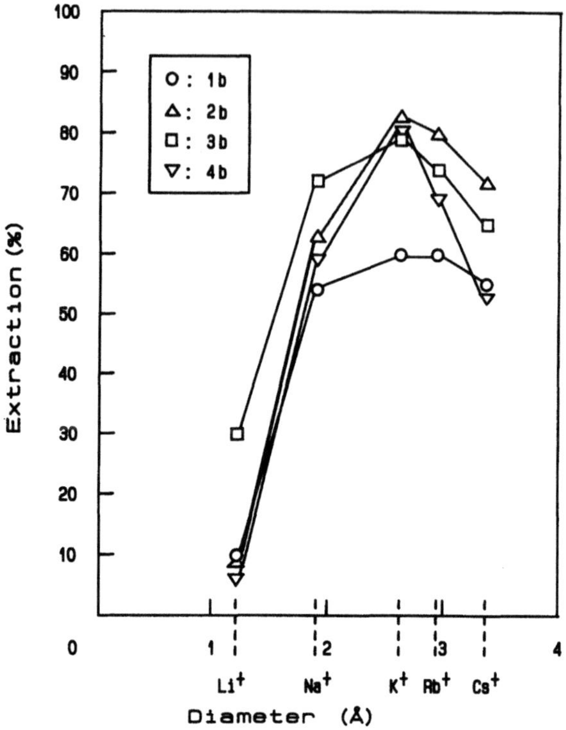

Fig. 2. Alkali-metal picrate extraction with the bis(crown ether)s containing both 15-crown and 18-crown rings.

Table I. Passive transport data[a)] from the basic phase to the acidic phase.

Ionophore	Transported Cations (%)[b)] Na+	K+	Selectivity (K+/Na+)
1a	4	10	2.5
2a	11	28	2.5
3a	14	22	1.6
4a	13	10	0.77
5	11	7	0.64
1b	5	18	3.6
2b	6	31	5.2
3b	10	29	2.9
4b	10	26	2.6

[a)]Phase 1 (basic): [PicK] = [PicNa] = 5×10^{-3} M, $[Me_4NOH] = 10^{-1}$ M / Membrane: [Ionophore] = [PicH] = 1.25×10^{-4} M / Phase 2(acidic): [HCl] = 5×10^{-3} M.
[b)]After 48 h.

Table II. Passive transport data [a] from the acidic phase to the basic phase.

Ionophore	Transported Cations (%)[b]		Selectivity
	Na^+	K^+	(Na^+/K^+)
1a	3	<1	>3
2a	7	3	2.3
3a	15	14	1.1
4a	17	10	1.7
5	21	12	1.8
1b	3	<1	>3
2b	9	2	4.5
3b	14	4	3.5
4b	16	6	2.7

[a]Phase 1 (basic): $[Me_4NOH] = 10^{-1}$ M / Membrane: [Ionophore] = [PicH] = 1.25×10^{-4} M / Phase 2(acidic): [PicK] = [PicNa] = [HCl] = 5×10^{-3} M.
[b]After 48 h.

3.4. ACTIVE TRANSPORT SYSTEM

Table III shows the quantity of cations transported after 48 h in the active transport system. Ionophores **2a**, **2b**, **3b**, and **4b** successfully transported K^+ and Na^+ in opposite directions. The ionophores consisting of 15-crown-5, monoaza-18-crown-6, and an appropriate connecting group (**2b**, **3b**) are especially effective. Ionophores **1a** and **1b** without an oxyethylene chain as the connecting group concentrated both K^+ and Na^+ in the acidic phase. In the acidic phase, the ionophore is protonated

Table III. Active transport data.[a].

Ionophore	Transported Cations (%)[b]			
	Basic Phase		Acidic Phase	
	Na^+	K^+	Na^+	K^+
1a	–	–	2	6
2a	5	–	–	19
3a	0	–	–	13
4a	0	–	–	5
5	5	0	–	–
1b	–	–	6	12
2b	10	–	–	31
3b	13	–	–	25
4b	7	–	–	24

[a]Phase 1 (basic): [PicK] = [PicNa] = 5×10^{-3} M, $[Me_4NOH] = 10^{-1}$ M / Membrane: [Ionophore] = [PicH] = 1.25×10^{-4} M / Phase 2(acidic): [PicK] = [PicNa] = [HCl] = 5×10^{-3} M.
[b]After 48 h.

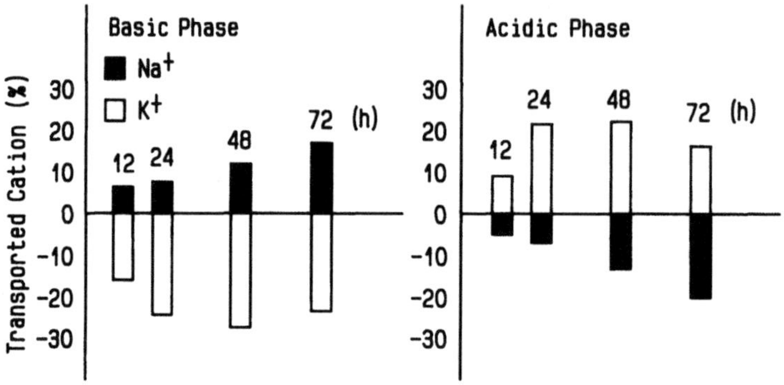

Fig. 3. Active transport of Na^+ and K^+ in opposite directions with **3b**. Standard transport conditions: Phase 1 (basic), [PicK] = [PicNa] = 5×10^{-3} M, [Me_4NOH] = 10^{-1} M / Liquid Membrane, [**3b**] = [PicH]′= 1.25×10^{-4} M / Phase 2 (acidic), [PicK] = [PicNa] = [HCl] = 5×10^{-3} M.

and has to complex with Na^+ in the form of an ammonium ion. This result clearly shows that an appropriate distance between the protonation site and the coordination site for Na^+ should be needed.

Figure 3 shows the quantity of K^+ and Na^+ transported using **3b** with the passage of time. In the basic phase, the quantity of K^+ decreased and the quantity of Na^+ increased. On the other hand, in the acidic phase, the reverse trend was observed. After 48 h, most of the protons were consumed and so the transport velocity of K^+ from the acidic to the basic phase increased. As a result, the concentration of K^+ in the acidic phase gradually decreased.

When the base was changed from tetramethylammonium hydroxide to lithium hydroxide in this transport system, the double uphill transport in opposite directions disappeared as shown in Table IV (Run 2). Both K^+ and Na^+ were concentrated in the acidic phase. At this point, the behaviour of picrate anion is interesting. In this case, picrate anion was concentrated in the basic phase. On the other hand, when tetramethylammonium ion was used as the base, picrate anion was concentrated in the acidic phase (Run 1). The control experiment without the ionophore also showed that picrate ion was concentrated in the acidic phase when

Table IV. Effect of tetramethylammonium cation on the active transport.[a)]

			Transported Ions (%)[b)]					
			Basic Phase			Acidic Phase		
Run No.	Base	Ionophore	Na^+	K^+	Pic^-	Na^+	K^+	Pic^-
1	Me_4NOH	**3b**	13	–	–	–	25	37
2	LiOH[c)]	**3b**	–	–	31	11	29	–
3	Me_4NOH	–[d)]	0	0	–	0	0	20

[a)]See the caption of Figure 3.
[b)]After 48 h.
[c)]Only base was changed from the standard transport conditions.
[d)]Ionophore was not present in the system.

tetramethylammonium ion was used (Run 3). Of course, in this case, the transport of both K^+ and Na^+ was not observed. These results suggest that tetramethylammonium ion plays an important role in transporting the picrate ion from the basic to the acidic phase. The transfer of the tetramethylammonium ion from the basic to the acidic phase was determined by ^1H NMR.

The proton concentrations in the acidic phase also affect the active transport (Table V). With the increase in the proton concentration, the transport of Na^+ from the acidic to the basic phase remarkably decreased. This result suggests that the ionophore tends to transport only H^+ without Na^+ from the acidic to the basic phase when the proton concentration is high. A suitable choice of transport conditions should be of importance to attain the double uphill transport.

Although the transport mechanism is complicated, the results obtained in this study may be explained as shown in Figure 4. In the basic interface, the ionophore **3b** complexes with K^+ rather than Na^+ as expected by the extraction data. The ionophore transports K^+ with a picrate ion across the liquid membrane to the acidic phase. In the acidic interface, the nitrogen atom of the ionophore is protonated to release K^+. After protonation, the ionophore still has the binding site, consisting of the 15-crown-5 ring and an electron-donating side-arm for Na^+. Thus, the ionophore can selectively uptake sodium picrate as a form of the ammonium salt from the acidic phase to the liquid membrane and the resulting complex moves to the basic phase across the membrane. Consequently, K^+ and Na^+ are concentrated in the acidic phase and the basic phase, respectively.

When this type of ionophore transports Na^+ from the acidic to the basic phase, the ionophore demands two picrate ion. According to this cycle, picrate ions should be concentrated in the basic phase. However, tetramethylammonium ion independently transports picrate ion from the basic to the acidic phase. The higher concentration of the picrate ion in the acidic phase is advantageous for the uptake process of Na^+ in the acidic phase.

Table V. Effect of proton concentration in acidic phase on the active transport [a] with **3b**.

		Transported Ions (%)[b]					
		Basic Phase			Acidic Phase		
Run No.	[HCl] in Phase 2[c]	Na^+	K^+	Pic^-	Na^+	K^+	Pic^-
1	5.0×10^{-3}	13	–	–	–	25	37
4	1.0×10^{-2}	6	–	–	–	35	22
5	2.0×10^{-2}	–	–	–	1	40	9
6	5.0×10^{-2}	–	–	32	17	38	–
7	1.0×10^{-1}	–	–	52	21	31	–

[a] See the caption of Figure 3.
[b] After 48 h.
[c] Only proton concentration in Phase 2 was changed from the standard transport conditions (Run 1).

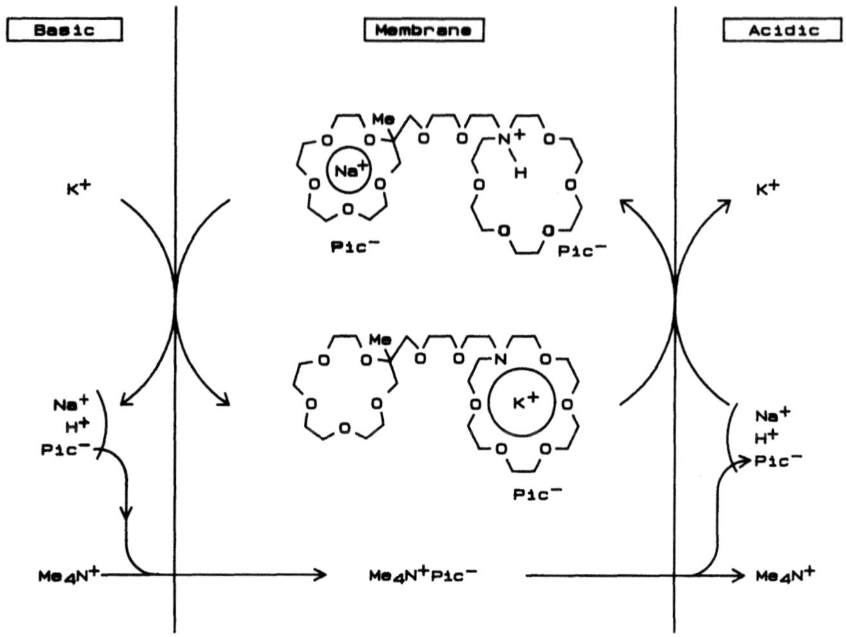

Fig. 4. Probable transport mechanism.

4. Conclusions

In this work, nine bis(crown ether)s were newly prepared and their complexing ability and transport ability were characterized. The function of the enzyme Na^+–K^+ ATPase was realized in an artificial membrane transport. To achieve the double uphill transport of Na^+ and K^+ in opposite directions, the selection of suitable transport conditions is clearly shown to be very important in addition to the design of the structure of the ionophore.

Acknowledgments

The Ministry of Education, Science and Culture, Japan is acknowledged for the partial support of this work through a Grant-in-Aid for Scientific Research on Priority Areas.

References

1. C. J. Pedersen: *J. Am. Chem. Soc.* **89**, 7017 (1967).
2. G. W. Gokel and S. H. Korseniowski: *Macrocyclic Polyether Syntheses*, Springer-Verlag, Heidelberg (1982).
3. M. Okahara, M. Miki, S. Yanagida, I. Ikeda, and K. Matsushima, *Synthesis* 854 (1977).
4. P.-L. Kuo, M. Miki, and M. Okahara: *J. Chem. Soc., Chem. Commun.* 504 (1978).
5. P.-L. Kuo, M. Miki, I. Ikeda, and M. Okahara: *Tetrahedron Lett.* 4273 (1978).
6. Y. Nakatsuji, Y. Watanabe, and M. Okahara: *Bull. Chem. Soc. Jpn.* **55**, 627 (1982).
7. Y. Nakatsuji, N. Kawamura, M. Okahara, and K. Matsushima: *Synthesis* 42 (1981).
8. I. Ikeda, T. Katayama, K. Tsuchiya, and M. Okahara: *Bull. Chem. Soc. Jpn.* **56**, 2473 (1983).
9. H. Maeda, T. Kikui, Y. Nakatsuji, and M. Okahara: *J. Org. Chem.* **47**, 5150 (1982).

10. T. Nakamura, Y. Nakatsuji, and M. Okahara: *J. Chem. Soc., Chem. Commun.* 219 (1981).
11. M. Okahara, P.-L. Kuo, S. Yamamura, and I. Ikeda: *J. Chem. Soc., Chem. Commun.* 586 (1980).
12. I. Ikeda, T. Katayama, M. Okahara, and T. Shono: *Tetrahedron Lett.* **22**, 3615 (1981).
13. A. Masuyama, Y. Nakatsuji, I. Ikeda, and M. Okahara: *Tetrahedron Lett.* **22**, 4665 (1981).
14. Y. Nakatsuji, T. Nakamura, M. Okahara, D. M. Dishong, and G. W. Gokel: *Tetrahedron Lett.* **23**, 1351 (1982).
15. Y. Nakatsuji, H. Kobayashi, and M. Okahara: *J. Chem. Soc., Chem. Commun.* 800 (1983).
16. Y. Nakatsuji, T. Mori, and M. Okahara: *J. Chem. Soc., Chem. Commun.* 1045 (1984).
17. M. Okahara and Y. Nakatsuji: *Top. Curr. Chem.* **128**, 37 (1985).
18. K. Matsushima, H. Kobayashi, Y. Nakatsuji, and M. Okahara: *Chem. Lett.* 701 (1983).
19. Y. Nakatsuji, M. Sakamoto, M. Okahara, and K. Matsushima: *Nippon Kagaku Kaishi* 430 (1987).
20. Y. Nakatsuji, R. Wakita, Y. Harada, and M. Okahara: *J. Org. Chem.* **54**, 2988 (1989).
21. S. L. Bonting and J. J. H. H. M. de Pont (eds.): *Membrane Transport In New Comprehensive Biochemistry*, Vol. 2, Elsevier/North Holland Biomedical Press, Amsterdam (1981).
22. A part of this work was published as a communication: Y. Nakatsuji, M. Sakamoto, and M. Okahara: *J. Chem. Soc., Chem. Commun.* 1101 (1988).
23. H. Maeda, Y. Nakatsuji, and M. Okahara: *J. Chem. Soc., Chem. Commun.* 471 (1981).
24. C. J. Pedersen: *J. Fed. Proc., Fed. Am. Soc., Exp. Biol* **27**, 1305 (1968).

Journal of Inclusion Phenomena and Molecular Recognition in Chemistry **12**: 213–222, 1992.

Effect of Solvent upon Competitive Liquid–Liquid Extraction of Alkali Metal Cations by Isomeric Dibenzo-16-crown-5-oxyacetic Acids[#]

WLADYSLAW WALKOWIAK,* EOK-GIU JEON,** HWANG HUH,† and RICHARD A. BARTSCH‡
Department of Chemistry and Biochemistry, Texas Tech University, Lubbock, Texas 79409-1061, U.S.A.

(Received: 18 October 1990; in final form: 3 December 1990)

Abstract. The selectivity and efficiency of competitive liquid–liquid extraction of alkali metal cations into organic solvents containing *sym*-(octyl)dibenzo-16-crown-5-oxyacetic acid (**2**) and *sym*-bis[4(5)-*tert*-butylbenzo]-16-crown-5-oxyacetic acid (**3**) have been determined. Solvents examined include: dichloromethane, chloroform, carbon tetrachloride, 1,2-dichloroethane, 1,1,1-trichloroethane, benzene, toluene, *p*-xylene, chlorobenzene, 1,2-dichlorobenzene, and 1,2,3,4-tetrahydronaphthalene. The Na^+/K^+ and Na^+/Li^+ extraction ratios are highest in chloroform. The extraction selectivity is found to correlate with the diluent parameter (DP) of the organic solvent.

Key words. Lipophilic crown ether carboxylic acids, solvent extraction of alkali metal cations, organic solvent variation in extraction.

Comments by Dr Richard A. Bartsch

Although I never met Charlie Pedersen, his work provided the foundation for our present research effort in metal ion separations with macrocyclic multidentate ligands. Without his discovery of synthetic routes to crown ethers and the initial exploration of metal salt interactions with such molecules, it is unlikely that our program, which uses crown ethers with pendant acidic groups to perform metal ion separations in solvent extraction, liquid membrane, and ion exchange resin systems, would have developed. Thanks to you Charlie for allowing me to have so much fun with my chemistry.

1. Introduction

The potential of crown ethers (macrocyclic polyethers) as the next generation of specific metal ion extracting agents was markedly enhanced by the introduction of crown ethers which bear proton-ionizable groups. Metal ion extraction by such chelating agents does not involve concomitant transfer of the aqueous phase anion into the organic medium. This factor is of immense importance to potential

[#] This paper is dedicated to the memory of the late Dr C. J. Pedersen.
* Present address: Institute of Inorganic Chemistry and Metallurgy of Rare Elements, Technical University of Wroclaw, 50-370 Wroclaw, Poland.
** Present address: Department of Chemistry, Korea Military Academy, Seoul 130–09, Korea.
† Present address: Department of Chemistry, University of Ulsan, Ulsan 690, Republic of Korea.
‡ Author for correspondence.

practical applications in which hard aqueous phase anions (chloride, nitrate, and sulfate) would be involved.

The influence of organic solvent variation upon the selectivity and efficiency of alkali metal cation extraction by 2-[(*sym*-dibenzo-16-crown-5)oxy]decanoic acid (**1**) has been examined [1]. For extractions conducted with lipophilic crown ether carboxylic acid **1** in chloroform, 1,1,1-trichloroethane, benzene, toluene, *p*-xylene and 1,2,3,4-tetrahydronaphthalene, formation of an oily second organic layer was noted for the last three solvents when the aqueous phase pH was ≥ 10. Despite this complication, it was noted that the extraction efficiency and Na^+ selectivity were much higher in chloroform than in other organic solvents.

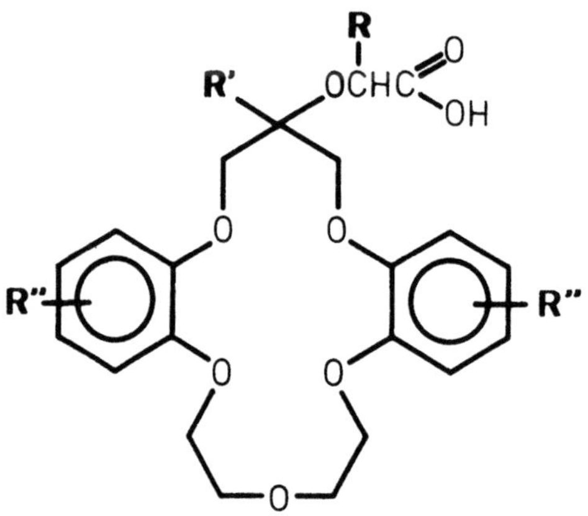

	R	R'	R''
1	C_8H_{17}	H	H
2	H	C_8H_{17}	H
3	H	H	$C(CH_3)_3$

Competitive liquid–liquid extractions of alkali metal cations by *sym*-(octyl)dibenzo-16-crown-5-oxyacetic acid (**2**) and *sym*-bis[4(5)-*tert*-butylbenzo]-16-crown-5-oxyacetic acid (**3**) in a wider range of hydrocarbon and chlorinated hydrocarbon solvents have now been conducted. Since the complicating formation of an oily second organic layer at highly alkaline pH was not observed for **2** or **3**, these results provide a better opportunity to determine the influence of organic solvent variation upon the selectivity and efficiency of alkali metal cation extraction by lipophilic crown ether carboxylic acids. Dibenzo-16-crown-5-oxyacetic acids **1**, **2**, and **3** are structural isomers which differ only in the lipophilic group attachment site(s). Therefore, the influence of varying the attachment site for the lipophilic group upon extraction selectivity and efficiency may also be assessed.

2. Experimental

2.1. REAGENTS

The sources of inorganic reagents were the same as reported previously [2]. Syntheses of crown ether carboxylic acids **2** and **3** have been published [3, 4]. Reagent grade chloroform (Fisher) was shaken four times with deionized water to remove the stabilizing ethanol and saturate the chloroform with water. Reagent-grade dichloromethane (Mallinkrodt), carbon tetrachloride (Fisher), 1,2-dichloroethane (MCB), 1,1,1-trichloroethane, 1,2-dichlorobenzene, 1,2,3,4-tetrahydronaphthalene (Aldrich), benzene, chlorobenzene, toluene, and *p*-xylene (Baker) were used.

2.2. APPARATUS AND PROCEDURE

The apparatus and procedure for competitive liquid–liquid extractions of alkali metal cations (0.25M in each) from aqueous solutions of their chlorides and hydroxides into 0.050M solutions of **2** in organic solvents were the same as those reported previously [1].

For the competitive liquid–liquid extraction of alkali metal cations by **3**, a new micro-extraction technique was developed. An aqueous solution of lithium, sodium, potassium, rubidium, and cesium chloride (1.00 mL, 0.25M in each with 0.25M cesium hydroxide added for pH adjustment) and an organic solution of **3** (1.00 mL, 0.050M) in a 5-mL, stoppered centrifuge tube was agitated with a vortex mixer (Scientific Industries, Model K-350-G) for four minutes at room temperature (21–23°C). After centrifuging for four minutes, the equilibrium pH of the aqueous layer was measured with a Fisher Accumet Model 620 pH meter and a Sargent Welch S3001005 ultra-semimicro combination electrode. A 0.50 mL sample of the organic layer was transferred to a 5-mL centrifuge tube and aqueous hydrochloric acid (0.50 mL, 0.20M) was added. The tube was stoppered and the mixture was agitated for four minutes with the vortex mixer to strip the alkali metal cations from the organic phase into the aqueous phase. The tube was centrifuged for four minutes and concentrations of alkali metal cations in the aqueous phase were determined with a Dionex Model 2000i ion chromatograph. The reproducibility of the extraction procedure has been demonstrated previously [1].

3. Results

Results for the competitive liquid–liquid extraction of alkali metal cations (0.25M in each) for aqueous solutions into 0.050M solutions of lipophilic crown ether carboxylic acid **2** in eleven organic solvents are presented in Figure 1. The solvent series includes dichloromethane, chloroform, carbon tetrachloride, 1,2-dichloroethane, 1,1,1-trichloroethane, benzene, toluene, *p*-xylene, 1,2,3,4-tetrahydronaphthalene, chlorobenzene, and 1,2-dichlorobenzene. Assuming formation of 1 : 1 crown ether carboxylate–metal ion complexes, the metal ion loading of all organic phases was quantitative when the equilibrium aqueous phase pH was 10. Extraction selectivity orders for the five alkali metal cations as well as the Na^+/K^+ and Na^+/Li^+ concentration ratios in the organic phases are given in Table I.

Organic phase concentrations of alkali metal cations as a function of the equilibrium aqueous phase pH for the isomeric lipophilic crown ether carboxylic acid **3** in eight organic solvents are shown in Figure 2. Organic phase metal ion loadings (82–94%), extraction selectivity orders, and selectivity ratios for Na^+/K^+ and Na^+/Li^+ are presented in Table II. In the absence of lipophilic crown ether carboxylic acid **2** or **3**, extraction of alkali metal cations into the organic phases was undetectable.

4. Discussion

4.1. ALKALI METAL CATION EXTRACTION SELECTIVITY AND EFFICIENCY

As would be predicted from comparison of the polyether cavity size and the diameter of the alkali metal cations [2], lipophilic dibenzo-16-crown-5 carboxylic

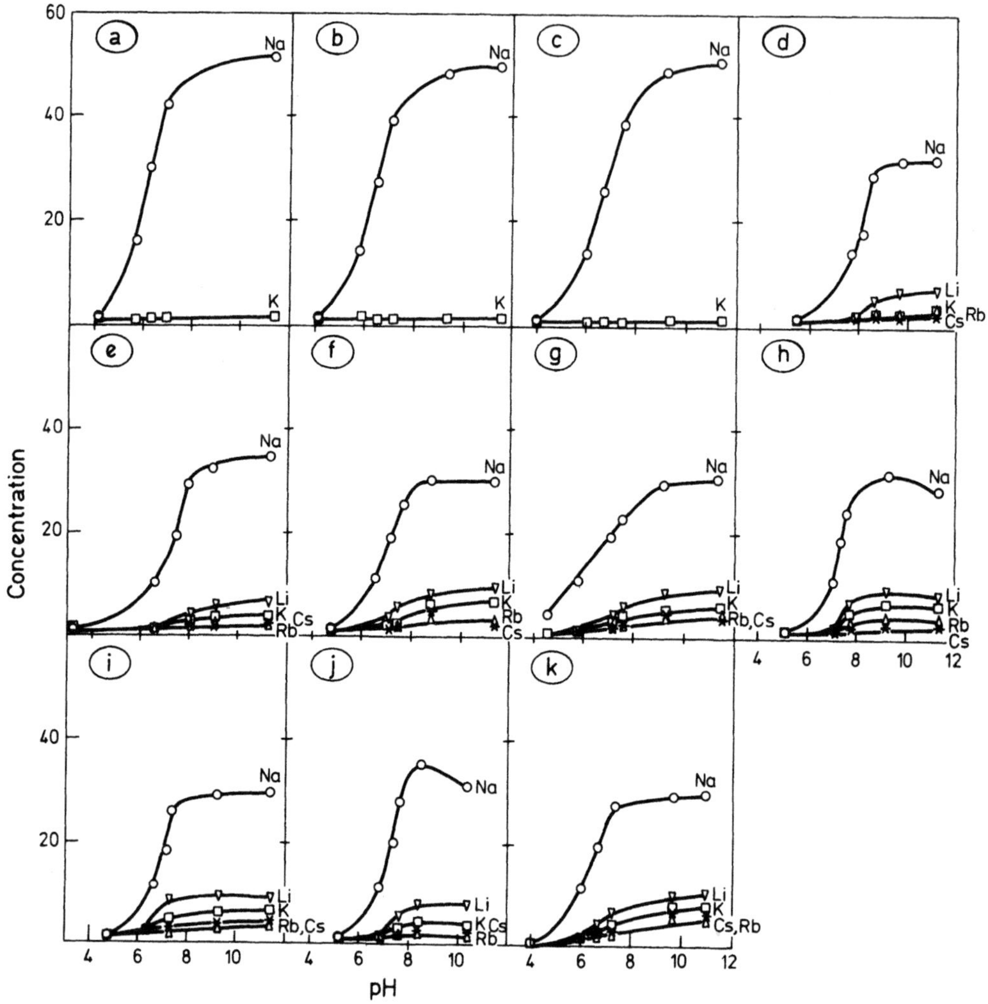

Fig. 1. Concentrations of metal cations (M × 10^3) in the organic phase vs. the equilibrium pH of the aqueous phase for competitive extraction of alkali metal cations (0.25M each) by 0.050M **2** in (a) chloroform, (b) dichloromethane, (c) 1,2-dichloroethane, (d) 1,2-dichlorobenzene, (e) chlorobenzene, (f) 1,1,1-trichloroethane, (g) benzene, (h) toluene, (i) *p*-xylene, (j) 1,2,3,4-tetrahydronaphthalene and (k) carbon tetrachloride.

acids **2** and **3** exhibit competitive extraction selectivity for Na^+. However, the degree of selectivity is found to be markedly influenced by the identity of the organic solvent and the positioning of the lipophilic group within the extractant molecule (Figures 1 and 2, Tables I and II).

For competitive solvent extraction of alkali metal cations into chloroform, dichloromethane, and 1,2-dichloroethane by lipophilic crown ether carboxylic acid **2**, the selectivity order was $Na^+ \gg K^+ > Li^+ \geqslant Rb^+$, Cs^+ with Na^+/K^+ ratios at pH = 10 of 71, 47 and 44, respectively (Table I). In contrast, for the remaining eight solvents of carbon tetrachloride, 1,1,1-trichloroethane, chlorobenzene, 1,2-dichlorobenzene, benzene, toluene, *p*-xylene, and 1,2,3,4-tetrahydronaphthalene, the

Table I. Selectivity in competitive extraction of alkali metal cations from aqueous solutions into organic solutions of lipophilic dibenzo-16-crown-5-oxyacetic acid **2** at pH = 10.0

Solvent	Extraction Selectivity Order	Selectivity Ratio	
		Na^+/K^+	Na^+/Li^+
dichloromethane	$Na^+ \gg K^+ > Li^+, Cs^+ > Rb^+$	47	85
chloroform	$Na^+ \gg K^+ > Li^+ > Rb^+ > Cs^+$	71	77
carbon tetrachloride	$Na^+ > Li^+ > K^+ > Rb^+, Cs^+$	4.4	2.9
1,2-dichloroethane	$Na^+ \gg K^+ > Li^+ > Cs^+ > Rb^+$	44	49
1,1,1-trichloroethane	$Na^+ > Li^+ > K^+ > Rb^+, Cs^+$	4.9	3.7
benzene	$Na^+ > Li^+ > K^+ > Rb^+, Cs^+$	5.2	3.5
toluene	$Na^+ > Li^+ > K^+ > Rb^+ > Cs^+$	5.1	3.4
p-xylene	$Na^+ > Li^+ > K^+Cs^+ > Rb^+$	5.1	3.4
chlorobenzene	$Na^+ > Li^+ > K^+ > Rb^+, Cs^+$	8.8	4.3
1,2-dichlorobenzene	$Na^+ > Li^+K^+, Rb^+, Cs^+$	13	5.7
1,2,3,4-tetrahydronaphthalene[a]	$Na^+ > Li^+ > K^+ > Cs^+ > Rb^+$	4.7	3.4

[a] At pH = 10.4.

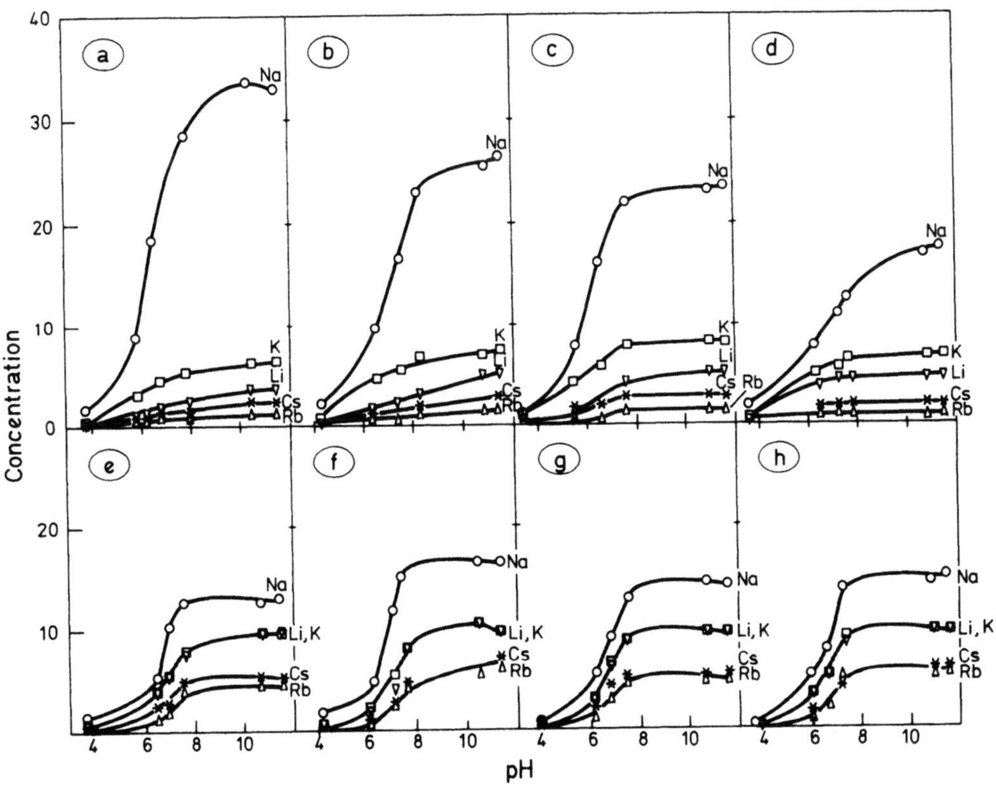

Fig. 2. Concentrations of metal cations (M × 10^3) in the organic phase vs. the equilibrium pH of the aqueous phase for competitive extraction of alkali metal cations (0.25M each) by 0.050M **3** in (a) chloroform, (b) dichloromethane, (c) 1,2-dichloroethane, (d) 1,2-dichlorobenzene, (e) chlorobenzene, (f) benzene, (g) 1,1,1-trichloroethane, and (h) carbon tetrachloride.

Table II. Organic phase loading and selectivity in competitive extraction of alkali metal cations from aqueous solutions into organic solutions of lipophilic dibenzo-16-crown-5-oxyacetic acid **3** at pH = 11.0.

Solvent	Organic Phase Loading, %	Extraction Selectivity Order	Selectivity Ratio	
			Na^+/K^+	Na^+/Li^+
dichloromethane	82	$Na^+ > K^+ > Li^+ > Cs^+ > Rb^+$	3.7	5.5
chloroform	94	$Na^+ > K^+ > Li^+ > Cs^+ > Rb^+$	5.2	8.9
carbon tetrachloride	86	$Na^+ > Li^+, K^+ > Rb^+, Cs^+$	1.6	1.6
1,2-dichloroethane	82	$Na^+ > K^+ > Li^+ > Cs^+ > Rb^+$	3.1	4.0
1,1,1-trichloroethane	85	$Na^+ > Li^+, K^+ > Rb^+, Cs^+$	1.7	1.7
benzene	94	$Na^+ > Li^+, K^+ > Rb^+, Cs^+$	1.7	1.7
chlorobenzene	84	$Na^+ > Li^+, K^+ > Cs^+ > Rb^+$	1.9	1.7
1,2-dichlorobenzene	91	$Na^+ > K^+ > Li^+ > Cs^+ > Rb^+$	1.9	1.7

second best-extracted metal ion changed to Li^+ and the Na^+/Li^+ ratios were much lower (2.9–5.7). For this latter grouping of solvents, 1,2-dichlorobenzene gave the highest Na^+ selectivity and carbon tetrachloride the lowest. Metal ion loading of the organic phases was quantitative for all eleven solvents.

For the lipophilic crown ether carboxylic acid **3**, the extraction selectivity order was $Na^+ > K^+ > Li^+ > Cs^+ > Rb^+$ in chloroform, dichlorometane, 1,2-dichloroethane, and 1,2-dichlorobenzene with Na^+/K^+ ratios of 5.2, 3.7, 3.1, and 1.9, respectively. For the other four solvents of carbon tetrachloride, 1,1,1-trichloroethane, chlorobenzene, and benzene, the extraction selectivity order changed to $Na^+ > Li^+$, $K^+ > Rb^+$, Cs^+ with Na^+/Li^+ ratios of 1.6–1.7. The organic phase metal ion loading varied from 82–96% for **3** (Table II) with no apparent correlation between extraction efficiency and selectivity.

4.2. VARIATION OF THE LIPOPHILIC GROUP ATTACHMENT SITE

Solvent extractions of alkali metal cations into organic solvents by the isomeric lipophilic dibenzo-16-crown-5-oxyacetic acids **1** [1], **2**, and **3** have now been conducted for the common solvents of chloroform, 1,1,1-trichloroethane, and benzene. For each of these three solvents, the extraction selectivity with **2** is much higher than with **1** or **3**. (The Na^+/K^+ ratios for **1**, **2**, and **3** when the equilibrium aqueous phase pH = 10–11 are 8.9 [1], 71, and 5.2, respectively.) Examination of CPK space-filling models reveals that when the lipophilic octyl groups attached to the central carbon of the three-carbon bridge in **2** extends away from the polar polyether ring then the carboxylic acid group is positioned directly over the crown ether cavity. Thus the enhanced extraction selectivity noted for crown ether carboxylic acid **2** may be attributed to pre-organization of the binding site [5, 6, 7].

4.3. EFFECT OF SOLVENT UPON EXTRACTION SELECTIVITY

Rationalization of the extraction selectivities observed for lipophilic crown ether carboxylic acids **2** and **3** should involve considerations of solvent properties and solvent–solute interactions. Five of the solvents investigated are haloalkanes, two are haloarenes, and four are alkylarenes. Selected physical properties of these eleven organic solvents are presented in Table III. In the first four columns, values for the molar volume [8], viscosity [9], dielectric constant [10], and dipole moment [8] are given. Since some amount of water is also present in the organic solvent during extraction, the solubility of water in the solvent [8] is listed in column five.

The solubility parameter (δ) as well as its hydrogen-bonding component (δ_h) are given in the next two columns. The solubility parameter is the square root of the cohesive energy density of the solvent and contains contributions from three types of interactions [11]. Thus

$$\delta^2 = \delta_d^2 + \delta_p^2 + \delta_n^2 \tag{1}$$

where δ_d, δ_p, and δ_h are dispersion, polar and hydrogen-bonding components, respectively, of the solubility parameter.

The penultimate column contains the value of the normalized parameter of solvent polarity, E_T^N. Reichardt [10] has recommended use of the normalized solubility parameter instead of E_T. The E_T^N value is defined according to Equation 2

$$E_T^N = \frac{E_T(\text{solvent}) - E_T(\text{TMS})}{E_T(\text{water}) - E_T(\text{TMS})} = \frac{E_T(\text{solvent}) - 30.7}{32.4} \tag{2}$$

using water and tetramethylsilane (TMS) as extreme reference solvents. The last column in Table III contains the diluent parameter, DP [11], of the solvent.

Some researchers have attempted to correlate extraction complex stability with the dielectric constant of the solvent [13] and the ion solution energy [14]. Correlations between solvent properties and extraction efficiency have been obtained by Iwachido, Masayoshi, Naito, and Toei [15]. In that paper, the extractability of potassium picrate from aqueous solutions into solutions of the non-ionizable crown ether 18-crown-6 in 57 organic solvents was investigated. Good correlation between the picrate distribution coefficient and the solubility parameter (δ) was noted. In addition, a good correlation between the picrate distribution coefficient and the solvent polarity parameter (E_T) was observed.

Shmidt [16] has introduced the empirical diluent parameter (DP) as a 'universal constant' for characterizing the diluent in extraction system that obey the relationship

$$\log D = k + (a)(DP) \tag{3}$$

in which D is the distribution coefficient and k and a are constants characteristic of the extraction system. Solvent properties which may be related to DP are those which describe the polarity of the solvent and its cohesiveness. The polarity is best represented by E_T and the cohesiveness of a solvent is best described by δ. In a recent study, Marcus [12] found the best correlation between DP, E_T and δ was

$$DP = -(10.6 \pm 0.93) + (0.241 \pm 0.026)E_T + (0.217 \pm 0.077)\delta \tag{4}$$

For 23 solvents (excluding chloroform) a correlation coefficient of 0.972 was obtained.

Returning now to the matter of selectivity in competitive alkali metal cation solvent extraction by the lipophilic dibenzo-16-crown-5-oxyacetic acids **1**, **2**, and

Table III. Selected properties of organic solvents used in competitive extractions of alkali metal cations by lipophilic dibenzo-16-crown-5-oxyacetic acids **2** and **3**.

Solvent	V^a	η^b	ε^c	μ^d	sol.[e]	δ^f	δ_h^f	E_T^{Ng}	DP^h
dichloromethane	64.5	0.499	8.93	1.14	0.198	9.88	3.0	0.309	–
chloroform	80.7	0.596	4.81(20°)	1.15	0.072(23°)	9.49	2.8	0.259	4.5
carbon tetrachloride	97.1	0.965	2.23	0.00	0.010(24°)	8.55	0.3	0.052	1.6
1,2-dichloroethane	79.4	0.887	10.37	1.86	0.150(20°)	9.78	2.0	0.327	3.5
1,1,1-trichloroethane	100.4	0.903	7.25(20°)	1.57	0.034	8.50	1.0	0.170	–
benzene	89.4	0.603	2.27	0.00	0.063	9.17	1.0	0.111	2.3
toluene	106.8	0.587	2.38	0.31	0.033	8.91	1.0	0.099	1.8
p-xylene	123.9	0.644	2.27(20°)	0.02	–	8.77	1.5	0.074	1.8
chlorobenzene	102.2	0.799	5.62	1.54	0.033	9.68	1.0	0.188	2.4
1,2-dichlorobenzene	113.0	1.324	9.93	2.27	0.309	10.05	1.6	0.225	2.7
1,2,3,4-tetrahydronaphthalene	136.8	2.202	2.77	0.60	–	9.50	1.4	0.093	–

[a] Molar volume in cm^3/mol from Ref. [8].
[b] Viscosity at 20°C in $(mN)s/m^2$ from Ref. [9].
[c] Dielectric constant at 25°C from Ref. [10].
[d] Dipole moment in Debye units (10^{-18} esu) from Ref. [8].
[e] Water solubility in the organic solvent at 25°C from Ref. [8].
[f] Solubility parameter in $(cal/cm^3)^{1/2}$ with its hydrogen-bonding component from Ref. [11].
[g] Normalized E_T^N values from Ref. [10].
[h] Diluent parameter from Ref. [12].

3, the Na^+/K^+ selectivity ratio for **2** decreases as the organic solvent is varied: chloroform $\gg$ dichloromethane, 1,2-dichloroethane $\gg$ 1,2-dichlorobenzene > chlorobenzene > benzene, toluene, *p*-xylene > 1,1,1-trichloroethane > 1,2,3,4-tetrahydronaphthalene > carbon tetrachloride (Table II). For **3**, the Na^+/K^+ selectivity ratio decreases as the diluent is varied: chloroform > dichloromethane > 1,2-dichloroethane > chlorobenzene, 1,2-dichlorobenzene > 1,1,1-trichloroethane, benzene > carbon tetrachloride. From data reported [1] for **1**, the Na^+/K^+ selectivity ratio when the equilibrium aqueous phase pH = 10 decreases as the solvent is varied: chloroform > 1,2,3,4-tetrahydronaphthalene, toluene > 1,1,1-trichloroethane > benzene > *p*-xylene. Although there is some variation in solvent ordering for the lipophilic dibenzo-16-crown-5-oxyacetic acids **1**, **2**, and **3**, the overall trends are quite similiar for the three structural isomers. Since the range of Na^+/K^+ selectivity ratios and the number of different solvents are larger for **2**, the comparison of extraction selectivity with solvent properties is most appropriate for this lipophilic crown ether carboxylic acid.

Comparison of the Na^+/K^+ selectivity ratios for **2** (Table I) with selected properties of the organic solvents (Table III) reveals no correlation with the molar volume, viscosity, dielectric constant, dipole moment, or water solubility in the organic solvent. The poor correlation of the solubility parameter (δ) with the Na^+/K^+ extraction selectivity ratio is improved when only the hydrogen-bonding component (δ_h) of the solubility parameter is considered. In this case, only the δ_h values for chloroform, xylene and 1,2,3,4-tetrahydronaphthalene fail to show ordering consistent with the observed selectivity. For the comparison between the Na^+/K^+ selectivity ordering and E_T^N values, good correlation is noted except for the E_T^N values of chloroform, xylene, and 1,1,1-trichloroethane.

The best correlation between Na^+/K^+ extraction selectivity and a solvent property is found for Shmidt's empirical diluent parameter (*DP*). Unfortunately, *DP* values have not been determined for dichloromethane, 1,1,1-trichloroethane and 1,2,3,4-tetrahydronaphthalene. However for the remaining eight solvents, the Na^+/K^+ extraction ratio and *DP* values have the same ordering. Particularly important is the correlation between the high Na^+/K^+ extraction selectivity observed for **2** in chloroform with the highest *DP* value for any of the solvents examined.

5. Conclusions

Selectivity in competitive liquid–liquid extraction of alkali metal cations by the lipophilic dibenzo-16-crown-5-oxyacetic acid isomers **1**, **2**, and **3** is strongly influenced by variation of the organic solvent and the lipophilic group attachment site. Pre-organization of the binding site in **2** by appropriate location of the lipophilic group produces excellent Na^+ selectivity with $Na^+/K^+ = 71$ and $Na^+/Li^+ = 77$ for extractions into chloroform. For **2** the Na^+/K^+ extraction selectivity ratio correlates well with the diluent parameter (*DP*) of the organic solvent. The extraction efficiencies observed for lipophilic dibenzo-16-crown-5-oxyacetic acids **2** and **3** are higher than those reported [1] and exhibit no apparent correlation with the organic solvent parameters.

Acknowledgements

This research was supported by the Division of Chemical Sciences of the Office of Basic Energy Sciences of the U.S. Department of Energy (Contract DE-AS05-80ER10604 and Grant DE-FG05-88ER13832). Dr. Hwang Huh received a postdoctoral fellowship from the Korea Science and Engineering Foundation.

References

1. W. A. Charewicz, W. Walkowiak, and R. A. Bartsch: *Anal. Chem.* **59**, 494 (1987).
2. J. Strzelbicki and R. A. Bartsch: *Anal. Chem.* **53**, 1894 (1981).
3. R. A. Bartsch, Y. Liu, S. I. Kang, B. Son, G. S. Heo, P. G. Hipes, and L. J. Bills: *J. Org. Chem.* **48**, 4864 (1983).
4. W. A. Charewicz, G. S. Heo, and R. A. Bartsch: *Anal. Chem.* **54**, 2094 (1982).
5. D. J. Cram: *Angew. Chem., Int. Ed. Engl.* **25**, 1039 (1986).
6. R. A. Bartsch: *Solv. Extr. Ion Exch.* **7**, (1989).
7. S. I. Kang, I.-W. Yang, M. J. Pugia, W. A. Charewicz, W. Walkowiak, and R. A. Bartsch: *Anal. Chem.* **62**, 2018 (1990).
8. J. A. Riddick, W. B. Bunger, and T. K. Sakano: *Organic Solvents. Physical Properties and Method of Purification* Volume II of Techniques of Chemistry, Fourth Edition, Wiley-Interscience (1986).
9. *Lange's Handbook of Chemistry* Thirteenth Edition, Ed. J. A. Dean, pp. 10–103, McGraw-Hill (1985).
10. C. Reichardt: *Solvents and Solvent Effects in Organic Chemistry* 2nd ed. pp. 364–372 and 408–410, VCH Verlagsgesellschaft (1988).
11. A. F. M. Barton: *Chem. Rev.* **25**, 731 (1975).
12. Y. Marcus: *Solv. Extr. Ion Exch.* **7**, 567 (1989).
13. J. D. Lamb, R. M. Izatt, and J. J. Christensen: *Coordination Chemistry of Macrocyclic Compounds* Ed. G. A. Melson, pp. 162–166, Plenum (1979).
14. J. D. Lamb, R. M. Izatt, C. S. Swain, and J. J. Christensen: *J. Am. Chem. Soc.* **102**, 475 (1980).
15. T. Iwachido, M. Masayoshi, H. Naito, and K. Toei: *Bull. Chem. Soc. Jpn.* **55**, 2378 (1982).
16. W. S. Shmidt: *Russ. Chem. Rev.* **47**, 929 (1978).

Journal of Inclusion Phenomena and Molecular Recognition in Chemistry **12**: 223–236, 1992.

Potassium Selective Chromoionophores

IAN P. DANKS and IAN O. SUTHERLAND*
Department of Chemistry, The University of Liverpool, P.O. Box 147, Liverpool, L69 3BX, U.K.

(Received 6 September 1990; in final form: 7 January 1991)

Abstract. The ionisable chromoionophores **5** were synthesised from simple precursors. These chromoionophores extract alkali metal cations from aqueous solutions at pH > 7 into dichloromethane. The ratio of extraction coefficients for the chromoionophore **5c** for K^+ and Na^+ indicate that it has potential for use in optical fibre sensors for K^+ in the presence of Na^+, Mg^{2+}, and Ca^{2+} at extracellular concentrations.

Key words. Chromoionophores, potassium, hemispherand, optical fibre sensors.

1. Introduction

The discovery of crown ethers [1] by the late Charles Pedersen[1] has led to the possibility of designing and synthesising ionophores that are specific for selected alkali metal cations. Thus the crown ether concept has been elegantly enlarged by the development of more rigid and selective systems such as the cryptands [2], hemispherands [3], and spherands [4], and, more recently, suitably bridged and functionalised calixarene derivatives [5]. The availability of these selective synthetic ionophores, which may readily be modified through synthesis has resulted in their exploitation in analytical devices [6] such as ion selective electrodes, selective chromogenic reagents for cations, and ion selective membranes for cation transport. The development [7] of other major areas in supramolecular and host-guest chemistry has also been inspired both directly and indirectly by the crown ethers and the chemistry of the next century will increasingly be based upon this new ability to design and control non-covalent interactions between synthetic compounds and between synthetic compounds and natural macromolecules.

The presence of metal cations in mmolar concentrations in both intracellular and extracellular fluids has led [8] to the development of reagents and analytical procedures for their estimation. Our interests have been in the development of optical fibre sensors and this has required the synthesis of ion selective chromoionophores to use as transducers at the tip of the optical fibre. The ionic concentrations in mammalian blood pose a particular problem for the measurement of potassium ion concentration (*ca* 4 mM) in the presence of a much higher concentration of sodium ions (*ca* 140 mM) and comparable concentrations of calcium (*ca* 1.8 mM) and magnesium (*ca* 1.5 mM) ions.

This paper is dedicated to the memory of the late Dr C. J. Pedersen.
* Author for correspondence.

2. Experimental

2.1. GENERAL

All reactions were performed under a nitrogen atmosphere using reagent grade materials in dried and purified solvents unless otherwise stated. The extraction experiments were carried out with ACS grade alkali and alkaline earth chlorides. Proton NMR spectra were recorded on Perkin Elmer R34 (220 MHz), Bruker WH250 (250 MHz), and Bruker AC200 (200 MHz) spectrometers, and ^{13}C NMR spectra were recorded on an AC200 (50.3 MHz) spectrometer. Visible and UV absorption spectra were recorded on Perkin Elmer Lambda 5 and Hewlett Packard 8452A spectrometers using quartz cells at ambient temperature.

2.2. MEASUREMENT OF EXTRACTION COEFFICIENTS

The hemispherands **5** were dissolved to known concentration (*ca* 6×10^{-5} M) in methylene chloride. A sample of the hemispherand solution (3 mL) was shaken with aqueous buffer (3 mL, pH 8 trishydroxymethylmethylamine hydrochloride (trisHCl) at 5×10^{-2} M) and the spectrum of the organic layer was recorded. A second sample of the hemispherand (3 mL) was shaken with an aqueous solution of MOH (3 mL, 1 M) and the spectrum was recorded. Similarly a third sample of the hemispherand solution was shaken with a buffered aqueous solution of MCl (3 mL, 1 M, pH 8, trisHCl buffer at 5×10^{-2} M) and the spectrum of the organic layer was recorded.

The extraction coefficient for the metal ion M^+ was calculated using the expression:–

$$K_e = [H^+]_{aq} \cdot (A - A_0)/[M^+]_{aq} \cdot (A_{00} - A)$$

where $[H^+]_{aq}$ is the proton concentration in the aqueous layer; $[M^+]_{aq}$ is the metal ion concentration in the aqueous layer; A is the absorbance of **5** at λ with MCl; A_0 is the absorbance of **5** at λ with pH 8 buffer; A_{00} is the absorbance of **5** at λ with MOH.

Due to the affinity of the hemispherands for alkali metal cations it was necessary to conduct extraction experiments using either acid washed pyrex glassware or polythene equipment. All measurements of K_e were carried out at least in duplicate and the value of K_e is independent of pH.

2.3. PREPARATION OF CHROMOIONOPHORES **5**

2-Methoxy-5-methylphenylboronic Acid **6**

A solution of butyl lithium in hexane (248 mmol) was added to TMEDA (37.5 mL, 248 mmol) in ether (30 mL) and left at room temperature for 20 min under an argon atmosphere. The above solution was added dropwise over 30 min to a stirred solution of 4-methylanisole (30 g, 245 mmol) in ether (150 mL) at room temperature. The solution was stirred at room temperature for 6 h then added to a solution of trimethyl borate (90 mL, 786 mmol) in THF (30 mL) at −78°C. The mixture was allowed to warm to room temperature and stirred for 2 h followed by partitioning between HCl (600 mL, 3 M aqueous solution) and ether (600 mL). The organic layer was washed with water and brine, dried ($MgSO_4$) and evaporated to give a semi-solid residue which gave the *boronic acid* **6** as a white crystalline powder (50–60% yield) which was used without further purification. δ ($CDCl_3$) 2.30 (*s*, 3 H, ArCH_3), 3.90 (*s*, 3 H, ArOCH_3), 6.40 (*s*, 2 H, B($OH)_2$), ABC system, δ_A7.65, δ_B7.25, δ_C6.82 (J_{AB}2 Hz, J_{BC}8 Hz, 3 aryl-*H*). (*Found*; C, 58.1; H, 6.7%; M^+, 166.0787. $C_8H_{11}BO_3$ *requires* C, 57.9; H, 6.7%; M^+, 166.0801).

2,2″-Dimethoxy-5,5″-dimethyl-2′-hydroxy-1,1,′:3,1″-terphenyl **8**

Tetrakis(triphenylphosphine) palladium(0) (2.30 g, 2 mmol) and Na_2CO_3 (25 mL, 2 M aqueous solution) were added to a solution of 2,6-dibromophenol **7** (5.00 g, 19.9 mmol) and 2-methoxyphenylboronic acid **6** (7.90 g, 47.6 mmol) in toluene (50 mL) and ethanol (10 mL). The mixture was refluxed with stirring for 16 h, cooled to room temperature, hydrogen peroxide (10 mL, 30% wt. aqueous solution) added, and the mixture stirred for a further 1 h. The resulting dark mixture was partitioned between water (300 mL) and ether (600 mL) and the organic layer washed with water and brine, dried ($MgSO_4$), and evaporated to give a brown viscous oil. Chromatography (dry flash, eluting solvent CH_2Cl_2) gave a major product (R_f 0.5 in CH_2Cl_2) as a pale yellow oil which crystallised from ethanol to give the *terphenyl derivative* **8** as a white powder, mp 112–114°C (yield 60–80%). δ ($CDCl_3$) 2.35 (*s*, 6 H, ArCH_3), 3.80 (*s*, 6 H, ArOCH_3), 6.45 (*s*, 1 H, O*H*), AB system, δ_A7.27, δ_B6.91 (J_{AB}10 Hz, 4 aryl-*H*), A_2B system, δ_A7.17, δ_B7.05 (J_{AB}8 Hz, 3 aryl-*H*), 7.19 (*s*, 2 aryl-*H*). (*Found*: C, 78.9; H, 6.5%; M^+, 334.1570. $C_{22}H_{22}O_3$ *requires* C, 79.0; H, 6.6%; M^+, 334.1569).

2,2″-Dimethoxy-5,5″-dimethyl-2′-methoxymethyleneoxy-1,1′:3′,1″-terphenyl **9**

Sodium hydride (80% suspension in oil, 0.54 g, *ca* 18 mmol) was added to a solution of the terphenyl derivative **8** (5.00 g, 15 mmol) in THF (50 mL) and the mixture stirred for 1 h at room temperature. Chloromethyl methyl ether (1.40 mL, 18 mmol) was added dropwise and the mixture stirred for a further 24 h. Water (30 mL) was added, and the product extracted with dichloromethane (100 mL, 2 × 50 mL) and the combined extracts washed with water and brine, dried ($MgSO_4$), and evaporated to give a discoloured semi-solid residue. Chromatopgraphy (dry flash, eluting solvent CH_2Cl_2) gave a major product (R_f 0.40 in CH_2Cl_2) as a pale yellow oil which crystallised from ethanol to give the *product* **9** as a white crystalline powder, m.p. 111–113°C (yield > 90%). δ ($CDCl_3$) 2.40 (*s*, 6 H, ArCH_3), 2.70 (*s*, 3 H, ArOCH_2OCH_3), 3.84 (*s*, 6 H, ArOCH_3), 4.45 (*s*, 2 H, ArOCH_2OCH_3), 6.94 (*d*, *J* 10 Hz, 2 aryl-*H*), 7.25 (*m*, 7 aryl-*H*). (*Found*: C, 76.3; H, 6.9%; M^+, 378.1835. $C_{24}H_{26}O_4$ *requires* C, 76.2; H, 6.9%; M^+, 378.1831).

3,3″-Diformyl-2,2″-dimethoxy-5,5″-dimethyl-2′methoxymethyleneoxy-1,1′:3′,1″-terphenyl **13**

A solution of butyl lithium in hexanes (19.8 mmol) was added to TMEDA (3.0 mL, 19.8 mmol) in ether (10 mL) and the solution left at room temperature under an atmosphere of argon for 20 min. The above solution was added over 30 min to a solution of the terphenyl **9** (3.00 g, 7.92 mmol) in ether (130 mL) and the stirred solution left for 6 h at room temperature. DMF (3.30 mL, 39.6 mmol) was added and the mixture stirred for 16 h before the addition of HCl(50 mL, 3M aqueous solution) and the extraction of the product into dichloromethane (3 × 100 mL). The combined extracts were washed with water and brine, dried ($MgSO_4$), and evaporated to give an off-white residue which was triturated with cold ether, filtered, and washed with light petroleum (bp 40–60°C) to give the *dialdehyde* **13** as a white powder, mp 148–150°C, which was used without further purification (yield 65–70%). δ ($CDCl_3$) 2.47 (*s*, 6 H, ArC*H*$_3$), 2.63 (*s*, 3 H, OCH_2OCH_3), 3.70 (*s*, 6 H, ArOC*H*$_3$), 4.48 (*s*, 2 H, OC*H*$_2$OCH$_3$), AB_2 system, δ_A7.40, δ_B7.53 (J_{AB}8 Hz, 3 aryl-*H*), AB system, δ_A7.77, δ_B7.56 (J_{AB}2 Hz, 4 aryl-*H*), 10.52 (*s*, 2 H, ArC*H*O). (*Found*: C, 72.3; H, 6.1%; M^+, 434.1724. $C_{26}H_{26}O_6$ *requires* C, 71.9; H, 6.0%; M^+, 434.1729).

3,3″-Bis(hydroxymethyl)-2,2″-dimethoxy-5,5″-dimethyl-2′-methoxymethyleneoxy-1,1′:3′1″-terphenyl **14**

Sodium borohydride (0.35 g, 9.20 mmol) was added to a solution of the above dialdehyde **13** (2.00 g, 4.60 mmol) in ethanol (25 mL). The solution was refluxed for 2 h, cooled, and evaporated to dryness. The residue was partitioned between HCl (50 mL, 3 M aqueous solution) and dichloromethane (100 mL) and the aqueous layer extracted wtih dichloromethane (2 × 50 mL). The combined organic layers were washed with water and brine, dried ($MgSO_4$), and evaporated to give the *diol* **14** (yield *ca* 100%) as a white foam which was used without further purification. δ ($CDCl_3$) 2.30 (*s*, 6 H, ArC*H*$_3$), 2.35 (*br.s*, 2 H, O*H*), 2.67 (*s*, 3 H, OCH$_2$OC*H*$_3$), 3.55 (*s*, 6 H, ArOC*H*$_3$), 4.50 (*s*, 2 H, OC*H*$_2$OCH$_3$), 4.82 (*s*, 4 H, 2 × ArC*H*$_2$OH), 7.25 (*s*, 4 aryl-*H*), AB_2 system, δ_A7.30, δ_B7.45 (J_{AB}8 Hz, 3 aryl-*H*). (*Found*: C, 70.7; H, 6.9%; M^+, 438. $C_{26}H_{30}O_6$ *requires* C, 71.2; H, 6.9%; M^+, 438).

3,3″-Bis(bromomethyl)-2,2″-dimethoxy-5,5″-dimethyl-2′-methoxymethyleneoxy-1,1′:3′,1″-terphenyl **10**

Dimethyl sulphide (1.2 mL, 16.5 mmol) was added dropwise to a suspension of *N*-bromosuccinimide (2.45 g, 13.8 mmol) in dichloromethane (15 mL) at 0°C. The mixture was stirred for 15 min. A solution of the above diol **14** (2.00 g, 4.6 mmol) in dichloromethane (30 mL) was added dropwise at 0°C, the solution was allowed to warm to room temperature and stirred for a further 16 h. Light petroleum (50 mL, bp 40–60°C) was added and the solution washed with ice cold water (3 × 20 mL) and ice cold brine (20 mL), dried ($MgSO_4$), and percolated through a wet flash chromatography column (15 cm long), further product was eluted with dichloromethane. The *dibromide* **10** (R_f0.65 in CH_2Cl_2) was recovered from the eluent as a colourless viscous oil (yield 60–70%). δ ($CDCl_3$) 2.40 (*s*, 6 H, ArC*H*$_3$), 2.67 (*s*, 3 H, OCH$_2$OC*H*$_3$), 3.62 (*s*, 6 H, ArOC*H*$_3$), 4.48 (*s*, 2 H, OC*H*$_2$OCH$_3$), 4.70 (*br.s*, 4 H, ArC*H*$_2$Br), 7.27 (*m*, 5 aryl-*H*), 7.47 (*d*, *J* 6 Hz, 2 aryl-*H*). (*Found*: C, 55.1; H, 5.0%; M^+, 566, 564, 562, $C_{26}H_{28}Br_2O_4$ *requires* C, 55.3; H, 5.0%; M^+, 566, 564, 562).

28,29-Dimethoxy-9,26-dimethyl-30-methoxymethyleneoxy-13,16,19,22-tetraoxatetracyclo [22.3.1^{2,6}.1^{7,11}]triaconta-1(28),2,4,6(30),7,9,11(29), 24,26-nonaene **11b**

A solution of the bis(bromomethyl)terphenyl **10** (0.65 g, 1.15 mmol) and triethyleneglycol (0.17 g, 1.15 mmol) in THF was added dropwise over 4 h, using a mechanical syringe, to a stirred suspension of sodium hydride (0.14 g, 80% suspension in oil, 4.6 mmol) and potassium bromide (0.27 g, 2.30 mmol) in refluxing THF (50 mL). Heating was continued for a further 12 h, the mixture was allowed to cool, water (30 mL) was added, and the product was extracted with dichloromethane (3 × 70 mL). The extracts were washed with water and brine, dried ($MgSO_4$) and evaporated to dryness. The residue was chromatographed (30 g wet flash silica mixed with 3 g KBr), major impurities were eluted first ($CHCl_3$ and $CHCl_3$–3% EtOH) followed by the required product ($CHCl_3$–15% EtOH) and KBr. The required fractions were combined, washed with HCl (50 mL, 3 M aqueous solution) and water, dried ($MgSO_4$), and evaporated. The residual solid was recrystallised from ethyl acetate to give the *macrocycle* **11b** as a white powder, mp 128–130°C (yield 50–60%). δ ($CDCl_3$) 2.35 (*s*, 6 H, $ArCH_3$), 2.50 (*s*, 3 H, OCH_2OCH_3), 3.55 (*s*, 6 H, $ArOCH_3$), 3.65 (*m*, 12 H, OCH_2), 4.20 (*s*, 2 H, OCH_2OCH_3), AB system, δ_A4.80, δ_B4.25 (J_{AB}11 Hz, 4 H, $ArCH_2O$), 7.10 (*s*, 2 aryl-*H*), 7.25 (*m*, 1 aryl-*H*), 7.35 (*m*, 4 aryl-*H*); ^{13}C NMR ($CDCl_3$) 20.3 ($ArCH_3$, 2C), 55.6 (OCH_2OCH_3), 61.3 ($ArOCH_3$, 2 C), 68.5 (OCH_2, 2 C), 69.1 (OCH_2, 2 C), 69.5 (OCH_2, 2 C), 70.6 (OCH_2, 2 C), 98.1 (OCH_2OCH_3), 123.7 (aryl-*C*), 129.7 (aryl-*C*, 2 C), 129.9 (aryl-*C*, 2 C), 131.6 (aryl-*C*, 2 C), 132.2 (aryl-*C*, 2 C), 132.3 (aryl-*C*, 2 C), 133.3 (aryl-*C* 2 C), 155.8 (aryl-*C*). (*Found*: C, 69.3; H, 7.3%; M^+, 552.2716. $C_{32}H_{40}O_8$ *requires* C, 69.5, H, 7.3%, M^+, 552.2723).

28,29-Dimethoxy-9,26-dimethyl-30-hydroxy-13,16,19,22-tetraoxatetracyclo-[22.3.1.1^{2,6}.1^{7,11}]triaconta-1(28),2,4,6(30),7,9,11(29),24,26-nonaene **15b**

The O-protected macrocycle **11b** (0.34 g, 0.61 mmol) was dissolved in dichloromethane (5 mL) and a trifluoroacetic acid (0.5 mL) water (0.5 mL) mixture was added. The suspension was stirred at room temperature for 16 h, diluted with water (20 mL) and extracted with dichloromethane (3 × 30 mL). The extract was dried ($MgSO_4$) and evporated and the residual solid crystallised from ethyl acetate to give the *phenolic hemispherand* **15b** as an analytically pure sample, m.p. 134–136°C (yield 60–70%). δ ($CDCl_3$) 2.37 (*s*, 6 H, $ArCH_3$), 3.50 (*s*, 6 H, $ArOCH_3$), 3.63 (*m*, 12 H, OCH_2), 4.35 (*br. s*, 2 H, $ArCH_2O$), 4.80 (*br. s*, 2 H, $ArCH_2O$), 7.10 *s* and *t*, *J* 10 Hz, 5 aryl-*H*), 7.40 (*d*, *J* 10 Hz, 2 aryl-*H*); ^{13}C NMR ($CDCL_3$) 20.65 ($ArCH_3$, 2 C), 61.55 ($ArOCH_3$, 2 C), 68.58 (OCH_2, 2 C), 68.82 (OCH_2, 2 C), 69.96 (OCH_2, 2 C), 70.46 (OCH_2, 2 C), 120.6 (aryl-*C*), 128.28 (aryl-*C*), 129.72 (aryl-*C*), 130.09 (aryl-*C*), 131.57 (aryl-*C*), 132.09 (aryl-*C*), 132.85 (aryl-*C*), 133.69 (aryl-*C*), 151.74 (aryl-*C*), 154.87 (aryl-*C*). (*Found*: C, 69.9; H, 7.1%; M^+, 508.2453. $C_{30}H_{36}O_7$ *requires* C, 70.8; H, 7.1%; M^+, 508.2461).

28,29-Dimethoxy-9,26-dimethyl-13,16,19,22-tetraoxatetracyclo-[22.3.1.1^{2,6}.1^{7,11}]triaconta-1(28),2,5,7,9,11(29),24,26-octaene-4,30-dione **12b**

A solution of the phenolic macrocycle **15b** (100 mg, 0.20 mmol) in chloroform (70 mL) was added carefully to a stirred solution of thallium trinitrate (0.27 g, 0.60 mmol) in methanol (80 mL) and ethanol (240 mL). The resulting orange solution was stirred at room temperature for 10 min, water (150 mL) was added and the solution stirred for a further 15 min. The organic layer was separated, washed twice with water, and evaporated to dryness. The residual orange solid was purified by chromatography on silica/KBr (10% w/w) and elution with dichloromethane/ethanol (0–10%). The major fraction was washed with aqueous HCl (100 mL, 3 M) and water and evaporated to given the *quinone* **12b** as orange crystals, m.p. 158–160°C (yield 65%). δ ($CDCl_3$) 2.32 (*s*, 6 H, ArCH_3), 3.52 (*s*, 4 H, OCH_2), 3.62 (*m*, 8 H, OCH_2), 3.80 (*s*, 6 H, ArOCH_3), 4.55 (*br. s*, 4 H, ArCH_2O), 6.80 (*s*, 2 H, =CH—), AB system, δ_A7.22, δ_B7.00 (J_{AB} 2 Hz, 4 aryl-*H*); ^{13}C NMR ($CDCl_3$) 21.0 (ArCH_3), 63.5 (ArOCH_3), 68.5 (OCH_2), 69.0 (OCH_2), 70.3 (OCH_2), 70.7 (OCH_2), 128.7, 130.0, 130.8, 132.8, 133.8, 149.0, 156.2 (12 aryl-*C* and 4 alkenyl-*C*), 187.7 (> C=O), (*Found*: C, 68.7; H, 6.5%; M$^+$, 522.2257. $C_{30}H_{34}O_8$ *requires* C, 69.0; H, 6.6%; M$^+$, 522.2254).

28,29-Dimethoxy-9,26-dimethyl-4-(2′,4′-dinitrophenylazo)-30-hydroxy-13,16,19,22-tetraoxatetracyclo[22.3.1.1^{2,6}.1^{7,11}]-triaconta-1(28),2,4,6(30),7,9,11(29),24,26-nonaene **5b**

Two mL of a solution of 2,4-dinitrophenylhydrazine (1.29 g, 6.5 mmol) and sulphuric acid (2.5 mL) in ethanol (20 mL) was added to a solution of the macrocyclic quinone **12b** (50 mg, 0.096 mmol) in chloroform (30 mL) and ethanol (50 mL). The solution was refluxed with stirring for 4 h, cooled, and left overnight. Dichloromethane (50 mL) was added and the solution was washed with water and evaporated to dryness, the residue was chromatographed (tlc, silica) and the major orange band extracted into dichloromethane and ethanol, washed with HCl (3 M, aqueous) and water and the solvent evaporated to give the *macrocyclic azophenol* **5b** which crystallised from ethyl acetate as an orange solid, m.p. 210–212°C (yield 85%). δ ($CDCl_3$) 2.40 (*s*, 6 H, ArCH_3), 3.55 (*s*, 6 H, ArOCH_3), 3.57 (*s*, 4 H, OCH_2), 3.60 (*m*, 8 H, OCH_2), *ca* 4.50 (*br. s*, 4 H, ArCH_2O), 7.21 (*s*, 4 aryl-*H*) 8.08 (*s*, 2 aryl-*H*), ABC system, δ_A8.78, δ_B8.53, δ_C7.90 (J_{AB}2 Hz, J_{BC}9 Hz, 3 aryl-*H*). (*Found*: C, 61.7; H, 5.3; N, 7.8%. $C_{36}H_{38}N_4O_{11}$ *requires* C, 61.5; H, 5.5; N, 8.0%); MS (CI source, carrier gas NH_3)$(M+NH_4)^+$, 720 (32%).

32,33-Dimethoxy-9,30-dimethyl-34-methoxymethyleneoxy-13,16,23,26-tetraoxapenta-cyclo[26.3.1.1^{2,6}.1^{7,11}.0^{17,22}] tetratriaconta-1(32),2,4,6(34),7,9,11(33),17(22),18,20,28,30-dodecaene **11c**

A similar method was used to that described for the preparation of the analogous macrocycle **11b**. The terphenyl derivative **10** (1.38 g, 2.45 mmol) was refluxed in THF with bis(2-hydroxyethyl)catechol (0.49 g, 2.45 mmol), sodium hydride (0.29 g, 80% dispersion in oil, 9.78 mmol), and potassium bromide (0.6 g, 5 mmol). Purification of the product gave the required *macrocycle* **11c** as a white foam (yield 35–40%) which crystallised from ethyl acetate as white crystals, m.p. 161–162°C. δ ($CDCl_3$) 2.30 (*s*, 6 H, ArC*H*$_3$), 2.43 (*s*, 3 H, OCH_2OCH_3), 3.72 (*s*, 6 H, ArOC*H*$_3$), 4.00 (*m*, 8 H, OC*H*$_2$), 4.20 (*s*, 2 H, OC*H*$_2$OCH$_3$), AB system, δ_A4.40, δ_B4.80 (J_{AB}12 Hz, ArC*H*$_2$O), 6.86 (*s*, 4 aryl-*H*), 7.15 (*s*, 4 aryl-*H*), 7.30 (*m*, 3 aryl-*H*); ^{13}C NMR ($CDCl_3$) 20.5 (Ar*C*H$_3$), 56.0 (ArCH$_2$O*C*H$_3$), 62.1 (ArO*C*H$_3$), 66.7 (O*C*H$_2$), 67.2 (O*C*H$_2$), 68.5 (O*C*H$_2$), 98.2 (O*C*H$_2$OCH$_3$), 111.6, 120.4, 123.9, 129.5, 129.6, 132.0, 132.1, 132.3, 132.6, 133.3, 148.1, 153.3, 156.1, (24 aryl-*C*). (*Found*: C, 71.8; H, 6.8%; M$^+$, 600.2726. $C_{36}H_{40}O_8$ *requires* C, 72.0; H, 6.7%; M$^+$, 600.2723).

32,33-Dimethoxy-9,30-dimethyl-34-hydroxy-13,16,23,26-tetraoxapentacyclo[26.3.1.1^{2,6}.1^{7,11}.0^{17,22}]tetratriaconta-1,(32),2,4,6(34),7,9,11(33),17(22),18,20,28,30-dodecaene **15c**

A similar method was used to that described for the preparation of the analogous phenolic macrocycle **15b**. The *phenol* **15c** was obtained as a white foam (*ca* 100% yield) which crystallised from ethyl acetate to give white crystals, m.p. 199–201°C. δ ($CDCl_3$) 2.35 (*s*, 6 H, ArC*H*$_3$), 3.45 (*s*, 6 H, ArOC*H*$_3$), 3.80 (*br. s*, 4 H, OC*H*$_2$), 4.00 (*br. s*, 4 H, OC*H*$_2$), 4.45 (*br. s*, 2 H, ArC*H*$_2$O), 4.75 (*br. s*, 2 H, ArC*H*$_2$O), 6.80 (*s*, 4 aryl-*H*), 7.15 (*s*, 4 aryl-*H*), AB$_2$ system, δ_A7.10, δ_B7.35 (J_{AB}8 Hz, 3 aryl-*H*); ^{13}C NMR ($CDCl_3$) 20.6 (Ar*C*H$_3$), 61.8 (ArO*C*H$_3$), 66.6 (*br.*, O*C*H$_2$*C*H$_2$O), 68.0 (*br.*, Ar*C*H$_2$O), 112.0, 120.5, 120.7, 128.2, 129.2, 129.4, 131.8, 132.7, 133.9, 148.0, 154.8 (24 aryl-*C*). (*Found*: M$^+$, 556.2453. $C_{34}H_{36}O_7$ *requires* M$^+$, 556.2461); the sodium salt was obtained analytically pure. (*Found*: C, 70.8; H, 6.2%. $C_{34}H_{35}NaO_7$ *requires* C, 70.6; H, 6.1%).

32,33-Dimethoxy-9,30-dimethyl-13,16,23,26-tetraoxapentacyclo[26.3.1.1.1^{2,6}.11^{7,11}.0^{17,27}]-tetratriaconta-1(32),2,5,7,9,11(33),17,(22),18,20,28,30-undecaene-4,34-dione **12c**

A similar method was used to that described for the macrocyclic quinone **12b**. The phenol **15c** (100 mg, 0.18 mmol) was oxidised using thallium trinitrate (0.24 g, 0.54 mmol), the product was purified by chromatography and crystallisation to give the *macrocyclic quinone* **12c** (94% yield) as orange crystals, m.p. 110–112°C. δ ($CDCl_3$) 2.25 (*s*, 6 H, ArC*H*$_3$), 3.80 (*s*, 6 H, ArOC*H*$_3$), ca 3.80 (*br.*, 4 H, OC*H*$_2$), 3.92 (*br.*, 4 H, OC*H*$_2$), 4.48 (*br. s*, ArC*H*$_2$O), 6.70 (*m*, 4 aryl-*H*), 6.72 (*s*, 2 H, —CH=), AB system, δ_A6.92, δ_B7.18 (J_{AB}2 Hz, 4 aryl-*H*). (*Found*: C, 71.1; H, 6.0%; M$^+$, 570.2254. $C_{34}H_{34}O_8$ *requires* C, 71.6; H, 6.0%; M$^+$, 570.2254).

32,33-Dimethoxy-9,30-dimethyl-4-(2′,4′-dinitrophenylazo)-34-hydroxy-13,16,23,26-tetraoxapentacyclo-[26.3.1.1^{2,6}.1^{7,11}.0^{17,22}]tetratriaconta-1(32),2,4,6(34),7,9,11(33),17(22),18,20,28,29-dodecaene **5c**

One mL of a solution of 2,4-dinitrophenylhydrazine (1.29 g, 6.5 mmol) and sulphuric acid (2.5 mL) in ethanol (20 mL) was added to a solution of the macrocyclic quinone **12c** (34 mg, 0.060 mmol) in chloroform (15 mL) and ethanol (25 mL). The solution was stirred and refluxed for 4 h, cooled to room temperature, and left overnight. Dichloromethane (50 mL) was added, the solution washed with water, and evaporated to dryness. The residue was purified by chromatography (tlc) and the major fraction was extracted with dichloromethane/5% ethanol, washed with HCl (3 M, aqueous solution) and water and evaporated to give the required *macrocyclic azophenol* **5c** which crystallised from ethyl acetate as orange crystals, m.p. 216–217°C (yield 100%). δ ($CDCl_3$) 2.39 (*s*, 6 H, ArCH_3), 3.58 (*s*, 6 H, Ar OCH_3), 3.88 (*m*, 4 H, OCH_2), 4.05 (*m*, 4 H, OCH_2), *ca* 4.60 (*br. s*, 4 H, ArCH_2O), 6.83 (*m*, 4 aryl-*H*), AB system, δ_A7.21, δ_B7.24 (J_{AB}2Hz, 4 aryl-*H*), ABC system, δ_A8.79. δ_B8.52, δ_C7.90 (J_{AB}2 Hz, J_{BC} 9 Hz, 3 aryl-*H*), 8.08(*s*, 2 aryl-*H*). (*Found*: C, 64.0; H, 5.0; N, 7.2%. $C_{40}H_{38}N_4O_{11}$ *requires* C, 64.0; H, 5.1; N, 7.5%); MS (CI source, carrier gas NH_3) $(M + NH_4)^+$, 768 (100%).

25,26-Dimethoxy-9,23-dimethyl-27-methoxymethyleneoxy-13,16,19-trioxatetracyclo[19.3.1.1^{2,6}.1^{7,11}]heptacosa-1(25),2,4,6(27),7,9,11(26),21,23-nonaene **11a**

A similar method was used to that described for the macrocycle **11b**. The terphenyl derivative **10** (1.36 g, 2.41 mmol) was reacted with 3-oxapentan-1,5-diol (0.26 g, 2.41 mmol) and sodium hydride (0.29 g, 80% dispersion in oil, 9.64 mmol) in refluxing THF. The product **11a** was isolated as before, but using silica/sodium chloride (10% ww) for chromatography, as white crystals, m.p. 143–145°C, from ethyl acetate (yield 60%). δ ($CDCl_3$) 2.35 (*s*, 6 H, ArCH_3), 2.40 (*s*, 3 H, OCH_2OCH_3), 3.38 (*s*, 6 H, ArOCH_3), 4.00 (*s*, 2 H, OCH_2OCH_3), 4.45 (*d*, *J* 12 Hz, 2 H, ArCH_2O), 4.75 (*d*, *J* 12 Hz, 2 H, ArCH_2O), 7.06 (*d*, *J* 2 Hz, 2 aryl-*H*), 7.15 (*d*, *J* 2 Hz, 2 aryl-*H*), 7.26 (*t*, *J* 8 Hz, aryl-*H*), 7.44 (*d*, *J* 8 Hz, 2 aryl-*H*). (*Found*: C, 70.7; H, 7.2%; M^+, 508.2458. $C_{30}H_{36}O_7$ *requires* C, 70.8; H, 7.1%; M^+, 508.2461).

25,26-Dimethoxy-9,23-dimethyl-27-hydroxy-13,16.19-trioxatetracyclo[19.3.1.1^{2,6}.1^{7,11}]heptacosa-1(25),2,4,6(27),7,9,11(26),21,23-nonaene **15a**

A similar method was used to that described for the phenolic macrocycle **15b**. The *product* **15a** (90–100% yield) was obtained as white crystals, m.p. 159–161°C. δ ($CDCl_3$) 2.38 (*s*, 6 H, ArCH_3), 3.45 (*m*, 4 H, OCH_2), 3.57 (*s*, 6 H, ArOCH_3), 3.72 (*m*, 4 H, OCH_2), AB system, δ_A4.98, δ_B4.40 (J_{AB} 12 Hz, 4 H, ArCH_2O), 7.10 (*s*, 2 aryl-*H*), 7.18 (*s*, 2 aryl-*H*), AB_2 system, δ_A7.25, δ_B7.55 (J_{AB}10 Hz, 3 aryl-*H*); ^{13}C NMR ($CDCl_3$) 20.7 (ArCH_3), 61.5 (ArOCH_3), 67.0 (OCH_2), 69.5 (OCH_2), 70.2 (OCH_2), 121.7, 128.1, 129.8, 130.1, 130.7, 131.9, 133.3, 151.5, 155.8 (18 aryl-*C*). (*Found*: C, 72.4; H, 7.0%; M^+, 464.2193. $C_{28}H_{32}O_6$ *requires* C, 72.4; H, 6.9%; M^+, 464.2199).

25,26-Dimethoxy-9,23-dimethyl-13,16,19-trioxatetracylco-[19.3.1.1^{2,6}.1^{7,11}]heptacosa-1(25),2,5,7,9,11(26),21,23-octaene-4,27-dione **12a**

A similar method was used to that described for the macrocyclic quinone **12b**. The phenol **15a** (126 mg, 0.27 mmol) was oxidised using thallium trinitrate (0.36 g,

0.81 mmol), the product was purified by chromatography and crystallisation to give the *macrocyclic quinone* **12a** (67% yield) as orange crystals, m.p. 147–149°C. δ ($CDCl_3$) 2.29 (*s*, 6 H, ArCH_3), 3.65 (*m*, 8 H, OCH_2) 3.80 (*s*, 6 H, ArOCH_3), 4.65 (*br. s*, 4 H, ArCH_2O), 6.95 (*s*, 2 H, —C*H*=), AB system, δ_A7.06, δ_B7.21 (J_{AB}2 Hz, 4 aryl-*H*). (*Found*: C, 69.8: H, 6.3%; M^+, 478.1984. $C_{28}H_{30}O_7$ *requires* C, 70.3; H, 6.3%; M^+, 478.1992).

25,26-Dimethoxy-9,23-dimethyl-4-(2′,4′-dinitrophenylazo)27-hydroxy-13,16,19-tri-oxatetracyclo[19.3.1.1^{2,6}.1^{7,11}]heptacosa-1(25),2,4,6(27),7,9,11(26),21,23-nonaene **5a**

One mL of a solution of 2,4-dinitrophenylhydrazine (1.29 g, 6.50 mmol) and sulphuric acid (2.5 mL) in ethanol (20 mL) was added to a solution of the macrocyclic quinone **12a** (80 g, 0.167 mmol) in chloroform (60 mL) and ethanol (90 mL). The solution was stirred and refluxed for 4 h, cooled, and left overnight. Dichloromethane (100 mL) was added and the solution washed with water and evaporated. The residue was purified by chromatography (tlc), the major band was extracted with dichloromethane/ethanol, the solution washed with HCl (3 M, aqueous solution) and water and evaporated to give the *azophenol* **5a** (*ca* 100% yield) which crystallised as orange crystals, m.p. 226–227°C, from ethyl acetate. δ ($CDCl_3$) 2.36 (*s*, 6 H, ArCH_3), 3.44 (*m*, 4 H, OCH_2), 3.60 (*s*, 6 H, ArOCH_3), 3.61 (*m*, 4 H, OCH_2), AB system, δ_A4.37, δ_B4.93 (J_{AB}12 Hz, 4 H, OCH_2Ar), AB system, δ_A7.10, δ_B7.18 (J_{AB}2 Hz, 4 aryl-*H*), 7.47 (*s*, O*H*), 8.18 (*s*, 2 aryl-*H*), ABC system, δ_A7.93, δ_B8.55, δ_C8.81 (J_{AB}9 Hz, J_{BC}2 Hz, 3 aryl-*H*). (*Found*: C, 61.3; H, 5.0; N, 7.8%. $C_{34}H_{34}N_4O_{10}$ *requires* C, 62.0; H, 5.2; N, 8.5%), MS (CI source, carrier gas NH_3) $(M + NH_4)^+$, 676 (12%).

3. Results and Discussion

3.1. OPTICAL FIBRE SENSORS FOR CATIONS

In our earlier work [9] it was demonstrated that the the chromoionophore **1** [10] can be immobilised on Amberlite XAD2 resin at the tip of an optical fibre. Such a device responds in the pH range 7–9 to mM concentrations of potassium salts in water but does not show the required selectivity for potassium, as compared with sodium and calcium, for the measurement of potassium ion concentration in extra-cellular fluids.

The response of chromoionophores, such as **1**, to ion capture is based upon the process summarised in Scheme 1, which is accompanied by a colour change, yellow to violet, of the azophenol chromophore. It was apparent from this work that a successful sensor for potassium should be based upon an ionophore with a binding energy for potassium ions similar to that of **1** but with a very much higher selectivity for potassium as compared with sodium and calcium. There are a number of reports of ionophores with high potassium selectivity, but for detection using optical fibres a correct value of the extraction coefficient for the concentration range to be detected is essential. Thus there is a possibility of saturation and loss of sensitivity if cation binding is too strong and an obvious loss of sensitivity at low concentrations if cation binding is too weak. The hemispherands **3** [11] and **4** [12]

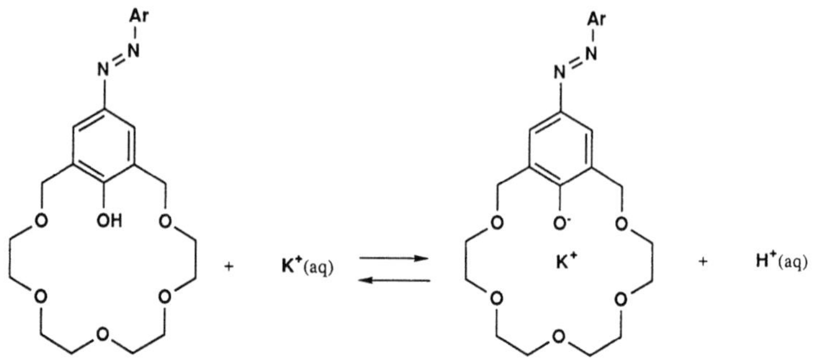

immobilised phase 1 immobilised phase 2

Scheme 1. Use of chromoionophores in optical fibre based cation sensors

appeared to be particularly suitable since these have binding constants for potassium in the same range as 18-crown-6 and in both cases the potassium/sodium selectivity is adequate (K_a's, based upon the extraction of potassium picrate from an aqueous solution by a solution of the ionophore in $CDCl_3$, are given below the structures with the ratio of K_a for K^+ and Na^+ given in parentheses). The hemispherand **3** was selected as the ionophore of choice since it is relatively readily synthesised, incorporation of a responding ionisable chromophore into this structure gave the chromoionophore **5b** as a synthetic target.

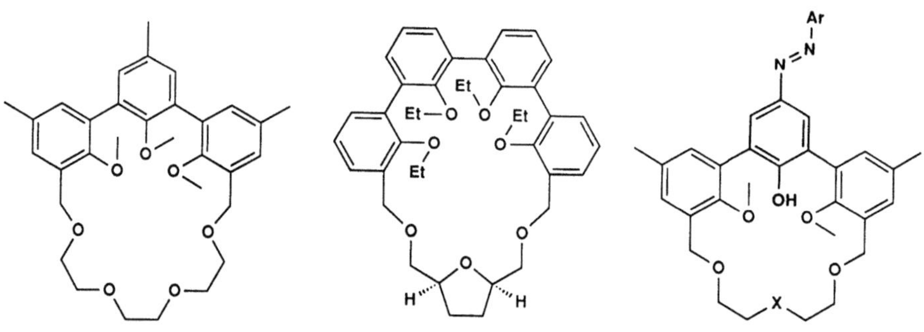

K_a for K^+ 10^8 (330)

3

K_a for K^+ 10^8 (2000)

4

5

a : X = O

b : X = OCH_2CH_2O

in 5 Ar =

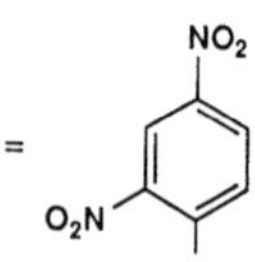

c : X =

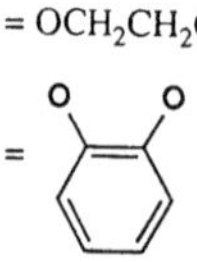

3.2. THE SYNTHESIS OF CHROMOIONOPHORES 5

The synthesis of **5b**, and of the closely related chromoionophores **5a** and **5c**, is summarised in Scheme 2, which is largely based upon established methodology but a few points deserve special comment. The synthesis of the terphenyl derivative **8** by Pd(0) catalysed coupling [13] proved to be a very reliable procedure and superior to a number of alternative methods [14]. The use of the methoxymethyl (MOM) [15] protecting group proved to be better than either protection by allyl or 2-methyoxyethoxymethyl (MEM), in particular it was retained during the formulation of the intermediate **9** so that the dialdehyde intermediate **13** was obtained in good yield. Reduction of the dialdehyde **13** to give the diol **14** was quantitative, but attempted preparation of the dibromide **10** from the diol **14** using PBr_3 in benzene resulted in extensive deprotection. A number of alternative procedures were investigated and eventually it was found that *N*-bromosuccinimide and dimethyl sulphide in CH_2Cl_2 at room temperature [16] gave consistently high yields of the required dibromide **10**.

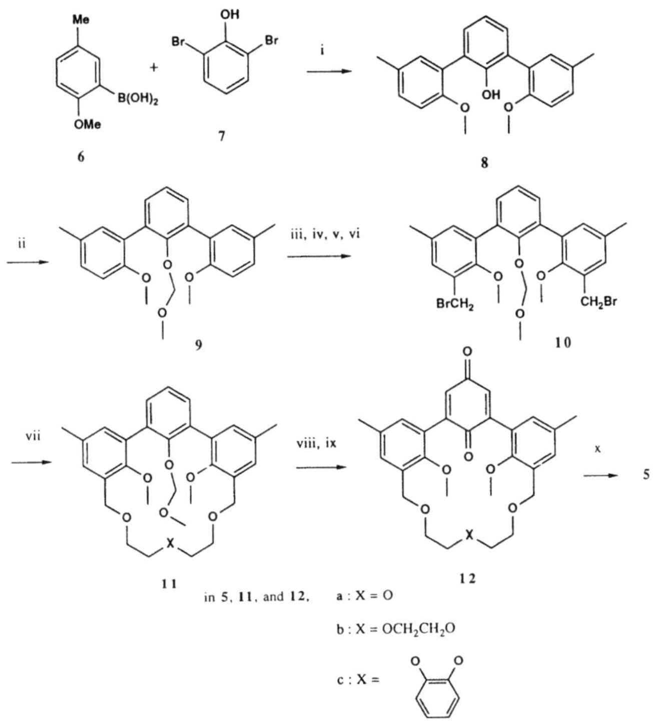

Scheme 2. Synthesis of hemispherand based chromoionophores **5**
Reagents: i, $Pd(PPh_3)_4$, EtOH, C_6H_6, Na_2CO_3/H_2O; ii, $MeOCH_2Cl$, NaH, THF; iii, BuLi, TMEDA, THF; iv, excess DMF; H_3O^+; v, $NaBH_4$; vi, 3 equiv NBS, 3 equiv Me_2S, CH_2Cl_2; vii, H-X-H, NaH, THF; viii, TFA, CH_2Cl_2/H_2O; ix, $Tl(NO_3)_3$; x, $ArNHNH_2$, H^+, EtOH, CH_2Cl_2.

The dibromide **10** reacted with diols under high dilution conditions to give hemispherands **11** in moderate to good yields. Sodium ions probably served as a template for the synthesis of **11a** (65% yield) and addition of potassium bromide to the reaction mixture gave increased yields (*ca* 50%) for the hemispherands **11b** and **11c**. The products **11** were purified by chromatography on silica gel mixed with a suitable metal salt (NaCl for **11a** and KBr for **11b** and **11c** at *ca* 1–2 g per 10 g of silica gel). The cyclic ionophores were eluted as complexes with the added metal salt after non-ionophoric impurities had been eluted by less polar solvents. This procedure has proved to be suitable for the purification of a wide range of

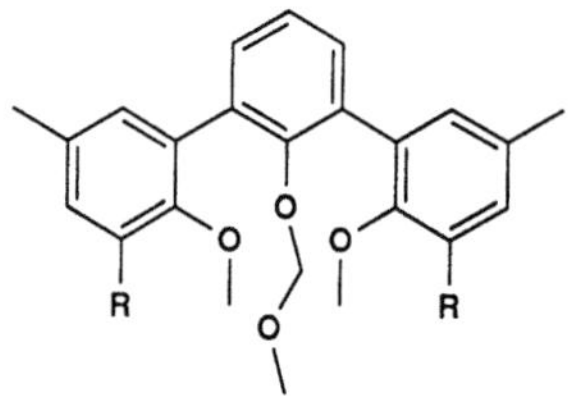

13 R = CHO

14 R = CH_2OH

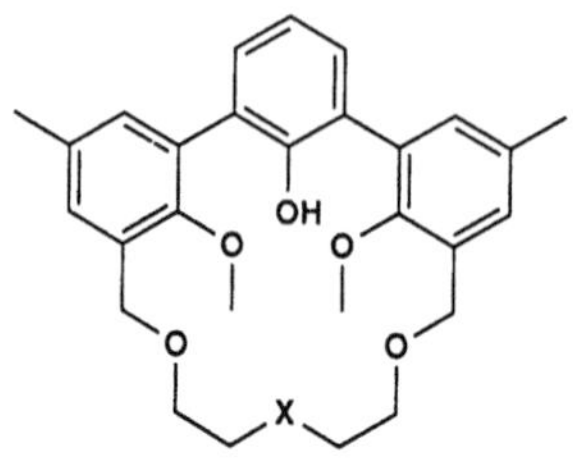

15

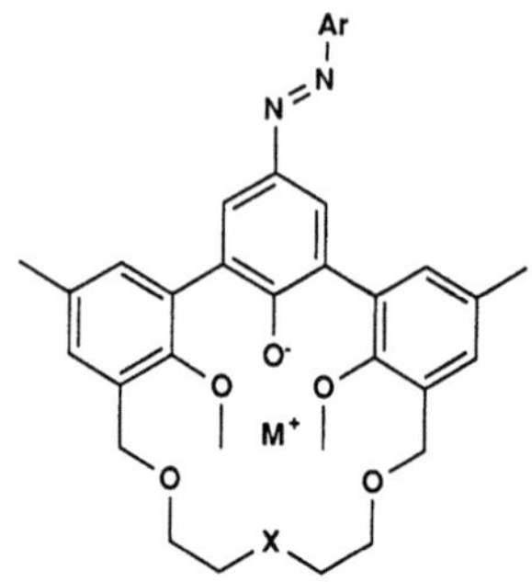

16

in **15** and **16**, a : X = O

b : X = OCH_2CH_2O

c : X = 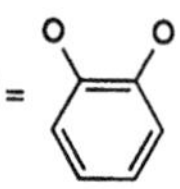

ionophores in other work. The parent ionophore **11** was regenerated from the metal salt complex by washing with water and crystallisation.

The MOM protecting group of **11** was removed quantitatively using TFA and the resulting phenol oxidised with thallium trinitrate [17] to give the quinones **12** in high yield. These reacted with arylhydrazines to give the required azophenols **5**. Alternative, direct coupling of the phenols with diazonium tetrafluoroborates proved to be a less satisfactory procedure. The route outlined in Scheme 2 is generally suitable for the sythesis of phenolic hemispherands **15** and hemispherand quinones **12**, the overall yields from the boronic acid **6** being greater than 10%.

3.3. EXTRACTION BEHAVIOUR OF CHROMOIONOPHORES 5

The azophenols **5** were examined for their selectivity in the extraction of metal cations from water into organic solvents in the pH range 7–9. The resulting values for K_e gave a very useful indication of their potential for use in optical fibre sensors. The results of this examination are given in Table I. Although the azophenols **5** extract all alkali metal ions into CH_2Cl_2 at high pH to form blue solutions of the salts **16**, with the absorption maxima indicated in the table, at lower pH they show

Table I. Extraction coefficients[a] for alkali metal cations and azophenols **5**.

Metal ion	$\log_{10}K_e(\pm 0.2)(\lambda_{max})$					
	5a[a]	$(\lambda_{max})^b$	**5b**[a]	$(\lambda_{max})^b$	**5c**[a]	$(\lambda_{max})^b$
Li^+	−9.34	(606)	−9.34	(594)	−9.91	(586)
Na^+	−7.06	(610)	−8.40	(606)	−9.57	(610)
K^+	−7.14	(612)	−6.99	(616)	−7.34	(612)
Rb^+	−8.38	(614)	−6.67	(620)	−6.63	(620)
Cs^+	−9.14	(614)	−6.53	(620)	−6.64	(620)

[a]Calculated with the expression $K_e = [H^+]_{aq}.[ML]_{org}/[M^+]_{aq}.[HL]_{org}$ for a solution of **5** at *ca* 6×10^{-5} M in CH_2Cl_2 and M^+Cl^- in H_2O at pH 8.
[b]For extraction of aqueous MOH (1 M) by a solution of **5** at *ca* 6×10^{-5} M.

greater selectivity. Thus, at pH 8 the hemispherand **5c** shows high selectivity for K^+ over Na^+ (K_{K^+}/K_{Na^+} *ca* 170) and Li^+ (K_{K^+}/K_{Li^+} ca 370) although Rb^+ and Cs^+ are extracted more efficiently than K^+. The more flexible hemispherand **5b** shows significantly less selectivity than **5c** for K^+ as compared with Na^+ (K_{K^+}/K_{Na^+} *ca* 26) and its structure can evidently adapt more readily to the smaller size of the sodium cation than can the more rigid benzo-fused structure of **5c**. Neither of the hemispherands **5b** and **5c** shows measurable extraction of Mg^{2+}, Ca^{2+}, or Sr^{2+}, presumably because the corresponding 1 : 1 complexes (*cf* **16**) would require the presence of a poorly solvated anion in the organic phase. However both azophenols extract Ba^{2+} although the resulting complex has an absorption spectrum that is quite distinct from those of the salts **16**. The reason for this is not clear but we note that ion pairing for $Ba^{2+} \cdot X^-$ occurs rather more readily than for the other doubly charged cations. The hemispherand derivative **5a** does not show useful K^+/Na^+ selectivity which is not unexpected in view of the reported K_a's for the corresponding metal picrates [18] by the related trimethoxy hemispherand.

Note

[1]This paper is submitted to honour the memory of the late Charles J. Pedersen. It would not have been possible to carry out this work without the inspiration of the many ideas and discoveries that have resulted from his discovery of the crown ethers.

References

1. C. J. Pedersen: *J. Am. Chem. Soc.* **89**, 7017 (1967); C. J. Pedersen and H. J. Frensdorff: *Angew. Chem., Int. Ed. Engl.* **11**, 16 (1972); C. J. Pedersen: *Angew. Chem., Int. Ed. Engl.* **27**, 1021 (1988).
2. B. Dietrich, J.-M. Lehn, and J.-P. Sauvage: *Tetrahedron Lett.* 2885, 2889 (1969); J.-M. Lehn: *Pure Appl. Chem.* **50**, 871 (1978); **51**, 579 (1979); J.-M. Lehn: *Angew. Chem., Int. Ed. Engl.* **27**, 89 (1988).
3. K. E. Koenig, G. M. Lein, P. Stuckler, T. Kaneda, and D. J. Cram: *J. Am. Chem. Soc.* **101**, 3553 (1979); D. J. Cram: *Angew. Chem., Int. Ed. Engl.* **25**, 1039 (1986); D. J. Cram: *Angew. Chem., Int. Ed. Engl.* **27**, 1009 (1988); J. A. Bryant, R. C. Helgeson, C. B. Knobler, M. P. deGrandpore, and D. J. Cram: *J. Org. Chem.* **55**, 4622 (1990).
4. D. J. Cram, T. Kaneda, R. C. Helgeson, S. B. Brown, C. B. Knobler, E. Maverick, and K. N. Trueblood: *J. Am. Chem. Soc.* **107**, 3645 (1985); D. J. Cram and G. M. Lein: *J. Am. Chem. Soc.* **107**, 3657 (1985); K. D. Stewart, M. Miesch, C. B. Knobler, E. F. Maverick, and D. J. Cram: *J. Org. Chem.* **51**, 4327 (1986); K. M. Doxsee, M. Feigel, K. D. Stewart, J. W. Canary, C. B Knobler, and D. J. Cram: *J. Am. Chem. Soc.* **109**, 3098 (1987); J. A. Bryant, S. P. Ho, C. B. Knobler, and D. J. Cram: *J. Am. Chem. Soc.* **112**, 5837 (1990).
5. C. D. Gutsche: *Top. Curr. Chem.* **123**, 1 (1984); C. D. Gutsche: *Prog. Macrocyclic Chem.* **3**, 93 (1987); C. D. Gutsche: *Calixarenes* (Vol. 1, Monographs in Supramolecular Chemistry, Ed. J. F. Stoddart), Royal Soc. Chem., Cambridge (1989).
6. J. Koryta (ed.): *Medical and Biological Applications of Analytical Devices*, Wiley, Chichester, (1980); Y. Takeda: *Top. Curr. Chem.* **121**, 1 (1984); M. Takagi and K. Ueno: *Top. Curr. Chem.* **121**, 39 (1984); J. van Gent, E. J. R. Sudholter, P. V. Lambeck, Th. J. A. Popma, G. J. Gerritsma, and D. N. Reinhoudt: *J. Chem. Soc. Chem. Commun.* 893 (1988); D. J. Cram, R. A. Carmack, and R. C. Helgeson: *J. Am. Chem. Soc.* **110**, 571 (1988); R. C. Helgeson, B. P. Czech, E. Chapoteau, C. R. Gebauer, A. Kumar, and D. J. Cram: *J. Am. Chem. Soc.* **111**, 6339 (1989); R. Y. Tsien: *Methods in Cell Biology* **30B**, 127 (1989); R. Y. Tsien: *Ann. Rev. Neurosci.* **12**, 227 (1989); F. Fages, J. P. Desvergne, H. Bouas-Laurent, J.-M. Lehn, J. P. Konopelski, P. Marsau, and Y. Barrons: *J. Chem. Soc., Chem. Commun.* 655 (1990); A. P. De Silva and H. Q. N. Gunaratne: *J. Chem. Soc., Chem. Commun.* 186 (1990).

7. D. J. Cram and K. N. Trueblood: *Top. Curr. Chem.* **98**, 43 (1981); F. Vögtle, H. Sieger, and W. M. Muller: *Ibid* **98**, 107 (1981); R. Hilgenfeld and W. Saenger: *Ibid* **101**, 1 (1982); I. Tabushi and K. Yamamura; *Ibid.* **113**, 145 (1983); C. D. Gutsche: *Ibid.* **123**, 1 (1984); F. Vögtle, W. M. Muller, and W. H. Watson: *Ibid* **125**, 131 (1984); F. P. Schmidtchen: *Ibid* **132**, 101 (1986); J. Franke and F. Vögtle: *Ibid.* **132**, 137 (1986); J. Rebek: *Ibid.* **149**, 189 (1988); F. Diederich: *Angew. Chem., Int. Ed. Engl.* **27**, 362 (1988); J. Rebek: *Molecular Recognition in Chemical and Biochemical Problems* (Ed. S. M. Roberts), Chem. Soc. Special Pub. No. 78, Cambridge (1989); Papers from 13th International Symposium on Macrocyclic Compounds, *Pure Appl. Chem.* **61** (1989).
8. L. Stryer: *Biochemistry* 2nd Edn., W. H. Freeman, New York (1981); J. Darnall, H. Lodish, and D. Baltimore: *Molecular Cell Biology* Scientific American Books, New York (1986).
9. J. F. Alder, D. C. Ashworth, R. Narayanaswamy, R. E. Moss, and I. O. Sutherland: *Analyst* **112**, 1191 (1987); R. E. Moss and I. O. Sutherland. *Anal. Proc.* **25**, 272 (1989).
10. T. Kaneda, Y. Sugihara, H. Kamiya, and S. Misumi: *Tetrahedron Lett.* **22**, 4407 (1981); K. Sugihara, T. Kaneda, and S. Misumi: *Heterocycles* **18**, 57 (1982).
11. P. ,J. Dijkstra, H. J. den Hertog, B. J. van Steen, S. Zijlstra, M. Skowronska-Ptasinska, D. N. Reinhoudt, J. van Erden, and S. Hardema: *J. Org. Chem.* **52** 2433 (1987).
12. S. P. Artz and D. J. Cram: *J. Am. Chem. Soc.* **106**, 2160 (1984).
13. N. Miyawa, T. Yanagi, and A. Suzuki: *Synth. Commun.* **11**, 513 (1981).
14. C. J. F. Du, H. Hart, and K. K. G. Ng: *J. Org. Chem.* **51**, 3162 (1986).
15. E. J. Corey, J. L. Gras, and P. Ulrich: *Tetrahedron Lett.* 809 (1976).
16. E. J. Corey, C. U. Kim, and M. Takeda: *Tetrahedron Lett.* 4339, (1972).
17. A. McKillop, D. H. Perry, and M. Edwards: *J. Org. Chem.* **41**, 282 (1976).
18. K. E. Koenig, G. M. Lein, P. Stuckler, T. Kaneda, and D. J. Cram: *J. Am. Chem. Soc.* **101**, 3553 (1979).

Journal of Inclusion Phenomena and Molecular Recognition in Chemistry **12**: 237–251, 1992.

Molecular Dynamics and the Structure of Macrocycles – Solvent Acetonitrile Interactions#

MEIZHEN XU and SERGIO PETRUCCI*
Weber Research Institute and Dept. of Chemistry, Polytechnic University, Long Island Center, Route 110, Farmingdale, NY 11735, U.S.A.

EDWARD M. EYRING*
Department of Chemistry, University of Utah, Salt Lake City, UT 84112, U.S.A.

(Received: 6 September 1990; in final form: 30 January 1991)

Abstract. Microwave complex permittivities, $\varepsilon^* = \varepsilon' - J\varepsilon''$, are reported in the 1–90 GHz frequency range for the macrocycles 18-crown-6 (18C6) and 15C5 added to acetonitrile in stoichiometric proportions, in the solvent CCl_4 at 25°C. Digitized infrared spectra of the 'CN stretch' ν_2 vibration of acetonitrile for the same systems are reported in the 2300–2200 cm^{-1} spectral region. The macrocycle 12C4 added to CH_3CN has also been investigated in the infrared. Both the dielectric relaxation and infrared results are interpreted in terms of macrocycle–acetonitrile interactions, probably involving a methyl-hydrogen to ethereal-oxygen interaction. These interactions with CH_3CN diminish in strength according to the sequence: 18C6 > 15C5 > 12C4.

Key words. Crown ethers, solvent-macrocycle interactions, microwave dielectric relaxation.

1. Introduction

The literature on macrocyclic ligands in solution [1, 2] abounds with information on their most salient interaction, the complexation of metal cations. The isomeric dynamics of macrocycles in aqueous and non-aqueous solvents have also been summarized for simple macrocycles [3]. These previous results focus on the macrocycle itself and on its dynamics in various solvents with the solvent present in excess. Hence, interactive changes of the macrocycles are masked by the more abundant solvent molecules.

In a pioneering effort, mainly utilizing NMR techniques, Mosier-Boss and Popov [4] focused on these macrocycle–solvent interactions by looking at the 18C6–acetonitrile system in 1 : 1 stoichiometric proportions, dissolved in the essentially inert solvent CCl_4. These interactions were expressed quantitatively by formation constants extracted [4] from the NMR data, and the existence of these interactions was confirmed by infrared data.

In the present work, microwave dielectric relaxation techniques were used to probe the effect of mixing the macrocycles (18C6 and 15C5) with acetonitrile (CH_3CN) in the solvent CCl_4 on the molecular relaxation dynamics of both the

This paper is dedicated to the memory of the late Dr C. J. Pedersen.

* Authors for correspondence

macrocycle and CH_3CN. Infrared data on the 'CN stretch' ν_2 vibration proved to be important for the characterization of these interactions.

It will be shown that interaction between CH_3CN and crown ethers diminish in strength in the order 18C6 > 15C5 > 12C4.

2. Experimental

The microwave dielectric instrumentation, the computer assisted infrared spectrometer, and the associated procedures have all been described before [5, 6]. 18C6 (Aldrich) was recrystallized from distilled, dry acetonitrile and dried *in vacuo*. 15C5 and 12C4 (Aldrich) were exposed to molecular sieves for several weeks before use. Acetonitrile (Aldrich, gold label) was refluxed for several hours over P_2O_5 before distilling in an all Pyrex distillation apparatus with no grease on the joints. CCl_4 was also distilled over P_2O_5 in an all Pyrex column.

3. Results and Discussion

3.1. MICROWAVE DIELECTRIC RELAXATION

Figure 1 reports the real part ε' and the loss coefficient ε'' of the complex dielectric permittivity $\varepsilon^* = \varepsilon' - J\varepsilon''$, in the frequency range 1–90 GHz. The system illustrated is the mixture CH_3CN 0.4M + 18C6 0.10M, in the solvent CCl_4 at 25°C. The spectrum profile can be interpreted by two Debye single relaxation dielectric processes, centered at $f_1 = 5$ GHz and $f_2 = 35$ GHz, respectively. Figure 2 reports ε' and the loss coefficient ε'' for CH_3CN 0.4M in CCl_4 at 25°C. The data can be interpreted, within experimental error, by a single Debye relaxation centered at $f_r = 30$ GHz with relaxation parameters comparable to those used for the higher frequency relaxation process of the mixture above. The data for the mixture are expressible by the function

$$\begin{aligned} \varepsilon &= \Delta\varepsilon_1' \frac{1}{1+(f/f_1)^2} + \Delta\varepsilon_2' \frac{1}{1+(f/f_2)^2} + \varepsilon_{\infty 2} \\ \varepsilon'' &= \Delta\varepsilon_1'' \frac{f/f_1}{1+(f/f_1)^2} + \Delta\varepsilon_2'' \frac{f/f_2}{1+(f/f_2)^2} \end{aligned} \tag{1}$$

with $\Delta\varepsilon_1' = \Delta\varepsilon_1'' = \varepsilon_0 - \varepsilon_{\infty 1}$ and $\Delta\varepsilon_2' = \Delta\varepsilon_2'' = \varepsilon_{\infty 1} - \varepsilon_{\infty 2}$ where ε_0 is the static permittivity, and $\varepsilon_{\infty 1}$ and $\varepsilon_{\infty 2}$ are the permittivity values for frequencies $f > f_1, f_2$.

It is important at this point to compare the results of the macrocycle + CH_3CN in CCl_4 with those of the macrocycle alone in CCl_4 and of CH_3CN alone in CCl_4. The attribution of the relaxation profile of the mixture CH_3CN 0.4M + 18C6 0.1M to two individual Debye processes, with the one at f_2 due to acetonitrile, is strengthened by the results of Figure 1C reporting the microwave dielectric relaxation parameter ε'' of 18C6 0.1M alone in CCl_4 and showing relaxation parameters comparable to those of the lower frequency relaxation component of Figure 3A. In Figure 3A the coefficient of the imaginary part ε'' of ε^* for the mixture CH_3CN 0.1M + 18C6 0.1M in CCl_4 is reported as a function of the frequency f. Figure 3B reports the dielectric spectrum of CH_3CN 0.1M in CCl_4 with parameters again comparable to those of the higher frequency relaxation in Figure 3A.

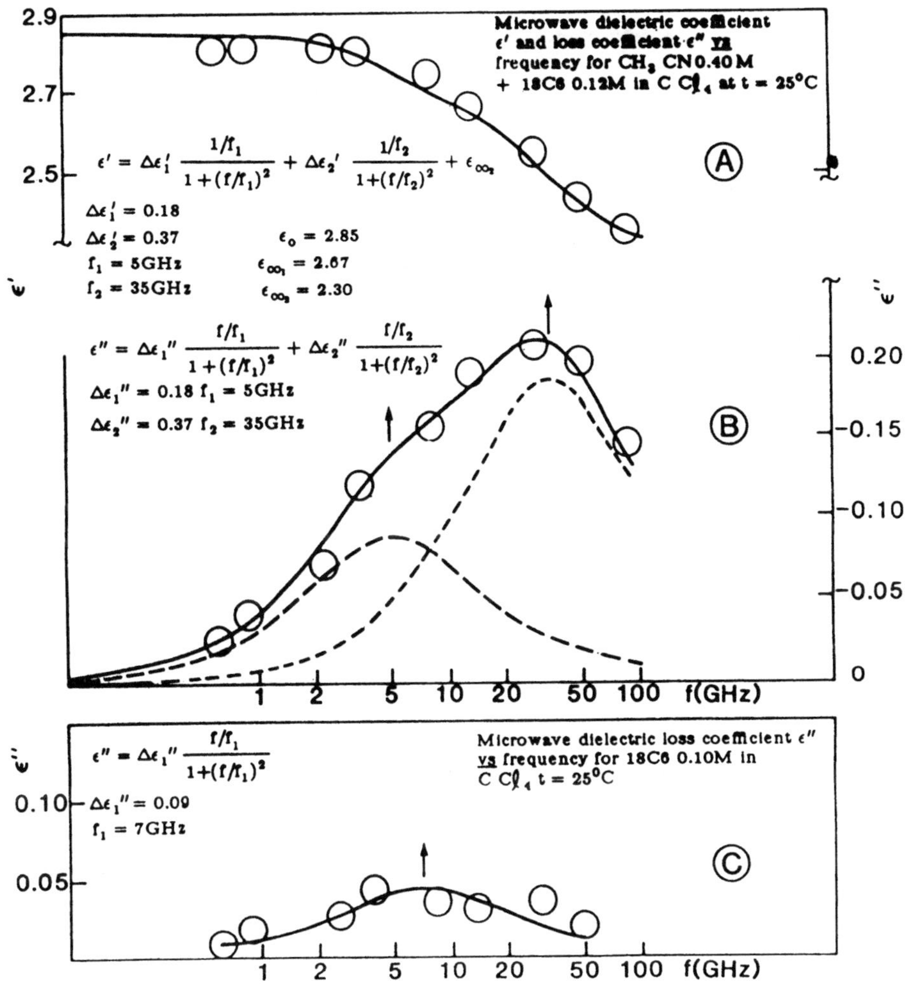

Fig. 1. (A) Microwave dielectric real coefficient ε' of the complex permittivity ε^* vs. frequency f for the mixture CH_3CN 0.4M + 18C6 0.12M in CCl_4 at 25°C. (B) Coefficient of the imaginary part ε'' of the complex permittivity ε^* vs. frequency f for the mixture CH_3CN 0.4M + 18C6 0.12M in CCl_4 at 25°C. (C) ε'' vs. f for 18C6 0.1M in CCl_4 at 25°C.

The relevant issue is that an interpretation of the spectral profile of the mixtures in terms of two relaxation processes, one due to CH_3CN and the other to 18C6, appears to be justified. In addition, the value of f_1 seems to decrease with increasing amounts of CH_3CN and the value of $\Delta\varepsilon_1''$ doubles in going from molar ratio $R = [CH_3CN]/[18C6] = 1$ to $R = 4$. Both these trends can be taken as an indication of the interaction between 18C6 and CH_3CN increasing $\Delta\varepsilon_1''$ through an increase of the apparent dipole moment of 18C6 and increasing the dielectric relaxation time of 18C6, $\tau_1(=2\pi f_1)^{-1}$, by increasing its moment of inertia.

We recall the Böttcher equation [7] correlating the concentration C of a dipolar species of apparent dipole moment μ to the relaxation strength $\Delta\varepsilon = \varepsilon_0 - \varepsilon_{\infty 1}$ for

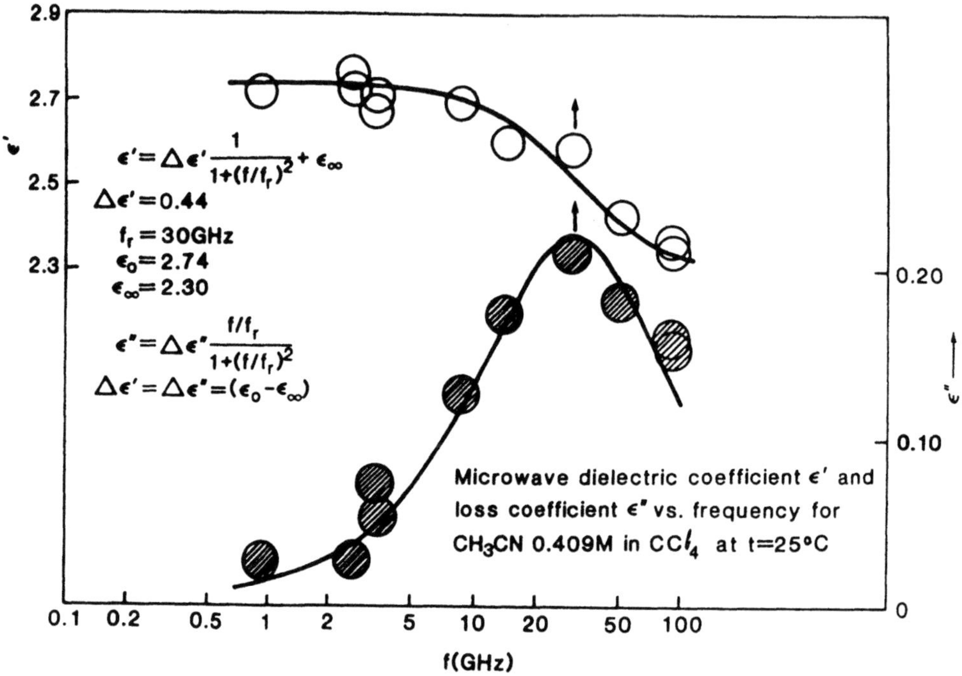

Fig. 2. Microwave dielectric coefficients ε' vs. f and ε'' vs. f for CH_3CN 0.41M in CCl_4 at 25°C.

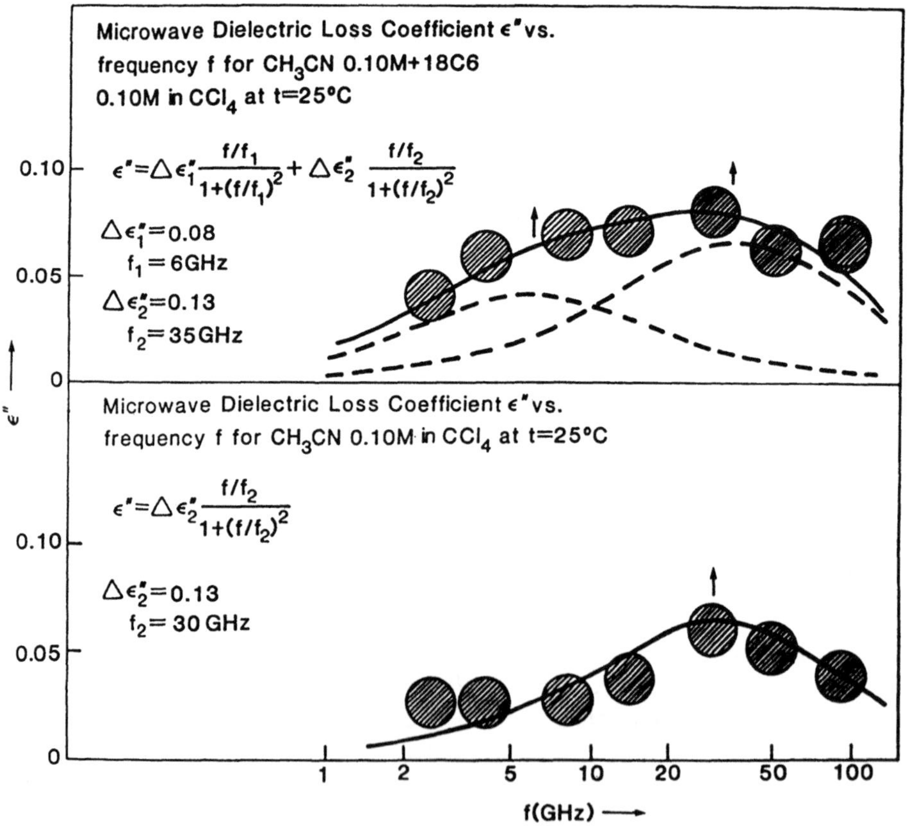

Fig. 3. Microwave loss coefficient ε'' vs. frequency f for the mixture CH_3CN 0.1M + 18C6 0.1M in CCl_4 at 25°C. Microwave loss coefficient ε'' vs. frequency f for CH_3CN 0.1M in CCl_4 at 25°C.

that species

$$\varepsilon_0 - \varepsilon_{\infty 1} = \frac{4\pi L \times 10^{-3}}{(1-\alpha f)^2} \frac{\mu^2}{3kT} \frac{3\varepsilon_0}{2\varepsilon_0 + 1} C \tag{2}$$

where L is Avogadro's number, α is the polarizability and f is the reaction field factor [7]. This can be rewritten, neglecting the polarizability α reacting field f factor term $(1-\alpha f)^2 = 0.9$ with respect to one, as

$$\phi(\varepsilon) = (\varepsilon_0 - \varepsilon_{\infty 1}) \frac{2\varepsilon_0 + 1}{3\varepsilon_0} = \frac{4\pi L \times 10^{-3}}{3kT} \mu^2 C \tag{3}$$

Thus the slope $d\phi(\varepsilon)/dC$ should increase with μ^2 for a given species.

Figure 4 shows a plot of $\phi(\varepsilon)$ vs. the molar ratio $R = [CH_3CN]/[18C6]$ for the system $CH_3CN + 18C6$ in CCl_4 at 25°C. The concentration of 18C6 is $C^0_{18C6} = 0.10$M, whereas the concentration of acetonitrile $C^0_{CH_3CN}$ varies from 0.1 to 0.5M. The plot has a positive slope indicating that the apparent dipole moment of 18C6 increases with R (as is also apparent by comparing the $\Delta\varepsilon''_1$ difference of Figures 1 and 3). These results reinforce the notion that 18C6 interacts with CH_3CN, affecting the rotational relaxation parameter of the species involved in the lower frequency relaxation process.

Table I reports all the dielectric parameters ε_0, $\varepsilon_{\infty 1}$, $\varepsilon_{\infty 2}$, f_1 and f_2 used to fit the dielectric spectra of the various R values investigated for the system $18C6 + CH_3CN$ in CCl_4 at 25°C.

Before leaving this system, consider Figure 5, which is a Böttcher plot of $\phi(\varepsilon)$ vs. $C^0_{CH_3CN}$ for acetonitrile alone dissolved in CCl_4. A curvature of the plot is clearly visible which indicates that some of the solvent exists in an apolar form. In fact,

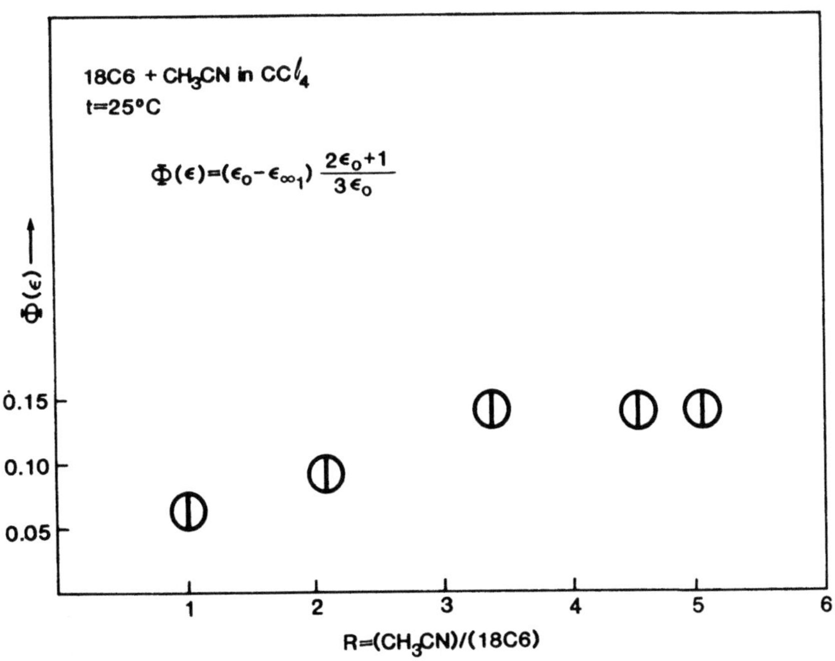

Fig. 4. Böttcher plot of $\phi(\varepsilon) = (\varepsilon_0 - \varepsilon_{\infty 1})(2\varepsilon_0 + 1)/(3\varepsilon_0)$ vs. the molar ratio $R = [CH_3CN]/[18C6]$ for the system $CH_3CN + 18C6$ in CCl_4 at 25°C.

Table I. Collected dielectric parameters ε_0, $\varepsilon_{\infty 1}$, $\varepsilon_{\infty 2}$, f_1 and f_2 used to interpret the dielectric spectra of the investigated systems CH_3CN and CH_3CN + macrocycles in CCl_4.[a]

C(M) CH_3CN	ε_0	$\varepsilon_{\infty 1}$	$\varepsilon_{\infty 2}$	f_1(GHz)	f_2(GHz)
0.100	2.37	2.24			30
0.200	2.51	2.24			30
0.303	2.64	2.28			30
0.409	2.74	2.30			30
0.503	2.85	2.30			30
18C6					
0.100	2.32	2.23		7	
18C6 + CH_3CN					
0.100 + 0.100	2.42	2.34	2.21	6	35
0.100 + 0.208	2.61	2.49_5	2.27_2	5	35
0.120 + 0.404	2.85	2.67	2.30	5	35
0.100 + 0.453	2.90	2.72	2.32	5	35
0.100 + 0.504	2.97	2.79	2.32	5	35
15C5					
0.100	2.29	2.22		9	
0.200	2.45	2.26		9	
0.300	2.61	2.26		8	
0.404	2.74	2.30		8	
0.501	2.85	2.30		8	
15C5 + CH_3CN					
0.100 + 0.100	2.53	2.43	2.25	4	30
0.200 + 0.202	2.78	2.60	2.28	4	30
0.301 + 0.301	3.01	2.72	2.28	4	25
0.406 + 0.404	3.35	2.95	2.40	3.8	23
0.501 + 0.499	3.62	3.08	2.40	4.3	23
0.106 + 0.539	3.06	2.87	2.35	5	30

[a] Concentrations are reported with three digits in order to avoid round-off errors in recalculations by others. The values of ε_0, $\varepsilon_{\infty 1}$ and $\varepsilon_{\infty 2}$ are affected by an average error of $\pm 1\%$; the values of f_1 and f_2 by an average error of $\pm 5\%$.

application of the dimerization equilibrium

$$2CH_3CN \rightleftarrows (CH_3CN)_2$$

$$2M \rightleftarrows D \quad \text{with} \tag{4}$$

$$K_d = \frac{[D]}{[M]^2} \quad \text{and} \quad C_0 = [M] + 2[D]$$

leads to the relation: $[M] = (-1 + \sqrt{1 + 8K_d C_0})/4K_d$.

Figure 5 presents the function $\phi(\varepsilon)$ vs. [M] using the minimum value of K_d that linearizes the plot [8] namely $K_d = 0.3_5$. This number is correct only to an order of magnitude, given the relative insensitivity of the linearization of the plot [8]. It is significant, however, that $K_d = 0.3_5$ is within the range of the standard error of the

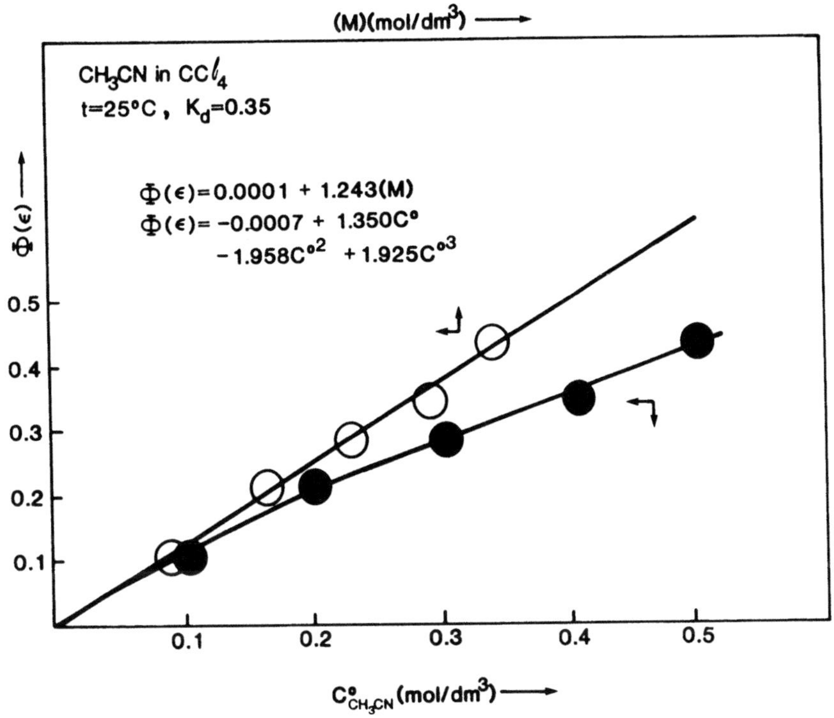

Fig. 5. Böttcher plot (dark circles) of $\phi(\varepsilon)$ vs. the total concentration of acetonitrile $C^0_{CH_3CN}$ for CH_3CN in CCl_4 at 25°C. The solid line corresponds to nonlinear regression giving 50% statistical weight to the origin. Same plot (open circles) of $\phi(\varepsilon)$ vs. the concentration of monomer CH_3CN (M), using a dimerization constant $K_d = 0.35$. The solid line corresponds to linear regression giving 50% statistical weight to the origin.

value $K_d = 0.5_6 \pm 0.2_7$ obtained by Popov by NMR techniques [4] for the dimerization of CH_3CN in CCl_4.

Equation 3 has been used successfully in a comparable concentration range for relatively strongly interactive systems such as dipolar ion-pairs in solution [8]. The correspondence with the K_d obtained by the NMR techniques [4] lends credence to the interpretation of the curvature of Figure 5 as due to dimerization of acetonitrile. The authors believe that the present interpretation based on the results of a relatively modern method such as NMR and a molecular interpretation (Equation 4) advantageously replaces previous ‘pre-NMR’ phenomenological relations based on the Kirkwood aggregation parameter g.

We were also interested in the possible interaction of another macrocycle, 15C5, with acetonitrile. Although the formation of weak adducts between 18C6 and CH_3CN in the solid state is well known, no quantitative information, to our knowledge, has been reported for 15C5 interacting with CH_3CN in an inert solvent. An inherent experimental advantage is the solubility of 15C5 in CCl_4 that is not limited to ~ 0.1M as in the case of 18C6.

Figure 6 shows a representative plot of ε'' vs. frequency for 15C5 0.4M + CH_3CN 0.4M (in molar ratio $R = 1$) in CCl_4 superimposed on the same plot of 15C5 0.4M alone in CCl_4 at 25°C. The solid lines interpret the loss data by two and one Debye

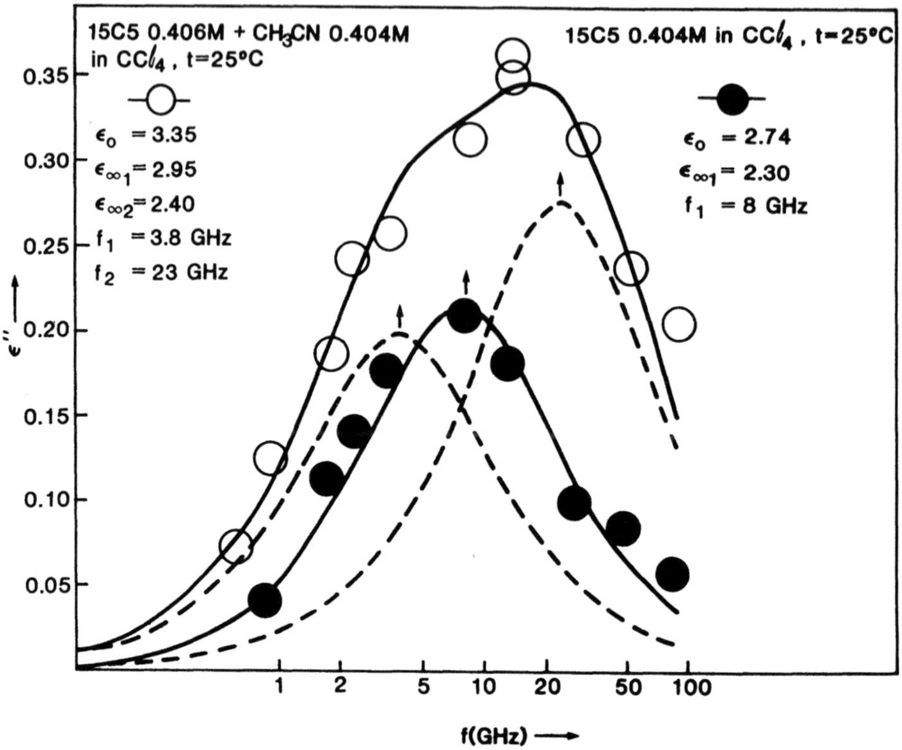

Fig. 6. Representative plot of ε'' vs. frequency f for CH_3CN 0.4M + 15C5 0.4M, at $R = 1$, (open circles) and of 15C5 0.4M in CCl_4 at 25°C (dark circles).

relaxation processes, respectively. While the lower frequency relaxation process drops considerably upon addition of CH_3CN to 15C5, the relaxation stength $\Delta\varepsilon''$ does not change appreciably. This is also shown in Figures 7A and 7B where the Böttcher function $\phi(\varepsilon)$ is plotted against the concentration of 15-crown-5, C^0_{15C5}. Linear regression, assigning 50% statistical weight to the intercept set equal to zero, gives for 15C5 alone

$$r^2 = 0.994, \quad \text{Intercept } -0.0045, \quad \text{Slope} = 0.87 \pm 0.07$$

and for $15C5 + CH_3CN$

$$r^2 = 0.996, \quad \text{Intercept } -0.0022, \quad \text{Slope} = 0.79 \pm 0.05$$

The two slopes are the same within experimental error and so are the two apparent dipole moments $\mu_1 = (3.7_7 \pm 0.1_6) \times 10^{-18}$ esu cm and $\mu_1 = (3.5_9 \pm 0.1_2) \times 10^{-18}$ esu cm for the two systems.

These results indicate that at molar ratio $R = 1$, CH_3CN solvates 15C5 as reflected by the lower relaxation frequency f_1, indicating a bulkier rotating entity when 15C5 is present. (This is true whether one retains for the relaxation time of the 15C5 moiety $\tau_1 = (2\pi f_1)^{-1}$, the decay of the polarization, or the microscopic relaxation time denoted τ_{1m}. The two quantities are related and, in the opinion of some [9], $\tau_1 \approx \tau_{1m}$.)

On the other hand, the interaction between 15C5 and CH_3CN at molar ratio $R = 1$ does not seem strong enough to alter the apparent dipole moment μ_1, namely

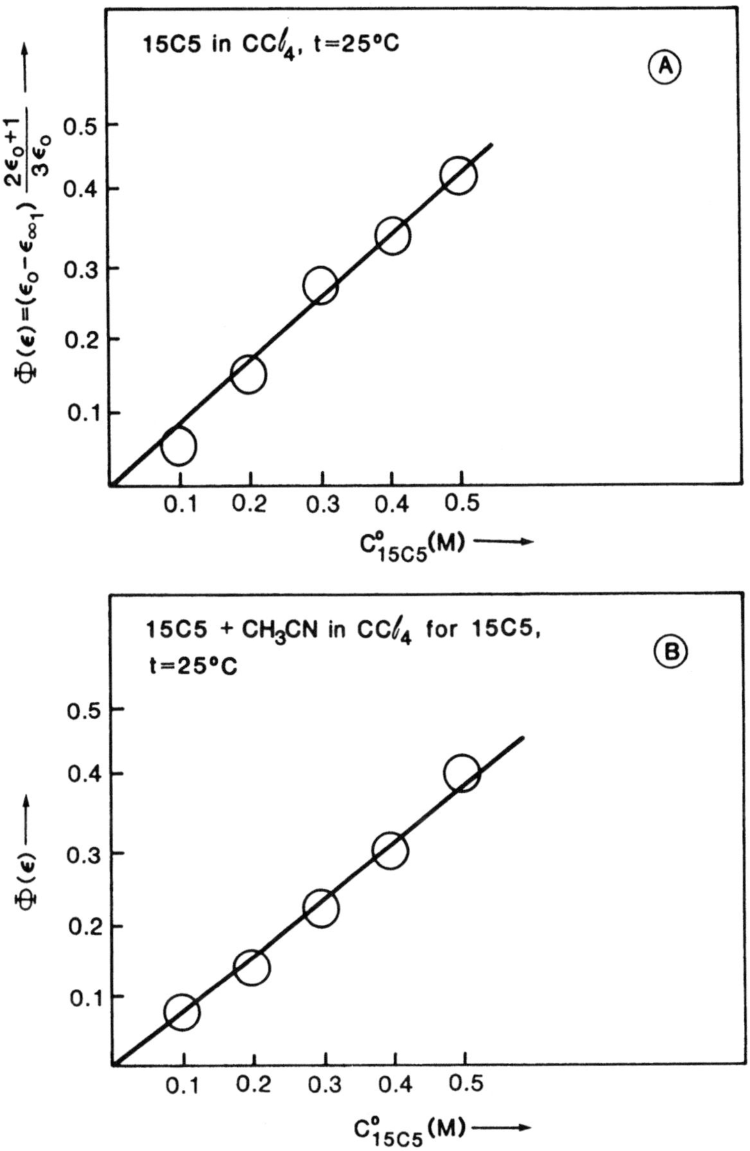

Fig. 7. (A) Böttcher plot for the system 15C5 in CCl_4 of $\phi(\varepsilon)$ vs. the concentration of 15C5, C^0_{15C5} at 25°C. (B) Böttcher plot for the systems $CH_3CN + 15C5$ at $R = 1$ of $\phi(\varepsilon)$ vs. the concentration of 15C5, C^0_{15C5}, at 25°C.

the formal charge separation or average molecular configuration of 15C5 as shown above. Table I reports all the dielectric parameters used to interpret the dielectric spectra of $15C5 + CH_3CN$ at molar ratio $R = 1$ and 25°C in CCl_4.

3.2. INFRARED SPECTRA

Figure 8A is the digitized spectrum corresponding to the ν_2 band of CH_3CN in CCl_4. The spectrum, taken relative to a matched cell filled with CCl_4, can be

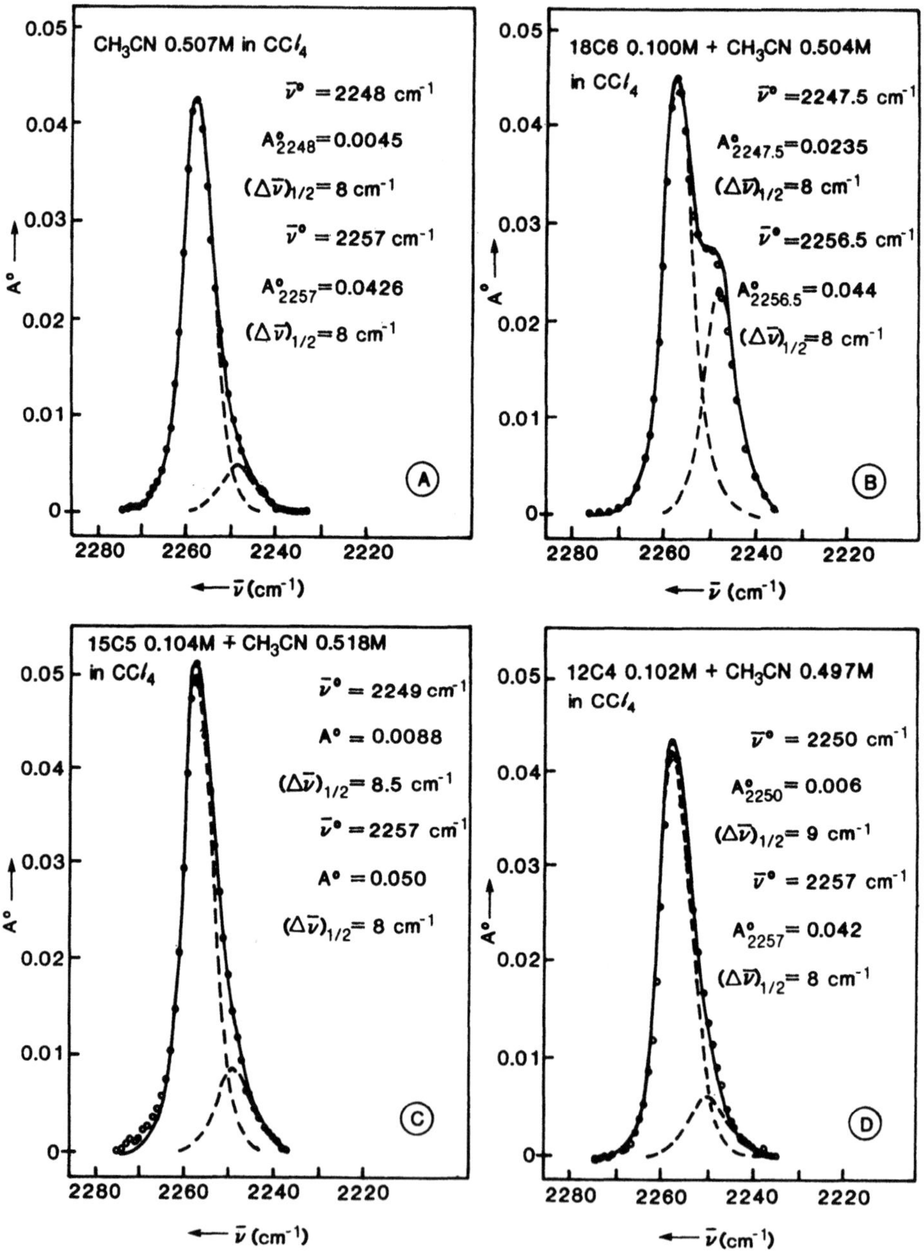

Fig. 8. (A) Representative digitized infrared spectrum of the ν_2 band of CH_3CN 0.5M in CCl_4. The spectral envelope can be deconvoluted into two Gaussian–Lorentzian components. (B) ν_2 band of CH_3CN for the system CH_3CN 0.5M + 18C6 0.1M in CCl_4. (C) ν_2 band of CH_3CN for the system 0.5M + 15C5 0.1M in CCl_4. (D) ν_2 band of CH_3CN for the system CH_3CN 0.5M + 12C4 0.1M in CCl_4.

deconvoluted into two Gaussian–Lorentzian product functions [10]

$$A = \sum_{j=1}^{2} A_j^0 \left[\exp\left(-\frac{(\bar{\nu} - \bar{\nu}_j^0)^2}{2\sigma_j^2} \right) \right] \left[1 + \frac{(\bar{\nu} - \bar{\nu}_j^0)^2}{\sigma_j^2} \right]^{-1} \tag{5}$$

(with σ_j^0 the variance) centered at the wavenumbers $\bar{\nu}_{2255}^0 = 2255\ \text{cm}^{-1}$ and $\bar{\nu}_{2248}^0 = 2248\ \text{cm}^{-1}$. A_j^0 and $(\Delta\bar{\nu}_j)_{1/2}$ are, respectively, the maximum absorbances at $\bar{\nu}_j^0$, and the widths of the band at half maximum absorbance ($A_j^0/2$).

Figures 8B, 8C and 8D are the digitized infrared envelopes related to the CN-stretch of acetronitrile in the mixtures 0.5M CH_3CN + 0.1M macrocycle (at $R = 5$). These spectral envelopes can also be interpreted by two Gaussian–Lorentzian bands centered at about the same wavenumbers as for CH_3CN in CCl_4. However, the maximum absorbances of the satellite band have increased relative to the one for CH_3CN alone in CCl_4 with the relative sequence $A_{2248}^0(18C6) > A_{2248}^0(15C5) > A_{2248}^0(12C4) > A_{2248}^0(CCl_4)$.

Figure 9A reports the quantities $A_{2248/l}^0$, namely the normalized maximum absorbance per unit optical path length vs. the molar ratio R for the three macrocycles investigated in CCl_4. Figure 9B presents the same quantities vs. the concentration $C_{CH_3CN}^0$. In this way even the data without added macrocycle can be viewed for comparison. Again it appears that the normalized maximum absorbance of the satellite band at $\bar{\nu}^0 = 2248\ \text{cm}^{-1}$ follows the trend 18C6 > 15C5 > 12C4 > CCl_4.

Raman studies of solutions of acetonitrile in CCl_4 have shown [11] that the deconvolution of the $\bar{\nu}_2$ band of CH_3CN is difficult because of interference attributed to hot bands [11], which occur at the low frequency side of the $\bar{\nu}_2$ band. Attribution of the 2248 cm^{-1} band for CH_3CN in CCl_4 to hot bands does not necessarily explain the nature of the increase in absorbance of the satellite band upon addition of the macrocycle in the order 12C4 < 15C5 < 18C6.

A possible alternate interpretation of the band at $\bar{\nu}^0 = 2248\ \text{cm}^{-1}$ for CH_3CN in CCl_4 could be found by invoking interaction of the protons of the methyl group of acetonitrile with the chlorine atoms of CCl_4. A model envisaging this specific interaction has indeed been proposed [4].

One could also argue that the CN stretching frequency of acetonitrile in CCl_4 added to macrocycles is affected (in the order 12C4 < 15C5 < 18C6) by interaction of the protons of the —CH_3 group with the oxygen atoms of the macrocycle [12], and models have been proposed to this end [4].

Perhaps the most convincing evidence of the existence of a specific additional source of the increase of the absorbance of the satellite band attributable to the macrocycles is shown in Figure 10. Here the maximum absorbance ratio $(A_{2248/l}^0)/(A_{2255/l}^0)$ is plotted against the molar ratio R. The absorbance ratio increases up to about $R = 2$ with the absorbance ratio for 18C6 being larger than those for the other two macrocycles. Thus $R \simeq 2$ apparently represents a 'saturation' of the interaction of the macrocycles with acetonitrile corresponding to a relative increase of A_{2248}^0 with respect to A_{2255}^0. It then follows that, after this 'saturation' occurs, further addition of acetonitrile does not affect the ratio of the absorbances, the individual values of the two bands increasing in the same order with respect to each other as in CH_3CN *alone* in CCl_4. Notice that a similar 'saturation' of the Böttcher dielectric function $\phi(\varepsilon)$ for $R > 3$ is visible in Figure 4.

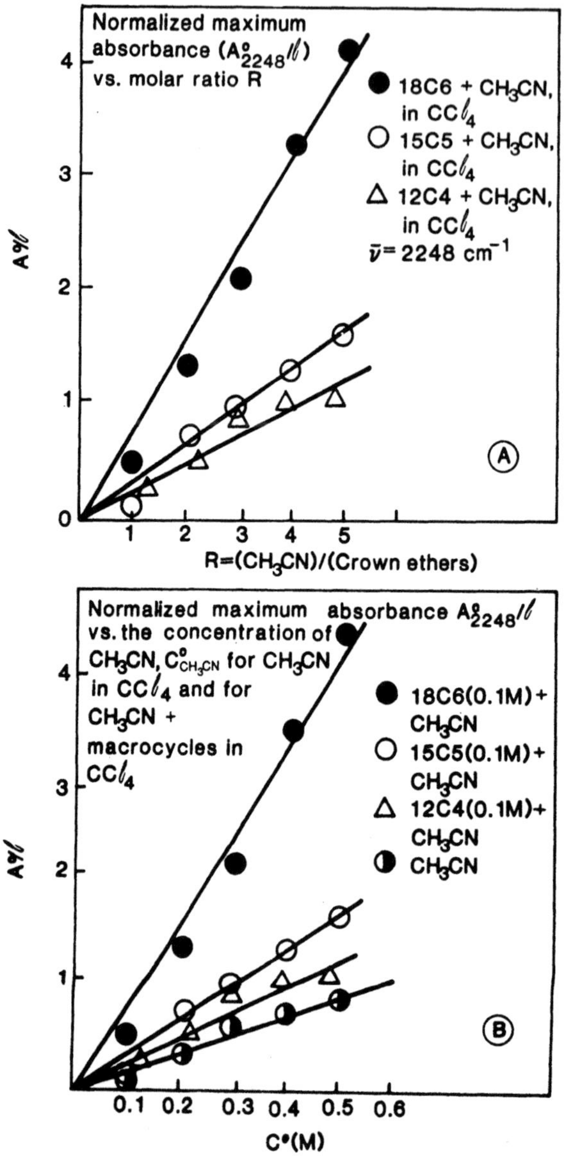

Fig. 9. (A) Normalized maximum absorbance A^0_{2248}/l per unit cell length vs. molar ratio R for the systems CH_3CN + macrocycles investigated in CCl_4 at 25°C. (B) Normalized maximum absorbance A^0_{2248}/l vs. the concentration of acetonitrile, $C^0_{CH_3CN}$, for CH_3CN in CCl_4 and for CH_3CN + macrocycles in CCl_4 at 25°C.

Table II reports all the infrared parameters A^0_j, $\bar{\nu}^0_j$, $(\Delta\bar{\nu}_j)_{1/2}$ of the Gaussian–Lorentzian bands used to interpret the infrared spectra of all the systems investigated.

Table III reports the numerical coefficients of the fitted equations related to the quantities A_0/l vs. either R or C_{CH3CN} for the mixtures shown in Figures 9A and 9B.

Table II. Collected infrared parameters $\bar{\nu}_j^0$, A_j^0, $(\Delta\bar{\nu}_j)_{1/2}$ for the Gaussian Lorentzian product functions used to deconvolute the spectral envelopes, related to the "CN stretch", for the systems CH_3CN and CH_3CN + Macrocycles in CCl_4.[a]

C(M) CH_3CN	$\bar{\nu}_{2257}^0$	A_{2257}^0	$(\Delta\bar{\nu}_{2257})_{1/2}$	$\bar{\nu}_{2248}^0$	A_{2248}^0	$(\Delta\bar{\nu}_{2248})_{1/2}$	l (cm)
0.100	2257.8	0.0217	7.3	2248	0.0017	8.0	0.0199
0.207	2257.5	0.0410	8.0	2248	0.0070	8.0	0.0217
0.302	2257.5	0.0177	8.0	2248	0.0030	8.0	0.00537
0.402	2257.2	0.0320	8.0	2248	0.0037	8.0	0.00544
0.507	2257.0	0.0426	8.0	2248	0.0045	8.0	0.00553
18C6 + CH_3CN							
0.100 + 0.100	2256.7	0.0260	8.5	2248	0.0100	8.0	0.01981
0.100 + 0.201	2257.0	0.0140	8.3	2248.2	0.075	8.0	0.00560
0.100 + 0.302	2256.8	0.0224	8.3	2248.5	0.0116	7.8	0.00556
0.100 + 0.407	2256.5	0.0333	8.0	2247.5	0.0180	7.8	0.00554
0.100 + 0.504	2256.5	0.0440	8.0	2247.5	0.0235	8.0	0.00567
15C5 + CH_3CN							
0.108 + 0.108	2256.6	0.0063	11	2248	0.00075	8.5	0.00557
0.109 + 0.229	2257	0.0163	8	2248.8	0.0040	8.5	0.00549
0.108 + 0.317	2257	0.0292	8	2248.5	0.0055	8.5	0.00551
0.103 + 0.410	2257	0.038	8	2248.5	0.0070	8.5	0.00538
0.104 + 0.518	2256.8	0.050	8	2248.8	0.0088	8.5	0.00545
12C4 + CH_3CN							
0.0976 + 0.119	2257.2	0.0078	7	2250	0.0014	9	0.00547
0.101 + 0.225	2257.1	0.0155	7	2250	0.0027	9	0.00542
0.107 + 0.325	2257.1	0.0265	7	2250	0.0050	9	0.00573
0.105 + 0.409	2257	0.0360	7	2250	0.0060	9	0.00576
0.102 + 0.497	2256.8	0.0420	8	2250	0.0060	9	0.00569
15C5 + CH_3CN							
0.108 + 0.108	2256.5	0.0063	11	2248	0.00075	8.5	0.00557
0.206 + 0.200	2256.5	0.0148	6.3	2251	0.0028	5.0	0.00546
0.301 + 0.304	2256.5	0.0262	8	2248.9	0.0060	8.0	0.00553
0.416 + 0.400	2256.0	0.0380	8	2248.5	0.0075	8.0	0.00576
0.501 + 0.508	2255.8	0.0480	8	2248	0.0115	8.0	0.00556
12C4 + CH_3CN							
0.502 + 0.497	2256	0.0500	8	2248	0.0070	9	0.00577

The last digit in the concentrations, the decimal figures in the $\bar{\nu}_j^0$ values and in the $(\Delta\bar{\nu}_j)_{1/2}$ values and the last digit in the A_j^0 values are affected by a significant error. These digits are reported here in order to avoid round-off errors in recalculations by others.

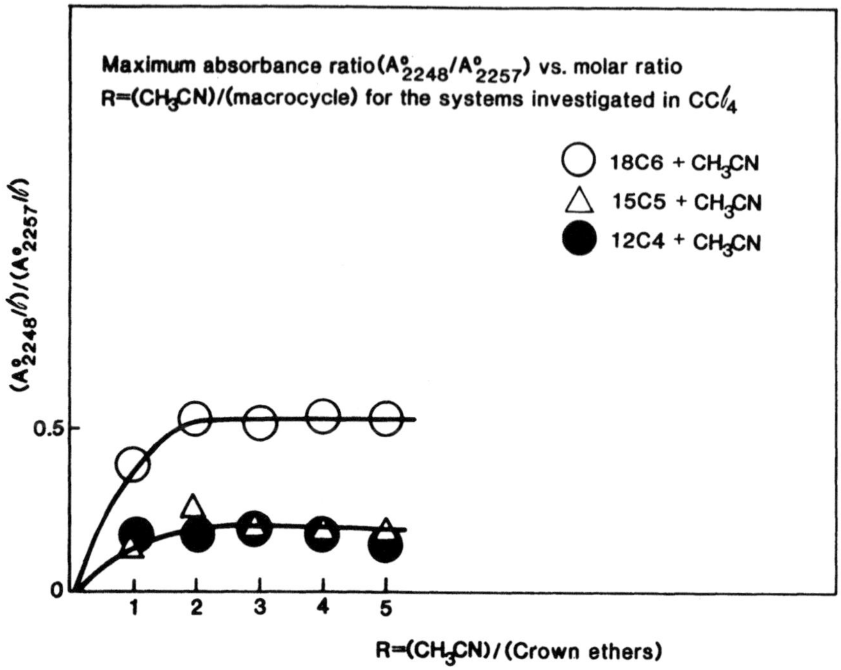

Fig. 10. Maximum absorbance ratio $(A^0_{2248}/l)/(A^0_{2251}/l)$ vs. molar ratio $R = [CH_3CN]/[\text{macrocycle}]$ for the systems investigated in CCl_4. The concentration of macrocycle is fixed at ~ 0.1M.

Table III. Fitted (A^0/l) values for the band at $\bar{\nu}^0 = 2248\ \text{cm}^{-1}$ vs molar ratio R and vs acetonitrile concentration for the macrocycles + acetonitrile mixtures investigated.

$A^0/l = a + bR$								
$A^0/l = \alpha + \beta R + \gamma R^2 + \delta R^3$								
$R = \dfrac{[CH_3CN]}{[18C6]}$	a	b	r^2	α	β	γ	δ	r^2
$18C6 + CH_3CN$	−0.0742	0.7969	0.990	−0.0003	0.3973	0.1394	−0.0108	0.999
$15C5 + CH_3CN$	−0.0185	0.3306	0.989					
$12C4 + CH_3CN$	0.0027	0.2422	0.975					
$A^0/l = a + bc$								
	a	b	r^2					
CH_3CN	−0.0082	1.6748	0.989					

Acknowledgement

The authors are grateful to the National Science Foundation for generous support through grant #CHE 8822333.

Notes and References

1. G. A. Melson (Ed.): *Coordination Chemistry of Macrocyclic Compounds*, Plenum Press, New York (1979).
2. R. M. Izatt and J. J. Christensen (Eds.): *Progress in Macrocyclic Chemistry*, Vol 2. John Wiley, New York, (1981).
3. G. W. Gokel, L. Echegoyen, M. S. Kim, E. M. Eyring, and S. Petrucci: *Biophys. Chem.* **26**, 225 (1987).
4. P. A. Mosier-Boss and A. I. Popov: *J. Am. Chem. Soc.* **107**, 6168 (1985).
5. H. Farber and S. Petrucci: *Dielectric Microwave Spectrometry* (The Chemical Physics of Solvation, Eds. R. R. Dogonadze, E. Kalman, A. A. Kornyshev and J. Ulstrup), Vol. B. Elsevier, Amsterdam (1986).
6. D. Saar and S. Petrucci: *J. Phys. Chem.* **90**, 3326 (1986).
7. C. F. Böttcher: *Theory of Electrical Polarization*, Elsevier, Amsterdam (1973).
8. M. Delsignore, H. Farber, and S. Petrucci: *J. Phys. Chem.* **89**, 4968 (1985).
9. M. Davies: *Dielectric Dispersion and Absorption and Some Related Studies in Condensed Phases* (Dielectric Properties and Molecular Behaviour, Eds. N. Hill, W. E. Vaughan, A. H. Price, and M. Davies), p. 298. Van Nostrand Reinhould Co., London (1969).
10. For a review of this function see: H. Maaser, M. Xu, P. Hemmes, and S. Petrucci: *J. Phys. Chem.* **91**, 3047 (1987); M. Inoue, M. Xu, and S. Petrucci: *J. Phys. Chem.* **91**, 4628 (1987).
11. J. Yarwood and R. Arndt: *The Study of Intermolecular Interactions in the Liquid Phase using Infrared and Raman Band Shapes and Intensities* (Molecular Association: Including Molecular Complexes, Ed. R. Foster) Vol 2, Chapter 4. Academic Press, New York (1979).
12. In the solid state the 18C6 · 2 CH_3CN complex appears to have the methyl protons of acetonitrile hydrogen-bonded to the ether oxygen atoms of 18C6 (see Ref. 6 in [4] above).

Journal of Inclusion Phenomena and Molecular Recognition in Chemistry **12**: 253–262, 1992.

Influence of Solvent Properties on Cation-Macrocycle Complexation: Cesium Cryptates[#]

LEE-LIN SOONG,† GEORGE E. LEROI* and ALEXANDER I. POPOV*
Department of Chemistry, Michigan State University, East Lansing, MI 48824-1322, U.S.A.

(Received: 8 October 1990; in final form: 30 November 1990)

Abstract. The chemical shift of the ^{133}Cs resonance has been measured as a function of the mole ratio of cryptands C222, C222B, C221, and C211 to the Cs^+ ion in *N*-methylformamide and dimethylacetamide solutions. The overall formation constants of the cesium cryptates were calculated from the NMR titration curves. From these and previously reported data it is clear that solvent properties such as dipole moment, donicity, dielectric constant and hydrogen bonding ability affect macrocyclic ligand-cation complexation equilibria through specific solvation of the cation, the ligand and the complex. A model of $M^+ \cdot L$ bonding in solution, involving a rather intimate electrostatic attraction between the cation and the local electron distribution of the macrocycle accompanied by little perturbation of the charge distributions of the separated partners, is presented. In addition to the specific interactions previously noted, the presence of the solvent mitigates the ion-multipole forces which would otherwise govern the long-range attraction.

Key words. Cesium cryptate stabilities, high dielectric solvents, ^{133}Cs NMR, electrostatic nature of complexation.

1. Introduction

Among the many factors which influence the formation of macrocyclic complexes of metal ions in solution, certainly not the least important is the ability of solvent molecules to solvate ions and thus to compete with the ligands for the coordination positions around a cation. Equally important, but often much less appreciated, is the interaction of some solvents with the macrocyclic ligands [1]. Thus both the cation and the ligand must be at least partially desolvated before the adduct can be formed. The two solvation mechanisms are usually different. Solvation of an alkali cation involves primarily ion–dipole forces (with some contribution from donor–acceptor interaction). On the other hand, solvation of macrocyclic ligands proceeds very largely through hydrogen bonding between the —OH, $—NH_2$, or even $—CH_3$ group of the solvent and the ether oxygens of crown ethers [1–3] and with oxygen and nitrogen atoms of cryptands [4].

In one of his early classical papers Pedersen suggested [5] that the formation of crown ether complexes of alkali ions results primarily from ion-dipole interactions between the cation and the local $[—CH_2—O—CH_2—]$ dipoles of the macrocyclic ligands. A later study showed that the ligand-cation vibration observed for a given M^+, L pair in the low frequency region of the infrared spectrum could not be

[#] This paper is dedicated to the memory of the late Dr C. J. Pedersen.
† Deceased.
* Authors for correspondence.

observed in the Raman spectrum [6]. The weakness of the Raman scattering indicates that this vibration is accompanied by little change in the polarizability, and thus supports Pedersen's assumption about the electrostatic nature of the cation–ligand bonds. However, it has also been observed that, in the absence of 'hard-soft' interactions, the stabilities of macrocyclic complexes often vary inversely with the donor abilities of the solvents employed [7] (as expressed by the Gutmann donicity values [8]), i.e. with their ability to solvate alkali cations and therefore inhibit the metal–ligand complexation reaction. Thus both electrostatic and donor–acceptor interactions play a role in the formation of a macrocyclic complex. On this basis, relative stabilities of macrocyclic complexes in solvents which have very nearly the same donating capacity and similar dipole moments will be governed by electrostatic forces and should vary inversely with the dielectric constants of the media.

In this study we have extended our previous NMR investigations [9] related to the influence of solvent properties on macrocyclic complexation to the reactions of the cesium ion with diazapolyoxa macrobicyclic ligands (cryptands; we follow nomenclature proposed by J.-M. Lehn [10] whereby *cryptand* is the free ligand, and *cryptate* is the complex) in two amide solvents, *N*-methylformamide and dimethylacetamide, which have very similar donor properties and dipole moments, but differ in the values of the dielectric constant (Table I) and in the ability to form hydrogen bonds. The ^{133}Cs chemical shift provides a sensitive probe of the cationic environment. From the variation of the chemical shift with relative concentration of a selected macrocyclic complexant, a value of the formation constant of the cesium-macrocycle complex can be calculated. Previous studies on cesium cryptates in dimethylformamide solutions [11–13] are included in the discussion of our results.

Table I. Solvent properties and formation constants of the Cs^+ ion cryptates at 25°C.

Solvent:		NMF	DMF	DMA
Donor number		25[a]	26.6[b]	27.8[b]
Dielectric constant[c] (25°C)		182.4	36.71	37.8
Dipole moment[c] (Debye)		3.86	3.86	3.72
Ligand	Cavity size (Å)[d]	log *K*		
C211	1.6	K ≈ 0	1.23 ± 0.07[e]	0.74 ± 0.06
C221	2.2	2.4 ± 0.2	3.33 ± 0.04[e]	3.07 ± 0.02
C222	2.8	2.35 ± 0.03	2.16 ± 0.03[f]	1.82 ± 0.01
C222B	≈2.8	1.75 ± 0.08	1.70 ± 0.03[f]	1.47 ± 0.01

[a] H.-H. Chuang, L. Soong, G. E. Leroi, and A. I. Popov: *J. Solution Chem.* **18**, 759 (1989).
[b] Ref. [8].
[c] J. A. Riddick and W. B. Burger: *Techniques of Chemistry, Vol II, Organic Solvents*, A. Weissberger, ed., Wiley-Interscience, New York, NY, 1970.
[d] Ref. [10].
[e] Ref. [13].
[f] Ref. [12].

2. Experimental Part

2.1. CHEMICALS

N-methylformamide [NMF] and *N,N*-dimethylacetamide [DMA] (both Aldrich) were refluxed over granulated barium oxide and then fractionally distilled at reduced pressure either from barium oxide or from calcium hydride. The middle fraction was collected in both cases. The water content of the solvents was measured with a gas chromatograph and was found to be always below 70 ppm. The solvents were stored over freshly activated 3 Å molecular sieves, in a dry-box under dry nitrogen atmosphere.

Cesium iodide (Alfa) was used as received except for drying at 120°C for two days. Cesium tetraphenylborate was prepared by the metathetical reaction between cesium chloride and sodium tetraphenylborate in a tetrahydrofuran-water mixture. The solid product was filtered and dried for two days, under vacuum, at 70°C. Cryptands C211, C221, C222 and C222B (all from MCB) were used as received except for drying under vacuum at room temperature for two days. Deuterium oxide (KOR Isotopes) was used as received. All solutions were prepared and stored in a dry-box under a dry nitrogen atmosphere.

2.2. MEASUREMENTS

Cesium-133 chemical shifts were measured on a Bruker WH-180 NMR spectrometer with a field strength of 43.2 kG. At this field the ^{133}Cs nucleus resonates at 23.62 MHz. The solutions were placed in 10 mm o.d. NMR tubes with a coaxial 4 mm o.d. tube containing 0.5M CsBr in D_2O, which served as a chemical shift reference and the lock. Measurements were made on a series of thermostatted (25 ± 1°C) solutions containing a constant amount of cesium salt and a variable amount of the ligand. Chemical shifts reported in this work are given vs. an infinitely dilute aqueous solution of the cesium ion.

Carbon-13 NMR measurements were carried out on a Bruker WM-250 spectrometer operating at a field of 58.7 kG and a frequency of 62.9 MHz. Benzene-d_6 was used as the external reference and lock. Chemical shifts are given with respect to Me_4Si. Both in ^{133}Cs and ^{13}C measurements, chemical shifts were corrected for the differences in the bulk diamagnetic susceptibility between the sample solvent and the reference solvent.

2.3. CALCULATIONS

Assuming that only a 1 : 1 complex is formed and that fast exchange conditions prevail, it can be readily shown [14] that in solutions containing both the free and the complexed cation, the observed chemical shift of the cesium ion is given by the equation:

$$\delta_{\text{obs}} = [(KC_{\text{M}} - KC_{\text{L}} - 1) + (K^2C_{\text{L}}^2 + K^2C_{\text{M}}^2 - 2K^2C_{\text{L}}C_{\text{M}} + 2KC_{\text{M}} + 2KC_{\text{L}} + 1)^{1/2}]\left[\frac{\delta_{\text{f}} - \delta_{\text{c}}}{2KC_{\text{M}}}\right] + \delta_{\text{c}} \quad (1)$$

where K is the concentration formation constant, C_L and C_M are the total concentrations of the ligand and of the cation respectively, δ_f is the chemical shift of the uncomplexed cesium ion, and δ_c is the chemical shift of the complexed cation.

A series of solutions was prepared containing a fixed amount of the cesium ion and a variable amount of the ligand. The titration curve of the ^{133}Cs chemical shift vs. [ligand]/[cation] mole ratio thus obtained was fitted to Equation 1 by using a non-linear least squares program, KINFIT [15]. Since in our solutions the chemical shift of the 'free' cesium ion varies with the concentration of the salt and with the nature of the anion, and thus δ_f is a variable, the values employed in the equation were obtained from measurements of ^{133}Cs chemical shifts as a function of concentration of the cesium salt, as exemplified in a previous publication [16].

3. Results and Discussion

Titration curves obtained in *N*-methylformamide and in dimethylacetamide are shown in Figures 1 and 2. Formation constants of cesium cryptates calculated from these data are listed in Table I, together with the values previously obtained for dimethylformamide [DMF] solutions [11–13]. It is seen that in all three solvents the cryptates are rather unstable (K rarely larger than 1000), and the apparent stability order is C221 > C222 > C222B > C211, irrespective of the solvent. It should be noted that these are *relative* stability constants, which reflect the competition between the ligand and the solvent for the cesium cation. The table also shows that the three solvents have very nearly the same dipole moments and very similar

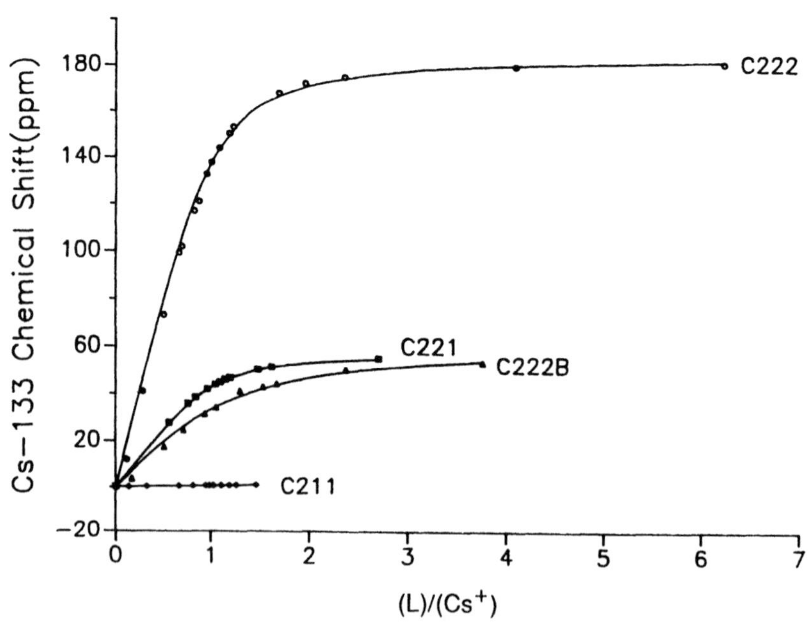

Fig. 1. Cesium-133 chemical shifts vs. cryptand/Cs^+ mole ratio: 0.05M CsI in NMF.

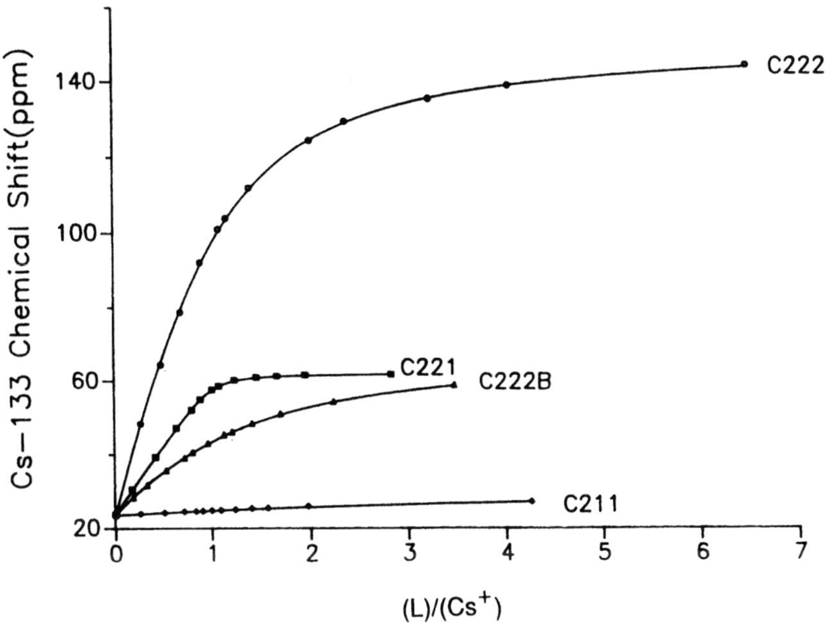

Fig. 2. Cesium-133 chemical shifts vs. cryptand/Cs^+ mole ratio: 0.05M CsI in DMA.

solvating abilities. However, the dielectric constant of NMF is much higher than those of the other two solvents.

As shown in Figures 1 and 2, the addition of the selected cryptands to amide solutions of a cesium salt results in paramagnetic (downfield) shifts of the ^{133}Cs resonance. These shifts represent, of course, the *difference* in the environment of the solvated and complexed cesium cation. The magnitude of the shift is not related to the stability of the complex, but reflects rather the tightness of the ligand-cation juxtaposition, i.e. the overlap of the lone pair orbitals of the oxygen and nitrogen atoms of the ligand and the outer *p*-orbital of the cation. Constraints on the $Cs^+ \cdot C222$ geometry suggest that the overlap will be particularly strong for this complex (*vide infra*). Thus, although cryptand C322 forms a much more stable complex with the cesium cation than does C222 ($\log K = 7.0$ vs. $\log K = 3.5$ in water–methanol mixtures [17]), the chemical shifts of the complexed cation in methanol solution are 22 ppm and 220 ppm, respectively [18]. Figure 2 illustrates that addition of C222 results in a larger ^{133}Cs paramagnetic shift than that observed with C221; however, in the latter case the chemical shift levels off shortly after the 1 : 1 mole ratio, which indicates qualitatively that, as in other solvents [12], $K(Cs^+ \cdot C221) > K(Cs^+ \cdot C222)$.

Formation constants (or free energies of formation) of macrocyclic ligand–cation complexes in solution, such as those reported in this paper *or elsewhere*, may be interpreted as representing the overall results of a series of reactions [19]. Of these, several elementary steps may have a strong effect on the quantity which is generally of primary interest: the *relative* stability of a given complex under a strictly defined set of conditions. Equations for the overall reaction and its component steps (which may occur in some concerted fashion) can be written:

$$M^+S_x + LS_y \rightleftharpoons M^+LS_z + (x + y - z)S$$

$$[i] \qquad M^+S_x \rightleftharpoons M^+ + xS$$

$$[ii] \qquad L_1S_y \rightleftharpoons L_1 + yS$$

$$[iii] \qquad L_1 \rightleftharpoons L_2$$

$$[iv] \qquad M^+ + L_2 \rightleftharpoons M^+L_2$$

$$[v] \qquad M^+L_2 + zS \rightleftharpoons M^+L_2S_z$$

These are: [i] desolvation (partial or complete) of the cation, [ii] desolvation (partial or complete) of the ligand, [iii] change in the conformation of the ligand (coincidental with [ii]?), [iv] the formation of the complex, and [v] solvation of the complex. It should be noted that in three of the five steps solvent molecules directly participate in the equilibria! Solvent properties such as the dielectric constant, dipole moment, donicity, and potentiality for hydrogen bonding will affect the reactions for each M^+, L pair differently.

In most previous studies of macrocyclic complexation reactions the influence of solvent–ligand interactions has been ignored. In the few cases where such interactions have been investigated, however, it has been clearly demonstrated that they may profoundly affect the extent of such reactions [1]. Carbon–13 NMR spectra of C221 in a number of solvents have been measured previously in our laboratory [4]; the C-4 and C-5 signals, which overlap in the spectrum of the *free* ligand, are split when direct solvent-ligand interaction occurs. In this work we have observed such a splitting for C221 in NMF, but not in DMA or DMF. The ^{13}C NMR spectrum of C221 in NMF (in the absence of cesium salt) is reproduced in Figure 3, which also shows the carbon atom numbering scheme for the ligand. Most likely the splitting reflects hydrogen bonding interactions between C221 and NMF, which are not possible in DMF or DMA. There is little doubt that similar interactions occur between NMF and the other cryptands.

Studies of C222 complexation with the Cs^+ ion in four different solvents, including DMF, have shown that the reaction occurs in two steps [11, 18]. Initially, the cation binds to the heteroatoms of two strands, forming an *exclusive*, partially-solvated, crown-like complex. The second step involves the penetration of the completely desolvated cation into the ligand cavity to form the *inclusive* [20] complex:

$$Cs^+S_x + C222 \rightleftharpoons Cs^+ \cdot C222S_y \ (\textit{exclusive}) + (x - y)S$$

$$Cs^+ \cdot C222S_y \ (\textit{exclusive}) \rightleftharpoons Cs^+ \cdot C222 \ (\textit{inclusive}) + yS$$

Thermodynamic parameters measured for the two reactions are given in the cited references. The values of the log K reported in Table I for the $Cs^+ \cdot C222$ cryptate are thus composites, which include both equilibria; therefore, they cannot be compared directly to formation constants calculated for cases in which only exclusive complexes are formed.

The situation for C222B is less clear. While the cavity dimensions of the free ligand should be very nearly the same as those of C222, the substitution of a benzene ring on one chain will tend to make the ligand less flexible, and may

C221

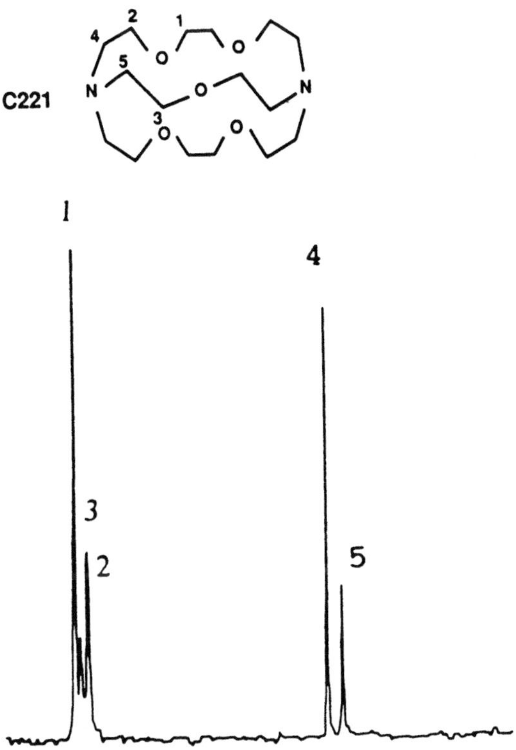

Fig. 3. Carbon-13 NMR spectrum of C221 in NMF at 25°C. The separation between the C4 and C5 signals is 0.81 ppm.

preclude or mitigate the formation of the inclusive complex with Cs^+. For the other two ligands studied in this work, C211 and C221, the cavities are much too small to allow the formation of an inclusive cesium cryptate.

Since the dipole moments and the donicities of the three solvents are essentially the same, their solvating abilities, *vis-à-vis* the Cs^+ ion, should also be approximately the same. Presuming that their interactions with the ligand and the complex are relatively similar, one would expect the overall complexation constant to be smaller in the high dielectric constant solvent, NMF. From Table I, this solvent dependence of log K is observed in the two cases where only a single-step complexation reaction, to form the exclusive complex, can occur. The same inverse dependence of the complexation constant on the solvent dielectric constant has been observed for $Cs^+ \cdot$ crown ether complexes in amide solvents [9].

4. Conclusions

The general subject of cation/macrocyclic ligand/solvent interactions has not been thoroughly discussed in the literature, especially with regard to the role of the solvent. It seems particularly appropriate in this volume to present an overview of the topic from a broader perspective than the new results reported in this paper. In particular, the nature of the interaction between an alkali metal cation and the macrocyclic ligand (step [iv], above) deserves deeper scrutiny than has been given in the literature heretofore. As mentioned above, in an early notable paper Pedersen suggested that the M^+—L bond is electrostatic [5], and the weakness of the scattering from this vibration in the experimental Raman spectra [6] is consistent with the small polarizability change to be expected under such circumstance. In our view, the important features of $M^+ \cdots L$ bonding *in solution* involve a rather intimate attraction between the positively-charged spherical ion and the local electron distribution of the macrocyclic complexant, the latter being dominated by the presence of non-bonding electron pairs. (The classical, long-range, ion-multipole forces which govern electrostatic interactions in the gas phase are mitigated by the charge distributions of intervening solvent molecules; the electrostatics specific to $M^+ \cdots L$ binding govern the interaction only when the partners are rather proximate.) In this 'close-binding' model, very little perturbation of the charge distributions of the separated partners accompanies the rather strong electrostatic interactions of the cation with the lone pairs of electrons on the macrocycle, which in many cases can adjust its conformation to partially envelop the cation and maximize the number of interacting electron pairs. The oxygen atoms in a strand of a macrocyclic ligand have two lone pairs of electrons, and thus high 'local dipole and polarizability'; a nitrogen atom with its single lone pair will be less electronegative, and will be a poorer partner for 'hard' acids, such as the alkali cations. Thus one would expect the aza-complexes to have smaller complexation constants than their oxygen counterparts with a particular alkali cation in a given solvent, as is indeed observed experimentally [21, 22].

Such a picture of electrostatic bonding has been recently propounded to account for the bonding between monopositive transition metal cations and neutral molecules in the gas phase [23–26]. For example, theoretical calculations reveal that the bonding between Sc^+, Ti^+, V^+, Cr^+, Ni^+ and Cu^+ ions with CO is essentially electrostatic, with very little charge transfer [23, 25, 26], unlike the covalent Dewar–Chatt scheme which has been accepted for the neutral species [27]. The attractive portion of the *ab initio* potential surfaces for the M^+—CO ground states is well represented by a classical electrostatic interaction potential, as is that for Sc^+—OC [23]. The metal–ligand binding in $LiCO^+$ and $ScNH_3^+$ ions [28, 29] is also clearly electrostatic, and rather strong.

In an alkali cation–macrocyclic ligand complex one expects the electrostatic attraction to be maximized, within geometrical limitations, until it balances the Pauli repulsion between the electrons on the metal and the ligand. The increased electron density thus predicted in the neighborhood of the metal ion is consistent with the large paramagnetic shifts observed in the alkali cation NMR spectra, relative to the corresponding gaseous atom, when $M^+ \cdot L$ complexes are formed [30]. For a given ligand, then, one might expect the interaction to be stronger for Li^+ than for Cs^+ because of the closer proximity of the cation to the negative charge distribution on the ligand; however, geometrical constraints related to the cation/cavity fit may well be dominant.

Although solvent molecules do not appear explicitly in reaction [iv], to the extent that the electrostatic interaction between M^+ and L occurs over long range, the bulk properties of the solvent will affect the equilibrium constant for this step via the conventional inverse dependence on the dielectric constant. If *specific* interactions of the solvent with the cation, the ligand and the complex (steps [i], [ii] and [v]) vary little from one solvent to another, for instance when solvents of similar donicity and functionality are intercompared, then the complexation constant measured for the overall reaction may reflect the influence of the dielectric properties of the solvent on the complexation equilibrium, [iv]. Indeed, some such correlations of cation/ligand stability constants with solvent dielectric constant have been suggested [9, 17, 19, 31] and the log K values we measure for $Cs^+ \cdot C211$ and $Cs^+ \cdot C221$ are inversely related to this property. However, we cannot reiterate too strenuously the danger of drawing such correlations for other systems without extreme caution. The dependence of the overall complexation constant on the properties of the solvent will usually be governed by differences in the interactions in steps [i], [ii] and/or [v], determined by differences in the solvent donicity, dipole moment, and/or its propensity for hydrogen bonding.

Acknowledgements

We appreciate enlightening discussions with Professors A. Mavridis and J. F. Harrison regarding the nature of electrostatic bonding. This work was supported in part by the U.S. National Science Foundation, under Grant CHE-8610421.

Notes and References

1. P. A. Mosier-Boss and A. I. Popov: *J. Am. Chem. Soc.* **107**, 6168 (1985), and references therein.
2. J. A. A. DeBoer, D. M. Reinhoudt, S. Harkema, G. J. van Hummel, and F. de Jong: *J. Am. Chem. Soc.* **104**, 4073 (1982).
3. R. D. Rogers and L. M. Green: *J. Incl. Phenomena* **4**, 77 (1986); R. L. Garell, J. C. Smyth, F. R. Fronczek, and R. D. Gandour: *Ibid.* **6**, 73 (1988).
4. E. Schmidt, J.-M. Tremillon, J.-P. Kintzinger, and A. I. Popov: *J. Am. Chem. Soc.* **105**, 7563 (1983).
5. C. J. Pedersen: *J. Am. Chem. Soc.* **89**, 7017 (1967).
6. Y. M. Cahen and A. I. Popov: *J. Solution Chem.* **4**, 599 (1975).
7. A. J. Smetana and A. I. Popov: *J. Solution Chem.* **9**, 183 (1980); Y.-C. Lee, J. Allison, and A. I. Popov: *Polyhedron* **4**, 441 (1981), and references therein.
8. V. Gutmann and E. Wychera: *Inorg. Nucl. Chem. Lett.* **2**, 257 (1966).
9. L. L. Soong, G. E. Leroi, and A. I. Popov: *Inorg. Chem.* **29**, 1366 (1990).
10. J.-M. Lehn: *Struct. Bonding* **16**, 1 (1973).
11. E. Mei, A. I. Popov, and J. L. Dye: *J. Am. Chem. Soc.* **99**, 6532 (1977).
12. E. Mei, L. Liu, J. L. Dye, and A. I. Popov: *J. Solution Chem.* **6**, 771 (1977).
13. L. Liu: M.S. Thesis, Michigan State University (1976).
14. E. T. Roach, P. R. Handy, and A. I. Popov: *Inorg. Nucl. Chem. Lett.* **9**, 359 (1973).
15. J. L. Dye and V. A. Nicely: *J. Chem. Educ.* **48**, 443 (1971).
16. L. L. Soong, G. E. Leroi, and A. I. Popov: *J. Solution Chem.* **18**, 759 (1989).
17. J.-M. Lehn and J.-P. Sauvage: *J. Am. Chem. Soc.* **97**, 6700 (1975).
18. E. Kaufmann, J. L. Dye, J.-M. Lehn, and A. I. Popov: *J. Am. Chem. Soc.* **102**, 2274 (1980).
19. J. D. Lamb, R. M. Izatt, and J. J. Christensen: *Progress in Macrocyclic Chemistry* (Vol 2, Eds. R. M. Izatt and J. J. Christensen), Ch 2, Wiley & Sons (1981).
20. The cavity of C222 (estimated from molecular models) is much smaller than the diameter of the cesium cation: 2.8 Å[10] vs. 3.7 Å [D. F. C. Morris: *Struct. Bonding* **4**, 63 (1968)]. Yet, analysis of the crystal structure of $Cs^+ \cdot C222 \cdot SCN^-$ shows that in the solid the Cs^+ ion is imbedded in the center of the three-dimensional macrocyclic cavity [D. Moras, B. Metz, and R. Weiss: *Acta Crystallogr.* **B29** 388 (1973)]. The fit is made possible by the flexibility of the ligand, which can, within reason, expand its cavity size by opening the $N-CH_2-CH_2-O$ and $O-CH_2-CH_2-O$ dihedral angles.

21. R. M. Izatt, J. S. Bradshaw, S. A. Nielsen, J. D. Lamb, J. J. Christensen, and D. Sen: *Chem. Rev.* **85**, 271 (1985).
22. H. K. Frensdorff: *J. Am. Chem. Soc.* **93**, 600 (1971).
23. A. Mavridis, J. F. Harrison, and J. Allison: *J. Am. Chem. Soc.* **111**, 2482 (1989).
24. J. Allison, A. Mavridis, and J. F. Harrison: *Polyhedron* **7**, 1559 (1988).
25. M. Merchan, I. Nebot-Gil, R. Gonzalez-Luque, and E. Orti: *J. Chem. Phys.* **87**, 1690 (1987).
26. C. W. Bauschlicher, Jr.: *J. Chem. Phys.* **84**, 260 (1986).
27. F. A. Cotton and G. Wilkinson: *Advanced Inorganic Chemistry*. Wiley & Sons (1980).
28. A. Mavridis and J. F. Harrison: unpublished results.
29. A. Mavridis, K. Kunze, J. F. Harrison, and J. Allison: *Gas-Phase Chemistry of First-Row Transition Metal Ions with Nitrogen-Containing Compounds: Theoretical and Experimental Investigations* (Bonding Energetics in Organometallic Compounds, Ed. T. J. Marks), *ACS Symp. Series, Vol. 428*, p. 263 (1990).
30. J. L. Dye, C. W. Andrews, and J. M. Ceraso: *J. Phys. Chem.* **79**, 3076 (1978).
31. E. Lada and M. K. Kalinowski: *Polyhedron* **8**, 2115 (1989).

Journal of Inclusion Phenomena and Molecular Recognition in Chemistry **12**: 263–274, 1992.

Li^+(TMPAND)Na^-: The First Alkalide Prepared from an Azacage Complexant#

JUDITH L. EGLIN, EVELYN P. JACKSON, KEVIN J. MOEGGENBORG, and JAMES L. DYE,*
Department of Chemistry, Michigan State University, East Lansing, Michigan 48824, U.S.A.

ANDREA BENCINI and MAURO MICHELONI
Department of Chemistry, University of Florence, via Maragliano 77, 50144 Florence, Italy

(Received: 5 October 1990; in final form: 15 December 1990)

Abstract. The new lithium-selective complexant 5,12,17-trimethyl-1,5,9,12,17-pentaazabicyclo[7.5.5]nonadecane (TMPAND) was used to synthesize Li^+(TMPAND)Na^-, the first alkalide prepared from an azacage complexant. This sodide was characterized by a variety of methods. Differential scanning calorimetry experiments showed a reversible, endothermic, solid-solid phase transition at an onset temperature of -75 ± 3°C and with $\Delta H = 3.3 \pm 1$ kJ/mol. ^{23}Na NMR spectra showed a peak at -61 ppm, characteristic of a sodium anion, and a second minor peak at -10 ppm, probably due to the interaction of Na^- with trapped electrons. The quadrupole coupling constant of the complexed lithium cation was found to be 0.19 MHz at -100°C, and ^{7}Li NMR spectra showed a discontinuity in the line width of the ^{7}Li NMR peak and in the quadrupole coupling constant at the phase transition.

Key words. Alkalide, electride, azacage, cryptand, sodium NMR, lithium NMR.

1. Introduction

Alkalides and electrides [1] are two unique classes of compounds with unusual properties that result from the presence of weakly bound electrons. Solubilization of an alkali metal by coordination with a crown ether [2], a discovery that grew out of the work of Pedersen *et al.* [3], led to the use of [2.2.2]cryptand [4] (C222) for the preparation of the first sodide, Na^+(C222)Na^- [5]. This alkalide is a stable crystalline salt with a sodium anion. More recently, crystalline electrides, analogous salts with electrons serving as anions, have been synthesized [6].

Since these initial discoveries, crystal structures have been obtained for 30 alkalides and four electrides. These compounds have been characterized and studied by a variety of techniques, including pressed powder d.c. conductivity, EPR spectroscopy, magnetic susceptibility, X-ray crystallography, powder X-ray diffraction, optical absorption spectroscopy of thin films, and magic angle spinning (MAS) and static NMR spectroscopy [7]. Such studies have yielded a great deal of insight into the nature of the trapped electron and the alkali metal anion [1d].

Alkalides and electrides are extremely sensitive to air, moisture and elevated temperature. They must be handled *in vacuo* or under an inert atmosphere, and at

This paper is dedicated to the memory of the late Dr C. J. Pedersen.
* Author for correspondence.

temperatures below about −40°C. These materials undergo irreversible decomposition due to their thermal instability, although the exact nature of the decomposition process (or products) is as yet unknown. The propensity to thermally decompose has limited the routine use of presynthesized alkalides and electrides as reagents in synthesis, in spite of their strong reducing ability. (The solvated electron is thermodynamically the most powerful reducing agent in a given solvent.) However, they can be used in crystalline form [8] or prepared *in situ* for use in reduction processes [9].

In pursuit of more thermally stable alkalides and electrides, the use of fully methylated aza analogs of crown ethers as complexing agents was investigated. These complexants [10] included hexamethylhexacyclen (HMHCY), pentamethylpentacyclen (PMPCY), and trimethyltricyclen [11] (TMTCY). They were expected to be more robust toward the highly reducing environment imposed by electron counterions and alkali metal anions [7f, 9a, 9b].

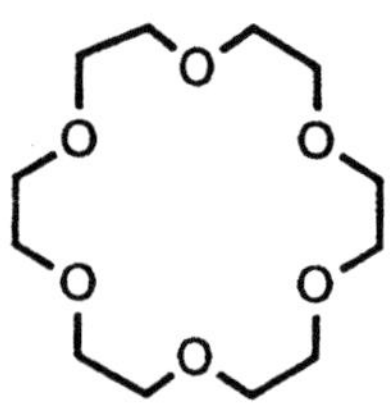

18crown6

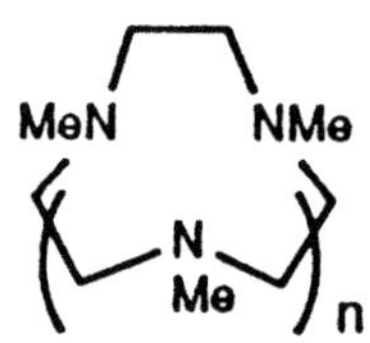

n=4 **HMHCY**

n=3 **PMPCY**

n=1 **TMTCY**

Several sodides were successfully prepared using HMHCY, and their relative stabilities were compared with the stabilities of the corresponding crown ethers. It was found by differential scanning calorimetry (DSC) that $Cs^+(HMHCY)Na^-$ and $K^+(HMHCY)Na^-$ melted without thermal decomposition at 8°C and 42°C, respectively; but on further temperature increase they underwent decomplexation at 37°C and 74°C, respectively [7f]. However, irreversible decomposition of the organic ligand did not occur below ~150°C. Under similar conditions in the DSC, alkalides that contain crown ethers or cryptands undergo irreversible thermal decomposition below ~90°C [12]. It is tempting to conclude from these data that alkalides made with HMHCY are stable at room temperature, when in fact they have a limited shelf-life (on the order of days) because of the thermodynamically favored decomplexation [7f].

These results indicated that the thermal instability of alkalides and electrides was complicated by competition between decomposition and *decomplexation*. With crown ether and cryptand complexants, irreversible decomposition generally occurs before the compound has an opportunity to decomplex. But when the aza crown ether HMHCY was used, decomplexation occurred before decomposition and at a lower temperature than was found for the oxa compounds. This decomplexation is not surprising since, in general, aza complexing agents have lower binding constants than their oxa counterparts [13]. Thus a *useful* increase in stability was not achieved with HMHCY.

In order to stabilize alkalides and electrides towards decomplexation without sacrificing chemical stability, aza complexants of increased binding strength were sought. The azacage complexant 5,12,17-trimethyl-1,5,9,12,17-pentaazabicyclo[7.5.5]nonadecane (TMPAND) selectively encapsulates a lithium cation in aqueous solution to form a stable 1 : 1 complex [14]. This fully methylated aza analog of [1.1.1]cryptand [15] (with two ethyl linkages replaced by *n*-propyl groups) was subsequently used to synthesize a sodide, $Li^+(TMPAND)Na^-$. The properties of this compound are described in this paper.

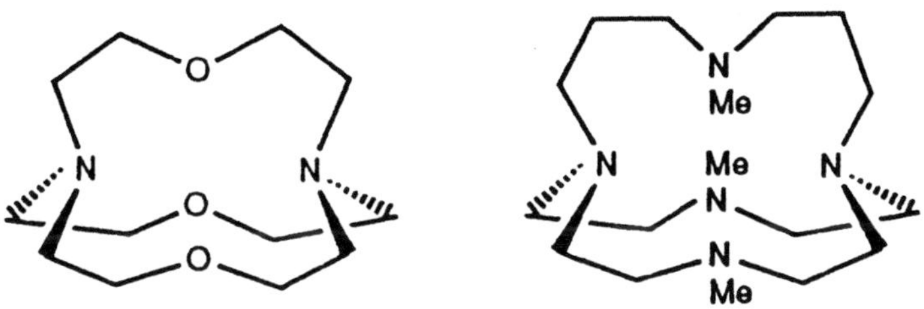

[1.1.1]cryptand TMPAND

2. Experimental

The synthesis of $Li^+(TMPAND)Na^-$ followed procedures that have been described in detail elsewhere [12]. In a helium-filled dry box, excess lithium and sodium were added to ~300 mg TMPAND in a modified, Kontes Airless-Ware® double tube ('H-cell'). After evacuation to $\sim 10^{-5}$ torr, purified methylamine was distilled onto the metals and complexant at −40°C to −50°C. The alkali metals dissolved to form a deep blue solution. The solution was packed in dry ice for at least 12 h, after which time it was deep purple in color, and bronze films formed on the sides of the glass apparatus when the solution was swirled. The methylamine was evaporated to dryness, followed by addition of dimethyl ether (Me_2O) to form a deep blue solution. This was poured through a frit to another glass bulb in order to separate the products from excess metal. After addition of trimethylamine (Me_3N) to induce crystallization, bronze-colored crystals formed. The Me_2O/Me_3N solvent mixture was evaporated, and pure Me_3N was added. The liquid was decanted through the frit and redistilled onto the compound to remove excess complexant. After several washes of this type, the Me_3N was removed by distillation and the compound was dried *in vacuo*. The polycrystalline powder was harvested in a nitrogen glove bag while the sample was kept cold on an aluminum support immersed in liquid nitrogen. Samples were stored in Cryotubes® under liquid nitrogen.

Optical absorption spectra of thin, solvent-free films of $Li^+(TMPAND)Na^-$ were measured from 300 to 1500 nm with a Model 260 Guided Wave Spectrophotometer at temperatures from −90°C to −10°C as recorded with a copper–constantan thermocouple. Thin films were made by first preparing a solution of the compound in Me_2O, then pouring the solution over the quartz cell window and rapidly evaporating the solvent.

Differential scanning calorimetry (DSC) was performed with a du Pont 910 Differential Scanning Calorimeter. Cold samples that had been at liquid nitrogen temperatures were quickly sealed hermetically in aluminum pans under a nitrogen atmosphere, and the samples were transported at liquid nitrogen temperature. They were loaded into the pre-cooled DSC instrument under an inert atmosphere.

Variable temperature ^{7}Li and ^{23}Na static NMR measurements were carried out for Li^+(TMPAND)Na^- at 155.5 and 105.8 MHz respectively on a Varian VXR-400 spectrometer equipped with a 45 to 165 MHz broadband probe by using single pulses of width 2 μs and a delay of 1 s. The NMR sample tube was loaded into the precooled probe and cold nitrogen gas was used to control the temperature to within 0.3°C after 3 to 10 minutes equilibration time. Chemical shifts are relative to aqueous Li^+ and Na^+ at infinite dilution.

The d.c. conductivity of packed powders was measured as a function of temperature with a computer-controlled Keithley 617 Programmable Electrometer. The temperature was measured with a computer-controlled Keithley 580 micro-ohmmeter and a four-probe carbon glass thermometer. Rectangular voltage pulses of alternate polarity 10 s in length and 100 mV in amplitude were applied to the sample to minimize polarization effects. To check for deviations from Ohm's law due to electrode effects, the electrodes were coated with sodium metal [16].

3. Results

3.1. OPTICAL SPECTRA

The optical absorption spectrum of Li^+(TMPAND)Na^- shown in Figure 1 has a major peak at ~700 nm. The transition is slightly temperature dependent, with a gradual and reversible red shift ($d\nu/dT \approx -5\ cm^{-1}\ deg^{-1}$). The peak also became broader at higher temperatures. A red shift of $-1.6\ cm^{-1}\ deg^{-1}$ and line broadening of about $1.1\ cm^{-1}\ deg^{-1}$ with increasing temperature had previously been observed with Na^+(C222)Na^- films [17].

3.2. THERMAL BEHAVIOR

Differential scanning calorimetry (DSC) studies of Li^+(TMPAND)Na^- show a reversible, endothermic, solid-solid phase transition with an onset temperature of -75 ± 3°C and a ΔH value of 3.3 ± 1 kJ/mol. Figure 2 shows this transition in addition to the exothermic decomposition peak ($\Delta H = -50 \pm 20$ kJ/mol) at ~95°C. By measuring the decomposition temperature at six ramp rates, an activation energy of 101 ± 5 kJ/mol was determined by the Kissinger method [18]. Extrapolation of these results to lower temperature may not be valid because other factors such as autocatalytic processes may control the rate of decomposition. Direct extrapolation of the high temperature data yields half-lives of about 70 h at 25°C and 2800 h at 0°C.

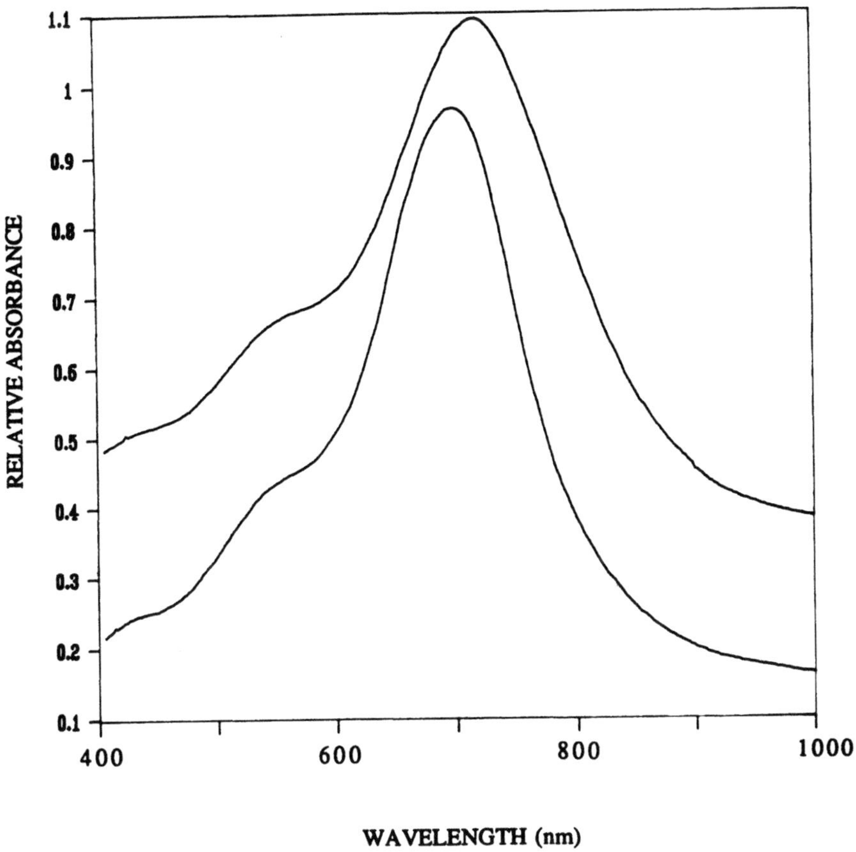

Fig. 1. Optical absorption spectra of thin films of $Li^+(TMPAND)Na^-$ obtained at $-10°C$ (*top line; offset by* $+0.2$ on the Relative Absorbance scale) and $-90°C$.

3.3. ^{23}Na AND ^{7}Li NMR SPECTRA

The major static ^{23}Na NMR peak of $Li^+(TMPAND)Na^-$ occurs at a chemical shift of -61 ppm with a width at half-height ($\Delta\nu_{1/2}$) of ~ 1500 Hz. This chemical shift is characteristic of sodides and indicates that the sodium anion is in a highly symmetric environment with minimal overlap of its filled *p* orbitals with the surroundings [19]. When the temperature of the sample was below the temperature of the phase transition, the ^{23}Na NMR peak at -61 ppm saturated readily and a second, rather broad peak of low intensity was observed at ~ -10 ppm ($\Delta\nu_{1/2} \approx$ 9500 Hz). This is a substantial paramagnetic shift from the typical sodide peak at -61 ppm. Figure 3 shows the ^{23}Na static NMR spectra above and below the phase transition. When the temperature was raised above that of the phase transition, the ^{23}Na peak at -10 ppm was obscured by the intense peak at -61 ppm. Evidently, the T_1 relaxation time of the major Na^- species decreases dramatically as the temperature is increased through the phase transition temperature.

The ^{7}Li static NMR spectrum of $Li^+(TMPAND)Na^-$ has a central ($-1/2 \rightarrow$ 1/2) transition at ~ -9 ppm independent of temperature. The line width, $\Delta\nu_{1/2}$,

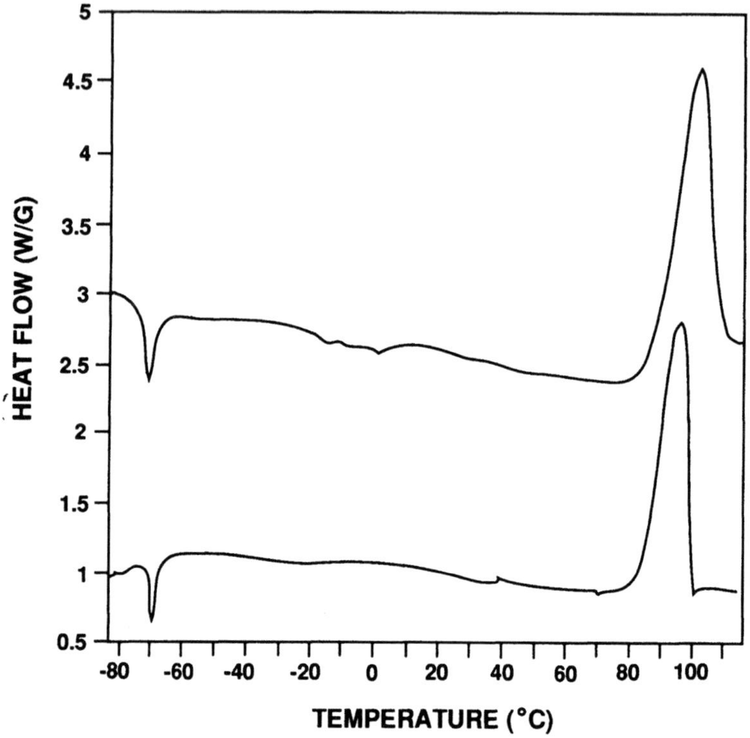

Fig. 2. DSC traces of Li^{+}(TMPAND)Na^{-} obtained at ramp rates of 12°C/min (*top line;* offset by 2 W/g) and 7°C/min. (Note that the area under a peak is inversely proportional to the ramp rate.)

decreases with increasing temperature from 7.7 kHz at −100°C to 6.5 kHz at a transition temperature of −75°C, and shows a discontinuity to 3.2 kHz at the transition temperature. Above the transition temperature, the line width continues to decrease, leveling off at 2.3 kHz above −60°C.

Although the crystal structure of Li^{+}(TMPAND)Na^{-} is not known,[1] that of Li^{+}(TMPAND)BPh_4^{-}, a model salt with the same cation, has been determined [14b]. A linewidth contribution of 8.2 kHz from the dipolar interaction was calculated with the Van Vleck equation [20] by using the lithium–nitrogen and lithium–proton distances obtained from the crystal structure of the model salt. This calculated linewidth is close to the experimental value of 7.7 kHz for Li^{+}(TMPAND)Na^{-} obtained at −100°C where molecular motion is minimized. The quadrupolar contribution to the linewidth is less than 80 Hz [21] and can be neglected.

In addition to the central transition, the ^{7}Li static NMR spectrum (Figure 4) shows the satellite transitions, whose separation gives the quadrupole coupling constant (QCC), assuming axial or nearly axial symmetry. The QCC for the ^{7}Li nucleus in different compounds reflects different values of the electric field gradient. The value of the QCC, 0.19 MHz, for Li^{+}(TMPAND)Na^{-} at −100°C is close to the value of 0.192 MHz obtained for the gas phase LiCl molecule [22]. The ^{7}Li QCC of the sodide depends on temperature as shown in Figure 5. The plot has a

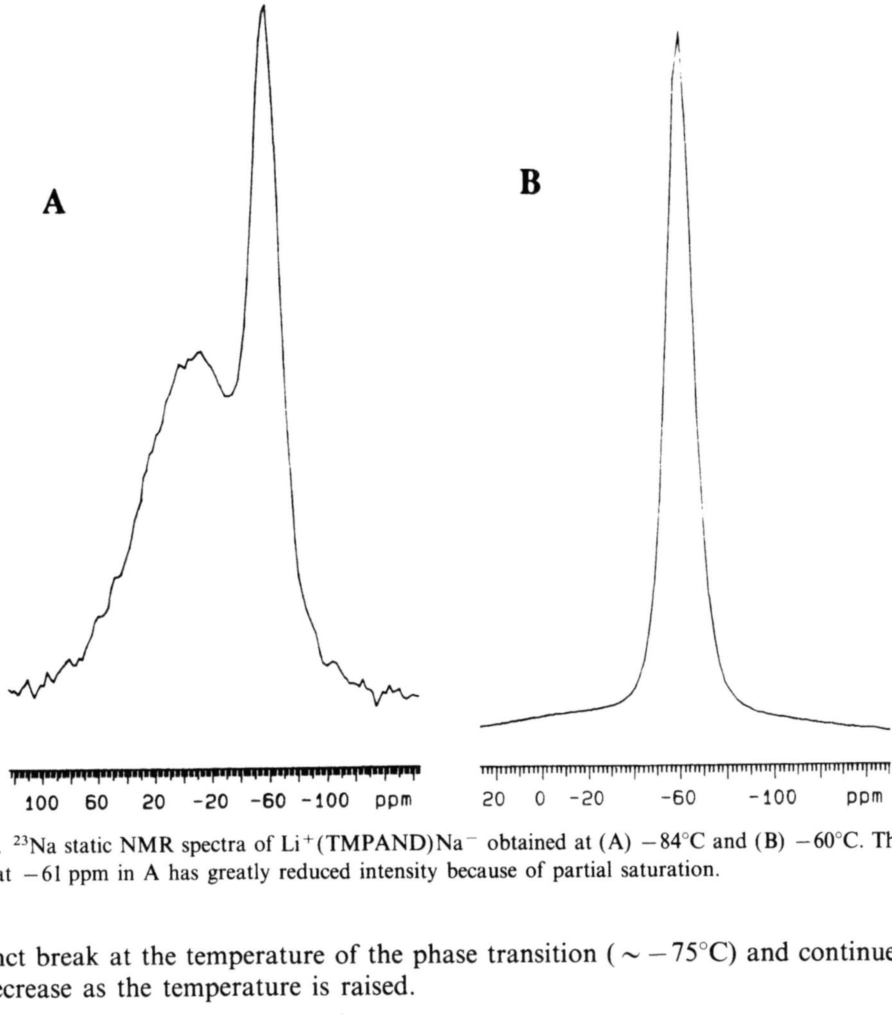

Fig. 3. ^{23}Na static NMR spectra of Li^+(TMPAND)Na^- obtained at (A) $-84°C$ and (B) $-60°C$. The peak at -61 ppm in A has greatly reduced intensity because of partial saturation.

distinct break at the temperature of the phase transition ($\sim -75°C$) and continues to decrease as the temperature is raised.

3.4. ELECTRICAL CONDUCTIVITY

The d.c. powder conductivity of Li^+(TMPAND)Na^- was studied as a function of temperature. Below the solid–solid phase transition temperature the sample resistance was beyond the measurement range of the instrument, thus providing no information about the conductivity of this phase, except that it is low ($<10^{-12}$ ohm^{-1} cm^{-1}). Above the transition temperature the sample behaved as a semiconductor with an apparent band gap of ~ 1.2 eV. A test for electrode effects was made by using a single sodium electrode and a steel counter-electrode [16]. Ohm's law was obeyed and there was no electrochemical cell behavior, suggesting that ionic conductivity is not important and that electrode effects are minimal.

4. Discussion

This work shows that the compound formed from lithium, sodium, and TMPAND is a sodide. The major peak at ~ 700 nm in the optical absorption spectrum is

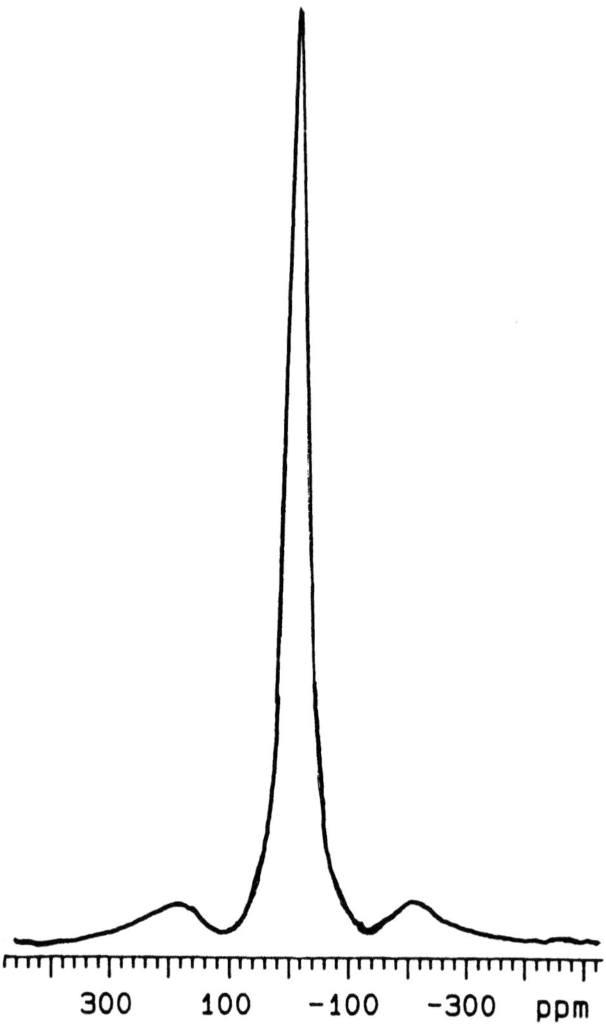

Fig. 4. ^{7}Li static NMR spectra of Li^+(TMPAND)Na^- obtained at -70°C.

typical of sodides, and it is attributed to a $3s^2 \rightarrow 3s3p$ transition of Na^- [1d, 1f]. No change was observed at or around the phase transition at -75°C, indicating that either Na^- is still present in the low temperature phase, or that this phase does not form in the thin films used.

Both the ^{7}Li and ^{23}Na NMR spectra are dramatically affected by the phase transition. The chemical shift range of ^{7}Li is small in diamagnetic compounds, so the constancy of the lithium chemical shift is not surprising, but does suggest the absence of *large* concentrations of unpaired electrons both above and below the phase transition.

Positive chemical shifts (paramagnetic direction) of ^{23}Na can have several origins: (1) overlap of electron pairs from surrounding molecules with the unoccupied $3p$ orbitals of sodium and mixing of the $2p$ and $3p$ states (Ramsey shift [23]); (2) interaction with paramagnetic species such as trapped electrons; (3) rapid exchange

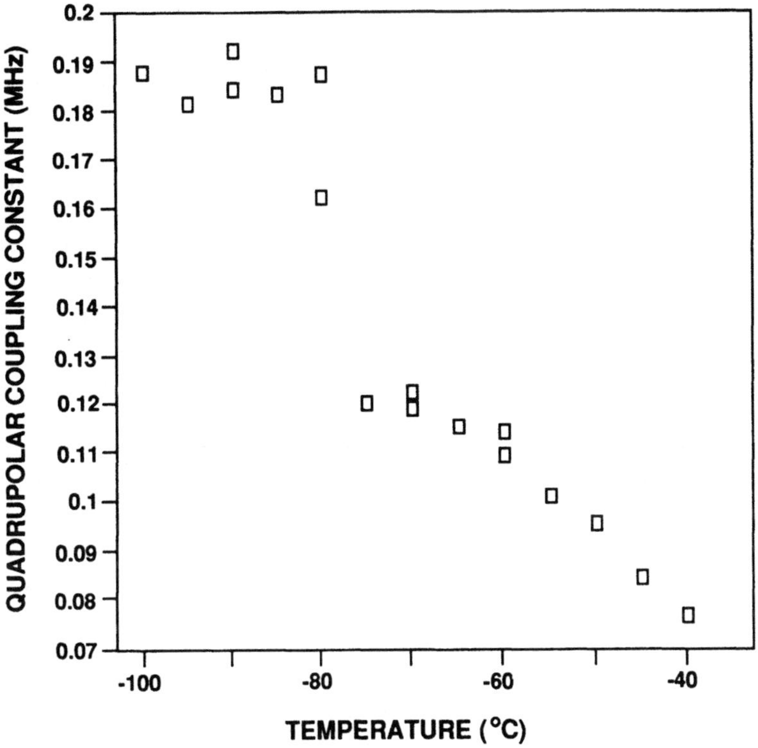

Fig. 5. Temperature dependence of the quadrupole coupling constant (QCC) calculated from the ^{7}Li static NMR spectrum of Li^{+}(TMPAND)Na^{-}.

with species such as Na atoms that exhibit large chemical shifts. Because Na^{-} is a very large anion (radius 2.6 Å [7c]) with two electrons in the 3*s* orbital that effectively screen the 2*p* electrons, the chemical shift of −61 ppm is essentially the same as that of Na^{-} in the gas phase and tends to be insensitive to the environment [19]. The chemical shift of the second ^{23}Na NMR peak is substantially more paramagnetic than the confirmed chemical shift of the sodium anion in any environment encountered to date. If the signal is from Na^{-} in a diamagnetic environment, the large chemical shift implies strong interaction of the *p* orbitals of Na^{-} with Li^{+} or with atoms of the complexant. EPR studies [24] show no change in the concentration of trapped electrons at the transition temperature, the optical spectrum is not altered, and the ^{7}Li chemical shift does not change. All of these observations make it unlikely that there is a significant difference in the trapped electron concentration above and below the phase transition. The most likely explanation for the shifted Na^{-} peak is the interaction between a small fraction of the sodium anions and nearby trapped electrons.

It appears that, below the phase transition temperature, Li^{+}(TMPAND)Na^{-} is locked in a conformation that increases the T_1 relaxation time of Na^{-} and gives the static value of the ^{7}Li linewidth and quadrupole coupling constant. Enough trapped electrons are present to yield a substantial paramagnetic shift of some of the sodium anions. According to this model, at the temperature of the phase transition a change occurs that results in substantially greater motion of the complexant. This could yield a sharp decrease in the value of the ^{7}Li QCC and its linewidth, coincident with a decrease in the longitudinal relaxation time of the major Na^- species.

The linewidth of the ^{7}Li NMR peak of the low temperature phase agrees with that calculated from a rigid lattice model and is primarily due to proton–lithium dipolar interactions. The abrupt decrease in width at the transition, and the continuing decrease as the temperature is raised, results from motion of the hydrogens of the complexant. Such motion, which reflects increased conformational freedom of the $—CH_2—$ groups, is commonly encountered in alkalides and electrides as indicated by both NMR line widths [1d] and powder X-ray diffraction studies [25].

In addition to preparing the alkalide, attempts were made to synthesize the corresponding lithium electride, Li^+(TMPAND)e^-. Paramagnetic solids were obtained, but the samples were not pure and included excess solvent. A potential difficulty with the use of azacage complexants is the slow rate of encapsulation of Li^+ into the cage. A similar problem was noted in the synthesis of the electride Li^+(C211)e^- when methylamine was used [26]. It was necessary to let the solution stand for several hours at −40°C to allow formation of the inclusive complex. Recent work [26b] used several dissolutions in Me_2O and subsequent evaporations to remove traces of methylamine. The current difficulty in preparing pure Li^+(TMPAND)e^- may be the formation of mixtures that contain both inclusive and exclusive complexes [27].

Unlike compounds synthesized with HMHCY [7f], decomplexation was not observed for Li^+(TMPAND)Na^-, and irreversible decomposition did not occur below 65°C in the DSC experiments. Both of these results point to good thermal stability. The high affinity of TMPAND for Li^+ results in an increased binding constant, which overcomes the propensity to decomplex, while retaining the resistance to decomposition. This chemical stability is not surprising, since fully methylated aza crown ethers were shown to be more stable to decomposition than their oxa analogs [7f, 9a, 9b], and we expect similar trends in the azacage/cryptand series. The mechanism of decomposition is not known, but a likely initial step is deprotonation of the methylene groups, a step that would be favored by the more acidic hydrogens of the oxa complexants. Other fully methylated azacage complexants become attractive candidates for the synthesis of more thermally stable alkalides and electrides.

Since thermal decomposition of electrides at temperatures higher than about −40°C is a continuing barrier to the study of their properties, the prospect of electrides that are stable to both decomposition and decomplexation at room temperature is an attractive one that warrants further study of the azacage systems.

Acknowledgements

This research was supported in part by U.S. National Science Foundation Solid-State Chemistry Grant DMR 87-14751 and by the Michigan State University Center for Fundamental Materials Research. We are indebted to Michael Wagner for helpful discussions about the NMR results.

Note

[1]A recent structure determination (R. H. Huang and J. L. Dye, unpublished results) shows the incorporation of one molecule of methylamine for every two $Li^+(TMPAND)Na^-$ units. In the present work, most or all of the methylamine may have been removed by the prolonged evacuation of the powdered sample.

References

1. For general reviews see: (a) J. L. Dye: *Science* **247**, 663 (1990); (b) J. L. Dye: *Pure Appl. Chem.* **61**, 1555 (1989); (c) J. L. Dye: *Proceedings of the The Robert A. Welch Foundation Conference on Chemical Research XXXII Valency*, 65 (1988); (d) J. L. Dye and M. G. DeBacker: *Ann. Rev. Phys. Chem.* **38**, 271 (1987); (e) J. L. Dye: *Scientific American* **257**, 66 (1987); (f) J. L. Dye: *Prog. Inorg. Chem.* **32**, 327 (1984).
2. J. L. Dye, M. G. DeBacker, and V. A. Nicely: *J. Am. Chem. Soc.* **92**, 5226 (1970).
3. (a) C. J. Pedersen: *J. Am. Chem. Soc.* **89**, 2495 (1967); (b) C. J. Pedersen: *J. Am. Chem. Soc.* **89**, 7017 (1967); (c) C. J. Pedersen: *J. Incl. Phenom.* **6**, 337 (1988); (d) C. J. Pedersen and H. K. Frensdorff: *Angew. Chem. Int. Ed. Engl.* **11**, 16 (1972).
4. J.-M. Lehn: *J. Incl. Phenom.* **6**, 351 (1988), and references therein.
5. J. L. Dye, J. M. Ceraso, M. T. Lok, B. L. Barnett, and F. J. Tehan: *J. Am. Chem. Soc.* **96**, 608 (1974).
6. (a) A. Ellaboudy, J. L. Dye, and P. B. Smith: *J. Am. Chem. Soc.* **105**, 6490 (1983); (b) S. B. Dawes, D. L. Ward, R. H. Huang, and J. L. Dye: *J. Am. Chem. Soc.* **108**, 3534 (1986).
7. See also: (a) R. H. Huang, D. L. Ward, and J. L. Dye: *Acta Crystallogr. C* **C46**, 1835 (1990); (b) J. L. Dye and R. H. Huang: *Chem. Brit.* **26**, 239 (1990); (c) R. H. Huang, D. L. Ward, and J. L. Dye: *J. Am. Chem. Soc.* **111**, 5707 (1989); (d) D. Issa and J. L. Dye: *Inorg. Chim. Acta* **160**, 111 (1989); (e) R. S. Bannwart, S. A. Solin, M. G. DeBacker, and J. L. Dye: *J. Am. Chem. Soc.* **111**, 5552 (1989); (f) M. E. Kuchenmeister and J. L. Dye: *J. Am. Chem. Soc.* **111**, 935 (1989); (g) D. L. Ward, R. H. Huang, and J. L. Dye: *Acta Crystallogr. C* **C44**, 1374 (1988); (h) J. Papaioannou and J. L. Dye: *Rev. Sci. Instrum.* **59**, 496 (1988); (i) R. Concepcion and J. L. Dye: *J. Am. Chem. Soc.* **109**, 7203 (1987); (j) F. J. Tehan, B. L. Barnett, and J. L. Dye: *J. Am. Chem. Soc.* **96**, 7203 (1974).
8. K.-L. Tsai and J. L. Dye: *J. Am. Chem. Soc.* **113**, 1650 (1991).
9. See for example: (a) A. G. M. Barrett, C. R. A. Godfrey, D. M. Hollinshead, P. Prokopiou, D. H. R. Barton, R. B. Boar, L. Joukhadar, J. F. McGhie, and S. C. Misra: *J. Chem. Soc., Perkin Trans I* 1501 (1981); (b) G. P. Pez, I. L. Mador, J. E. Galle, R. K. Crissey, and C. E. Forbes: *J. Am. Chem. Soc.* **107**, 4098 (1985); (c) Z. Jedlinski, M. Kowalczuk, and A. Misiolek: *J. Chem. Soc., Chem. Commun.* 1261 (1988); (d) T. Ohsawa, T. Kobayashi, Y. Mizuguchi, T. Saitoh, and T. Oishi: *Tetrahedron Lett.* **26**, 6103 (1985); (e) D. J. Mathre and W. C. Guida: *Tetrahedron Lett.* **21**, 4773 (1980).
10. The IUPAC names are: 1,4,7,10,13,16-hexaaza-1,4,7,10,13,16-hexamethylcyclooctadecane (HMHCY, hexamethylhexacyclen); 1,4,7,10,13-pentaaza-1,4,7,10,13-pentamethylcyclopentadecane (PMPCY, pentamethylpentacyclen); and 1,4,7-triaza-1,4,7,-timethylcyclononane (TMTCY, trimethyltricyclen).
11. For an example of the use of TMTCY as a complexing agent see: K. Wieghardt, S. Brodka, K. Peters, E. M. Peters, and A. Simon: *Z. Naturforsch. B* **42**, 279 (1987).
12. J. .L. Dye: *J. Phys. Chem.* **88**, 3842 (1984).
13. J. J. Christensen, D. J. Eatough, and R. M. Izatt: *Chem. Rev.* **74**, 351 (1974).
14. (a) A. Bencini, A. Bianchi, M. Ciampolini, E. Garcia-Espana, P. Dapporto, M. Micheloni, P. Paoli, J. A. Ramirez, and B. Valtancoli: *J. Chem. Soc., Chem. Commun.* 701 (1989); (b) A. Bencini, A. Bianchi, A. Borselli, M. Ciampolini, E. Garcia-Espana, P. Dapporto, M. Micheloni, P. Paoli, J. A. Ramirez, and B. Valtancoli: *Inorg. Chem.* **28**, 4279 (1989).
15. (a) J. Cheney and J.-M. Lehn: *J. Chem. Soc., Chem. Commun.* 487 (1972); (b) J. Cheney, J. P. Kintzinger, and J.-M. Lehn: *Nouv. J. Chim.* **2**, 411 (1978).
16. (a) K. J. Moeggenborg: Ph.D. Dissertation, Michigan State University (1990); (b) K. J. Moeggenborg, J. Papaioannou, and J. L. Dye: *Chem. Mater* **3**, 514 (1991).
17. J. L. Dye, M. R. Yemen, M. G. DaGue, and J.-M. Lehn: *J. Chem. Phys.* **68**, 1665 (1978).
18. H. E. Kissinger: *J. Res. Nat. Bur. Stand.* **57**, 217 (1956).

19. (a) J. M. Ceraso and J. L. Dye: *J. Chem. Phys.* **61**, 1585 (1974); (b) J. L. Dye: *J. Phys. Chem.* **88** 3842 (1984); (c) P. P. Edwards, A. Ellaboudy, D. M. Holton, and N. C. Pyper: *Annu. Rep. NMR Spectroscopy* **20**, 315 (1988).
20. J. H. Van Vleck: *Phys. Rev.* **74**, 1168 (1948).
21. D. Freude, T. A. Carpenter, J. Haase, J. Klinowski, and G. Roniker: *Chem. Phys. Lett.* **119**, 365 (1985).
22. R. A. Logan, R. E. Cote, and P. Kusch: *Phys. Rev.* **86**, 280 (1952).
23. N. F. Ramsey: *Phys. Rev.* **77**, 567 (1950); **78**, 699 (1950); **83**, 540 (1951); **86**, 243 (1952).
24. M. A. Atamian: unpublished results, this laboratory.
25. S. Doeuff, K.-L. Tsai, and J. L. Dye: *Inorg. Chem.* **30**, 849 (1991).
26. (a) J. S. Landers, J. L. Dye, A. Stacy, and M. J. Sienko: *J. Phys. Chem.* **85**, 1096 (1981); (b) R. H. Huang, M. K. Faber, J. Kim, and J. L. Dye: unpublished results, this laboratory.
27. (a) E. Mei, A. I. Popov, and J. L. Dye: *J. Am. Chem. Soc.* **99**, 6532 (1977); (b) E. Kauffman, J. L. Dye, J.-M. Lehn, and A. I. Popov: *J. Am. Chem. Soc.* **102**, 2274 (1980).

Journal of Inclusion Phenomena and Molecular Recognition in Chemistry **12**: 275–289, 1992.

Evaluation of a Direct ¹H NMR Method for Determining log *K* and Δ*H* Values for Crown Ether–Alkylammonium Cation Complexation [#]

C. Y. ZHU, J. S. BRADSHAW, J. L. OSCARSON, and R. M. IZATT*
Departments of Chemistry and Chemical Engineering, Brigham Young University, Provo, Utah 84602, U.S.A.

(Received: 10 September 1990; in final form: 15 December 1990)

Abstract. A direct ¹H NMR method for determining log *K* and Δ*H* values for crown ether–ammonium cation complexation using milligrams of sample was tested and evaluated for accuracy and precision by comparing the results with those obtained using a titration calorimetric method. Log *K* values for the interactions of a non-chiral crown ether, diketopyridino-18-crown-6 (K_2P18C6), with α-phenylethylammonium ($PhEt^+$) perchlorate in 50%–50% and 90%–10% (v/v) mixtures of deuterated methanol (CD_3OD) and deuterated chloroform ($CDCl_3$) at four temperatures and, with α-(1-naphthyl)ethylammonium ($NapEt^+$) perchlorate in 50%CD_3OD–50%$CDCl_3$ (v/v) at 25°C were determined by a direct ¹H NMR method. Values of Δ*H* for the interactions of K_2P18C6 with $PhEt^+$ in the two solvents were calculated from the temperature dependence of log *K*. Log *K* values for the interactions of a chiral crown ether, dimethyldiketopyridino-18-crown-6 (M_2K_2P18C6), with (R) and (S) enantiomers of $NapEt^+$ in pure CD_3OD at 25.0°C were also determined by the NMR method. The results were compared with those determined by a calorimetric method at 25.0°C in 50%–50% and 90%–10% (v/v) mixtures of plain methanol and chloroform, in 100% plain methanol, and in a 50%–50% mixture of partially deuterated methanol (with deuterium substitution on the methanol OH group, CH_3OD) and deuterated chloroform. The log *K* values determined by both methods were found to be in good agreement, but the standard deviations associated with the NMR log *K* values were two to three times greater. The agreement of the Δ*H* values determined by the two methods was poor, differing by approximately 10 kJ/mol with the NMR method giving more negative values. The standard deviations associated with the NMR Δ*H* values were approximately ten times greater than those for the calorimetric values. Ion-pairing was observed for the interaction of perchlorate ion with both free and bound $PhEt^+$ in 50%methanol–50%chloroform mixture. It is concluded that the NMR procedure is satisfactory for the determination of log *K*, but not Δ*H* values.

Key words. Crown ether-ammonium cation complexation, formation constants (*K*) and enthalpy changes (Δ*H*), direct ¹H NMR method, titration calorimetric method.

1. Retrospect

In February 1968, the late Professor James J. Christensen and one of us (RMI) visited Charlie Pedersen and Karl Frensdorf in their du Pont laboratory. This visit resulted from a conversation concerning Charlie's work on alkali metal coordination which we had had with George Eisenman (then at the University of Chicago). We had become interested in alkali metal coordination as a result of a post-doctoral stay by Jim at R. J. P. William's laboratory at Oxford two years earlier. Charlie

[#] This paper is dedicated to the memory of the late Dr C. J. Pedersen.
* Author for correspondence.

kindly provided us with a sample of dicyclohexano-18-crown-6 (mixed isomers) which became the basis for our first study of log K, ΔH, and ΔS values for cyclic polyether–alkali metal binding, *Science* **164**, 443–444 (1969).

This contact changed the direction of our research and, through collaboration at BYU with Jerald S. Bradshaw and N. Kent Dalley, led to extensive studies of factors determining cation selectivities by a large number of new cyclic polyethers and their derivatives. We at BYU owe much to Charlie for the direction and impetus he gave to our research program. Our associations with him after 1968 were limited, but always cordial. He was invited each year to the symposia on macrocyclic chemistry which were held at BYU beginning in 1977. Unfortunately, he was unable to attend due to the illness of his wife. In May 1989, one of us (RMI) visited him at his home in Salem, New Jersey. At the entrance to the city, a sign proclaimed Salem to be the home of Charles J. Pedersen, Nobel Prize Laureate. Charlie was not well, but took great interest in providing a guided tour of his home and garden.

2. Introduction

Crown ethers form complexes with inorganic and organic cations [1–4] and with neutral molecules [5, 6]. It is important to know the values of the equilibrium constant (K) and enthalpy change (ΔH) for the formation of these complexes. Large numbers of log K and ΔH values for crown ether–cation interactions determined by various methods have been reported [7, 8] since Pedersen described the synthesis of these macrocycles [9].

Among the various methods that have been used to determine log K and ΔH values, calorimetry is useful in being able to determine both of these quantities from a single experiment provided that log K values lie within a certain range [10–12]. Log K values determined in this way are comparable in accuracy and precision to those obtained by conventional potentiometric procedures [13, 14]. Calorimetry is the preferred method for ΔH determination since the heat of reaction is measured directly. For many systems, a major disadvantage of the calorimetric method is that a large amount of sample, usually tens or hundreds of milligrams, is required for the determination. This requirement is not easily met in the case of many crown ether compounds which are difficult to prepare.

NMR methods have been used for the determination of K values for the interaction of crown ethers with metal cations [15, 16], organic cations [17, 18], and neutral molecules [19]. Among the various NMR methods available, that involving ^{1}H NMR [18] is the simplest and, probably, the most accurate if it is applicable to the system under study and is used properly. Advantages of the direct ^{1}H NMR method are that a small amount (usually milligrams) of sample is required, useful structural information can often be obtained, and the experiment is much less time consuming than any of the other NMR methods. Despite the advantages that the ^{1}H NMR method has, its accuracy and precision in determining K and ΔH values relative to those of the calorimetric or other conventional methods have not been examined.

It has been common to obtain ΔH values from the temperature dependence of K using the Van 't Hoff equation. Various indirect methods [20–22] including the ^{1}H

NMR method [19] have been used to determine ΔH values. In these determinations, the temperature dependence of ΔH is assumed to be negligible. It is expected that ΔH values obtained through indirect methods should have larger standard deviations than those determined by calorimetric measurements. Also, the temperature dependence of ΔH is not always negligible [23, 24]. Thus, the accuracy of ΔH values obtained from indirect methods is open to question. A major objective of the present work was to provide an evaluation of our ^{1}H NMR method for determining log K and ΔH values by comparing these values with those obtained using the calorimetric method under experimental conditions which are identical except for deuterium substitution in the solvent.

The direct ^{1}H NMR method provides a desirable means for the determination of log K values for the interaction of chiral crown ethers with chiral alkylammonium cations or protonated amino acids. Those systems showing significant chiral recognition [25] are of particular interest. It is desirable to characterize these interactions and ^{1}H NMR provides a promising means to do this since only a few milligrams of each crown ether compound are used for each log K determination. In the present work, our ^{1}H NMR method was evaluated by determining log K and ΔH values for the interactions of diketopyridino-18-crown-6 (K_2P18C6) with α-phenylethylammonium ($PhEt^+$) and α-(1-naphthyl)ethylammonium ($NapEt^+$) cations in two deuterated methanol-chloroform mixtures with different volume ratios. In addition, log K values were determined for the interactions of (R,R)-dimethyldiketopyridino-18-

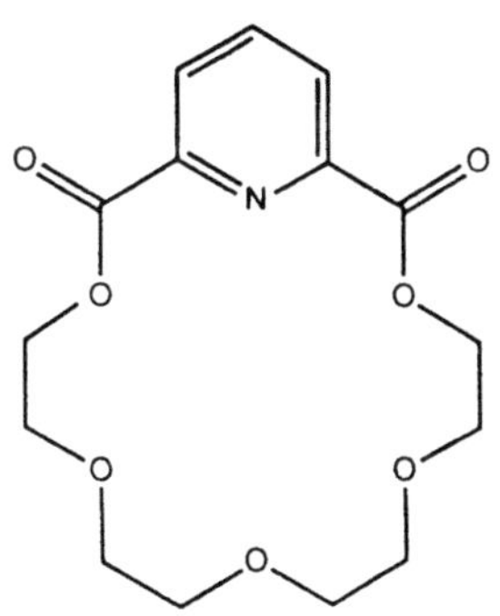

Diketopyridino-18-Crown-6 (K_2P18C6)

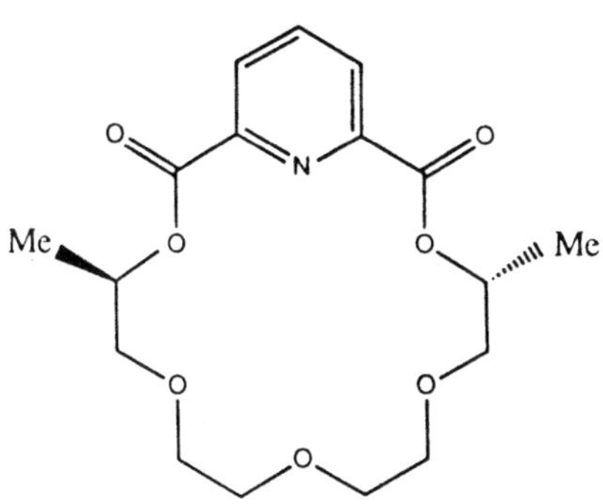

(R,R)-Dimethyldiketopyridino-18-Crown-6 (M_2K_2P18C6)

crown-6 (M_2K_2P18C6) with both the (R) and (S) enantiomers of $NapEt^+$ in pure CD_3OD. The method was evaluated by comparing the results with those determined by a titration calorimetric method in analogous non-deuterated and partially deuterated solvents.

3. Experimental

3.1. MATERIALS

K_2P18C6 and M_2K_2P18C6 were prepared as reported earlier [26, 27]. Each compound was purified by recrystallization in hot diisopropyl ether and drying *in vacuo*.

(*R*) and (*S*) enantiomers of $PhEt^{+}ClO_4^{-}$ and $NapEt^{+}ClO_4^{-}$ were prepared according to Zhu *et al.* [28]. The purities of K_2P18C6, M_2K_2P18C6, $PhEt^{+}ClO_4^{-}$ and $NapEt^{+}ClO_4^{-}$ were better than 99% as evidenced by their ^1H NMR spectra. $CDCl_3$, CH_3OD and CD_3OD (Aldrich, 98%) were used as purchased. $CHCl_3$ and CH_3OH (Fisher, HPLC grade) were used without further publication.

3.2. ^1H NMR EXPERIMENTS

All ^1H NMR spectra were recorded on a Varian Gemini 200 MHz apparatus with a temperature accessory.

For each log *K* determination at a given temperature, a sample containing a few milligrams of a macrocycle in a known volume of solvent was first loaded into the probe and a spectrum was taken. The sample was then unloaded, added to the sample tube with a small amount of the solid ammonium salt, reloaded into the probe, and another spectrum was taken. This process was repeated until no significant change was observed in successive ^1H NMR spectra. Usually eight to twelve spectra were taken for each log *K* determination. The crown ether concentrations were about 0.01–0.015M and the ammonium salt concentrations varied from 0.0M to approximately 0.06M for each of the experiments. In each experiment, an accurately weighed quantity of the crown ether was dissolved in a known volume of solvent at 25.0°C. The analytical balance used was calibrated for accuracy using a standard weight from the National Institute of Standards and Technology. The salt concentrations were calculated based on the integral ratio of a particular ammonium salt signal to a particular crown ether signal in the spectra. In order to obtain a quantitative integration, the time delay between the two pulses for each NMR acquisition was set long enough to allow sufficient relaxation of the signals of interest. The NMR parameter settings were kept the same for all of the experiments and Me_4Si (TMS) was used as the internal standard for all of the samples.

Under conditions of fast exchange on the NMR time scale, the average chemical shift of a crown ether signal is the weighted average of the chemcial shifts of the same signals for the free and complexed crown ether

$$\delta_{\text{ave}} = X_f\delta_f + (1 - X_f)\delta_c \tag{1}$$

where δ_{ave} = average chemical shift of the crown ether signal of interest, δ_f = chemical shift of the same signal for the free crown ether, and X_f = the mole fraction of free crown ether. By using a non-linear least-squares treatment, the best fit of the expermental data points can be achieved through the minimization of the function

$$U = \Sigma[\delta_{\text{obs},i} - X_{f,i}\delta_f - (1 - X_{f,i})\delta_c]^2 \tag{2}$$

where δ_{obs} = observed average chemical shift of the crown ether signal of interest. Since X_f is a function of log *K*, *U* is a function of log *K* also. The log *K* value that results in the minimum *U* value is taken to be the correct value. The log *K* values were calculated this way using the program EQDD on a VAX 11/780 computer. The ΔH values for the interactions were obtained from the temperature dependence of log *K* according to the relationship

$$2.303 \log K = -\Delta H/RT + C \tag{3}$$

which is obtained from the integration of the Van 'tHoff equation assuming that ΔH is a constant over the temperature range, where C is an integration constant. Linear least-squares treatment of the log $K - T^{-1}$ data pairs based on Equation 3 gives the following expression for ΔH

$$\Delta H = (2.303\mathrm{R})\frac{\Sigma[(T_{\mathrm{av}}^{-1} - T_i^{-1})(\log K_i - \log K_{\mathrm{av}})]}{\Sigma(T_i^{-1} - T_{\mathrm{av}}^{-1})^2} \tag{4}$$

where T_{av}^{-1} and log K_{av} = the averages of T_i^{-1} values and log K_i values, respectively.

3.3. CALORIMETRIC EXPERIMENTS

Calorimetric determinations were carried out on a Tronac model 450 isoperibol calorimeter using procedures described elsewhere [29]. The calorimeter was calibrated and tested according to Christensen *et al.* [10]. The calculations were done on a VAX 11/780 computer using the programs developed earlier in our laboratory [11].

4. Results and Discussion

K_2P18C6 forms complexes with $PhEt^+$ and $NapEt^+$ in 90%MeOH–10%H_2O [28]. These complexes are expected to be more stable in methanol–chloroform mixtures because of the decreased solvent polarity. The use of methanol–chloroform mixtures as the solvents in this work was based on solubility considerations since the complexes of K_2P18C6 with either $PhEt^+ClO_4^-$ or $NapEt^+ClO_4^-$ did not dissolve in methanol at concentrations larger than approximately 0.01M. However, the use of these mixed solvents increases the likelihood that ion-pair formation is present due to the decreased solvent polarity. In order to evaluate the contribution of ion-pairing to the overall complexation reaction, the interactions of K_2P18C6 with $PhEt^+ClO_4^-$ in two methanol-chloroform mixtures with different volume ratios were studied and compared.

The ^{1}H NMR spectra of the samples containing K_2P18C6 and either $PhEt^+$ or $NapEt^+$ showed that the pyridine signals of the macrocycle were a doublet and a triplet. Both of these signals were well resolved in the range 8.10–8.40 ppm, and were well separated from the phenyl signals of $PhEt^+$ or the naphthyl signals of $NapEt^+$. Therefore, the pyridine signals of K_2P18C6 were used as the probe for monitoring the complexation. The methyl signal of either $PhEt^+$ or $NapEt^+$ was found to be a well isolated sharp doublet at about 1.5 ppm. The integral of the methyl signal was used for the calculation of the ammonium salt concentrations. In the presence of $PhEt^+$, downfield shifts of up to 0.17 ppm (34 Hz) for the pyridine signals of K_2P18C6 were observed. However, in the presence of $NapEt^+$, upfield shifts of up to 0.19 ppm (38 Hz) for the same signals were observed. This difference in the direction of the chemical shift change is understood by considering the difference between the proposed structures of the K_2P18C6–$PhEt^+$ and K_2P18C6–$PhEt^+$ complexes [28] as illustrated in Figures 1A and 1B, respectively. The orientation of the phenyl group of $PhEt^+$ opposite to that of the pyridine ring of K_2P18C6 is the cause for the downfield shift of the pyridine signals in the case of K_2P18C6–$PhEt^+$ complexation since the deshielding zone of the phenyl group

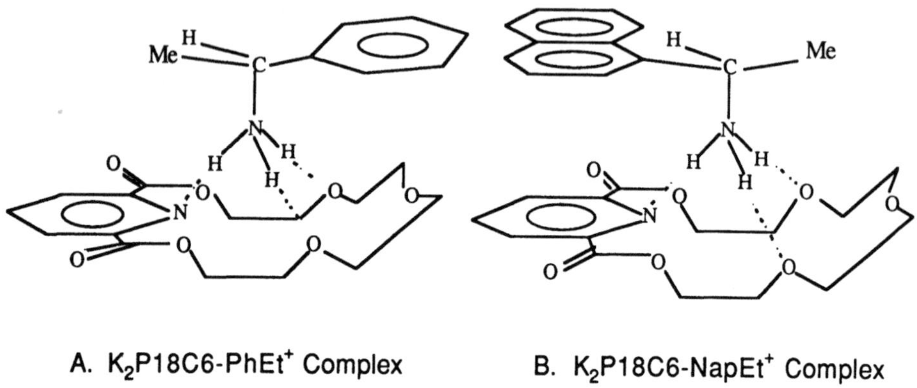

Fig. 1. A schematic representation of the molecular structures of the diketopyridino-18-crown-6 complexes with α-phenylethylammonium cation (A) and with α-(1-naphthyl)ethylammonium cation (B) [28].

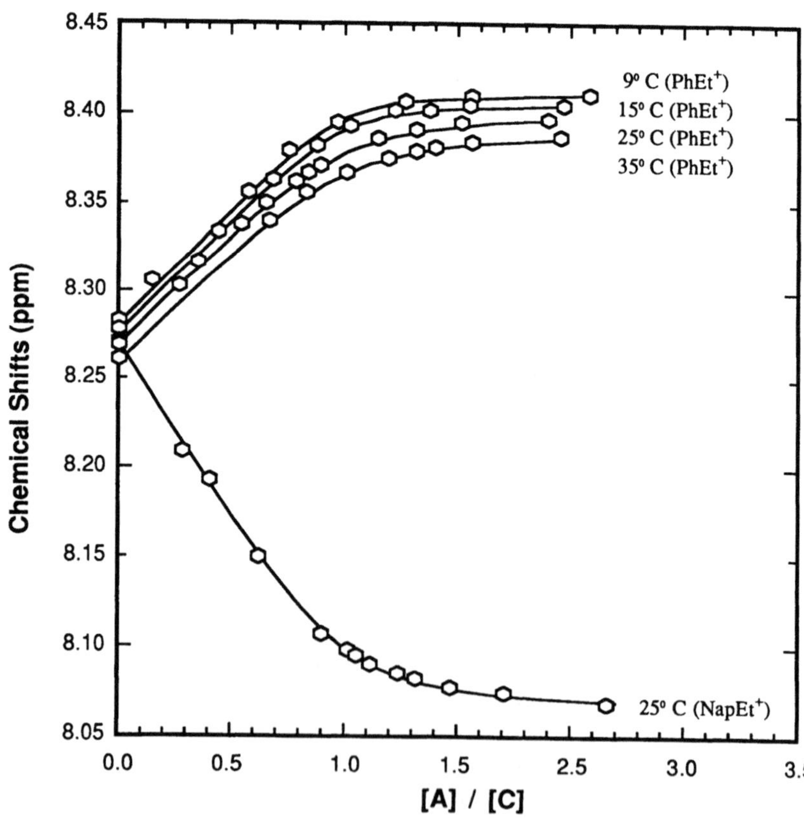

Fig. 2. Observed chemical shifts of the 3,5 protons on the pyridine moiety of diketopyridino-18-crown-6 (K_2P18C6) in 50%CD_3OD–50%$CDCl_3$ at different temperatures as a function of the molar ratio, [A]/[C], of the α-phenylethylammonium ($PhEt^+$) or α-(1-naphthyl)ethylammonium ($NapEt^+$) perchlorate to K_2P18C6, where [A] and [C] represent the concentrations of $PhEt^+$ (or $NapEt^+$) and K_2P18C6, respectively.

covers the pyridine ring. The overlap between the naphthyl group of $NapEt^+$ and the pyridine ring is the cause for the upfield shift of the pyridine signals in the case of K_2P18C6–$NapEt^+$ complexation since the shielding zone of the naphthyl group covers the pyridine ring. The plots of the observed and simulated pyridine chemical shifts of K_2P18C6 vs. the molar ratios of either $PhEt^+$ or $NapEt^+$ to K_2P18C6 in 50%CD_3OD–50%$CDCl_3$ (v/v) at different temperatures are shown in Figure 2.

(S,S)-M_2K_2P18C6 shows good chiral recognition toward chiral $NapEt^+$ by favorably complexing the (R) rather than the (S) enantiomer [30]. (R,R)-M_2K_2P18C6 is expected to show equal recognition towards chiral $NapEt^+$, but in favor of the (S) rather than the (R) enantiomer. The calorimetrically determined log K values for the interactions of (S,S)–M_2K_2P18C6 with both (R) and (S) enantiomers of $NapEt^+$ in methanol have been reported [30]. In the present work, the log K values for the interactions of (R,R)–M_2K_2P18C6 with (R)- and (S)-$NapEt^+$ in CD_3OD were determined by the ^1H NMR method, and compared with those for (S,S)–M_2K_2P18C6 determined by calorimetry. In the ^1H NMR spectrum of M_2K_2P18C6 with the presence of either the (R) or (S) form of $NapEt^+$, an upfield shift of the pyridine signals and a downfield shift of the methyl signal of M_2K_2P18C6 were observed. The methyl signal of M_2K_2P18C6 was a nice doublet and was well separated from the methyl signal of $NapEt^+$. The integral ratio of the two methyl signals was used to calculate the concentration of the salt. The plots of the observed and simulated pyridine chemical shifts of M_2K_2P18C6 vs. the molar ratios of either the (R) or (S) form of $NapEt^+$ to M_2K_2P18C6 in CD_3OD at 25.0°C are shown in Figure 3.

Analysis of the chemical shift–molar ratio plots showed that the maximum slope change of each curve happened where the molar ratio was about 1 : 1, indicating that 1 : 1 complexation dominated in each system studied. Similar analysis of the heat–molar ratio plots generated by calorimetry for the same systems in the corresponding non-deuterated and partially deuterated solvents also suggested the domination of 1 : 1 complexation. The simulation of the NMR or calorimetric data for each system by assuming 1 : 1 complexation was good with reasonably small statistical deviation. The log K, ΔH and ΔS values obtained by both ^1H NMR and calorimetric methods are listed in Tables I and II.

In Table I, the NMR ΔH values were calculated from the temperature dependence of log K according to Equation 3 where ΔH is assumed to be temperature independent. Values of ΔS at 25.0°C were calculated according to the relationship

$$2.303RT \log K = \Delta H - T\Delta S \tag{5}$$

Plots of log K vs. T^{-1} for K_2P18C6–$PhEt^+$ interactions in the two solvents are shown in Figure 4.

It is seen from Tables I and II that the log K values determined by the NMR method at 25.0°C are in good agreement with those determined by calorimetry at the same temperature. The method should prove useful in quantitating chiral recognition for new chiral systems in terms of Δ log K values determined by the NMR method. In addition, only very small amounts of valuable sample are used. However, as seen in Table I, the ΔH and ΔS values obtained by the NMR method are in poor agreement with those determined by calorimetry.

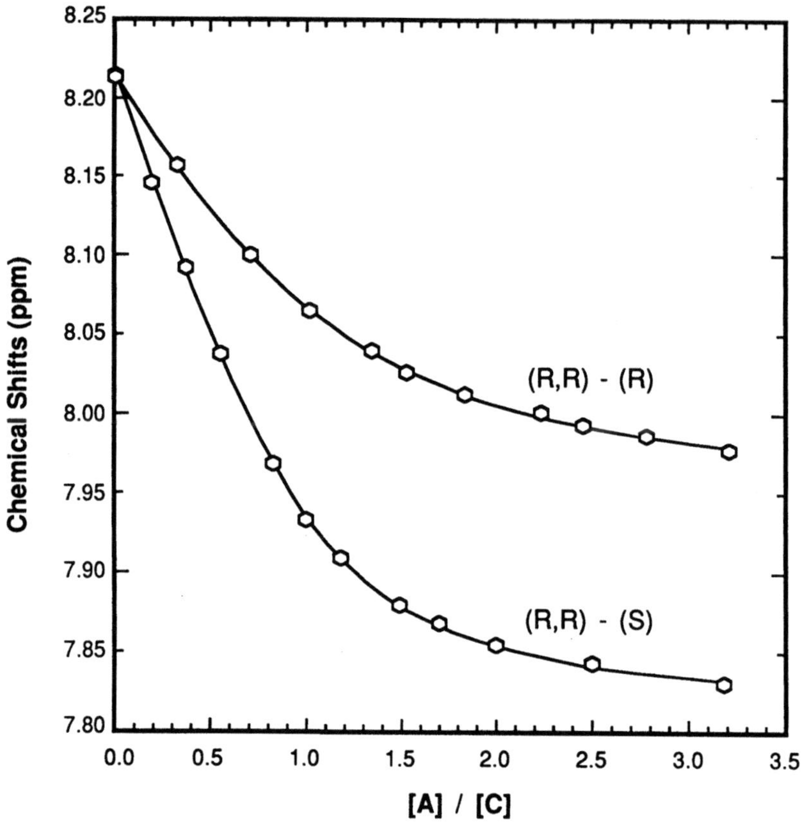

Fig. 3. Observed chemical shifts of the 3,5 protons on the pyridine moiety of (R,R)-dimethyldiketopyridino-18-crown-6 (M_2K_2P18C6) in pure CD_3OD at 25°C as a function of the molar ratio, [A]/[C], of the (R) on (S) enantiomer of α-(1-naphthyl)ethylammonium ($NapEt^+$) perchlorate of $NapEt^+$ and M_2K_2P18C6, respectively.

As mentioned earlier, one of the advantages of calorimetry is that the reaction heat is measured directly, and the log K and ΔH values are obtained simultaneously from a single experiment. Each of the calorimetrically determined log K and ΔH values listed in Tables I and II is the average of at least three values obtained from parallel experiments at the same conditions. The standard deviation of each log K or ΔH value is the standard deviation of the parallel values from the average. Each ΔS value is calculated from log K and ΔH values according to Equation 5, and the standard deviation of each ΔS value is the weighted sum of those of the log K and ΔH values.

In the case of the NMR method, only a log K value can be obtained from a single experiment and the ΔH value is calculated from the variation of log K with temperature according to Equations 3 and 4. Therefore, the standard deviation of each ΔH value is determined by, first, the standard deviation of the log K values used for the ΔH calculation, second, the number of log $K - T^{-1}$ data pairs used and the temperature range that these data pairs cover, and third, the validity of the assumption that ΔH is independent of temperature in the temperature range covered. So, the standard deviations of the ΔH values obtained by the NMR

Table I. Log K, ΔH and ΔS values for the interactions of diketopyridino-18-crown-6 (K_2P18C6) with α-phenylethylammonium ($PhEt^+$) perchlorate in different solvents and at different temperatures.

Method	Solvent[a]	Temp.(°C)	Log K	ΔH(kJ/ mol)	ΔS(J/K mol)
NMR	1 : 1M/C(D)	9.0	3.73		
		15.0	3.62		
		25.0	3.33	−44.4[b]	−84.5[b]
		35.0	3.10		
Cal.[c]	1 : 1M/C	25.0	3.42 ± 0.01	-30.5 ± 0.5	-36.8 ± 1.6
NMR	9 : 1M/C(D)	15.0	3.18		
		25.0	2.99		
		35.0	2.76	−37.2[b]	−68.2[b]
		42.0	2.64		
Cal.[c]	9 : 1M/C	25.0	2.96 ± 0.01	-29.2 ± 0.5	-41.3 ± 1.6

[a] 1 : 1M/C(D) = 50%CD_3OD–50%$CDCl_3$ (v/v), 1 : 1M/C = 50%CH_3OH–50%$CHCl_3$ (v/v), 9 : 1M/C(D) = 90%CD_3OD–10%$CDCl_3$ (v/v), and 9 : 1M/C = 90%CH_3OH–10%$CHCl_3$ (v/v).
[b] The NMR ΔH values were calculated from the temperature dependence of log K according to the integrated form of the Van 'tHoff equation. The linear correlation coefficients for log K vs. $1/T$ were greater than 0.997.
[c] Cal. = Titration calorimetry.

method are expected to be greater than those of the ΔH values determined by calorimetry. As seen in Figure 4, each of the log K vs. T^{-1} plots is linear, indicating that no visible temperature dependence of ΔH exists. If the temperature dependence of ΔH is ignored, the standard deviation of an NMR ΔH value can be evaluated based on the following equation

$$\delta \Delta H = (2.303R)\frac{\Sigma|T_{\text{av}}^{-1} - T_i^{-1}|\,|\delta \log K_i|}{\Sigma(T_{\text{av}}^{-1} - T_i^{-1})^2} \tag{6}$$

where $\delta \log K_i$ = the standard deviation of each log K value.

In evaluating the standard deviation of the NMR log K values, the reproducibility of these values determined for the same system by separate experiments was examined. The log K values obtained from six parallel experiments for K_2P18C6–$PhEt^+$ complexation at 25°C in 50%CD_3OD–50%$CDCl_3$ (3.33, 3.31, 3.39, 3.42, 3.40, and 3.36, respectively) were in good agreement with each other, and the standard deviation was ± 0.04. This standard deviation is a little greater than that of a calorimetric log K value which is typically ± 0.01–0.02. In comparison with the calorimetric results, the greater standard deviation of an NMR log K value is expected since the deviations of the sample concentrations are greater due to the use of very small amounts of sample and to the limitation of the balance precision. In addition, the smaller number of data points used for an NMR log K calculation may also increase the standard deviation of the result. The standard deviation of an NMR log K value is also affected by the magnitude of the maximum chemical shift change for the signal of interest induced by the complexation reaction and the standard deviations of the chemical shift values for the same signal at different ammonium salt-crown ether molar ratios. The larger the maximum chemical shift change and the smaller the standard deviation of each

Table II. Log K, ΔH and ΔS values for the interactions of diketopyridino-18-crown-6 (K_2P18C6) with α-(-naphthyl)ethylammonium ($NapEt^+$) perchlorate, and of (R,R)- and (S,S)-dimethyldiketopyridino-18-crown-6 (M_2K_2P18C6) with (R) and (S) enantiomers of $NapEt^+ClO_4^-$ at 25.0°C.

Ligand	Cation	Method	Solvent[a]	Log K	ΔH(kJ/mol)	ΔS(J/K mol)
K_2P18C6	NapEt	NMR	1 : 1M/C(D)	3.35		
		Cal.[b]	1 : 1M/C	3.49 ± 0.01	−29.8 ± 0.5	−33.3 ± 1.5
M_2K_2P18C6	NapEt					
(*RR*)	(*R*)	NMR	CD_3OD	2.08		
(*SS*)	(*S*)	Cal.[b]	CH_3OH	2.06 ± 0.01[c]	−26.4 ± 0.5[c]	−49.8 ± 1.2[c]
(*RR*)	(*S*)	NMR	CD_3OD	2.50		
(*SS*)	(*R*)	Cal.[b]	CH_3OH	2.47 ± 0.01[c]	−27.6 ± 0.5[c]	−45.2 ± 1.1[c]

[a] See footnote [a] of Table I.
[b] See footnote [c] of Table I.
[c] Ref. [30].

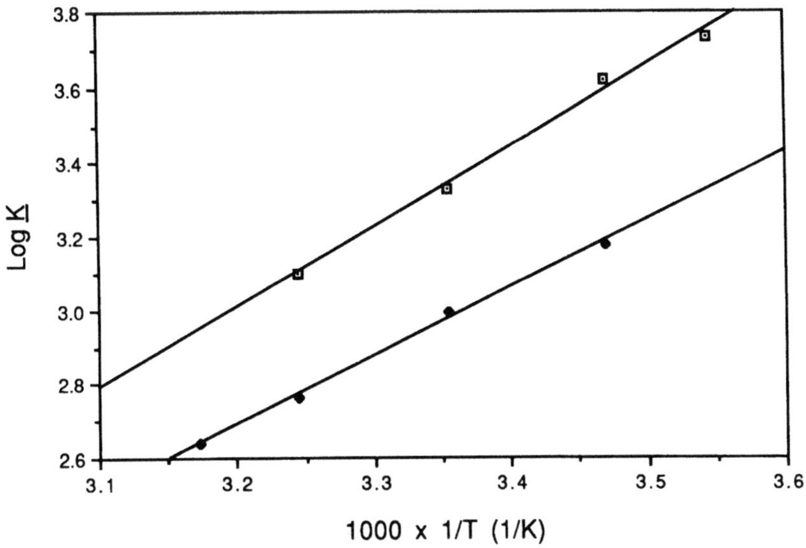

Fig. 4. Plot of log K for the interaction of diketopyridino-18-crown-6 with α-(1-naphthyl)ethylammonium ($NapEt^+$) perchlorate in 50%CD_3OD–50%$CDCl_3$ (solid circles) and in 90%CD_3OD–10%$CDCl_3$ (solid triangles).

chemical shift value, the smaller the standard deviation of the NMR log K value. Since the maximum chemical shift changes induced by the complexation reactions for all of the systems studied are of a comparable magnitude (0.17–0.35 ppm) and are much greater than the standard deviation of each chemical shift value (0.001 ppm), it is expected that the standard deviations of all of the NMR log K values reported in the present work are similar to that found for the K_2P18C6–$PhEt^+$ interaction, i.e., ± 0.04 log K unit.

As mentioned earlier, the standard deviation of an NMR ΔH value is dependent on those of the log K values used. If the standard deviation of each log K value in Table I for the K_2P18C6–$PhEt^+$ interaction in 50%CD_3OD–50%$CDCl_3$ is taken to be ± 0.04, the standard deviation of the ΔH value will be ± 8.4 kJ/mol according to Equation 6. The standard deviation of ΔS ($\pm$ 17 J/K mol) is even greater than that of ΔH. It is also interesting to note that the standard deviation of each calculated ΔH value is dependent on the number of log K–T^{-1} data pairs used and the temperature range covered by these data pairs. Calculations based on Equation 6, where ΔH is assumed to be temperature independent, showed that the wider the temperature range, the smaller the standard deviation of the ΔH value. In a given temperature range, the more the number of data pairs, the smaller the standard deviation of the ΔH value.

Since it is difficult to differentiate experimental data accurately, the poor agreement between the NMR and calorimetric ΔH values is not surprising. Even small, random experimental errors in the log K values can lead to large errors in calculating the slope of the curve going through the log K values when plotted versus T^{-1}. The slope of the log K vs. T^{-1} plot is usually evaluated by assuming some functional relationship between log K and T^{-1}. The assumption made about this functionality influences the answer obtained. In this work, it was assumed that ΔH was constant, hence the slope of log K versus T^{-1} was constant. This appears to be a reasonable assumption from observation of the data, but if it were assumed that the relationship was quadratic, a different ΔH value would be obtained. The researcher must decide which function is the best. Including too many terms in the function can lead to a curve which is somewhat wavy resulting in unreasonable ΔH values at some temperatures. Inclusion of too few terms does not allow enough temperature dependence in the ΔH term. This problem can be overcome somewhat by taking many log K values over a wide temperature range and fitting the data to a smooth curve with the fewest possible parameters in the fitting equation. Even with great care the ΔH values obtained from log K values will not usually be as good as those obtained from calorimetric data.

Our conclusion is that the direct ^{1}H NMR method can produce acceptable log K values in comparison with those determined by the calorimetric method. However, the ΔH and ΔS values obtained by the NMR method have large standard deviations and care must be taken before discussing any chemical reactions in terms of these values. The standard deviation problem for ΔH and ΔS exists in other indirect methods [20–22] also.

It is important to mention a possible limitation of the ^{1}H NMR method. Since the proton of interest whose signal is used as a probe for the complexation is, in most cases, not directly involved in the bond breaking and forming processes and is far from the reaction center, the proton chemical shift change induced by the complexation reaction is not necessarily proportional to the stability of the complex. The formation of a stable complex may result in only a very small chemical shift change due to the special structural feature of the complex in which the proton of interest remains basically unaffected upon complexation. If this happens, the NMR method will fail to provide the log K value for the particular system. However, if the formation of a complex does induce a significant chemical shift change for a proton signal of interest, the NMR method will provide a reliable log K value no matter how far the proton of interest is from the reaction center. The factor that reflects the complex stability is not the maximum chemical shift change induced by the complexation reaction, but the curvature of the chemical shift vs. molar ratio plot (Figures 2 and 3), provided the conditions of fast exchange on the NMR time scale are met, i.e., Equation 1 is valid.

It is seen from Table I that the log K values for the interactions of K_2P18C6 with $PhEt^+$ in 50%methanol–50%chloroform are greater than those in 90%methanol–10%chloroform at corresponding temperatures, respectively. This is expected since the 50%–50% methanol–chloroform mixture has the smaller solvent polarity and dielectric constant. Both NMR and calorimetric ΔH values for the same interaction in 50%methanol–50%chloroform are more negative than those in 90%methanol–10%chloroform. The calorimetric ΔS value for the same interaction in 50%methanol–50%chloroform is less negative than that in 90%methanol–10%chloroform, but the relative magnitudes of the two NMR ΔS values are reversed. Since the NMR ΔH and ΔS values have much greater standard deviations than the calorimetric values, the poor agreement between the NMR and calorimetric ΔH or ΔS values is not surprising. However, the use of deuterated solvent in the NMR measurements may account for part of this difference. Since the hydrogen on

Table III. Log K, ΔH and ΔS values for the interaction of K_2P18C6 with α-phenylethylammonium perchlorate in 50%–50% (v/v) CH_3OH–$CHCl_3$, CH_3OD–$CDCl_3$, and CD_3OD–$CDCl_3$ at 25.0°C.

Solvent	Method	Log K	ΔH(kJ/mol)	ΔS(J/K mol)
CH_3OH–$CHCl_3$	Cal.[a]	3.42 ± 0.02	-30.5 ± 0.5	-36.8 ± 1
CH_3OD–$CDCl_3$	Cal.[a]	3.42 ± 0.02	-33.2 ± 0.8	-45.9 ± 2
CD_3OD–$CDCl_3$	NMR	3.33 ± 0.04	-44.4[b]	-84.5[b]

[a] See footnote [c] of Table I.
[b] The estimated deviations of the NMR ΔH and ΔS values are ± 8.4 kJ/mol and ± 17 J/K mol, respectively.

the solvent methanol OH group is capable of fast exchange with those on an ammonium cation, the substitution of deuterium for hydrogen on the methanol OH group may affect the interaction of an ammonium cation with a crown ether ligand. In order to evaluate the difference that deuterium substitution on the methanol OH group could make, the calorimetric measurement of log K and ΔH values for the K_2P18C6–$PhEt^+$ interaction in 50%CH_3OD–50%$CDCl_3$ was made and the results are compared with those in 50%CH_3OH–50%$CHCl_3$ (Table III). It is found that the log K values for the reaction in the two solvents are in excellent agreement, while the ΔH and ΔS values are different. The ΔH and ΔS values in 50%CH_3OD–50%$CDCl_3$ are about 2.7 kJ/mol and 9.1 J/K mol more negative than those in 50%CH_3OH–50%$CHCl_3$ (Table III), respectively. Apparently, the substitution of deuterium for the hydrogen on the methanol hydroxy group has an appreciable effect on the interaction of K_2P18C6 with $PhEt^+$. Since the deuterium on the hydroxy group of a deuterated methanol molecule can be involved in a fast exchange with the hydrogens of an ammonium cation [31], this fast exchange will result in deuterium substitution for part or all of the protons in the N^+—H—N or N^+—H—O hydrogen bonds formed between a crown ether ligand and an ammonium cation. The substitution of deuterium for hydrogen in a hydrogen bond was found to increase the bond strength [32]. This is consistent with our observation of a more negative ΔH value for the K_2P18C6–$PhEt^+$ interaction in 50%CH_3OD–50%$CDCl_3$ than that in 50%CH_3OH–50%$CHCl_3$. The substitution of protons from N^+—H—N or N^+—H—O hydrogen bonds by deuterium atoms may be an entropy reducing process, as we observed a more negative ΔS value for the same interaction in 50%CH_3OD–50%$CDCl_3$ than in 50%CH_3OH–50%$CHCl_3$.

The use of methanol–chloroform mixtures as solvents in the present work increases the likelihood of ion-pairing due to the reduced solvent polarity and dielectric constant in comparison with pure methanol. Therefore, care has been taken in the analysis of both the NMR and calorimetric data to look for the possibility of ion-pairing. It was found that the fit of either the NMR or the calorimetric data assuming 1 : 1 complexation for the reactions in the 50%methanol–50%chloroform mixture generally gave larger statistical deviations than in the 90%methanol–10%chloroform mixture. NMR spectra of $PhEt^+ClO_4^-$ in deuterated 50%methanol–50%chloroform showed that the methyl chemical shift changed as the concentration of the salt changed. Since the extent of ion-pairing depends on the salt concentration, the weighted average of the chemical shifts for the free and ion-paired cations will also depend on the salt concentration. Therefore, the observed chemical shift change as a result of the salt concentration change provided clear evidence for the presence of ion-pairing. A similar phenomenon was not observed in deuterated 90%methanol–10%chloroform. The ion-pairing constant for $PhEt^+ClO_4^-$ ($\log K_{ip}$) in the 50%–50% mixture was found to be 2.08 using a ^{1}H NMR method similar to the one used for $\log K$ determination. The calculation of the ion-pairing constant from the ^{1}H NMR chemical shift data of the salt as a function of the salt concentration used a basic equation similar to Equation (1)

$$\delta_{ave} = X_f\delta_f + (1 - X_f)\delta_{ip} \tag{7}$$

where δ_{ave} = average chemical shift of the ammonium cation signal of interest, δ_f = chemical shift of the same signal for the free, non ion-paired ammonium cation, and X_f = the mole fraction of free ammonium cation. The δ_f value is known by extrapolating δ_{ave} values at zero salt concentration, where X_f is assumed to be equal to 1.

By incorporating the calculated ion-pairing constant of $PhEt^+ClO_4^-$ into the calculation of $\log K$, the fit of both the NMR and calorimetric data for the 50%–50% mixture was significantly improved, while no significant effect was found for the 10%–90% mixture. This indicates the presence of ion-pairing in the 50%–50% mixture, but not in the 90%–10% mixture. By taking ion-pairing into account, the $\log K$ value for the $K_2P18C6–PhEt^+$ interaction and the ion-pairing constant, $\log K'_{ip}$ for the $(K_2P18C6–PhEt^+)–ClO_4^-$ interaction were calculated. The $\log K$ value thus calculated was very close to that calculated assuming 1 : 1 complexation only and the $\log K'_{ip}$ value, 2.04, was also very close to that ($\log K_{ip} = 2.08$) for the $PhEt^+–ClO_4^-$ interaction. The $\log K_{ip}$ and $\log K'_{ip}$ values are the same within their expected standard deviations. During the interaction of K_2P18C6 with $PhEt^+$, the contribution from ion-pair dissociation of $PhEt^+ClO_4^-$ is mostly cancelled out by the contribution from ion-pair formation of the complexed ammonium cation with the anion. Therefore, the ion-pairing effect does not affect the observed complexation constant when 1 : 1 complexation is assumed.

Acknowledgement

Appreciation is expressed to the Office of Naval Research for financial support of this research and Dr. Peter Huszthy for purifying the crown ether compounds.

References

1. F. de Jong and D. N. Reinhoudt: *Adv. Phys. Org. Chem.* **17**, 279 (1980).
2. C. J. Pedersen in *Synthetic Multidentate Macrocyclic Compounds* (R. M. Izatt and J. J. Christensen, eds.); Academic Press, New York (1978).
3. J. D. Lamb, R. M. Izatt, J. J. Christensen, and J. D. Eatough in *Coordination Chemistry of Macrocyclic Compounds* (G. A. Melson, ed.); Plenum Press, New York (1979).
4. J. P. Behr, J. M. Lehn, and P. Vierling: *J. Chem. Soc. Chem. Commun.* 621 (1976).
5. J. A. A. de Boer, D. N. Reinhoudt, S. Harkema, G. J. Van Hummel, and F. de Jong: *J. Am. Chem. Soc.* **104**, 4073 (1982).

6. F. Vogtle, H. Sieger, and W. M. Muler: *Top. Curr. Chem.* **98**, 107 (1981).
7. R. M. Izatt, J. S. Bradshaw, S. A. Nielsen, J. D. Lamb, and J. J. Christensen: *Chem. Rev.* **85**, 271 (1985).
8. R. M. Izatt, K. Pawlak, J. S. Bradshaw, and R. L. Bruening: *Chem. Rev.* in press.
9. C. J. Pedersen: *J. Am. Chem. Soc.* **89**, 2495 (1967).
10. J. J. Christensen, J. Ruckman, D. J. Eatough, and R. M. Izatt: *Thermochim. Acta* **3**, 203 (1972).
11. D. J. Eatough, J. J. Christensen, and R. M. Izatt: *Thermochim. Acta* **3**, 219 (1972).
12. J. J. Christensen, D. P. Wrathall, J. L. Oscarson, and R. M. Izatt: *Anal. Chem.* **40**, 1713 (1968).
13. H. K. Frensdorf: *J. Am. Chem. Soc.* **93**, 606 (1971).
14. T. Yamabe, K. Hori, and K. Akagi: *Tetrahedron* **35**, 1065 (1979).
15. M. Shamsipur, G. Rounaghi, and A. I. Popov: *J. Solution Chem.* **9**, 701 (1980).
16. C. Erk: *J. Mol. Liquids* **40**, 143 (1989).
17. F. de Jong, D. N. Reinhoudt, C. J. Smit, and R. Hius: *Tetrahedron Lett.* 4783 (1976).
18. J. A. A. de Boer and D. N. Reinhoudt: *J. Am. Chem. Soc.* **107**, 5347 (1985).
19. H. S. Gold and M. R. Rice: *Anal. Data* **29**, 637 (1982).
20. F. Wada, Y. Wada, K. Kikukawa, and H. Matsuda: *Bull. Chem. Soc. Jpn.* **54**, 458 (1981).
21. H. Shizuka, K. Kameta, and T. Shinozaki: *J. Am. Chem. Soc.* **107**, 3956 (1985).
22. R. A. Alberty, R. M. Smith, and R. M. Bock: *J. Biol. Chem.* **193**, 425 (1951).
23. R. M. Izatt, D. P. Nelson, J. H. Rytting, B. L. Haymore, and J. J. Christensen: *J. Am. Chem. Soc.* **93**, 1619 (1971).
24. R. M. Izatt, H. D. Johnston, and J. J. Christensen: *J. Chem. Soc. Dalton Trans* 1152 (1972).
25. J. S. Bradshaw, P. Huszthy, C. W. McDaniel, C. Y. Zhu, N. K. Dalley, R. M. Izatt, and S. Lifson: *J. Org. Chem.* **55**, 3129 (1990).
26. R. M. Izatt, J. D. Lamb, R. E. Asay, G. E. Maas, J. S. Bradshaw, J. J. Christensen, and S. S. Moore: *J. Am. Chem. Soc.* **99**, 6134 (1977).
27. B. A. Jones, J. S. Bradshaw, and R. M. Izatt: *J. Heterocycl. Chem.* **19**, 551 (1982).
28. C. Y. Zhu, R. M. Izatt, J. S. Bradshaw, and N. K. Dalley: *J. Incl. Phenom.* in press.
29. R. M. Izatt, J. D. Lamb, N. E. Izatt, B. E. Rossiter, Jr., J. J. Christensen, and B. H. Haymore: *J. Am. Chem. Soc.* **101**, 6273 (1979).
30. R. B. Davidson, J. S. Bradshaw, B. A. Jones, N. K. Dalley, J. J. Christensen, and R. M. Izatt: *J. Org. Chem.* **49**, 353 (1984).
31. J. W. Akitt: in *Multinuclear NMR* (J. Mason, ed.); Plenum Press, New York (1987).
32. H. Wolff: in *The Hydrogen Bond* (P. Schuster, G. Zundel, and C. Sandorfy, eds.); North-Holland Publishing Company, Amsterdam (1976).

Journal of Inclusion Phenomena and Molecular Recognition in Chemistry **12**: 291–304, 1992.

Cation and Anion Coordination Chemistry of Palladium(II) with Polyazacycloalkanes. Thermodynamic and Structural Studies[#]

A. BENCINI, A. BIANCHI,* M. MICHELONI,* and P. PAOLETTI,*
Department of Chemistry, University of Florence, Via Maragliano 75, I-50144 Florence, Italy.

P. DAPPORTO and P. PAOLI
Department of Energetics, University of Florence, Italy.

E. GARCIA-ESPAÑA*
Department of Inorganic Chemistry, University of Valencia, Spain.

(Received: 19 October 1990; in final form: 26 November 1990)

Abstract. The interaction of $PdCl_4^{2-}$ with the macrocyclic ligands of the series [3*k*]aneN$_k$ has been studied both in solution and in the solid state. [18]aneN$_6$ and [21]aneN$_7$ form both mono- and binuclear Pd^{2+} complexes, whose stability constants have been determined in 0.5 mol dm^{-3} NaCl at 298.15 K. [21]aneN$_7$ also forms, in solution, a trinuclear species in which an amino group deprotonates to bridge two Pd^{2+} ions, as observed in the solid state. The crystal structure of the complexes [Pd$_2$([18]aneN$_6$)Cl$_2$][ClO$_4$]$_2$ and [Pd$_3$([21]aneN$_7$)Cl$_3$][ClO$_4$]$_2 \cdot H_2O$ have been solved by single crystal X-ray analysis. $C_{12}H_{30}N_6Cl_4O_8Pd_2$: monoclinic, space group *C*2/*m*, $a = 10.876(2)$, $b = 18.117(2)$, $c = 7.043(2)$ Å, $\beta = 113.78(2)°$, $V = 1270(12)$ Å^3, $Z = 2$, $D_{calc} = 1.92$ g cm^{-3}, $\mu = 16.94$ cm^{-1}. $R = 0.063$, $R_w = 0.059$. $C_{14}H_{36}N_7Cl_5O_9Pd_3$: orthorhombic, space group *Pcab*, $a = 13.125(7)$, $b = 13.213(3)$, $c = 33.570(5)$ Å, $V = 5822(3)$ Å^3, $Z = 8$, $D_{calc} = 2.15$ g cm^{-3}, $\mu = 21.20$ cm^{-1}. $R = 0.074$, $R_w = 0.061$. In very acidic solutions the polyammonium cations (H$_k$[3*k*]aneN$_k$)$^{k+}$ interact with $PdCl_4^{2-}$ forming second sphere coordinated species. These reactions have been followed by a microcalorimetric technique in 2 mol dm^{-3} HCl solutions. The slowness of the reactions of (H$_{10}$[30]aneN$_{10}$)$^{10+}$ with $PdCl_4^{2-}$ has been interpreted in terms of inclusion of the anion into the receptor's cavity as shown by the crystal structure of [(PdCl$_4$)(H$_{10}$[30]aneN$_{10}$)][PdCl$_4$]$_2$Cl$_4$: triclinic, space group $P\bar{1}$, $a = 7.760(3)$, $b = 11.448(4)$, $c = 13.399(11)$ Å, $\alpha = 96.31(8)°$, $\beta = 104.50(6)°$, $\gamma = 92.30(3)°$, $Z = 1$. $R = 0.046$ and $R_w = 0.039$.

Key words. Macrocyclic chemistry of Pd^{2+}, deprotonation of secondary amino group, supramolecular chemistry, anion coordination chemistry, anion inclusion, crystal structures.

Supplementary Data. relating to this article are deposited with the British Library as Supplementary Publication No. SUP 82117 (45 pages).

1. Introduction

Since 1967, when Pedersen's first papers appeared [1, 2] reporting that alkali metal ions bind crown ethers to form highly structured complexes, the field of synthetic macrocyclic compounds has undergone spectacular growth. The growth has largely been due to the synthesis of a great number and variety of synthetic macrocycles

[#] This paper is dedicated to the memory of the late Dr C. J. Pedersen.
* Authors for correspondence.

showing remarkably selective behavior [3]. This field has attracted scientists working in many areas, including reaction catalysis, transport processes, industrial applications, model systems and others. A variety of crown ethers have been shown to bind ammonium groups as well as metal–amine complexes giving supramolecular species with second sphere coordination [4]. Among synthetic macrocyles polyazacycloalkanes or azacrowns, in which nitrogen donor atoms have replaced oxygen donor atoms of the analogous crown ethers, have been much studied. Many papers have been published especially dealing with tetraaza macrocycles [5]. More recently much attention has been devoted to azamacrocycles having more than six nitrogen donor atoms (*large polyazacycloalkanes*). The possibility that these compounds bind to more than one metal ion in the macrocyclic framework, as well as the recent development of anion coordination chemistry justify this interest [6–8]. Indeed polyammonium species formed by these macrocycles can both form adducts with anionic complexes and selectively bind nucleotide anions, catalyzing their dephosphorylation [9, 10]. In the former case the chemical properties of the supramolecular species have been studied in an attempt to understand whether the coordination of the complexed anions takes place inside or outside the macrocyclic cavity [7, 11].

In the present paper we report some studies on Pd(II) ion which is able to form both polynuclear cationic complexes and anionic 'supercomplexes' with polyazacycloalkanes of different size [12, 13]. The polyazamacrocycles employed in the present study are the $[3k]aneN_k$ $(k = 6–11)$ ligands.

2. Experimental

2.1. CRYSTAL PREPARATION

Yellow crystals of $[Pd_2([18]aneN_6)Cl_2](ClO_4)_2$ were obtained by mixing aqueous solutions of $[18]aneN_6$ and $[PdCl_4]^{2-}$. Crystals of $[Pd_3([21]aneN_7)Cl_3][ClO_4]_2 \cdot H_2O$ were obtained in the same way. Crystals of $[(PdCl_4)(H_{10}[30]aneN_{10})](PdCl_4)_2Cl_4$ suitable for X-ray analysis, were obtained by slow evaporation, at room temperature, of a solution containing $[30]aneN_{10} \cdot 10\,HCl$ (1 mmol) and K_2PdCl_4 (3 mmol) in 50 mL of HCl 2 mmol dm^{-3}.

2.2. CRYSTAL DATA AND STRUCTURES REFINEMENT

All intensity measurements were done on an Enraf-Nonius CAD4 automatic four-circle diffractometer with graphite monochromated MoK_α radiation ($\lambda = 0.7106$ Å).

For $C_{12}H_{30}N_6Cl_4O_8Pd_2$: $M = 741.01$, monoclinic, space group $C2/m$, $a = 10.876(2)$, $b = 18.117(2)$, $c = 7.043(2)$ Å, $\beta = 113.78(2)°$, $V = 1270(12)$ Å^3, $Z = 2$, $D_{calc} = 1.92$ g cm^{-3}, $\mu = 16.94$ cm^{-1}. 1272 Intensities were collected in the range $5 \leqslant 2\theta \leqslant 50°$ and corrected for Lorentz and polarization effects. An absorption correction was applied once the structure was solved [14]. The 1062 reflections having $I \geqslant 3\sigma(I)$ were considered observed and were used in the structure analysis. The structure was solved by the heavy-atom method, which gave the position of the palladium atom. Successive Fourier syntheses showed all non-hydrogen atoms. The refinement, performed by means of the full-matrix least-squares method [15], gave $R = 0.063$ and $R_w = 0.059$. Hydrogen atoms were included in calculated positions with an overall temperature factor U of 0.05 Å^2. Anisotropic temperature factors were used for all non-hydrogen atoms except for the oxygen atoms of the perchlorate anion, which shows some degree of disorder. Two different models were found for the perchlorate ion; therefore population parameters of 0.5 were used for these oxygen atoms.

For $C_{14}H_{36}N_7Cl_5O_9Pd_3$: $M = 942.93$, orthorhombic, space group $Pcab$ $a = 13.125(7)$, $b = 13.213(3)$, $c = 33.570(5)$ Å, $V = 5822(3)$ Å, $Z = 8$, $D_{calc} = 2.15$ g cm^{-3}, $\mu = 21.20$ cm^{-1}, 4275 reflections were collected in the range $5 \leqslant 2\theta \leqslant 45°$. The intensities were corrected for Lorentz and polarization effects, and an absorption correction was applied once the structure was solved [14]. The 1899 unique reflections having $I \geqslant 3\sigma(I)$ were considered observed and were used in the structure analysis. The structure was solved by the heavy-atom method, which gave the position of the palladium atoms. Successive Fourier syntheses showed all non-hydrogen atoms. Refinement was performed by means of the full-matrix least-squares method [15] to obtain a final R factor of 0.074 and $R_w = 0.061$. Hydrogen atoms were included in calculated positions with an overall temperature factor U of 0.05 Å^2. Isotropic temperature factors were used for all atoms except for the palladium and chlorine.

$[(PdCl_4)(H_{10}[30]aneN_{10})](PdCl_4)_2Cl_4$: $M = 1327.15$, triclinic, space group $P\bar{1}$, $a = 7.760(3)$, $b = 11.448(4)$, $c = 13.399(11)$ Å, $\alpha = 96.31(8)°$, $\beta = 104.50(6)°$, $\gamma = 92.30(3)°$, $Z = 1$. A total of 4061 reflections ($5 \leqslant 2\theta \leqslant 50°$) were collected of which 2956 having $I \geqslant 3\sigma(I)$ were used in the structure solution and refinement. The structure was solved by the heavy atom technique. A ΔF Fourier synthesis showed two different positions for the N(3) atom. The population parameters of these positions were refined. The rather high thermal parameters of the attached carbon atoms are indicative of an extensive disorder around the N(3) atom which affects the values of the bond angles. Final refinement with anisotropic thermal parameters for all non-hydrogen atoms converged at $R = 0.046$ and $R_w = 0.039$.

2.3. EMF MEASUREMENTS

All the potentiometric titrations were carried out at 298.15 K, by using the equipment that has been fully described [16]. The measurements have been performed in 0.5 mol dm^{-3} NaCl in order to avoid the dissociation of the tetrachloro complex $PdCl_4^{2-}$. Furthermore, competition between the azamacrocycles and the chloride ligands for Pd^{2+} favors the determination of the equilibrium constants of these macrocyclic Pd^{2+} complexes [17]. In order to reach the equilibrium in these complex formation reactions, several minutes were allowed to elapse between each emf reading. The protonation constants of the ligands [18]aneN_6 and [21]aneN_7, redetermined under these experimental conditions, are shown in ref. [18]. The computer program SUPERQUAD [19] has been used to process the data and determine the equilibrium constants.

2.4. MICROCALORIMETRY

The enthalpy changes associated to the interaction of $PdCl_4^{2-}$ with the cations $(H_k[3k]aneN_k)^{k+}$ have been determined by means of a BATCH 10700 LKB microcalorimeter. These reactions were carried out in 2 mol dm^{-3} HCl solutions as the formation of Pd^{2+} complexes were strongly competitive with $PdCl_4^{2-}$ second sphere interaction in less acidic solution. Ratios, $R = [(H_k[3k]aneN_k)^{k+}]/[PdCl_4^{2-}]$, varying from 0.2 to 5 were used. The heat of the reactions were corrected for the heat of dilution of the reagents by blank experiments.

2.5. SPECTROPHOTOMETRIC MEASUREMENTS

Electronic spectra were recorded on a Perkin Elmer Model Lambda 9 instrument.

3. Results and Discussion

3.1. METAL COMPLEXES

Crystal Structure of $[Pd_2([18]aneN_6)Cl_2][ClO_4]_2$ *and* $[Pd_3([21]aneN_7)Cl_3][ClO_4]_2 \cdot H_2O$

The structure of the compound $[Pd_2([18]aneN_6)Cl_2][ClO_4]_2$ consists of a dinuclear $[Pd_2([18]aneN_6)Cl_2]^{2+}$ cation and disordered uncoordinated perchlorate anions. In the dinuclear complex, which possess a $2/m$ symmetry, each palladium ion shows a square-planar geometry involving three nitrogen atoms and one chloride ion. Each

Table I. Positional parameters ($\times 10^4$) and their estimated standard deviations in parentheses for $(Pd_2[18]aneN_6)Cl_2(ClO_4)_2$.

Atom	x/a	y/b	z/c
Pd[a]	1526(1)	0	846(2)
Cl(1)[a]	1018(4)	0	3717(6)
Cl(2)[a]	5000	1852(3)	5000
N(1)[a]	2252(12)	0	−1329(17)
N(2)	1701(9)	1123(4)	718(16)
C(1)	3017(11)	699(7)	−1104(20)
C(2)	2106(14)	1286(7)	−919(23)
C(3)	675(11)	1632(6)	960(24)
O(1)[a]	4329(31)	1176(22)	4987(60)
O(2)[a]	4634(25)	1863(15)	2679(45)
O(3)[a]	4315(25)	2469(18)	4082(45)
O(4)[a]	3674(32)	1880(20)	4914(49)

[a]Atom with occupancy factor 0.5.

Table II. Selected bond distances (Å) and angles (°) for $(Pd_2[18]aneN_6)Cl_2(ClO_4)_2$.

Bond distances		Bond Angles	
Pd—N(1)	1.99(1)	N(2)—Pd—N(2′)	166.5(4)
Pd—N(2)	2.049(9)	N(1)—Pd—N(2)	83.9(3)
Pd—Cl(1)	2.301(5)	Cl(1)—Pd—N(2)	95.6(3)
		Cl(1)—Pd—N(1)	171.4(4)

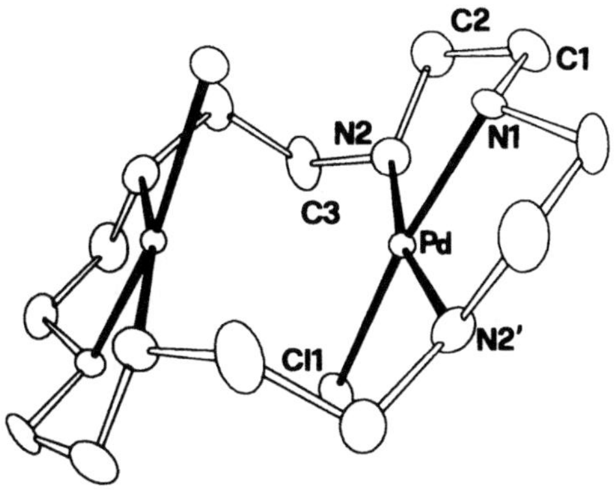

Fig. 1. ORTEP drawing of the $[Pd_2([18]aneN_6)Cl_2]^{2+}$cation.

metal ion lies 0.14 Å out of the plane identified by the donor atoms N(1), N(2), N(2′), Cl(1). The two palladium atoms are 3.04 Å apart and point towards each other with respect to the two planes containing the donor atoms (see Figure 1).

In $[Pd_3([21]aneN_7)Cl_3][ClO_4]_2 \cdot H_2O$ the ligand and three chlorine atoms bind the three palladium atoms in a square planar fashion. An amino group of the macrocyclic ligand, which has undergone deprotonation on complex formation, bridges two palladium atoms. The coordination planes containing Pd(1) and Pd(2) are nearly parallel, the angle between them being 1.0°. The third square plane containing N(3), N(4), N(5), Cl(3) and Pd(3) forms an angle of 92.0° with the other ones. The Pd(1) and Pd(2) atoms lie significantly out from their mean coordination planes, 0.084 Å and 0.048 Å, respectively, pointing towards each other (see Figure 2). The distance Pd(1)—Pd(2) [3.057(4) Å] is the shortest among the three contact interactions between the three metal atoms, the other being 3.232(4) Å for Pd(1)—Pd(3) and 3.478(4) Å for Pd(2)—Pd(3). The macrocyclic geometry is

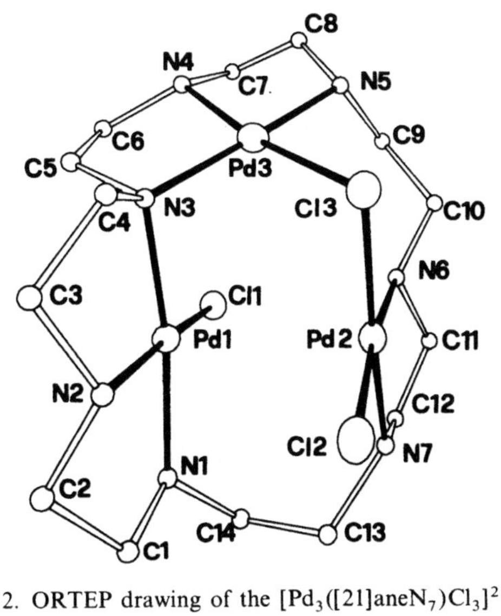

Fig. 2. ORTEP drawing of the $[Pd_3([21]aneN_7)Cl_3]^{2+}$ cation.

Table III. Positional parameters (× 10^4) with their estimated standard deviations in parentheses for $[Pd_3([21]aneN_7)Cl_3](ClO_4)_2 \cdot H_2O$.

Atom	x/x	y/a	z/c	Atom	x/a	y/b	z/c
Pd(1)	1594(2)	3162(2)	4152(1)	C(8)	5781(26)	−1685(28)	3204(11)
Pd(2)	4482(2)	2063(2)	3779(1)	N(5)	5089(24)	−940(23)	3325(9)
Pd(3)	664(2)	287(2)	1155(1)	C(9)	4957(29)	−241(29)	2981(11)
Cl(1)	−2120(6)	6694(7)	6516(3)	C(10)	4225(26)	670(25)	3052(10)
Cl(2)	946(6)	7731(7)	5619(3)	N(6)	4717(19)	1554(19)	3212(8)
Cl(3)	880(6)	5473(7)	5975(3)	C(11)	4581(26)	2460(24)	2970(9)
Cl(4)	657(10)	3141(10)	596(4)	C(12)	5015(28)	3375(27)	3142(11)
Cl(5)	1985(14)	4087(15)	2272(6)	N(7)	4738(19)	3417(18)	3541(7)
N(1)	6760(18)	3362(19)	4156(9)	C(13)	5330(22)	4269(24)	3765(10)
C(1)	6537(27)	3751(24)	4545(10)	C(14)	6397(24)	4088(25)	3831(10)
C(2)	6739(23)	2947(21)	4874(9)	O(1)	474(28)	2218(30)	428(11)
N(2)	6245(17)	2105(20)	4726(7)	O(2)	539(31)	3910(27)	342(11)
C(3)	6498(24)	1137(23)	4934(10)	O(3)	145(40)	3273(42)	912(15)
C(4)	6087(26)	285(27)	4676(10)	O(4)	1618(30)	3096(33)	655(12)
N(3)	6456(20)	313(20)	4276(8)	O(5)	2425(34)	4814(36)	2474(16)
C(5)	7454(25)	−196(25)	4281(10)	O(6)	1210(39)	3705(40)	2357(16)
C(6)	7786(30)	−666(32)	3900(12)	O(7)	1809(48)	4188(46)	1889(17)
N(4)	6858(21)	−1129(20)	3706(10)	O(8)	−2792(34)	6457(35)	7758(14)
C(7)	6759(28)	−1394(30)	3284(12)	O(9)	3426(19)	3008(19)	1233(8)

determined by the coordination spheres around the palladium atoms, which are able to arrange the nitrogen donor atoms on the three different coordination planes. The Pd—N distances in the present structures are very similar to those found in other reported structures of Pd^{2+} complexes with saturated tri- and tetraazamacrocycles [20], although some shorter distances have been observed.

3.1.2. *Solution Chemistry*

The formation of Pd^{2+} complexes by the macrocyclic ligands $[18]aneN_6$ and $[21]aneN_7$ has been followed in solution by means of potentiometric and spectrophotometric techniques. For $[18]aneN_6$ mono- and dinuclear species have been observed while for $[21]aneN_7$ mono-, di- and trinuclear species are formed. The logarithms of the stability constants determined for these systems are presented in Table VII. The electronic spectra indicate that Pd^{2+} is coordinated in these complexes in a square planar mode. Therefore, in the mononuclear complexes, two nitrogens of $[18]aneN_6$ and three of $[21]aneN_7$ remain uncoordinated. The decrease of the stability constants when passing from $[Pd([18]aneN_6)]^{2+}$ [$\log K = 29.2$] to $[Pd([21]aneN_7)]^{2+}$ [$\log K = 24.5(1)$] can be attributed to the presence of a larger chelate ring in the latter which lowers its stability. Furthermore, two protonated complexes for $[18]aneN_6$ and three for $[21]aneN_7$ have been found, in agreement with the number of uncoordinated nitrogens in each complex.

The molar conductivity of pure water solutions containing $[Pd_2([18]aneN_6)Cl_2][ClO_4]_2$ in the dilution range 10^{-1}–10^{-4} mol dm^{-3} showed that no appreciable dissociation of the chloride ligands takes place in the cation

Table IV. Selected bond distances (Å) and angles (°) for $[Pd_3([21]aneN_7)Cl_3](ClO_4)_2 \cdot H_2O$.

Bond Distances				Bond Angles			
Pd(1)—Cl(1)	2.3556(88)	Pd(2)—N(6)	2.0422(254)	N(2)—Pd(1)—N(3)	87.61(1.)	Cl(2)—Pd(2)—N(7)	92.84(0.74)
Pd(1)—N(1)	2.0249(250)	Pd(2)—N(7)	1.9887(243)	N(1)—Pd(1)—N(3)	167.99(1.10)	Cl(2)—Pd(2)—N(6)	173.04(0.75)
Pd(1)—N(2)	2.0105(237)	Pd(3)—Cl(3)	2.3407(89)	N(1)—Pd(1)—N(2)	81.04(1.08)	Cl(2)—Pd(2)—Cl(3)	89.15(0.33)
Pd(1)—N(3)	2.0657(260)	Pd(3)—N(3)	1.9521(263)	Cl(1)—Pd(1)—N(3)	97.98(0.76)	N(4)—Pd(3)—N(5)	81.83(120)
Pd(2)—Cl(2)	2.2756(92)	Pd(3)—N(4)	1.9769(277)	Cl(1)—Pd(1)—N(2)	173.42(0.75)	N(3)—Pd(3)—N(5)	168.78(1.17)
Pd(2)—Cl(3)	2.3063(94)	Pd(3)—N(5)	2.0874(299)	Cl(1)—Pd(1)—N(1)	93.13(0.79)	N(3)—Pd(3)—N(4)	88.88(1.13)
				N(6)—Pd(2)—N(7)	84.02(1.01)	Cl(3)—Pd(3)—N(5)	94.65(0.89)
				Cl(3)—Pd(2)—N(7)	176.62(0.76)	Cl(13)—Pd(3)—N(4)	171.18(0.83)
				Cl(3)—Pd(2)—N(6)	93.70(0.75)	Cl(3)—Pd(3)—N(3)	95.42(0.80)

Table V. Positional parameters ($\times 10^4$) for $[C_{20}H_{60}N_{10}(PdCl_4)](PdCl_4)_2Cl_4$ with their e.s.d. in parentheses.

Atom	x/a	y/b	z/c	Atom	x/a	y/b	z/c
Pd(1)	0	5000	0	N(31)[b]	−931(47)	1952(30)	6031(27)
Pd(2)	0	0	0	N(4)	−2153(10)	4999(6)	6655(6)
Pd(3)	5000	0	5000	N(5)	−4277(10)	7250(7)	7244(5)
Cl(1)	2782(3)	5280(2)	1119(2)	C(1)	5816(13)	3041(8)	11417(7)
Cl(2)	1269(3)	4926(2)	−1376(2)	C(2)	4572(11)	2651(7)	9437(6)
Cl(3)	73(3)	−1963(2)	226(2)	C(3)	5098(11)	2002(8)	8544(7)
Cl(4)	2683(3)	−55(2)	−417(2)	C(4)	1963(13)	1547(10)	7415(8)
Cl(5)	3630(4)	−175(2)	3251(2)	C(5)	849(15)	1371(11)	6292(9)
Cl(6)	4376(4)	1935(2)	5167(2)	C(6)	−1139(15)	3038(11)	6404(8)
Cl(7)	3674(3)	4634(2)	6462(2)	C(7)	−1324(13)	4196(8)	6006(7)
Cl(8)	1690(4)	8164(3)	6316(2)	C(8)	−2067(14)	6234(8)	6424(8)
N(1)	6094(9)	2595(6)	10381(5)	C(9)	−2381(12)	7118(8)	7278(7)
N(2)	3834(10)	2113(7)	7534(6)	C(10)	−4475(12)	7835(9)	8264(7)
N(3)[a]	342(14)	2418(10)	5890(9)				

[a]Occupancy factor 0.76.
[b]Occupancy factor 0.24.

$[Pd_2([18]aneN_6)Cl_2]^{2+}$. This result agrees with the fact that the electronic spectra of solid $[Pd_2([18]aneN_6)Cl_2][ClO_4]_2$ and its solution (ε 1330 at $\lambda = 342$ nm, in 0.5 mol dm^{-3} NaCl) are essentially the same. The very high stability of the complex $[Pd_2([18]aneN_6)Cl_2]^{2+}$ ($\log K = 51.8$), which is formed in very acidic solutions, is accompanied by a marked kinetic inertness towards acid dissociation. Some months are required for the complete dissociation of $[Pd_2([18]aneN_6)Cl_2]^{2+}$ in 2 mol dm^{-3} HCl solution. Also $[21]aneN_7$ forms a very stable dinuclear species with estimated stoichiometry $[Pd_2([21]aneN_7)Cl]^{3+}$ ($\log K > 52$). This complex can react with a further $PdCl_4^{2-}$ ion forming a trinuclear species whose electronic spectrum (ε 2400 at $\lambda = 377$ nm, in 0.5 mol dm^{-3} NaCl) is quite similar to the reflectance spectrum of the $[Pd_3([21]aneN_7)Cl_3][ClO_4]_2 \cdot H_2O$ solid compound. The titration curves show that the binding of the third Pd^{2+} causes the release of a further equivalent of H^+ ion. These results suggest that the overall structure of $[Pd_3([21]aneN_7)Cl_3]^{2+}$ observed in the solid state, in which a nitrogen deprotonates to bridge two Pd^{2+} ions, is also likely to be maintained in solution.

3.2. ANION COORDINATION

3.2.1. *Crystal Structure of* $[(PdCl_4)(H_{10}[30]aneN_{10})](PdCl_4)_2Cl_4$

The crystal structure (see Figures 3 and 4) consists of $[(PdCl_4)(H_{10}[30]aneN_{10})]^{8+}$ cations, where the $(PdCl_4)^{2-}$ anion (**A**) is enclosed into the $(H_{10}[30]aneN_{10})^{10+}$ cavity, and of isolated $(PdCl_4)^{2-}$ (**B**), (**C**) and Cl^- counterions. All the tetrachloropalladate(II) anions are centrosymmetric, the metal atom residing on an inversion centre in each case. The structural features of $(PdCl_4)^{2-}$ are not influenced by the inclusion into the decacharged $(H_{10}[30]aneN_{10})^{10+}$ macrocycle, as shown by the small deviation

Table VI. Bond distances (Å) and angles (°) for $[(PdCl_4)(H_{10}[30]aneN_{10})](PdCl_4)_2Cl_4$ with their e.s.d. in parentheses.

Bond distances				Bond Angles			
Pd(1)—Cl(1)	2.287(2)	N(3)—C(6)	1.629(17)	Cl(1)—Pd(1)—Cl(2)	89.63(0.09)	N(1)—C(2)—C(3)	106.21(0.68)
Pd(1)—Cl(2)	2.293(3)	N(31)—C(5)	1.536(38)	Cl(3)—Pd(2)—Cl(4)	90.08(0.08)	N(2)—C(3)—C(2)	112.46(0.71)
Pd(2)—Cl(3)	2.301(2)	N(31)—C(6)	1.319(37)	Cl(5)—Pd(3)—Cl(6)	90.13(0.09)	N(2)—C(4)—C(5)	112.77(0.85)
Pd(2)—Cl(4)	2.288(3)	N(4)—C(7)	1.471(13)	C(1)—N(1)—C(2)	117.44(0.67)	C(3)—N(2)—C(4)	113.85(0.70)
Pd(3)—Cl(5)	2.300(2)	N(4)—C(8)	1.483(12)	N(3)—C(5)—C(4)	114.02(1.05)	N(31)—C(5)—C(4)	118.20(1.56)
Pd(3)—Cl(6)	2.285(2)	N(5)—C(9)	1.475(12)	C(5)—N(3)—C(6)	112.25(0.94)	N(31)—C(6)—C(7)	136.44(1.75)
N(1)—C(1)	1.496(12)	N(5)—C(10)	1.501(13)	C(5)—N(31)—C(6)	123.53(2.70)	N(3)—C(6)—C(7)	104.80(0.85)
N(1)—C(2)	1.508(10)	C(2)—C(3)	1.491(13)	C(7)—N4—C8	113.06(0.72)	N(4)—C(7)—C(6)	110.40(0.79)
N(2)—C(3)	1.481(10)	C(4)—C(5)	1.522(14)	C(9)—N(5)—C(10)	110.87(0.70)	N(4)—C(8)—C(9)	113.38(0.75)
N(2)—C(4)	1.529(13)	C(6)—C(7)	1.482(16)	N(4)—C(9)—C(8)	114.48(0.79)		
N(3)—C(5)	1.399(17)	C(8)—C(9)	1.515(14)				

Table VII. Logarithms of the equilibrium constants of Pd^{2+} complexes with [18]aneN_6 and [21]aneN_7 determined in 0.5 mol dm^{-3} NaCl solution at 298.15 K.

Reaction	[18]aneN_6	[21]aneN_7
	log *K*	
$Pd^{a} + L$	29.2(1)[b]	24.55(6)
Pd + L + H	37.47(7)	34.52(6)
Pd + L + 2H	42.40(7)	42.63(4)
Pd + L + 3H		47.13(4)
2Pd + L + 2Cl	51.8(8)	
2Pd + L + Cl		> 52
PdL + H	8.3	10.0
PdLH + H	4.9	8.1
$PdLH_2 + H$		4.5

[a]Charges have been omitted for simplicity.
[b]Values in parentheses are standard deviations on the last significant figure.

from orthogonality of the Cl(1)—Pd(1)—Cl(2) angle. Furthermore Pd—Cl bond distances differ by less than 0.016 Å for the three anions. The decaprotonated macrocycle is centrosymmetric, the centre of symmetry being coincident with that of the $(PdCl_4)^{2-}$ anion. The macrocycle displays an S-shaped conformation which produces an elongated macrocyclic cavity, of approximate dimensions 9 × 11 Å (see Figure 3). The plane of the tetrachloropalladate(II) anion is perpendicular to the direction of main elongation of the cavity, with the chlorine atoms standing out of the macrocyclic frame. Such arrangement allows the chlorine atom Cl(2) of (**A**) (see Figure 3) to interact via a hydrogen bond with the N(4) protonated nitrogen atom of the receptor [Cl(2)···HN(4) 2.308(7) Å]. The $(PdCl_4)^{2-}$ anion (**A**) further interacts via hydrogen bonds with the nitrogen atoms of a symmetry related supercomplexed species [Cl(1)···HN(4) 2.411(9) Å and Cl(2)···HN(1) 2.611(7) Å]. Hydrogen bond interactions with the macrocycle are also formed by the chlorine atoms of the $(PdCl_4)^{2-}$ anion (**B**) [Cl(3)···HN(1) 2.699(8) Å and Cl(4)···HN(1) 2.024(8) Å], as well as by those of (**C**) [Cl(5)···HN(2) 2.415(8) Å, Cl(6)···HN(2) 2.399(8) Å and Cl(6)···HN(4) 2.589 Å]. The hydrogen bond framework is completed by several other interactions involving the chloride counterions Cl(7) and Cl(8).

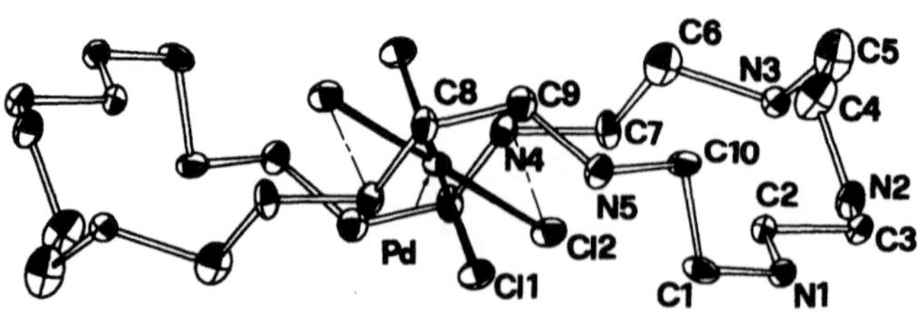

Fig. 3. ORTEP drawing of the $[(PdCl_4)(H_{10}[30]aneN_{10})]^{8+}$ 'super complex'.

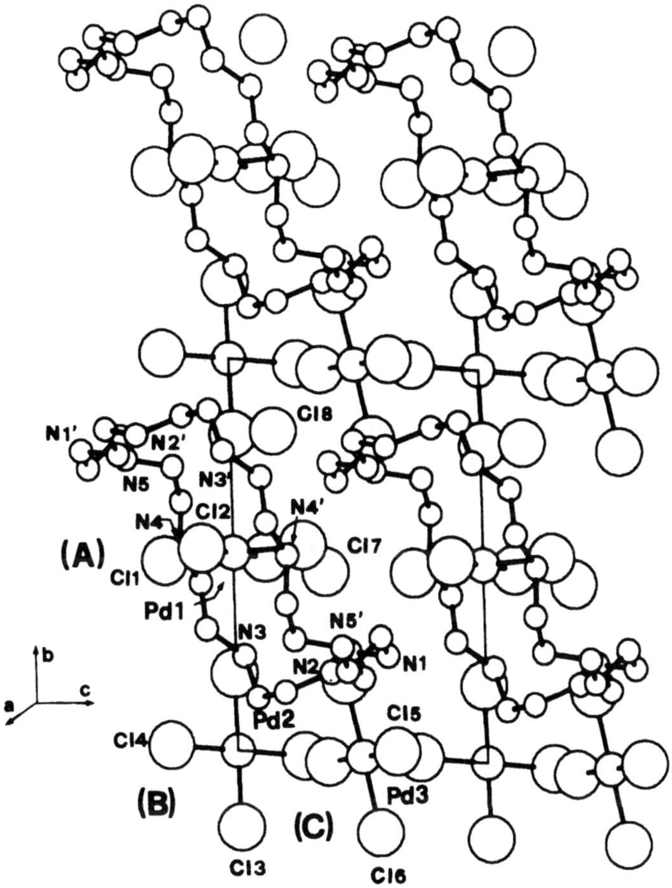

Fig. 4. Crystal packing of $[(PdCl_4)(H_{10}[30]aneN_{10})](PdCl_4)_2Cl_4$.

3.2.2. *Solution Chemistry*

The interaction between $PdCl_4^{2-}$ and the polyammonium cations $(H_k[3k]aneN_k)^{k+}$ has been studied by direct microcalorimetry in 2 mol dm^{-3} HCl solutions. These experimental conditions have been chosen in order to avoid dissociation of the $PdCl_4^{2-}$ anion and to prevent the formation of palladium(II) complexes of the macrocyclic ligands. All the reactions are exothermic. In the cases of $(H_6[18]aneN_6)^{6+}$ and $(H_7[21]aneN_7)^{7+}$, the reactions are fast for all the experimental ratios, $R = [(H_k(3k]aneN_k)^{k+}]/[PdCl_4^{2-}]$, between 0.2 and 5. The enthalpy changes for both reactions are the same within experimental error ($\Delta H^\circ = -1.5(1)$ kcal mol^{-1}). The reaction of $(H_8[24]aneN_8)^{8+}$ with $PdCl_4^{2-}$ is also fast, with a $\Delta H^\circ = -1.6(1)$ kcal mol^{-1} for $1 \leqslant R \leqslant 5$. However for $0.2 \leqslant R \leqslant 1$, a slight increment of the reaction enthalpy is observed. For $1 \leqslant R \leqslant 5$ the three larger polyammonium cations $(H_9[27]aneN_9)^{9+}$, $(H_{10}[30]aneN_{10})^{10+}$ and $(H_{11}[33]aneN_{11})^{11+}$ react slowly with enthalpy changes of $-2.9(1)$, $-3.9(1)$ and $-3.1(1)$ kcal mol^{-1} respectively. In these cases, remarkable increases of the reaction heats were observed for R values lower than 1. The reactions of $PdCl_4^{2-}$ with

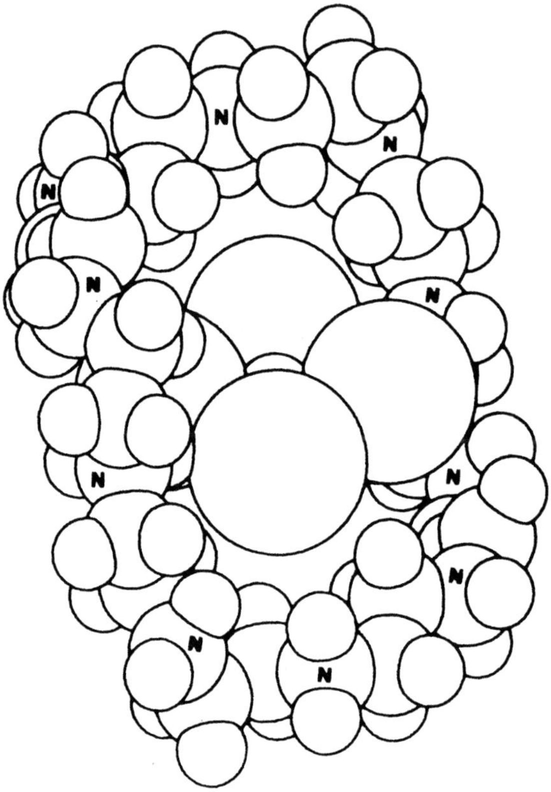

Fig. 5. Space-filling representation of the $[(PdCl_4)(H_{10}[30]aneN_{10})]^{8+}$ cation.

$[(PdCl_4)(H_k[3k]aneN_k)^{(k-2)+}$ ($k = 8-11$) are fast. These results suggest that the two smaller polyammonium receptors $(H_6[18]aneN_6)^{6+}$ and $(H_7[21]aneN_7)^{7+}$ form just 1 : 1 species while, $(H_8[24]aneN_8)^{8+}$ and even more clearly $(H_9[27]aneN_9)^{9+}$, $(H_{10}[30]aneN_{10})^{10+}$ and $(H_{11}[33]aneN_{11})^{11+}$ interact with more than one $PdCl_4^{2-}$ anion. The slowness of the reactions of the larger macrocycles can be explained in terms of the structure of the solid compound $[(PdCl_4)(H_{10}[30]aneN_{10})](PdCl_4)_2Cl_4$ (see Figure 5). In this structure a $PdCl_4^{2-}$ anion is included in the cavity of the decaprotonated cation $(H_{10}[30]aneN_{10})^{10+}$ being disposed along the shortest axes of the macrocycle. If this inclusion is also maintained in solution the conformational energy of the receptor, needed to accommodate the guest species inside the cavity so as to form a compact structure, would be greater. Indeed, the reactions of $PdCl_4^{2-}$ with the polyammonium receptors $(H_6[18]aneN_6)^{6+}$ and $(H_7[21]aneN_7)^{7+}$, whose cavities are too small to include $PdCl_4^{2-}$, are fast. If the observed slowness is indicative of inclusion we may conclude that $(H_9[27]aneN_9)^{9+}$ is the smallest receptor able to incorporate $PdCl_4^{2-}$ inside the cavity. For $(H_{11}[33]aneN_{11})^{11+}$ a faster reaction was observed; this can be explained in terms of the greater cavity size allowing the anion to be included more snugly.

4. Conclusions

Polyazacycloalkanes have been shown to be good receptors both for transition metal ions and anionic species. In the present paper we have presented the case of a substrate ($PdCl_4^{2-}$) able, as a function of pH, to interact with these receptors providing either covalent or non-covalent interactions. The second sphere coordination takes place for those acid pH values where the macrocyclic nitrogens do not compete with the chloride anions for the direct coordination to the metal. When the pH is raised, the nitrogen donors of the macrocycle remove, at least partially, the chloride anions from the first coordination sphere to interact directly with the Pd^{2+} ions. For the first time, the inclusion of a complex anion inside the cavity of a macrocyclic receptor has been undoubtedly provided by X-ray analysis. On the other hand, the matching between a metal ion presenting rigid stereochemical requirements and ligands constrained by their cyclic topology results in some rare chemical events such as the deprotonation of an amine group to bridge two metal ions, that has been observed for the first time.

Notes and References

1. C. J. Pedersen: *J. Am. Chem. Soc.* **89**, 2495 (1967).
2. C. J. Pedersen: *J. Am. Chem. Soc.* **89**, 7017 (1967).
3. G. A. Melson: *Coordination Chemistry of Macrocyclic Compounds*; Plenum Press, New York (1979). (b) M. Hiraoka: *Crown Compounds. Their Characteristics and Applications*, Elsevier, Amsterdam (1982). L. F. Lindoy: *The Chemistry of Macrocyclic Ligand Complexes*, Cambridge University Press (1989).
4. H. M. Colquhoun, J. F. Stoddart and D. J. Williams: *Angew. Chem. Int. Ed. Engl.* **25**, 487 (1986).
5. R. M. Izatt, J. D. Bradshaw, S. A. Nielsen, J. D. Lamb and J. J. Christensen: *Chem. Rev.* **85**, 271 (1985).
6. (a) A. Bianchi, S. Mangani, M. Micheloni, V. Nanini, P. Orioli, P. Paoletti and B. Seghi: *Inorg. Chem.* **24**, 1182 (1985). (b) M. Micheloni, P. Paoletti and A. Bianchi: *Inorg. Chem.* **24**, 3702 (1985). (c) A. Bencini, A. Bianchi, E. Garcia-España, M. Giusti, M. Micheloni and P. Paoletti: *Inorg. Chem.* **26**, 681 (1987). (d) A. Bencini, A. Bianchi, E. Garcia-España, M. Giusti, S. Mangani, M. Micheloni, P. Orioli and P. Paoletti: *Inorg. Chem.* **26**, 1243 (1987). (e) A. Bencini, A. Bianchi, E. Garcia-España, S. Mangani, M. Micheloni, P. Orioli and P. Paoletti: *Inorg. Chem.* **27**, 1104 (1988). (f) A. Bencini, A. Bianchi, P. Dapporto, E. Garcia-España, M. Micheloni and P. Paoletti: *Inorg. Chem.* **28**, 1188 (1989). (g) A. Bencini, A. Bianchi, E. Garcia-España, M. Micheloni and P. Paoletti: *Inorg. Chem.* **28**, 2480 (1989). (h) A. Bencini, A. Bianchi, M. Castello, M. Di Vaira, J. Faus, E. Garcia-España, M. Micheloni and P. Paoletti: *Inorg. Chem.* **28**, 347 (1989). (i) A. Bencini, A. Bianchi, M. Castello, P. Dapporto, J. Faus, E. Garcia-España, M. Micheloni, P. Paoletti and P. Paoli: *Inorg. Chem.* **28**, 3175 (1989).
7. (a). A. Bencini, A. Bianchi, E. Garcia-España, M. Giusti, S. Mangani, M. Micheloni, P. Orioli and P. Paoletti: *Inorg. Chem.* **26**, 3902 (1987). (b) A Bianchi, M. Micheloni, P. Orioli, P. Paoletti and S. Mangani: *Inorg. Chim. Acta.* **146**, 153 (1988) (c) E. Garcia-España, M. Micheloni, P. Paoletti and A. Bianchi: *Inorg. Chim. Acta* **L9**, 102 (1985).
8. K. B. Mertes and J. M. Lehn: *Comp. Coord. Chem.* **2**, 915 (1987) and references therein.
9. M. W. Hosseini, J. M. Lehn, L. Maggiora, K. B. Mertes and M. P. Mertes: *J. Am. Chem. Soc.* **109**, 537 (1987).
10. A. Bencini, A. Bianchi, E. Garcia-España, E. C. Scott, L. Morales, B. Wang, M. P. Mertes, K. B. Mertes and P. Paoletti: *Bioorg. Chem.*, in press.
11. (a) F. Pina, L. Moggi, M. F. Manfrin, V. Balzani, M. W. Hosseini and J. M. Lehn: *Gazz. Chim. Ital* **119**, 65 (1989). (b) M. F. Manfrin, L. Moggi, V. Castelvetro, V. Balzani, M. W. Hosseini and J. M. Lehn: *J. Am. Chem. Soc.* **107**, 6888 (1985).

12. A. Bencini, A. Bianchi, P. Dapporto, E. Garcia-España, M. Micheloni, P. Paoletti and P. Paoli: *J. Chem. Soc., Chem. Commun.* 1382 (1990).
13. A. Bencini, A. Bianchi, P. Dapporto, E. Garcia-España, M. Micheloni, P. Paoletti and P. Paoli: *J. Chem. Soc., Chem. Commun.* **753** (1990).
14. N. Walker, D. D. Stuart: *Acta Crystallogr., Sect. A.* **39**, 158 (1983).
15. G. M. Sheldrick: SHELX-76, *A Program for Crystal Structure Determination*, University of Cambridge: Cambridge England, 1976.
16. A. Bianchi, L. Bologni, P. Dapporto, M. Micheloni, P. Paoletti: *Inorg. Chem.* **23**, 1201 (1984).
17. The equilibrium constants for the formation of Pd^{2+}-chloro complexes have been taken from: W. F. Rittner, A. Gulko, and G. Schmukler: *Talanta* **17**, 807 (1970).
18. Basicity constants for the ligands [18]aneN_6 and [21]aneN_7 determined in 0.5 mol dm^{-3} NaCl at 298.15 K: [18]aneN_6, $\log K_1 = 10.10$, $\log K_2 = 9.67$, $\log K_3 = 8.88$, $\log K_4 = 5.17$, $\log K_5 = 2.55$, $\log K_6 = 1.77$; [21]aneN_7, $\log K_1 = 9.88$, $\log K_2 = 9.45$, $\log K_3 = 8.94$, $\log K_4 = 7.26$, $\log K_5 = 4.99$, $\log K_6 = 2.41$, $\log K_7 = 1.87$.
19. P. Gans, A. Sabatini and A. Vacca: *J. Chem. Soc., Dalton Trans.* 1195 (1985).
20. (a) K. Toriumi, M. Yamashita, H. Ito and T. Ito: *Acta Crystallogr.* **C42**, 963 (1986). (b) A. Blake, R. O. Gould, T. I. Hyde, M. Schroder: *J. Chem. Soc., Chem. Commun.* 431 (1987) (c) A. Blake, L. M. Gordon, A. J. Holder, T. I. Hyde, G. Reid and M. Schroder: *J. Chem. Soc., Chem. Commun.* 1452 (1988). (d) P. V. Berndardt, G. A. Lawrance, W. C. Patalinghug, B. W. Skelton, A. H. White, N. F. Curtis and A. Siriwardena: *J. Chem. Soc., Dalton Trans.* 2853 (1990).

Journal of Inclusion Phenomena and Molecular Recognition in Chemistry **12**: 305–312, 1992.

The Crystal Structure of Dibenzo-14-Crown-4, A Preorganized Basis of Square Pyramid Coordination for Lithium Ions[#]

N. KENT DALLEY* and WEIMING JIANG
Department of Chemistry, Brigham Young University, Provo, Utah, 84602, U.S.A.

URIEL OLSHER
Department of Chemical Services, The Weizmann Institute of Science, Rehovot 76100, Israel.

(Received: October 25, 1990, in final form: 7 January 1991)

Abstract. Dibenzo-14-crown-4 (DB14C4) has a high selectivity for Li^+. The rigidity of the molecule caused by the two benzene rings suggests that this is a preorganized ligand for metal ions and particularly for Li^+. A single crystal structure study of the molecule was performed. The crystal data are: space group $P2_1$, $a = 12.811(2)$, $b = 5.106(1)$, $c = 12.816(3)$Å, $\beta = 115.44(1)°$, $V = 757.0(2)$Å^3 with $Z = 2$. The structure was refined to $R = 0.049$ *and* $R_w = 0.058$ using 1804 unique data with $F < 4\sigma(F)$. The conformation of the free ligand was found to be similar to that of the complexed ligand and the conformational parameters of the free ligand and its derivatives are compared to Li^+ complexes of the ligand and its derivatives.

Key words. X-ray crystal structure, dibenzo-14-crown-4, preorganized ligand, Li^+ binding.

Supplementary Data relating to this article are deposited with the British Library as Supplementary Publication No. SUP 82126 (12 pages).

1. Introduction

Previously it was shown that small crown ethers with a ring size of 12–16 atoms containing four ether oxygens exhibit high selectivity toward Li^+ [1–8]. The fourteen membered ring crown-ethers dibenzo-14-crown-4 [3] and 14-crown-4 [4–6] exhibit the highest selectivity towards Li^+ among all the crown ethers which have been tested. Molecular structure studies of Li^+ complexes with crown ethers [9–19], linear related compounds [20], cryptands [21] and naturally occurring macromolecules [22, 23] reveal that Li^+ occurs in various coordinations (tetra, penta, hexa, hepta and octa), and the Li^+—ligand bond distances vary from 1.86 to 2.37Å [24]. In many cases the Li^+ is five coordinated and the most favorable coordination geometry is that of a square pyramid (SP) in which the macrocyclic ring provides the basis for a square pyramid, while the apical site is occupied by the anion [9, 10, 12–19, 24] or solvent molecule [11, 19, 22, 23, 24].

The most effective ion-dipole interactions in five-fold coordination geometry can be achieved in an arrangement of a square pyramid (SP) or a trigonal bipyramid (TBP) configuration of the dipoles around the Li^+ [25, 26] (see Figure 1). The

[#] This paper is dedicated to the memory of the late Dr C. J. Pedersen.
*Author for correspondence

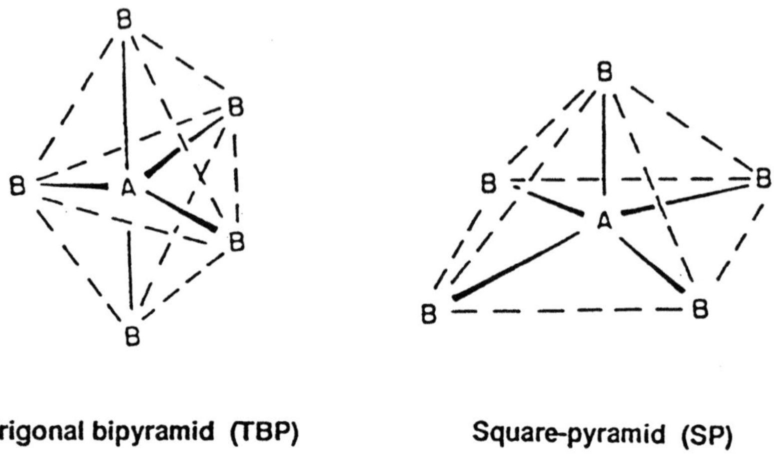

Fig. 1. Common types of five-fold coordination.

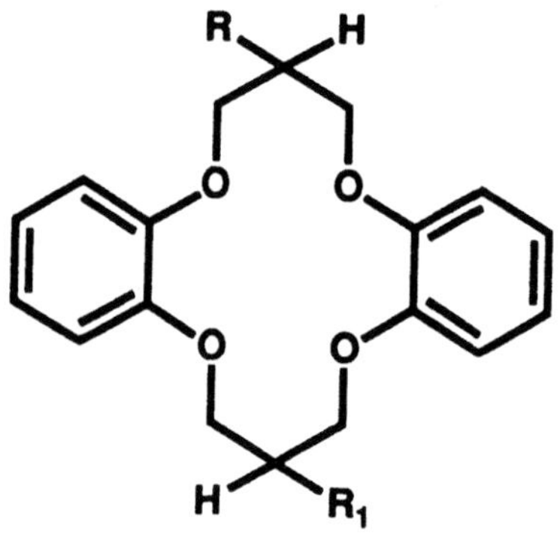

	R	R_1
1	H	H
2	OH	H
3	OH	OH (trans)
4	CH_2OH	H
5	OCH_2CO_2H	H

Fig. 2. A schematic drawing of compounds **1**–**5**.

square pyramid conformation is dictated by the rigidity and the conformation of the macrocyclic ring.

The dibenzo-14-crown-4 (DB14C4) (1) (see Figure 2) molecule provides an 'ideal base' of square-pyramidal coordination in LiNCS [9], LiPicrate, $LiClO_4$ and LiI [19, 24] complexes. The goal of this research is to find out whether DB14C4 is a preorganized basis of square pyramid coordination for lithium ions or whether conformational changes take place in DB14C4 upon complexation.

2. Experimental

The synthesis of DB14C4 has been described by Pedersen [27]. Single crystals of DB14C4 were grown from a 1:1 (volume) solution of CH_2Cl_2 and methanol. Crystal and intensity data were obtained using a Nicolet R3 automated diffractometer which utilized graphite monochromated MoK_α radiation ($\lambda = 0.71073$Å). The lattice parameters and orientation matrix were obtained using a least-squares procedure

Table I. Crystal data and experimental conditions.

Formula	$C_{18}H_{20}O_4$
Formula Weight	300.3
F(000)	320
Crystal Size (mm)	0.25 × 0.30 × 0.40
μ (cm^{-1})	0.86
Space group	$P2_1$
a(Å)	12.811(2)
b(Å)	5.106(1)
c(Å)	12.816(3)
β(deg)	115.44(1)
V(Å^3)	757.0
Z	2
D_x(g/cc)	1.32
Max sin θ/λ	0.70
Observed unique data	1804
Unobserved data($F < 4\sigma(F)$)	524
R	0.049
R_w	0.058
G (in weights)	7.8×10^{-4}
$GOOF$	1.33
largest peaks in difference maps (eÅ^{-3})	0.21, −0.18

involving 25 carefully centered data ($11.0° < 2\theta < 27.7°$). A cell reduction calculation yields the result that the unit cell could be described as C-centered orthorhombic but the diffraction symmetry ruled out that possibility. Intensity data were collected using a variable scan rate (4 to 60 deg/min) $\theta - 2\theta$ scanning mode. Crystal data and experimental information are summarized in Table I.

The trial structure was obtained using direct methods. All non-hydrogen atoms were refined anisotropically. Positions for hydrogen atoms were obtained from difference maps. The hydrogen atoms were allowed to ride on the carbon to which they are bonded with a C—H distance of 0.96Å during the refinement procedure and only their isotropic thermal parameters were refined. The R values were $R = 0.049$ and $R_w = 0.058$. Weights were based on counting statistics and were of the form $w = 1/\sigma^2(F) + GF^2$ with G being refined to bring the goodness of fit value to approximately 1.0. Atomic scattering factors were obtained from the *International Tables for X-Ray Crystallography* Vol. 4. [28] All programs used in the solution and characterization of the structure are contained in the program package SHELXTL [29].

3. Results and Discussion

The conformation of DB14C4 and the atom labels are shown in the SHELXTL drawing of the molecule, Figure 3. Figure 4 shows a view of the molecule in which the angle between the planes, the hinge angle, is emphasized. Atomic parameters are listed in Table II. The atom groups C(2), C(3)—C(10), C(11) and C(11), C(12)—C(1), C(2) (see Figure 3) are essentially planar with the dihedral angle

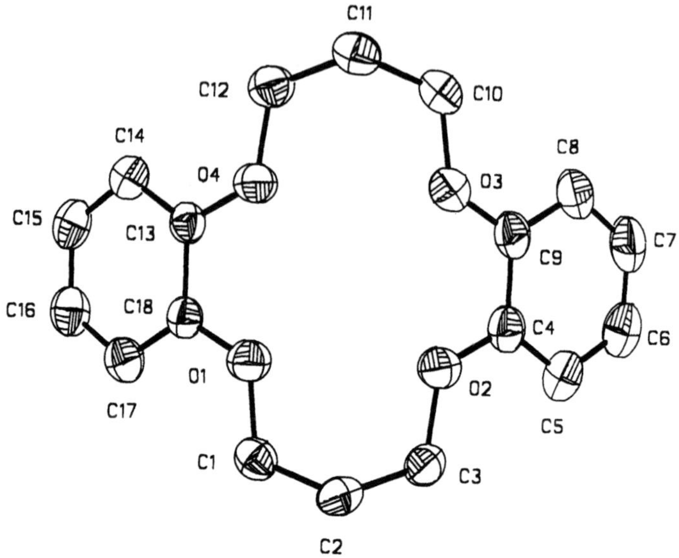

Fig. 3. A SHELXTL drawing of dibenzo-14-crown-4 with atom labels. Thermal ellipsoids are drawn at the 50% probability level.

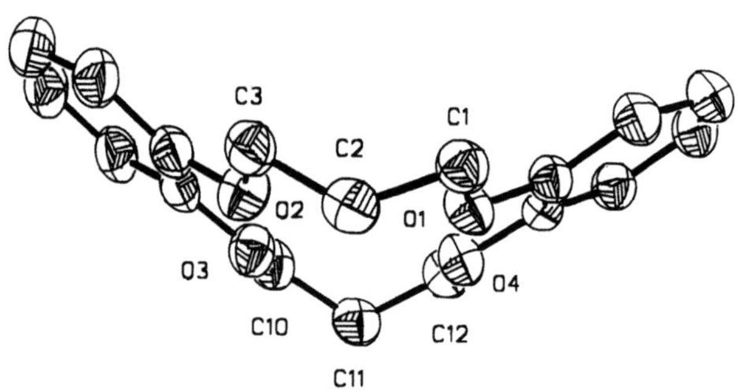

Fig. 4. A view of the dibenzo-14-crown-4 molecule which emphasizes the hinge angle between the two major planes of the molecule.

between these least squares planes being 122.8°. The four oxygen atoms form a plane, with the largest deviation of any oxygen from their least-squares plane being 0.02Å. The oxygens form a distorted rectangle, with dimensions O(1)—O(2) 2.860Å, O(2)—O(3) 2.575Å, O(3)-O(4) 2.916Å and O(4)-O(1) 2.563Å. The O—O—O angles range from 88.6° to 91.1°. The diagonal O—O distances are O(1)-O(3) 3.886Å and O(2)-O(4), 3.843Å. The bond lengths and angles in the ether portion of the molecule are typical of crown molecules with the average aliphatic C—C distance, 1.509(4) Å and C—O distance 1.429(5) Å being slightly shorter than those found in non-crown ethers. The $C_{benzene}$—O distances are not included in the C—O average. All bond lengths and angles are contained in Table III. The average C—O—C—C torsion angle is 176(2)° (*trans*) and the average O—C—C—C torsion angle is 73(1)° (*gauche*). These values indicate little strain in the ring.

Table II. Positional (× 10^4) and thermal (Å^2 × 10^3) parameters for the atoms with e.s.d. values in parenthesis.

Atoms	x/a	y/b	z/c	U
O(1)	5373(1)	9060	2493(1)	45(1)[a]
C(1)	4968(2)	7297(6)	1542(2)	41(1)[a]
H(1)C(1)	4431	6059	1595	40(7)
H(2)C(1)	4579	8419	863	50(8)
C(2)	5988(2)	5833(6)	1547(2)	42(1)[a]
H(1)C(2)	5751	4533	981	56(8)
H(2)C(2)	6386	4766	2250	50(8)
C(3)	6791(2)	7412(7)	1206(2)	42(1)[a]
H(1)C(3)	7286	6093	1026	72(9)
H(2)C(3)	6339	8653	505	57(8)
O(2)	7429(2)	9153(5)	2137(2)	48(1)[a]
C(4)	8182(2)	10852(7)	1999(2)	44(1)[a]
C(5)	8443(2)	10899(8)	1051(3)	54(1)[a]
HC(5)	8199	9389	480	70(9)
C(6)	9224(3)	12761(8)	1008(3)	61(1)[a]
HC(6)	9436	13043	345	54(8)
C(7)	9733(2)	14500(8)	1889(3)	62(1)[a]
HC(7)	10172	15823	1823	59(8)
C(8)	9480(2)	14484(8)	2838(3)	53(1)[a]
HC(8)	9690	16038	3405	91(9)
C(9)	8698(2)	12677(7)	2898(2)	44(1)[a]
O(3)	8371(1)	12496(5)	3783(2)	48(1)[a]
C(10)	8657(2)	14651(7)	4576(2)	52(1)[a]
H(1)C(10)	9445	14820	4799	78(10)
H(2)C(10)	8271	16799	4109	123(14)
C(11)	8087(2)	14169(8)	5377(2)	51(1)[a]
H(1)C(11)	8302	12458	5762	47(7)
H(2)C(11)	8431	15663	5951	68(9)
C(12)	6798(2)	14616(7)	4866(2)	47(1)[a]
H(1)C(12)	6509	14564	5375	60(8)
H(2)C(12)	6627	16153	4586	57(9)
O(4)	6244(1)	12503(5)	4092(1)	45(1)[a]
C(13)	5069(2)	12594(6)	3493(2)	37(1)[a]
C(14)	4346(2)	14358(7)	3680(2)	45(1)[a]
HC(14)	4769	16011	4274	74(9)
C(15)	3158(2)	14263(7)	2991(2)	52(1)[a]
HC(15)	2653	15437	3148	59(8)
C(16)	2700(2)	12446(8)	2128(2)	52(1)[a]
HC(16)	1855	12111	1662	55(8)
C(17)	3409(2)	10659(7)	1930(2)	46(1)[a]
HC(17)	3070	9210	1275	49(7)
C(18)	4593(2)	10714(6)	2608(2)	38(1)[a]

[a] Equivalent isotropic U_{eq} defined as one-third of the trace of the orthogonalized U_{ij} tensor.

A comparison of the conformational features of 1 and LiNCS–1 (see Table IV) shows that there are only small differences between the complexed and uncomplexed ligand. The dihedral angles between the two major planes in the free and complexed ligand differ by only about 2°. The ether oxygens are essentially planar in both compounds and the diagonal oxygen interatomic distances in the

Table III. Bond lengths (Å) and angles (deg) in **1** with e.s.d. values in parentheses.

1	2	3	1—2(Å)	1—2—3(deg)
C(1)	O(1)	C(18)	1.422(3)	118.0(2)
O(1)	C(1)	C(2)		108.3(2)
C(1)	C(2)	C(3)	1.503(4)	115.6(3)
C(2)	C(3)	O(2)	1.511(4)	107.8(2)
C(3)	O(2)	C(4)	1.429(3)	118.0(2)
O(2)	C(4)	C(5)	1.365(4)	124.9(3)
O(2)	C(4)	C(9)		115.1(3)
C(5)	C(4)	C(9)	1.391(5)	120.0(3)
C(4)	C(5)	C(6)		119.4(3)
C(5)	C(6)	C(7)	1.398(5)	120.4(4)
C(6)	C(7)	C(8)	1.362(5)	120.9(4)
C(7)	C(8)	C(9)	1.388(5)	119.9(3)
C(4)	C(9)	C(8)	1.406(4)	119.4(3)
C(4)	C(9)	O(3)		115.5(3)
C(8)	C(9)	O(3)	1.388(5)	125.1(3)
C(9)	O(3)	C(10)	1.371(4)	117.3(3)
O(3)	C(10)	C(11)	1.435(4)	107.8(3)
C(10)	C(11)	C(12)	1.513(5)	116.0(2)
C(11)	C(12)	O(4)	1.510(4)	108.1(3)
C(12)	O(4)	C(13)	1.431(4)	117.5(2)
O(4)	C(13)	C(14)	1.366(3)	125.7(2)
O(4)	C(13)	C(18)		114.9(3)
C(14)	C(13)	C(18)	1.385(4)	119.5(2)
C(13)	C(14)	C(15)		119.8(3)
C(14)	C(15)	C(16)	1.395(3)	120.5(3)
C(15)	C(16)	C(17)	1.368(5)	120.5(2)
C(16)	C(17)	C(18)	1.385(5)	119.9(3)
O(1)	C(18)	C(13)	1.364(3)	115.1(2)
C(17)	C(18)	O(1)	1.387(3)	125.1(2)
C(13)	C(18)	C(17)	1.409(4)	119.8(3)

two molecules are similar. The fact that the conformation of the ligand in the solution and solid states are similar suggests that the ligand is preorganized in such a way that it will selectively complex a lithium ion.

Table IV also gives conformational features of derivatives of 1 and Li^+ complexes of those derivatives. A schematic drawing of compounds 1–5 is presented in Figure 2. The similarity of the values in the Table indicates a rigidity of the basic unit of the molecule. The values also suggest some trends in conformational differences in the various derivatives. For example, substitutents on the hinge carbons (C(2) and C(11) in the title compound) affect the hinge angle; the angle increases as OH groups are added (see compounds 2 and 3). Molecule 4 in which a CH_2—OH group is substituted on a hinge carbon has a larger angle between the major planes than either 2 or 3. Finally the hinge angle is always several degrees smaller in the complex than in the uncomplexed ligand except for the unsubstituted ligand where it is less than 2° smaller. The fact that the hinge angle in complexes of 4 is greater when the anion is NO_3^- instead of SCN^- may be significant but more data are required before it can be suggested that there is an anion effect.

Table IV. Comparison of complexed and uncomplexed DB14C4 and derivatives of that ligand.

Ligand	Dihedral angle between two major planes (deg) (hinge angle)	O(1)—O(3)(Å)	O(2)—O(4)(Å)	Average deviation of ether oxygen from planarity (Å)	Average Li—O distance (Å)	Ref.
1 uncomplexed	122.8	3.886	3.843	0.021		this study
LiNCS–**1** complex	121	3.77	3.77	0.000[a]	2.043	9, 24
2 uncomplexed	129.7	3.813	3.813	0.000[a]		30
LiNCS–**2** A molecule	125.8	3.679	3.852	0.087	2.003	31
B molecule	126.1	3.779	3.760	0.012	2.015	31
3 uncomplexed	132.5	3.787	3.808	0.012		30
LiNCS—**3**	124.4	3.792	3.802	0.017	2.029	31
4 uncomplexed	135.2	3.793	3.805	0.011		30
LiNCS–**4**	122.3	3.788	3.736	0.009	2.021	31
$LiNO_3$–**4**	126.3	3.699	3.824	0.056	2.038	31
Li–**5**[b]	118.6	3.70	3.83	0.046	2.047	11

[a] Required by the symmetry of the molecule.
[b] Ligand also is the anion.

The similarity of the conformations of 1 and its derivatives to the conformations of the respective ligands when complexing Li^+, is further evidence for the preorganization of the DB14C4 structural unit for the complexation of Li^+.

Acknowledgement

Appreciation is expressed to the State of Utah Centers of Excellence Program and the U.S. Department of Energy, Basic Energy Sciences (Grant No. DE-FG02-86ER-13463) for financial support for this work.

References

1. V. P. Y. Gadzekpo and G. D. Christian: *Anal. Letters* **16,** 1371 (1983).
2. U. Olsher and J. Jagut-Grodzinski: *J. Chem. Soc Dalton Trans.* 501 (1981).
3. U. Olsher: *J. Am. Chem. Soc.* **104,** 4006 (1982).
4. K. Kimura, H. Yano, S. Kitazawa, and T. Shono: *J. Chem. Soc. Perkin Trans. 2* 1945 (1986).
5. K. Kimura, H. Oishi, T. Miura and T. Shono: *Anal. Chem.* **59,** 2331 (1987).
6. K. Kimura, H. Sakamoto, S. Kitazawa and T. Shono: *J. Chem. Soc., Chem. Commun.* 669 (1985).
7. C. S. Chen, H. E. Chao, S. J. Wang and S. C. Wu: *Inorg. Chim. Acta* **145,** 85 (1988).
8. K. M. Aalmo and J. Krane: *Acta Chem. Scand.* **A36,** 227 (1982).
9. G. Shoham, W. N. Lipscomb and U. Olsher: *J. Chem. Soc. Chem., Commun.* **208** (1983).
10. G. Shoham, W. N. Lipscomb and U. Olsher: *J. Am. Chem. Soc.* **105,** 1247 (1983).
11. G. Shoham, D. W. Christianson, R. A. Bartsch, G. S. Heo, U. Olsher and W. N. Lipscomb: *J. Am. Chem. Soc.* **106,** 1280 (1984).
12. P. Groth: *Acta Chem. Scand.* **A35,** 460 (1981).
13. P. Groth: *Acta Chem. Scand.* **A36,** 109 (1982).
14. J. Dale, J. Eggestad, S. B. Fredriksen and P. Groth: *J. Chem. Soc., Chem., Commun.* 1391 (1987).
15. M. Czugler and E. Wever: *J. Chem. Soc., Chem. Commun.* 472 (1981).
16. V. W. Bhagwat, H. Manohar and N. S. Poonia: *Inor. Nucl. Chem. Lett.* **16,** 373 (1980).
17. M. VanBeylen, B. Roland, G. S. D. King and J. Aerts: *J. Chem. Research* (M) 4201 (1985).
18. P. P. Power: *Acc. Chem. Res.* **21,** 147 (1988).
19. U. Olsher and G. Shoham: Unpublished Results.
20. K. Ueno, K. Hiratani, K. Taguchi, T. Okado and H. Sugihara: *Chem. Lett.* 949 (1987).
21. D. Moras and R. Weiss: *Acta Crystallogr.* **B29,** 400 (1973).
22. I. L. Karle: *J. Am. Chem. Soc.* **96,** 4000 (1974).
23. I. L. Karle, J. Karle, T. Wieland, W. Burgermeister, H. Faulstich and B. Witkop: *Proc. Nat. Acad. Sci. USA* **70,** 1836 (1973).
24. U. Olsher, R. M. Izatt, J. S. Bradshaw and N. K. Dalley: *Chem Rev.* **91,** 137 (1991).
25. F. A. Cotton and G. Wilkinson: *Advanced Inorganic Chemistry*, Interscience; New York, pp. 131–134 (1966).
26. J. M. Lehn: *Struct. Bonding* **16,** 1 (1973).
27. C. J. Pedersen: *J. Am. Chem. Soc.* **89,** 7017 (1967).
28. *International Tables for X-ray Crystallography* Vol. 4, J. A. Ibers and W. C. Hamilton, (eds.), Kynock Press, Birmingham, England, 1974, p. 99. (Distributed by Kluwer Academic Publishers Group, Dordrecht and Boston.)
29. G. M. Sheldrick: *SHELXTL.* An Intergrated System for Solving, Refining and Displaying Crystal Structures from Diffraction Data, University of Gottingen, Federal Republic of Germany, 4th revision, 1983.
30. U. Olsher, N. K. Dalley and R. A. Bartsch: unpublished results.
31. U. Olsher, F. Frolow, N. K. Dalley, J. Weiming, Z. Y. Yu, J. M. Krobelock, and R. A. Bartsch: *J. Am. Chem. Soc. 113,* 6570 (1991).

Journal of Inclusion Phenomena and Molecular Recognition in Chemistry **12**: 313–332, 1992.

On the Solid-State Conformations of 18-crown-6 Complexes

THOMAS M. FYLES
Department of Chemistry, University of Victoria, Victoria, B.C. Canada V8W 3P6

and

RICHARD D. GANDOUR
Department of Chemistry, Louisiana State University, Baton Rouge, LA 70802-1804, U.S.A.

(Received: 28 September 1990; in final form: 15 December 1990)

Abstract. A procedure for displaying macrocylic torsion angles as a map on polar coordinates is discussed with reference to the solid-state conformations of 18-crown-6 and its complexes. The maps aid in comparisons of related structures, in the perception of pseudo-symmetry elements, and in the classification of the conformations of 18-crown-6. Only four conformational groups are found in the 1 : 1 complexes of 18-crown-6 with sodium, potassium, rubidium, cesium, thallium(I), calcium and strontium cations. The relationship of donor number, mean cavity radius and effective ionic radius combined with skeletal drawings of the donors and the polar map of the torsion angles provide a composite picture of the structures and insight into the balance between cation-donor interaction energy and conformational energy.

Key words. 18-crown-6, conformations, solid-state complexes, torsion angle.

1. Introduction

Crown ethers and conformations – the two have been tied from Pedersen's first report [1]. Even the 'crown' nomenclature suggests a particular molecular shape. Pedersen was always careful to point out that "some of the molecules might resemble the drawings, [but] they are not intended to represent their actual conformations" [1]. Even so, it is clear that he believed from the beginning that the 18-crown-6 ring system would adopt a 'crown' conformation as implied from planar drawings [2, 3].

The first X-ray structures of crown ether complexes [4], showed that Pedersen's caution was warranted: dibenzo-18-crown-6 forms a 'shallow cup' for cations, and dibenzo-24-crown-8 twists to encircle two potassium ions. An ideal 'crown' conformation was first reported by Dunitz, Dobler and their co-workers in 1974 [5] as found in the K^+, Rb^+ and Cs^+ complexes of unsubstituted 18-crown-6. They also reported the C_i conformation for the free ligand, and an 'irregular conformation' for the Na^+ complex [5]. But it is the highly symmetric, D_{3d} conformation that immediately became the paradigm of crown ether conformations.

Crown ether conformations control the complexation reactions. Thermodynamic stability is enhanced when the ligand is 'preorganized' to reduce the energy of

This paper is dedicated to the memory of the late Dr C. J. Pedersen.

reorganization [6]. Kinetic barriers to complexation are principally due to conformational changes which precede interactions with cations [7]. Thus control of ligand conformation, by use of conformationally defined sub-units, has become a standard feature in the design of new complexing agents [8].

Inherent conformational properties are important, but for flexible ligands such as crown ethers, 'the guest organizes the host' [9]. This is seen in the solid-state structures of lariat ethers [10] and of polycarboxylate crown ethers derived from (+)-tartaric acid [11]. In such cases, the ligand mean cavity radius adapts to match the effective ionic radius of the guest. The host optimizes the number of donors to the cation and alters its conformation to best surround different guest cations.

Although the D_{3d} conformation is commonly observed in the solid-state structures of 18-crown-6 and its derivatives, 'irregular' conformations are not rare. Even among the 'irregular' conformations there are marked conformational regularities. The ethylenedioxy units show a preference for an *anti*, *gauche*, *anti* sequence of torsion angles for the bonds of the —O—C—C—O— unit [4, 12, 13]. Additional substituents with well defined conformational preferences, such as *R*,*R*-(+)-tartaric acid, reinforce the preference for a *gauche* relationship between oxygens [14]. Yet even in such conformationally restricted cases, 'irregular' conformations are common [11].

Our goal in this paper is to explore the conformations found in the solid-state structures of 18-crown-6 and its derivatives with special emphasis on 'irregular' or distorted conformations. We outline a procedure for displaying torsion angle information in a form that permits immediate recognition of similarities and differences between related structures. We will show that there are four main groups of conformations, each with its own defined donor array for binding cationic guests. The idea of 'irregular' conformation will be replaced by a proposal for the perception and naming of conformational regularities. We will then examine the solid-state complexes of 18-crown-6, and will show that although the guest certainly organizes the host, the conformational preferences of the host are completely respected.

2. Polar Maps of Endocyclic Torsion Angles

A specific conformation of 18-crown-6 is defined by the 18 endocyclic torsion angles and the C—C and C—O bond lengths. The bond lengths vary little [5, 12, 13] and the torsion angles cluster about $\pm 60°$ (O—C—C—O) and 180° (C—O—C—C). Stoddart [14] and Dale [15] exploited these regularities to define idealized 'bird's-eye' perspective drawings to represent conformations of macrocycles. More recent computer generated perspective and stereo drawings convey the exact conformation of a particular solid-state or molecular mechanics structure. Even so, the *perception* of the conformation, and discovery of its relationship to other conformations is a challenging task. An alternative procedure, discussed in detail by Weiler [16], summarizes the torsion angle information as a map on polar coordinates. The three dimensional perception problem is reduced to *pattern recognition* in two dimensions.

2.1. MAKING A POLAR COORDINATE MAP

Endocyclic torsion angles are determined with the convention that a positive torsion angle exhibits a clockwise rotation between the front and the rear three

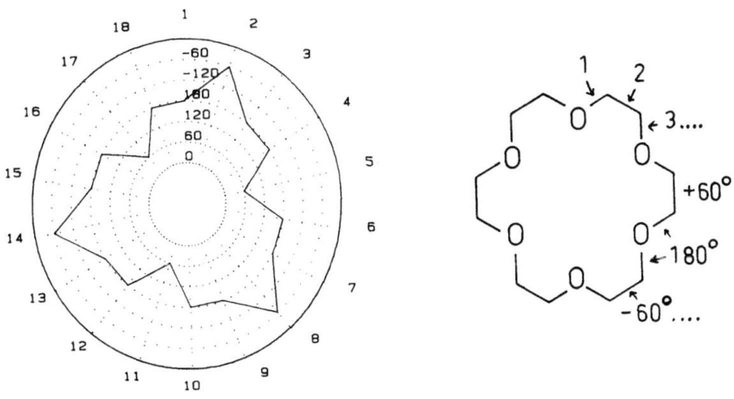

Fig. 1. Polar coordinate map of the torsion angles of an ideal D_{3d} conformation showing the bond numbering convention.

atom planes. The macrocycle *bonds* are numbered sequentially: the starting point and direction can be chosen randomly. The torsion angle of each bond is then plotted by bond number on a polar coordinate system. For 18-crown-6, each bond accounts for 20° (one-eighteenth) of 360°, and the resultant map is an octadecagon. The torsion angles are plotted from the center at 0° to the outer edge (360°); negative torsion angles are plotted as 360° plus the negative torsion angles. This plotting convention differs from Weiler [16], but is useful for crown ethers. It diminishes the small differences about *anti* torsion angles (+175° vs. −175°) and accentuates the equally small differences between *eclipsed* torsion angles (+5° vs. −5°). The former are very common in crown ethers, while the latter occur very rarely [13]. Figure 1 illustrates the procedure for an 'ideal' D_{3d} conformation consisting of torsion angles set to 180° and ±60°. The map, like the actual conformation has three-fold symmetry. The three mirror planes of the map correspond to the three two-fold axes of the D_{3d} symmetry group. The starting point, and direction of bond numbering can be chosen arbitrarily. Rotations and reflections of the resultant map will give equivalent maps for overlay comparisons between structures. Pattern recognition is not impaired by the arbitrary starting point.

2.2. USING POLAR COORDINATE MAPS

One of the uses of the polar coordinate maps is *perception of symmetry elements*. This was alluded to above and will be generally true: a mirror plane in a polar coordinate map will always locate a two-fold rotation axis in the conformation represented by the map [16]. Similarly, perpendicular rotation axes or inversion centers are simply identified. However, some three-dimensional symmetry operations, such as rotation/reflection, S_n, have no two-dimensional counterparts, and the polar map will not assist in perceiving such elements. Real structures are rarely perfectly symmetric. The maps are very useful for detecting pseudo symmetries as individual torsion angles can be selectively ignored while the remaining features are examined.

More importantly, the maps are useful for *comparisons of structures*. For example, the structure of the NaI complex of 2,3,11,12-tetraphenyl 18-crown-6 dihydrate [17] shows the Na^+ ion at the center of the macrocycle, coordinated to

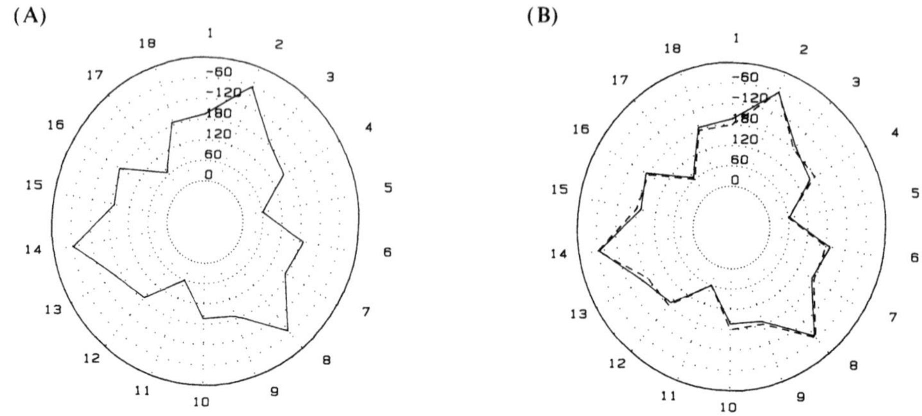

Fig. 2. (A) Polar coordinate map of the torsion angles of the *cis-anti-cis*-2,3,11,12-tetraphenyl-18-crown-6 complex of NaI dihydrate [17,CENPEQ]; (B) Overlay comparison with an "ideal" D_{3d} map (- - -).

two water molecules in axial positions. The water—Na^+—water axis is tilted due to interactions with the phenyl substituents. The polar map of the torsion angles, Figure 2a, is clearly related to the D_{3d} map of Figure 1, with lowered symmetry; compare bonds 3 and 18, which should be mirror images of one another. An overlay of the map with the 'ideal' D_{3d} case is given in Figure 2b. Several minor adjustments of torsion angle, localized in the C—O—C—C units, rather than the O—C—C—O units, can be found. The macrocycle in this structure is pseudo D_{3d}; the maps reveal how the minor distorsions are accommodated.

The central thrust of this paper is an analysis of conformations which are not pseudo-D_{3d}. Consider how the differences can be detected with reference to the maps of Figure 3. Figure 3a refers to the Cs^+ complex of the 18-crown-6 hexaacid derived from three units of *R*,*R*-(+)-tartaric acid [11]. The map lacks three-fold symmetry, but has an approximate mirror plane through bonds 5 and 14. The

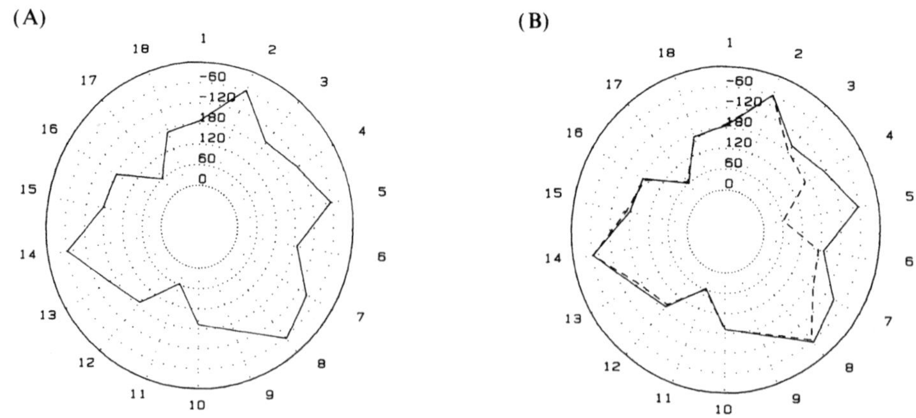

Fig. 3. (A) Polar coordinate map of torsion angles of the Cs^+ salt complex of an 18-crown-6 hexaacid [11]; (B) Overlay comparison of the Cs^+ (——) and Tl^+ (- - -) polar maps of the same ligand.

related Tl^+ complex of the same crown ether is pseudo-D_{3d}. Close inspection of the stereo drawings showed that the two structures differed, but it was difficult to localize the essential differences in the conformation. An overlay of the maps for the Cs^+ and Tl^+ complexes is given in Figure 3b. It is trivial to detect that the principal differences occur in bonds 4, 5 and 7. Bond 5 is an O—C—C—O torsion angle which has switched sign from *gauche*$^+$ in the Tl^+ complex to *gauche*$^-$ in the Cs^+ complex. This is accompanied by changes in C—O—C—C torsion angles at bonds 4 (*anti* to $-128°$) and 7 (*anti* to $-90°$). Reinspection of the structure shows that these motions reorganize the pendent carboxyl groups to enable intermolecular chelation of a Cs^+ ion in an adjacent complex [11].

The essential point in using these maps is that complex information is summarized in a form which draws the chemist's attention to critical substructures. Our uses of the maps have concentrated on analyses of pairs of structures [11, 18], as outlined in the two examples above. This will be possible for any pair of related macrocycles, of any ring size, and will permit easy perception of conformational relationships.

2.3. CONFORMATIONAL GROUPS IN 18-CROWN-6 STRUCTURES

Polar maps can also be used for *perception of groups of conformations*. In our survey of the maps drawn from available crystal structure data, we were struck by recurring patterns in the maps. Some maps, like the Cs^+ example in Figure 3, had pseudo-mirror planes. Others were axially symmetric. Still others lacked any symmetry elements. We were eventually able to recognize six different types of map patterns and symmetries.

The clue to organizing the patterns was provided by the example of Figure 3. The Cs^+ structure is related to the Tl^+ structure by a *gauche*$^+$ to *gauche*$^-$ interconversion in a single ethylenedioxy unit (plus other angle adjustments discussed in detail below). The sequence of the *gauche* torsion angles of the ethylenedioxy units in the Cs^+ structure is $g^+, g^-, g^-, g^-, g^+, g^-$, compared to the D_{3d} sequence $g^+, g^-, g^+, g^-, g^+, g^-$.

We had already generated a map for the 'ideal' D_{3d} conformation with $\pm 60°$ and $180°$ torsion angles (Figure 1). What does the map for the Cs^+ torsion angle sequence look like when 'ideal' values are used? What other sequences of *gauche* torsion angles occur in 18-crown-6 structures? What do the maps of those look like when 'ideal' values are used? The answers are given in Figures 4–6.

There are two sequences involving three positive and three negative torsion angles (Figure 4): the D_{3d} sequences $g^+, g^-, g^+, g^-, g^+, g^-$, (Figure 4a), and another sequence, $g^+, g^-, g^-, g^+, g^+, g^-$, given in Figure 4b. There are four sequences involving four torsion angles of like sign and two torsion angles of opposite sign. They occur in pairs, as the sequences $g^+, g^-, g^-, g^-, g^+, g^-$, (as in the Cs^+ structure of Figure 3, 'ideal' map Figure 5a), and $g^-, g^+, g^+, g^+, g^-, g^+$, (Figure 5b) are enantiomers. The maps of the remaining sequence ($g^+, g^-, g^-, g^+, g^-, g^-$,) and its enantiomer ($g^-, g^+, g^+, g^-, g^+, g^+$,) are given in Figure 6.

Two points require comment. Firstly, these torsion angle sequences are the ones that *commonly* occur. The coverage of our analysis is given in Section 2 and

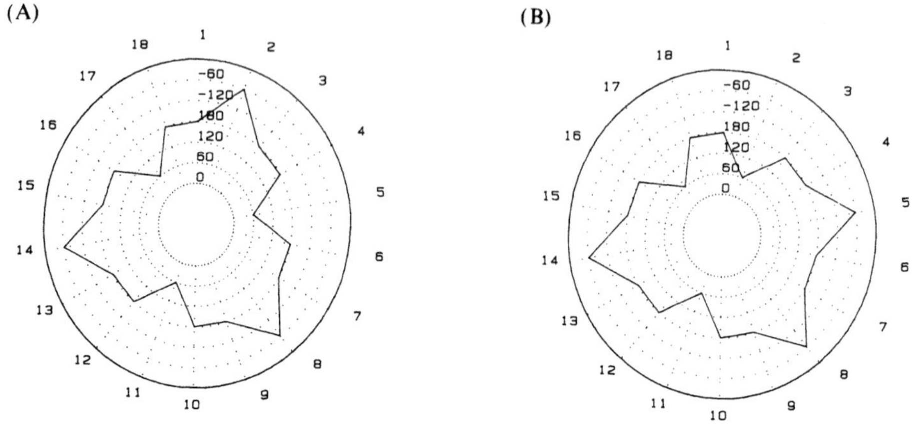

Fig. 4. Polar coordinate map of torsion angles of 'ideal' conformations: (A) D_{3d}; (B) $S_2(g^+, g^+, g^-, g^+, g^-, g^-)$.

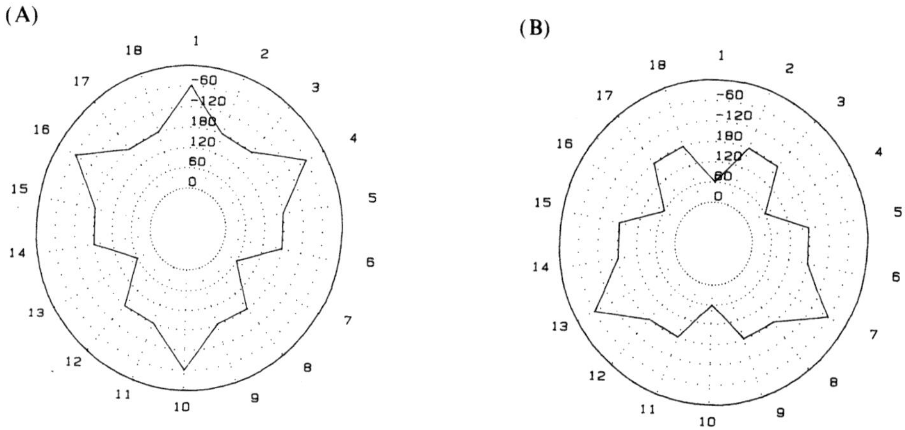

Fig. 5. Polar coordinate map of torsion angles of 'ideal' conformations: (A) $C_2(g^+, g^-, g^-, g^-, g^+, g^-)$; (B) $C_2(E^+)(g^-, g^+, g^+, g^+, g^-, g^+)$.

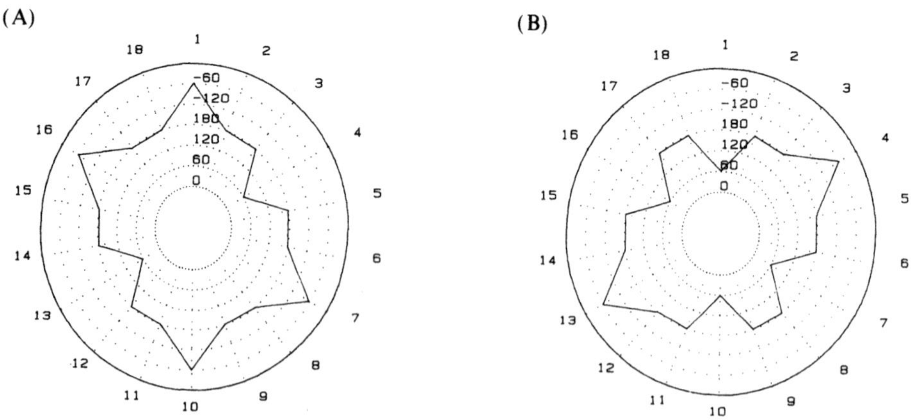

Fig. 6. Polar coordinate map of torsion angles of 'ideal' conformations: (*A*) $C_2(A-)(g^+, g^-, g^-, g^+, g^-, g^-,)$; (B) $C_2(A+)(g^-, g^+, g^+, g^-, g^+, g^+)$.

examples drawn from real structures are discussed. There are several other combinations of signs for the *gauche* O—C—C—O units that can be written ($g^+, g^+, g^-, g^-, g^-, g^-$, for example), but real examples were not found within the group considered.

Secondly, the maps of the pairs of enantiomers are not mirror images of one another. This is an artifact of the plotting convention. The Weiler convention [16] places $+180°$ at the center and $-180°$ at the other edge and gives the result that enantiomeric maps will be superimposible by rotation. As noted above, we choose to minimize the difference about the *anti* torsion angles, with the result that maps of enantiomers are different. The symmetry elements in the plane are the same for the two enantiomers. Both maps of Figure 5 have a single mirror plane; both maps of Figure 6 have two mirror planes. These planes correspond to C_2 axes in the structures.

The 'ideal' maps illustrated define conformational regularities in 18-crown-6 structures. We propose to name the conformations found in real structures by matching the experimental torsion angle map to one of the ideal cases illustrated in Figures 4–6. This is current practice for pseudo-D_{3d} cases, such as Figure 2, and it will be possible for other map patterns as well. Our naming proposal draws on the symmetry elements of the 'ideal' cases. Both enantiomeric sets in Figures 5 and 6 display C_2 axes, but the axes differ in orientation. The Figure 5 axis lies in the plane of the macrocycle, while the Figure 6 axes include one perpendicular to the plane of the macrocycle. The groups are named accordingly: Figure 5 defines the group $C_2(E)$ (for an equatorial C_2 axis) and Figure 6 defines the group $C_2(A)$ (for an axial C_2 axis). Within each group, the enantiomers are distinguished by the sign of the *gauche* angle in excess i.e. $g^+, g^-, g^-, g^-, g^+, g^-$, Figure 5a, is $C_2(A-)$ and its enantiomer is $C_2(A+)$.

The remaining group (Figure 4b) is defined as S_2 as the real structures in this class display a pseudo-rotation reflection axis. Note that S_2 *structures* cannot exist as enantiomers, but the *maps* lack a plane symmetry element and therefore could appear as mirror images. This is another artifact of the plotting conventions: the two different numbering directions will result in mirror image maps from a single structure of S_2 symmetry. The perception of the S_2 class is not impaired by this artifact, and rotation/inversion of the map will lead to maps of the same handedness for overlay comparisons.

The pseudo symmetries of the groups can be discerned in the stereo pairs given in Figure 7. The 18-crown-6 conformation are excised from X-ray structures of examples of each class. The $C_2(A)$ and $C_2(E)$ structures have an excess of *gauche*$^-$ torsion angles i.e. they are $C_2(A-)$ and $C_2(E-)$.

Recall that a sign change in one of the *gauche* torsion angles converted the hexaacid Tl^+ structure (D_{3d}) to the Cs^+ structure ($C_2(E-)$, Figure 3). The other groups are similarly related by single *gauche* sign changes. The interconversion relationships are illustrated in Figure 8. These could represent interconversion pathways for 18-crown-6 in solution [15]. We use them for describing structures near the limit of one group as a 'transition' structure between two pseudo-symmetry classes. This is illustrated in Figure 9 for the Sr^{2+} complex of the *anti*-diastereomer of an 18-crown-6 diacid dianilide [19]. The conformation lies near the extreme of S_2, with significant deviations at bonds 6 and 17. The latter is a regular feature, discussed in the next section; the former is a striking deviation from typical values ($-1°$, nearly perfectly *eclipsed*). Had this angle distorted even further to 5°, the map would then be an extreme of the $C_2(A+)$ group (Figure 9b).

2.4. 'IDEAL' VS. REAL CONFORMATIONS

The 'ideal' maps of Figures 4–6 were generated by assuming that the macrocycle could achieve $\pm 60°$ and 180° torsion angles in a prescribed sequence. Unfortu-

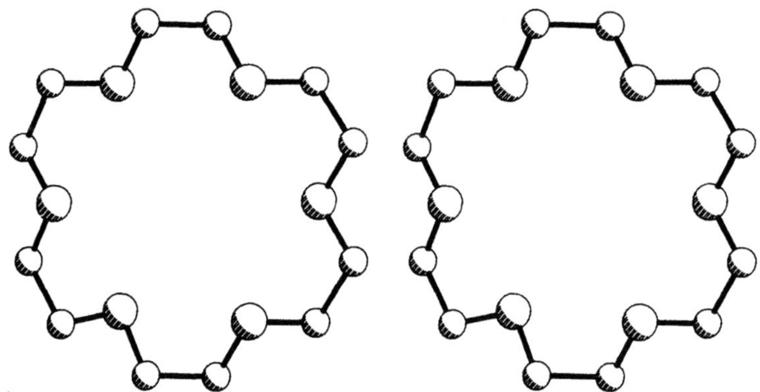

(A)

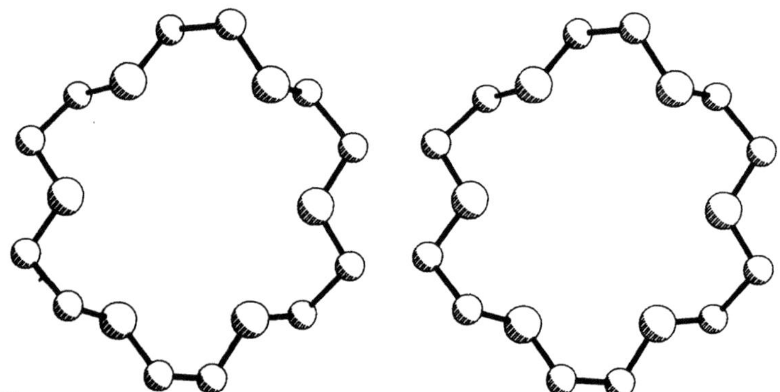

(B)

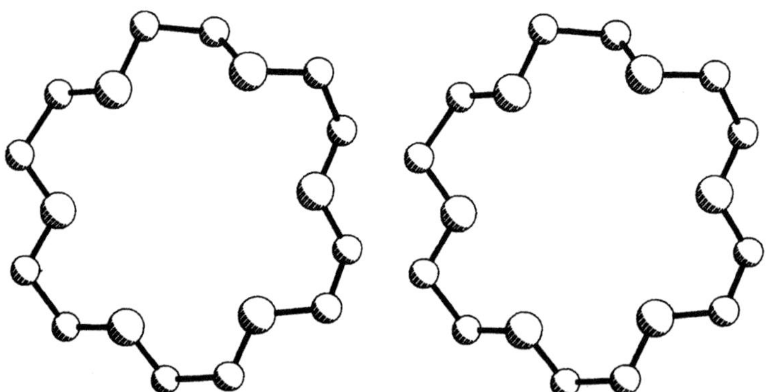

(C)

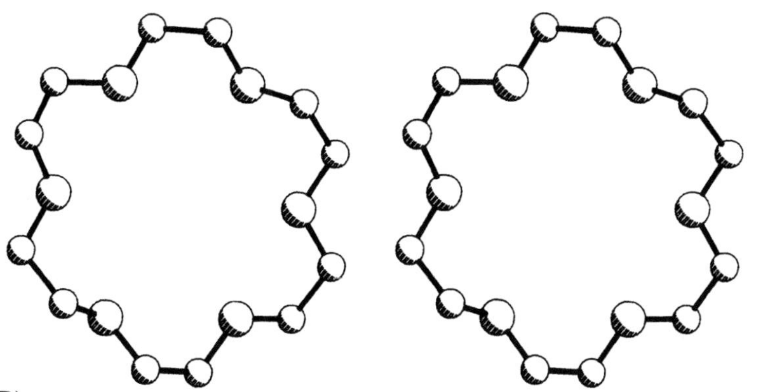

(D)

Fig. 7. Stereo drawings of 18-crown-6 conformations taken from representative structures: (A) D_{3d} (HEXH); (B) S_2 (TETNA); (C) $C_2(E-)$(HEXCS); (D) $C_2(A-)$ (TETCS).

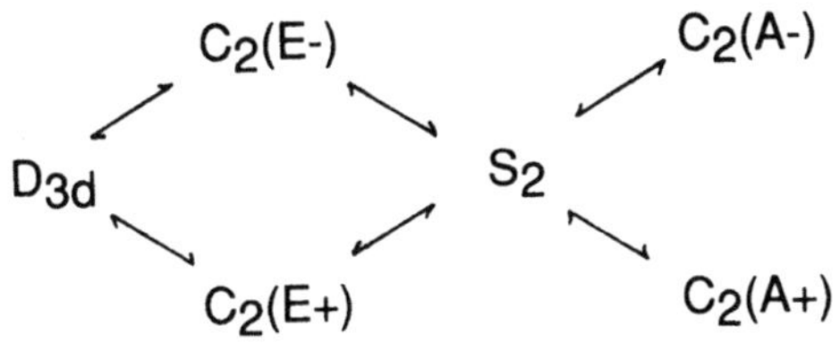

Fig. 8. Interconversions between conformational groups produced by single *gauche*$^+$ to *gauche*$^-$ changes in the O—C—C—O torsion angles.

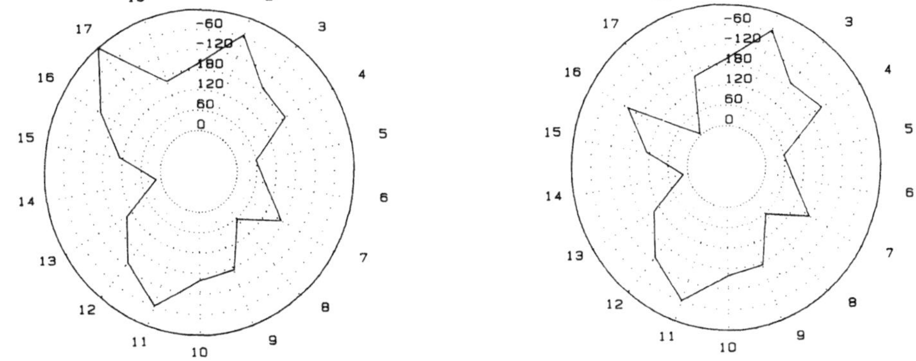

Fig. 9. (A) Polar coordinate map of the torsion angles of CAHXOB illustrating an extreme of the S_2 conformational group. (B) torsion angle map of CAHXOB with bond 17 switched from $-1°$ to $+5°$, illustrating an extreme of the $C_2(A+)$ conformational group.

nately this is not physically realistic. We 'built' and optimized the bond lengths and bond angles of an oligoethylene glycol chain using a molecular mechanics package [20], and then imposed the 'ideal' torsion angles on the chain using the torsion angle input features of the program. The distance from the first to the seventh oxygen was determined. For an 'ideal' D_{3d} sequence, the separation was zero, as expected. For the other groups, the separation was 610–640 pm, clearly indicating that the 'ideal' conformation cannot exist. Distorsions are inevitable. A close inspection of the maps of real conformations, in overlap with their 'ideal', shows that the main distorsions are localized in key regions. When these key distorsions are included in the molecular model, the first and seventh oxygens approach to within 30 pm.

The structure used for the stereoscopic drawing of the S_2 class is the Na^+ salt complex of the 18-crown-6 tetraacid derived from two tartaric acid units [11]. The map and its overlay with the 'ideal' S_2 map are given in Figure 10. The main difference in torsion angle between the two structures is at bond 1 (using the orientation of the 'ideal' map as illustrated in Figures 4 and 10b). The 'ideal' *anti* torsion angle for a C—O—C—C unit is 90° in the real structure [11]. The same bond in the map in Figure 9, also S_2 (bond 6), is 86°. From molecular modelling, a value between 75° and 105° brings the first and seventh oxygens to within 50 pm of one another, with the closest approach at 90°. Not just any distortion from 'ideal' will work to bring the ends close; *only distorsions of bond 1 will close the macrocycle.* All the examples in this class exhibit distortion and closely overlay the map in Figure 10 (or its mirror image) and take values for the bond 1 torsion angle ranging from 73° to 112°.

The other classes behave similarly; there are critical locations for deviation from the 'ideal' values and the deviations take well defined values. In every case, the deviation applied to the molecular model brings the first and seventh oxygens close to one another. Real examples in the two C_2 classes are illustrated in Figures 11 and 12. In general, two or three *gauche* angles of like sign in the O—C—C—O units

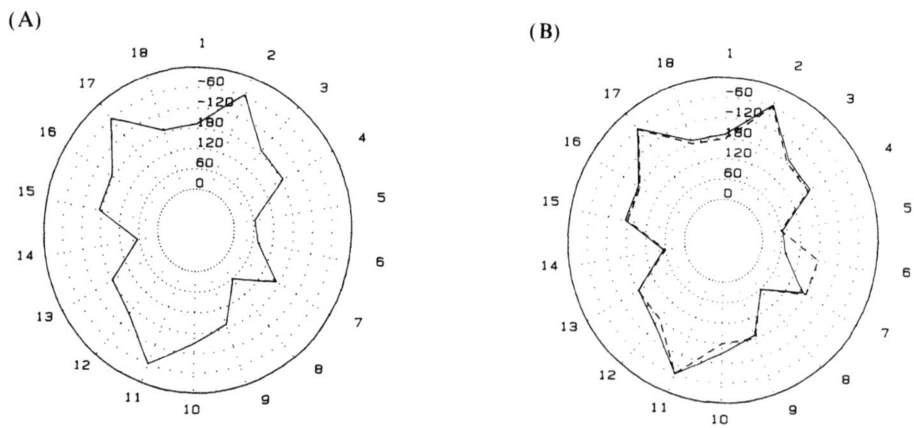

Fig. 10. (A) Polar coordinate map of TETNA torsion angles. (B) Overlay of the TETNA map with an ideal S_2 map. Note the deviation at bond 6. The S_2 map of Figure 4B has been inverted and rotated to overlay the experimental map of Figure 10A.

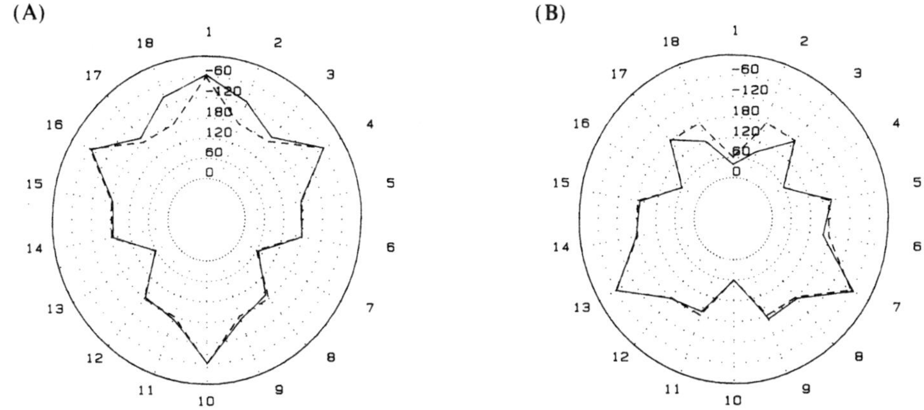

Fig. 11. Comparison of ideal and real polar coordinate maps: (A) Overlay of torsion angle maps of TETH (——) and 'ideal' $C_2(E-)$ (- - -); (B) Overlay of torsion angle maps of NACRWB (——) and 'ideal' $C_2(E+)$ (- - -).

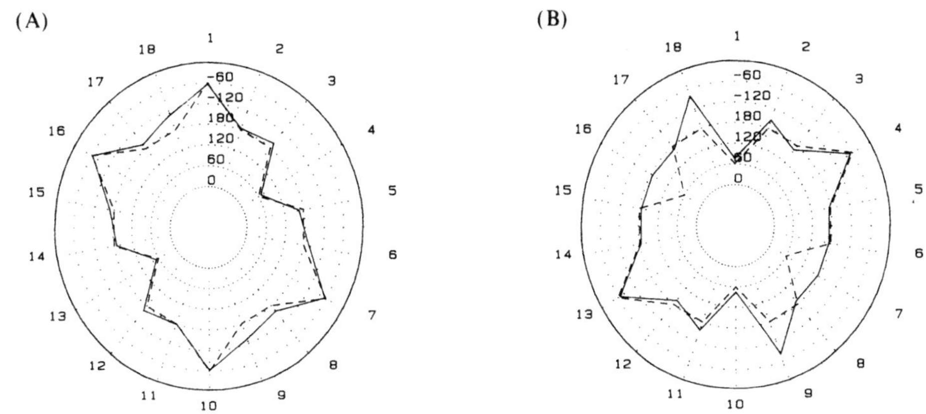

Fig. 12. Comparison of ideal and real polar coordinate maps: (A) Overlay of torsion angle maps of TETCS (——) and 'ideal' $C_2(A-)$ (- - -); (B) Overlay of torsion angle maps of 18C6 (——) and 'ideal' $C_2(A+)$ (- - -).

provoke deviations in the adjacent C—O—C—C units. The distorsions take an *anti* torsion angle to a value about $\pm 90°$.

These are major and systematic 'first-order' adjustments to the zeroth order solution presented by the 'ideal' map. There are small 'second-order' adjustments throughout the structures. These will optimize interactions with guests, relieve strain, improve contacts in the crystal, or achieve any of the other optimizations that are important in the solid state. The value of the pattern recognition approach is that it ignores the minor detail in favor of the larger picture.

It would be possible to redefine 'ideal' to incorporate the additional regular distorsions observed. The resultant maps would then closely resemble the real cases in Figures 10–12. But this is not necessary for pattern recognition purposes. Assigning a structure to one of the conformational groups is easily done by inspection of the map in conjunction with Figures 4–6. Once an assignment is made, detailed comparison with structures from the same group will reveal how both the systematic, and the specific distorsions from 'ideal' are accommodated.

3. Structures of 18-Crown-6 Complexes in the Solid State

We turn to a survey of the structures of 18-crown-6 complexes in the solid state. A search of the Cambridge Crystallographic Database in August 1988 using '18-crown-6' with Group 1 and 2 cations gave a core of references. Structures with $R > 0.10$, structures in which the crown ether was disordered, and 2:1 or 1:2 complexes were rejected. The remaining structures were supplemented with structures from our own work on 18-crown-6 tetra- and hexa-acids [11]. The final dataset is given in Tables I–III. The set is complete for 18-crown-6 and its derivatives which do not involve ring fusion with the macrocycle (benzo-, cyclohexano-, polycyclic structures). The coverage is similarly complete for Na^+, K^+, Rb^+, Cs^+, Tl^+, Ca^{2+}, Sr^{2+} and Ba^{2+} as single guests. Ammonium salts and neutral guests have not been systematically examined, although a few examples are included in Table III.

Our analysis considers two aspects of the structures. Firstly we consider the coordination geometry about the cation and the relationship of cavity size to the effective ionic radius of the guest cation. For each complex we calculate the mean cavity radius by the procedure of Mathieu *et al.* [21], which averages the differences between the metal–donor distance and the covalent radius of donor oxygen of all donors from the ligand. The mean cavity radius can then be compared to the effective ionic radius of the cation for various coordination numbers [23]. The donor geometry around the cation can be determined from skeletal drawings, and classified by analogy between the six oxygen positions of the 18-crown-6 with the conformations of cyclohexane [10, 11]. These analyses emphasize the coordination from the guest's perspective. The host's perspective is the macrocycle conformation, as revealed by polar maps of torsion angles. Here the host's role in establishing the coordination environment is the issue. We can then compare the macrocycle conformation with the cavity radius and the skeletal donor atom arrangement.

3.1. POTASSIUM 18-CROWN-6 COMPLEXES

The structural data are summarized in Table I and give the overwhelming sense of a close kinship between the complexes. The coordination numbers range from 6 to 9, but the majority of these are accommodated by a D_{3d} macrocyclic conformation. This places the donors in a chair skeletal arrangement. The exceptions are 18-crown-6 tetraamides derived from (+)-tartaric acid [19]. The tartaro units prefer an *anti* conformation of the carboxyl groups [22]; this places the carboxyl substituents in axial positions on the macrocycle and sets a *gauche*$^-$ conformation between the ether oxygens of each tartaro unit [11]. The substituent placement and the conformational preference combine to exclude a D_{3d} conformation. No D_{3d} conformations are found for any of the 18-crown-6 tetraacid complexes; this K^+ structure is not anomalous in that context.

Table I. Summary of K^+ structures[a].

Structure	Conformational Group	Skeletal Configuration	Donor Number	Cavity Radius (Å)	Other Donors
KCROFE	D_{3d}	CH	6	1.37	
CULJOI	D_{3d}	PL/CH	6	1.37	
FAPNAL	D_{3d}	CH	7	1.38	$I(PhSbI_2)_4^-$
BONSAY	$C_2(E-)$	PL	7	1.39	C=O
DECDUK	D_{3d}	CH	7	1.39	W
DEDFUN	D_{3d}	CH	7	1.40	$(SCN)_4Co^-$
KTHOXD	D_{3d}	CH	7	1.40	SCN^-
DEJWOE	D_{3d}	PL/CH	7	1.41	O_2NAr^-
DUMZAM	D_{3d}	CH	7	1.44	NO_3^-
DAXNAR	D_{3d}	CH	7	1.50	$H_3Ru(PPh_3)_3^-$
CIBFEY	D_{3d}	CH	7	1.50	$H\text{-}MoCp_2^-$
DIKXEA	D_{3d}	CH	8	1.41	$Fe_4(AuPEt_3)(CO)_{13}^-$
DIJXOJ	D_{3d}	CH	8	1.41	$P(CN)_2^-$
DIJLAJ	D_{3d}	PL/CH	8	1.41	CuI_2^-
HOXOKM	D_{3d}	CH	8	1.41	W, $Mo_6O_{19}^{2-}$
CEXJIY	D_{3d}	CH	8	1.45	C=O
DEBVUB	D_{3d}	CH	8	1.51	C=O
BONSAY	$C_2(E-)$	PL	8	1.51	C=O
CRKEAC	D_{3d}	CH	8	1.53	$C{=}O^-$
HOXOMK	D_{3d}	CH	8	1.53	W, MoO_4^{2-}
HEXK	D_{3d}	PL	9	1.48	W, CO_2^-, C=O
DEJWUK	D_{3d}	CH	9	1.53	O_2NAr^-, W

[a] Abbreviations: CH = chair; PL = planar; TW BT = twist boat; BT = boat; HF CH = half chair; RE CH = rectangular chair; * = no group assigned; W = water; THF = tetrahydrofuran.

The range of cavity size for a given donor number is larger than anticipated by an earlier analysis [10]. The group averages of cavity radius (*CR*) compare well with the K^+ effective ionic radii (*IR*) for different donor numbers [23]: 6 coord., $CR = 137$ pm, $IR = 138$ pm; 7 coord., $CR = 142$ pm, $IR = 146$ pm; 8 coord., $CR = 147$ pm, $IR = 151$ pm; 9 coord, $CR = 151$ pm, $IR = 155$ pm. In general, smaller cavity radii than expected are associated with soft, highly delocalized donors contributed by the counterion. The Shannon ionic radii were derived from oxide and chalconide structures and thus would apply best to hard donor sites [23]. A soft donor to K^+ would require the crown ether sites to compensate with increased interaction and shorter donor-cation separations. Larger cavity radii than expected (DAXNAR [24] and CIBFEY [25]) are associated with very large counterions that block potential sites for additional donors without actually contributing a formal donor atom for the K^+. Both these structures also involve a metal hydride coordinated to K^+. There are too few examples to generalize further on factors leading to larger cavity sizes.

There is no Relationship Between the Conformational Symmetry and the Cavity Size. In K^+ complexes, D_{3d} can contract to 137 pm and expand to 153 pm. The changes are accommodated by minor adjustments in the *anti* torsion angles of the C—O—C—C fragments as discussed above (Figure 2). There is a relationship between the D_{3d} conformational group and the chair skeletal arrangement; only D_{3d} gives the chair donor atom set.

3.2. SODIUM 18-CROWN-6 COMPLEXES

The sodium complexes, summarized in Table II, show a much broader spectrum of structures. The cavity sizes range from 113 pm to 137 pm and follow the donor

Table II. Summary of Na^+ structures (see Table I for abbreviations).

Structure	Conformational group	Skeletal Congfiguration	Donor Number	Cavity Radius (Å)	Other Donors
DIDCEY	*	BT	6	1.13	
NATHOD	$C_2(E+)$	BT	7	1.15	W
TETNA	S_2	TW BT	7	1.16	C=O
CENPOA	S_2	HF CH	7	1.16	I^-
NACRWB	$C_2(E+)$	HF CH	8	1.23	$W_2(CO)_{10}SH^-$
HEXNA	D_{3d}	PL	8	1.28	W CO_2^-
BOYYUJ	D_{3d}	CH	8	1.34	THF
NACRWA	D_{3d}	Pl CH	8	1.35	$W(CO)_5SH^-$
NACNPB	D_{3d}	CH	8	1.36	THF
XOCCPN	D_{3d}	PL CH	8	1.36	THF
CENPEQ	D_{3d}	CH	8	1.36	W
NACNPA	D_{3d}	CH	8	1.37	THF

numbers. However, the cavity radii are consistently larger than the effective ionic radii [23]; 6 coord., $CR = 113$ pm, $IR = 102$ pm; 7 coord., $CR = 116$ pm, $IR = 112$ pm; 8 coord., $CR = 130$ pm, $IR = 118$ pm. 18-Crown-6 is indeed 'too large for Na^+', even though the cavity size can adapt through a large range. This is the sign that the macrocycle is not completely flexible, nor can it be completely organized by the guest cation. Truter [4] suggests that these larger Na^+—O distances result from all potential donors of the macrocycle interacting with the cation. In other words, the macrocycle *enforces* a high coordination number for the complex. D_{3d} conformations giving chair or planar donor skeletal arrangements are associated with the larger cavity radii found in the 8 coordinate complexes. As the donor number and the cavity radius decreases, the skeletal array adopts half chair, twist boat and ultimately boat arrangements to permit shorter Na^+—O contacts. This is an energy balance between the Na^+—O interaction and the conformational energy of the macrocycle. The energy to achieve a conformation with shorter Na^+—O contacts must be too large, because the observed cavity radii are consistently larger than the effective ionic radii.

The K^+ structures suggested that soft donor sites would lead to closer cation–macrocycle contacts, and smaller cavity sizes. A similar effect is found in the Na^+ data (NACRWB [26]), but the data set is too small to establish a trend. The DIDCEY structure [27] represents an extreme for Na^+ to 18-crown-6 interaction. The counterion is a bis-mercapto Fe(III) tetraphenylporphyrin complex which has no localized donor site for additional interaction with the Na^+. Consequently, the 18-crown-6 must provide the full stabilization for the cation. The macrocycle is distorted to the point where none of the conformational groups is appropriate to describe the conformation (Figure 13). This may be an example of a conformational group derived from four consecutive O—C—C—O torsion angles of like sign followed by two of the opposite sign, but the 'ideal' map is so remote that generalization is not warranted. The DIDCEY structure represents an uneasy compromise between the conformational energy of the 18-crown-6 and the Na^+—O interaction. Crystal packing might play a significant role in such a balanced system.

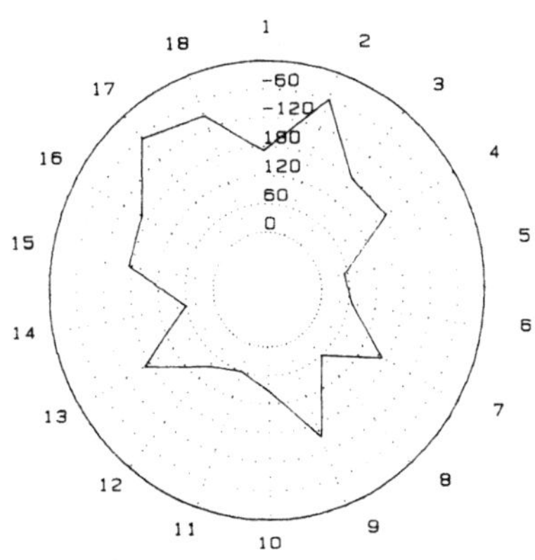

Fig. 13. Polar coordinate map of the torsion angles of DIDCEY [27].

3.3. OTHER 18-CROWN-6 COMPLEXES

The remaining complexes are summarized in Table III. The data set for any cation is too small to permit generalizations, but the main themes developed above can be detected. Donor number and cavity size are related, and the effective ionic radii of the guests follow the cavity radii closely. The ideal D_{3d} conformation is poorly represented, but when it occurs, the skeletal donor array is in a chair arrangement. There is no relationship between any other conformational group and a particular skeletal array.

4. Extensions and Exceptions

Exceptions? By definition there will be exceptions [28]. The DIDCEY structure discussed above is one example. There will always be exceptions to any procedure that attempts to describe a real system using idealized limiting structures. It is precisely the deviations from ideality that provide the most insight into the molecular recognition process in the structures discussed.

There are certainly other 18-crown-6 structures with polar coordinate maps which do not fall within the classes defined above. Our coverage excluded 2:1 complexes, with hydrogen bonding guests, 18-crown-6 fused to additional rings etc. There may well be additional regularities among the excluded structures. The *perception* of those regularities will certainly be aided by the polar maping procedure outlined here.

We have not explored the question of relative energy of the various conformational groups. Molecular mechanics calculations have established that the solid-state structures of 18-crown-6, and its Na^+ and K^+ complexes are the minimum energy structures [29–31]. The distribution of structures among the conformational groups suggests that the D_{3d} conformation is generally a low energy structure with the other conformational groups somewhat higher in energy. It may be appropriate to extend the skeletal analogy with cyclohexane to include energetics as well; the chair (D_{3d}) in a global well, with several closely related conformations at higher but similar energy. This proposal is supported by the results of Uiterwijk *et al.* [32]. They describe a diamond-lattice method for enumeration of the number of conformation of crown ethers, including 18-crown-6. They then compute relative conformational energies using average values of attractive and repulsive energies. The D_{3d} conformation is a minimum in their analysis; and 'ideal' S_2 conformation lies 9.1 kJ/mol higher while an 'ideal' $C_2(A)$ conformation is 11.7 kJ/mol above D_{3d}. More recently the same group has examined the energetics as a function of dielectric constant [33]. The energy differences between 'ideal' conformation are attenuated for realistic values of dielectric constant, but the conformations examined lie in distinct minima with 'steep walls' [33]. A fuller exploration is warranted.

We have chosen to examine 18-crown-6 structures because these were relevant to our current projects. The polar maps of other macrocycles are easily generated and show interesting patterns. Analogy with cyclohexane may not be appropriate, but there will certainly be more to uncover about conformations and molecular recognition. Polar coordinate maps and cavity size/donor number relationships are the tools for the task.

Table III. Summary of other structures considered (See Table I for abbreviations).

Structure	Guest	Conformational Group	Skeletal Configuration	Donor Number	Cavity Radius (Å)	Other Donors
RBTHXD	Rb^+	D_{3d}	CH	8	1.62	NCS^-
CSTHX	Cs^+	D_{3d}	CH	8	1.74	NCS^-
TETCS	Cs^+	$C_2(A-)$	PL	9	1.76	CO_2^-, W, C=0
HEXCS	Cs^+	$C_2(E-)$	TW BT	10	1.80	CO_2^-, W, C=0
DUBCIM	Cs^+	$C_2(A+)$	TW BT	12	1.95	
TETTL	Tl^+	S_2	HF CH	8	1.61	CO_2^-, C=0
HEXTL	Tl^+	D_{3d}	PL	10	1.61	CO_2^-, W, C=0
CAHXOA	Ca^{2+}	S_2	TW BT	9	1.37	CO_{2-}, W
DUJJOH	Ca^{2+}	$C_2(A+)$	TW BT	10	1.12	NO_3^-
CAHXOB	Sr^{2+}	$C_2(A-)$	PL	9	1.47	CO_2^-, Cl^-
TETH	W	$C_2(P-)$	HF CH			
HEXH	W_2	D_{3d}	CH			
18C6	NONE	$C_2 2(A+)$	RE CH			
SUCCIN	$(HO_2CCH_2)_2$	$C_2 2(A+)$	TW BT			
TETEN	$(^+H_3NCH_2)_2$	$C_2(A-)$	TW BT			

Acknowledgement

The ongoing support of the Natural Sciences and Engineering Research Council is gratefully acknowledged.

References

1. C. J. Pedersen: *J. Am. Chem. Soc.* **89,** 7017 (1967).
2. C. J. Pedersen and H. K. Frensdorff: *Angew. Chem. Int. Edn. Engl.* **11**, 16 (1972).
3. C. J. Pedersen: *Science* **241**, 536 (1988).
4. M. R. Truter: *Struct. Bonding* **16**, 71 (1973).
5. J. D. Dunitz, M. Dobler, P. Seiler and R. P. Phizackerley: *Acta Crystallogr.* **B30**, 2733 (1974).
6. (a) R. M. Izatt, D. J. Eatough, and J. J. Christensen: *Struct. Bonding* **16**, 161 (1973); (b) R. M. Izatt, J. S. Bradshaw, S. A. Nielsen, J. D. Lamb and J. J. Christensen: *Chem. Rev.* **85**, 271 (1985); (c) L. F. Lindoy: *The Chemistry of Macrocyclic Ligand Complexes*, Cambridge University Press, 1989.
7. (a) P. B. Chock: *Proc. Natl. Acad. Sci. U.S.A.* **69**, 1939 (1972); (b) C. C. Chen and S. Petrucci: *J. Phys. Chem.* **86**, 2601 (1982); (c) Ref. [6b].
8. (a) J. M. Lehn: *Struct. Bonding* **16**, 1 (1973); (b) D. J. Cram: *Angew. Chem. Int. Edn.* **25**, 1039 (1986).
9. D. J. Cram and K. N. Trueblood: *Top. Curr. Chem.* **98**, 43 (1981).
10. R. D. Gandour, F. R. Fronczek, V. J. Gatto, C. Minganti, R. A. Schultz, B. D. White, K. A. Arnold, A. D. Mazzocchi, S. R. Miller and G. W. Gokel: *J. Am. Chem. Soc.* **108**, 4078 (1986).
11. P. J. Dutton, T. M. Fyles and S. J. McDermid: *Can. J. Chem.* **66**, 1097 (1988); P. J. Dutton, F. R. Fronczek, T. M. Fyles and R. D. Gandour, crystal structures of TETH and HEXH *J. Am. Chem. Soc.*, **112,** 8984 (1990); crystal structures of salt complexes submitted.
12. M. Dobler: *Ionophores and Their Structures*, Wiley & Sons, (1981).
13. L. Parentau and F. Brisse: *Can. J. Chem.* **67**, 1293 (1989).
14. J. F. Stoddart: *Quart. Rev.* **8**, 85 (1979).
15. J. Dale: *Top. Stereochem.* **9**, 199 (1977).
16. J. P. Ounsworth and L. Weiler: *J. Chem. Ed.* **64**, 568 (1987).
17. G. Weber, G. M. Sheldrick, T. Burgemeister, F. Dietl, A. Mannschreck and A. Merz: *Tetrahedron* **40**, 855 (1984).
18. F. R. Fronczek, T. M. Fyles, R. D. Gandour, P. J. Hocking and P. D. Wotton: *Can. J. Chem.* **69,** 12 (1991).
19. J. P. Behr, J. M. Lehn, D. Moras and J. C. Thierry: *J. Am. Chem. Soc.* **103**, 701 (1981).
20. J. J. Gajewski and W. K. Gilbert: *PC Model*, Serena Software, Bloomington, Indiana.
21. F. Mathieu, B. Metz, D. Moras, R. Weiss: *J. Am. Chem. Soc.* **100**, 4412 (1978).
22. J. Gawronski, K. Gawronska and U. Rychlewska: *Tetrahedron Lett.* **30**, 6071 (1989) and references therein; M. Egli and M. Dobler: *Helv. Chim. Acta* **72**, 1136 (1989).
23. R. D. Shannon: *Acta Crystallogr.* **A32**, 751 (1976).
24. A. S. C. Chan and H. S. Shieh: *J. Chem. Soc., Chem. Commun.* 1378 (1985).
25. J. A. Bandy, A. Berry, M. H. Green, R. N. Perutz, K. Prout, and J. N. Verpeaux: *J. Chem. Soc., Chem. Commun.* 729 (1984).
26. M. K. Cooper, P. A. Duckworth, K. Hendrick and M. McPartlin: *J. Chem. Soc., Dalton Trans.* 2357 (1981).
27. P. Doppelt, J. Fischer and R. Wiess: *Croat. Chem. Acta* **57**, 507 (1984).
28. R. B. Woodward and R. Hoffman: *The Conservation of Orbital Symmetry*, Verlag Chemie, Weinheim (1970), p. 173.
29. R. Perrin, C. Decoret, G. Bertholon and R. Lamartine: *Nouv. J. Chim.* **7**, 263 (1983).
30. G. Wipff, P. Weiner and P. Kollman: *J. Am. Chem. Soc.* **104**, 3249 (1982).
31. G. Ranghino, S. Romano, J. M. Lehn and G. Wipff: *J. Am. Chem. Soc.* **107**, 7873 (1985).
32. J. W. H. M. Uiterwijk, S. Harkema, B. W. van de Waal, F. Gobel and H. T. M. Nibbeling: *J. Chem. Soc., Perkin Trans. 2* 1843 (1983).
33. J. W. H. M. Uiterwijk, S. Harkema, and D. Feil: *J. Chem. Soc., Perkin Trans. 2* 721 (1987).

Crystal Structure References

18C6 $C_{12}H_{24}O_6$; J. D. Dunitz and P. Seiler: *Acta Crystallogr.* **B30**, 2739 (1974).

BONSAY $2(C_{24}H_{44}N_4O_{10}) \cdot 3KBr \cdot 7H_2O$; A.-C. Dock, D. Moras, J.-P. Behr and J.-M. Lehn: *Acta Crystallogr.* **C39**, 1001 (1983).

BOYYUJ $C_{12}H_{24}O_6 \cdot Na^+ \cdot (Nb(C_{14}H_{12})(C_5H_5)_2^-) \cdot 2(C_4H_8O)$; S. I. Bailey, L. M. Engelhardt, W.-P. Leung, C. L. Rastom, I. M. Ritchie and A. H. White: *J. Chem. Soc., Dalton Trans.* 1747 (1985).

CAHXOA $C_{28}H_{32}N_2O_{12}^{2-} \cdot Ca^{2+} \cdot 3H_2O$; J.-P. Behr. J.-M. Lehn, D. Moras and J.-C. Thierry: *J. Am. Chem. Soc.* **103,** 701 (1981).

CAHXOB $C_{28}H_{32}N_2O_{12}^{2-} \cdot Ca^{2+} \cdot 2Cl^-$; J.-P. Behr, J.-M. Lehn, D. Moras and J.-C. Thierry: *J. Am. Chem. Soc.* **103**, 701 (1981).

CENPEQ $C_{36}H_{40}O_6 \cdot NaI \cdot 2H_2O \cdot CH_2Cl_2$; G. Weber, G. M. Sheldrick, T. Burgemeister, F. Dietl, A. Mannschreck and A. Merz: *Tetrahedron* **40**, 855 (1984).

CENPOA $C_{36}H_{40}O_6 \cdot NaI$; G. Weber, G. M. Sheldrick, T. Burgemeister, F. Dietl, A. Mannschreck and A. Merz: *Tetrahedron* **40**, 855 (1984).

CEXJIY $C_{12}H_{24}O_6 \cdot K^+ \cdot C_8H_4O_2P$; C. L. Liotta, M. L. McLaughlin, D. G. Van Derveer and B. A. O'Brien: *Tetrahedron Lett.* **25**, 1665 (1984).

CIBFEY $C_{12}H_{24}O_6 \cdot K^+ \cdot (Mo(C_5H_5)_2H^-)$; J. A. Bandy, A. Berry, M. L. H. Green, R. N. Perutz, K. Prout and J.-N. Verpeaux: *J. Chem. Soc., Chem. Commun.* 729 (1984).

CRKEAC $C_{12}H_{24}O_6 \cdot K^+ \cdot C_6H_9O_3^-$; C. Cambillau, G. Bram, G. Corset, C. Riche and C. Pascard-Billy: *Tetrahedron* **34**, 2675 (1978).

CSTHX $C_{12}H_{24}O_6 \cdot CsNCS$; M. Dobler and R. P. Phizackerley: *Acta Crystallogr.* **B30**, 2748 (1974).

CULJOI $C_{12}H_{24}O_6 \cdot K^+ \cdot C_{19}Cl_{15}^-$; C. Miravitilles, E. Molins, X. Solans, G. Germain and J. P. DeClerq: *J. Incl. Phenom.* **3**, 27 (1985).

DAXNAR $C_{12}H_{24}O_6 \cdot K^+ \cdot (H_3Ru(P(C_6H_5)_3)_3^-)$; A. S. C. Chan and H.-S. Sheih: *J. Chem. Soc., Chem. Commun.* 1378 (1985).

DEBVUB $2(C_{12}H_{24}O_6 \cdot K^+) \cdot Cu^{2+} \cdot 2(S_2C_2O_2^{2-})$; M. G. Kanatzidis, N. C. Baenziger and D. Coucouvanis: *Inorg. Chem.* **24**, 2680 (1985).

DECDUK $C_{12}H_{24}O_6 \cdot K^+ \cdot$ (*closo*-$(P(C_6H_5)_3)_2 \cdot RhC_2B_9H_{11}^-$); J. A. Walker, C. B. Knobler and M. F. Hawthorn: *Inorg. Chem.* **24**, 2688 (1985).

DEDFUN $(C_{12}H_{24}O_6 \cdot K^+)_2 \cdot (Co(SCN)_4^{2-})$; Y. Fan, Z. Zhou, X. Wang, J. Zhang and J. Han: *Kexue Tonbao* **30**, 910 (1985).

DEJWOE $C_{12}H_{24}O_6 \cdot K^+ \cdot C_{12}N_3O_9^-$; M. P. Egorov, V. K. Bel'skii, E. S. Petrov, M. I. Terekhova and I. P. Beletskaya: *Zh. Org. Khim.* **20**, 1855 (1984).

DEJWUK $C_{12}H_{24}O_6 \cdot K^+ \cdot C_{12}N_3O_9^- \cdot 2H_2O$; M. P. Egorov, V. K. Bel'skii, E. S. Petrov, M. I. Terekhova and I. P. Beletskaya: *Zh. Org. Khim.* **20**, 1855 (1984).

DIDCEY $C_{12}H_{24}O_6 \cdot Na^+ + \cdot (C_{44}H_{28}N_4Fe(SC_6HF_4)_2^-) \cdot C_6H_6$; P. Doppelt, J. Fischer and R. Weiss: *Croat. Chem. Acta* **57**, 507 (1984).

DIJLAJ $C_{12}H_{24}O_6 \cdot K^+ \cdot CuI_2^-$; N. P. Rath and E. M. Holt: *J. Chem. Soc., Chem. Commun.* 311 (1986).

DIJXOJ $C_{12}H_{24}O_6 \cdot K^+ \cdot (P(CN)_2^-)$; A. Schmidtpeter, G. Burget, F. Zwaschka and W. S. Sheldrick: *Z. Anorg. Allg. Chem.* **527**, 17 (1985).

DIKXEA $C_{12}H_{24}O_6 \cdot K^+ \cdot (Fe_4(AuP(C_2H_5)_3)(CO)_{13}^-) \cdot CH_2Cl_2$; C. P. Horowitz, E. M. Holt, C. P. Brock and D. F. Shriver: *J. Am. Chem. Soc.* **107**, 8136 (1985).

DUBCIM $(C_{12}H_{24}O_6)_2 \cdot Cs^+ \cdot e^-$; S. B. Dawes, D. L. Ward, R. H. Huang and J. L. Dye: *J. Am. Chem. Soc.* **108**, 3534 (1986).

DUJJOH $C_{16}H_{32}O_6 \cdot Ca(NO_3)_2$; R. B. Dyer, D. H. Metcalf, R. G. Ghirardelli, R. A. Palmer and E. M. Holt: *J. Am. Chem. Soc.* **108**, 3621 (1986).

DUMZAM $C_{14}H_{24}O_6 \cdot KNO_3$; R. B. Dyer, R. G. Ghirardelli, R. A. Palmer and E. M. Holt: *Inorg. Chem.* **25**, 3184 (1986).

FAPNAL $C_{12}H_{24}O_6 \cdot K^+((C_6H_5SbI)_4I^-)$; J. von Seyerl, O. Scheidsteger, H. Berke and G. Huttner: *J. Organomet. Chem.* **311**, 85 (1986).

HEXCS $C_{18}H_{23}O_{18}^- \cdot 2H_2O$; P. J. Dutton, F. R. Fronczek, T. M. Fyles, R. D. Gandour and V. V. Suresh: submitted.

HEXH $C_{18}H_{24}O_{18} \cdot 4H_2O$; P. J. Dutton, F. R. Fronczek, T. M. Fyles and R. D. Gandour: *J. Am. Chem. Soc.* **112,** 8984 (1990).

HEXK $C_{18}H_{23}O_{18}^- \cdot K^+ \cdot 3H_2O$; P. J. Dutton, F. R. Fronczek, T. M. Fyles, R. D. Gandour and V. V. Suresh: submitted.

HEXNA $C_{18}H_{23}O_{18}^- \cdot Na^+ \cdot 3H_2O$; P. J. Dutton, F. R. Fronczek, T. M. Fyles, R. D. Gandour and V. V. Suresh: submitted.

HEXTL $C_{18}H_{23}O_{18}^- \cdot Tl^+ \cdot 2H_2O$; P. J. Dutton, F. R. Fronczek, T. M. Fyles, R. D. Gandour and V. V. Suresh: submitted.

HOXOKM $(C_{12}H_{24}O_6 \cdot K^+)_2 \cdot Mo_6O_{19}^{2-} \cdot H_2O$; O. Nagano and Y. Sasaki: *Acta Crystallogr.* **B35**, 2387 (1979).

HOXOMK $(C_{12}H_{24}O_6 \cdot K^+)_2 \cdot MoO_4^{2-} \cdot 5H_2O$; O. Nagano: *Acta Crystallogr.* **B35**, 465 (1979).

KCROFE $C_{12}H_{24}O_6 \cdot K^+ \cdot (C_{44}H_{28}N_4Fe(SC_6H_5)_2^-)$; M. P. Byrn and C. E. Strouse: *J. Am. Chem. Soc.* **103**, 2633 (1981).

KTHOXD $C_{12}H_{24}O_6 \cdot KNCS$: P. Seiler, M. Dobler and J. D. Dunitz: *Acta Crystallogr.* **B30**, 2744 (1974).

NACNPA $C_{12}H_{24}O_6 \cdot Na^+ \cdot PBr_2(CN)_2^- \cdot (C_4H_8O)$; W. S. Sheldrick, A. Schmidpeter, F. Zwaschka, K. B. Dillon, A. W. G. Platt and T. C. Waddington: *J. Chem. Soc., Dalton Trans.* 415 (1981).

NACNPB $C_{12}H_{24}O_6 \cdot Na^+ \cdot PBr(CN)_3^- \cdot 2(C_4H_8O)$; W. S. Sheldrick, A. Schmidpeter, F. Zwaschka, K. B. Dillon, A. W. G. Platt and T. C. Waddington: *J. Chem. Soc., Dalton Trans.* 415 (1981).

NACRWA $C_{12}H_{24}O_6 \cdot Na^+ \cdot (W(CO)_5SH^-)$; M. K. Cooper, P. A. Duckworth, K. Hendrick and M. McPartlin: *J. Chem. Soc., Dalton Trans.* 2357 (1981).

NACRWB $C_{12}H_{24}O_6 \cdot Na^+ \cdot (W_2(CO)_{10}SH^-)$; M. K. Cooper, P. A. Duckworth, K. Hendrick and M. McPartlin: *J. Chem. Soc., Dalton Trans.* 2357 (1981).

NATHOD $C_{12}H_{24}O_6 \cdot NaNCS \cdot H_2O$; M. Dobler, J. D. Dunitz and P. Seiler: *Acta Crystallogr.* **B30**, 2741 (1974).

RBTHXD $C_{12}H_{24}O_6 \cdot RbNCS$; M. Dobler and R. P. Phizackerley: *Acta Crystallogr.* **B30**, 2746 (1974).

SUCCIN $C_{12}H_{24}O_6 \cdot C_4H_6O_4 \cdot H_2O$; L. Parenteau and F. Brisse: *Can. J. Chem.* **67**, 1293 (1989).

TETCS $C_{16}H_{23}O_{14}^- \cdot Cs^+ \cdot 2H_2O$; P. J. Dutton, F. R. Fronczek, T. M. Fyles, R. D. Gandour and V. V. Suresh: submitted.

TETEN $C_{16}H_{24}O_{14} \cdot C_2N_2H_8 \cdot 3H_2O$; J. J. Daly and P. Schönholzer: *Helv. Chim. Acta* **64**, 1444 (1981).

TETH $C_{16}H_{24}O_{14} \cdot 2H_2O$; P. J. Dutton, F. R. Fronczek, T. M. Fyles and R. D. Gandour: *J. Am. Chem. Soc.* **112,** 8984 (1990).

TETNA $C_{16}H_{23}O_{14}^- \cdot Na^+ \cdot 2H_2O$; P. J. Dutton, F. R. Fronczek, T. M. Fyles, R. D. Gandour and V. V. Suresh: submitted.

TETTL $C_{16}H_{23}O_{14}^- \cdot Tl^+ \cdot 2H_2O$; P. J. Dutton, F. R. Fronczek, T. M. Fyles, R. D. Gandour and V. V. Suresh: submitted.

XOCCPN $(C_{12}H_{24}O_6 \cdot Na^+)_2 \cdot (P(CN)_2 \cdot 2(C_4H_8O)$; W. S. Sheldrick, J. Kroner, F. Zwaschka and A. Schmidpeter: *Angew. Chem. Int. Ed. Engl.* **18**, 934 (1979).

Journal of Inclusion Phenomena and Molecular Recognition in Chemistry **12**: 333–339, 1992.

Structure and Thermodynamic Aspects of Macrobicyclic Polyether–Metal Ion Interactions[#]

N. KENT DALLEY,* WEIMING JIANG, GENG WU, JERALD S. BRADSHAW, HAOYUN AN, KRZYSZTOF E. KRAKOWIAK and REED M. IZATT
Department of Chemistry, Brigham Young University, Provo, UT 84602, U.S.A.

(Received: 12 November 1990; in final form: 15 January 1991)

Abstract. The crystals structure of a K^+-diptychand-15C5-18C6 iodide complex has been determined from X-ray data. The complexed bicyclic molecule crystallizes in the triclinic space group $P\bar{1}$ with $a = 9.995(4)$ Å, $b = 10.097(4)$ Å, $c = 13.725(6)$ Å, $\alpha = 90.12(3)°$, $\beta = 93.62(4)°$, $\gamma = 97.56(3)°$. The structure was solved using heavy atom methods and refined to $R = 0.032$ for 3262 independent reflections. In the crystal structure, the K^+ lies between the two crown ether rings, and is coordinated by the nine donor atoms of the ligand molecule. The complexation properties of the ligand with K^+ and Na^+ were studied by titration calorimetry in 90% (v/v) $MeOH/H_2O$ solution. The studies indicate the formation of 1 : 1 ligand : metal ion complexes in both solid state and solution.

Key words. X-ray crystal structure, macrobicycle, diptychand, equilibrium constant, K^+ binding.

Supplementary Data relating to this article are deposited with the British Library as Supplementary Publication No. Sup. S2118 (26 pages).

1. Introduction

Crown compounds, first popularized by Pedersen [1], have been shown to have a high degree of selectivity for certain metal cations, organic cations, and organic molecules. This selectivity is based primarily on the preorganization and size of the cavity and types of donor atoms. It has been found that the presence of one or two nitrogen atoms in the ring of a bicyclic macrocycle may be desirable for selective complexation of alkali metal ions [2, 3]. Also, nitrogen atoms are Brønsted–Lowry bases. Thus, complexation by the aza-crowns is pH dependent which allows the metal ion stripping process to take place by decreasing the pH of the solution. The monocyclic aza-crowns show low selectivities towards the alkali metal ions but cryptands, which contain two nitrogen atoms, have high selectivity for these ions. However, the cryptands are difficult and expensive to prepare and show slow kinetics towards releasing metal ions.

In order for a complexing agent to have practical use for separating metal ions, it must have high selectivity and labile kinetics, be inexpensive, and the complexed metal ions must be easily stripped at the end of the separation process. Macrocycles with pendant arms containing donor atoms (the lariat crown ethers) have many of these desired properties [4–7]. However, most of the lariat crowns have a low degree of preorganization and the stoichiometry of complexation is often uncertain. This paper reports a study of the structure and complexation properties of one of

[#] This paper is dedicated to the memory of the late Dr C. J. Pedersen.
* Author for correspondence.

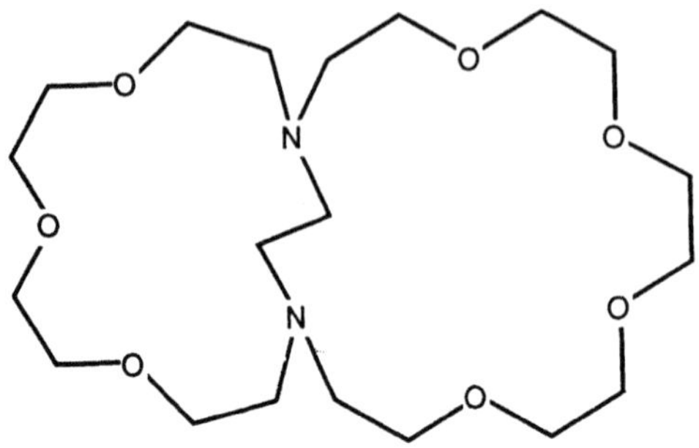

Fig. 1. Structural formula of diptychand–15C5–18C6.

a series of macrobicyclic polyethers containing nitrogen atom bridgeheads [8] which appears to have several advantages over lariat crown ethers and the cryptands as complexing agents for alkali metal ions. These new compounds are more rigid and preorganized than the lariat crowns and have an optimum balance of high selectivity, labile kinetics, low cost, ease of synthesis, and pH stripping capability.

2. Experimental

The synthesis of the macrobicyclic polyether receptor, diptychand–15C5–18C6 (see Figure 1) has been reported [8]. Complexation between dipytchand–15C5–18C6 and K^+ or Na^+ has been studied by a calorimetric titration technique in a 90% methanol/10% H_2O (v/v) solvent at 25°C. The calorimetric experimental procedures have been described [9]. The computer program EQDH [10] was used to process the calorimetric data and calculate the log K and ΔH values.

Crystals of KI–diptychand–15C5–18C6 were prepared by dissolving equal molar quantities of the ligand and KI in a warm methanol solution and then allowing the solvent to evaporate. The material obtained was recrystallized from warm methylene chloride and, after slow evaporation of the solvent, single crystals of the title compound suitable for an X-ray study were obtained. A crystal was mounted on a Nicolet R3 diffractometer which utilized graphite monochromated Mo$K\alpha$ radiation ($\lambda = 0.71073$ Å). The lattice parameters and orientation matrix were obtained using a least-squares procedure involving 21 carefully centered reflections with $10.27 < 2\theta < 35.67°$. Crystal data and X-ray crystallographic experimental conditions are summarized in Table I. Single crystal data were collected to a sin θ/λ limit of 0.59. The structure of the macrobicycle–K^+ complex was solved using the heavy-atom method. The position of the iodide ion was obtained from a Patterson map. Positions of the remaining atoms were then obtained from difference Fourier maps phased by the iodide ion. An extra peak located in the difference map was identified as an oxygen of a water of crystallisation. The positions of all hydrogen atoms were obtained from difference maps. The positional and isotropic thermal parameters of these atoms were included in the refinement. The C—H bonds range from 0.91(3) Å to 1.12(4) Å while the H—C—H angles range from 103(3)° to 120(3)°. The Ow—H distances were 0.75(3) Å and 1.06(11) Å and the H—Ow—H angle was 85(8)°. All non-hydrogen atoms were refined anisotropically. An empirical extinction correction was applied to the data but no absorption correction was

Table I. Details of the crystallographic analysis of K^+–diptychand–15C5–18C6.

Formula	$[K(C_{20}H_{40}N_2O_7)]I \cdot H_2O$
F.W.	604.55
Crystal size (mm)	0.20 × 0.35 × 0.35
Space group	$P\bar{1}$
a(Å)	9.995(4)
b(Å)	10.097(4)
c(Å)	13.725(6)
α(°)	90.12(3)
β(°)	93.62(4)
γ(°)	97.56(3)
V(Å^3)	1370.2(9)
Z	2
D_x(g cm^{-3})	1.465
Number of unique data collected	4000
Unique observed data $F > 3\sigma(F)$	3263
Scan rate (°/min)	3.9–29.3
2θ range (°)	4–45
R	0.032
R_s	0.031
Goodness of fit	1.36
Residuals in final ΔF map (e/Å^3)	general ±0.35 near I ±0.58

made. Calculations were carried out using the SHELXTL program package [11]. Atomic scattering factors were obtained from The International Tables for X-ray Crystallography [12].

3. Results and Discussion

The bicyclic ligand coordinates the potassium cation with all nine donor atoms as shown in Figures 2 and 3. The positional and isotropic thermal parameters of the non-hydrogen atoms are given in Table II. The K^+—O distances range from 2.772(2) Å to 2.898(3) Å and K^+ – N distances are 2.910(3) Å and 2.926(3) Å.

The nine-fold coordination can best be described as follows. The K^+ is coordinated by four nearly planar atoms N(1), O(4), O(6), and O(9) with the largest deviation of a donor atom from the least-squares plane of these atoms being 0.13 Å. The K^+ is 0.21 Å above this plane. Three more donor atoms, O(5), O(7), and O(9), form a triangular plane above the K^+ and this plane forms a dihedral angle of 21° with the four donor atom plane. An additional two donor atoms complex the K^+ below the four atom plane. The line determined by these donor atoms, N(2) and O(3), is nearly parallel to the least-squares plane of the four atoms. Figure 3 shows the relationship of K^+ to the donor atoms when viewed in a direction nearly parallel to the four atom plane.

Donor atom—K^+ interatomic distances and donor atom—K^+—donor atom angles are listed in Table III. Torsion angles about C—C bonds in the ligand are given in Table IV. The K^+—I^- $(-x, -y, 1-z)$ distance of 6.429(2) Å indicates that the I^- does not interact with the K^+. The water molecule in the crystal structure also does not interact with the K^+ as the Ow—K^+ $(1-x, 1-y, 1-z)$

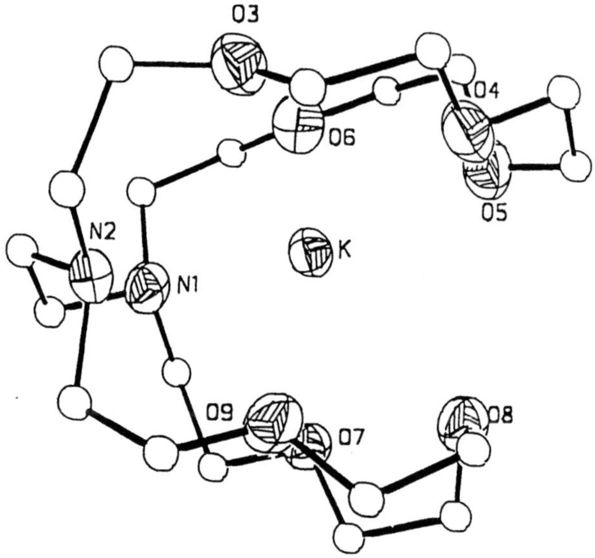

Fig. 2. Computer drawing of the K^+ complex showing the conformation of the two rings.

distance is 6.351(2) Å. However, it likely interacts with the I^- as indicated by two fairly short Ow—I^- interatomic distances. They are Ow—I^- $(1-x, 1-y, 1-z)$, 3.707(2) Å and Ow—I $(1+x, y, z)$, 3.882(2) Å. The symmetry translation in parenthesis indicates the symmetry relationship of the second atom listed to the atom coordinates contained in Table II.

The conformation of the uncomplexed diptychand–15C5–18C6 determined by molecular mechanics (MM) calculations is similar to the conformation of the ligand in the K^+-complex as determined in this X-ray study. This fact suggests that the

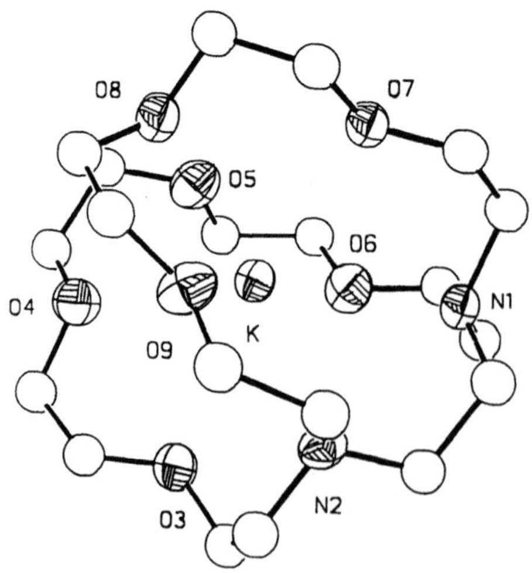

Fig. 3. A view of the K^+ complex viewed nearly parallel to the four donor atom plane consisting of N(1), O(4), O(6), and O(9).

Table II. Atom coordinates (× 10[4]) and isotropic temperature factors (Å[2] × 10[3]) for non-hydrogen atoms with e.s.d. values in parenthesis

atom	x/a	y/b	z/c	U^a
I	1201(1)	3300(1)	6760(1)	60(1)
K	3058(1)	1601(1)	1925(1)	35(1)
N(1)	3574(3)	−1057(2)	2527(2)	43(1)
N(2)	2416(3)	877(3)	3921(2)	41(1)
O(3)	384(2)	1937(2)	2506(2)	49(1)
O(4)	1676(2)	3664(3)	1159(2)	56(1)
O(5)	2853(3)	2039(3)	−104(2)	60(1)
O(6)	2066(3)	−513(2)	663(2)	55(1)
O(7)	5779(2)	1075(2)	2066(2)	44(1)
O(8)	5150(2)	3698(2)	1679(2)	47(1)
O(9)	4137(2)	3232(2)	3485(2)	51(1)
C(1)	3382(4)	−1257(4)	3573(3)	55(1)
C(2)	2248(4)	−596(4)	3966(3)	52(1)
C(3)	1165(4)	1381(4)	4176(3)	52(1)
C(4)	47(4)	1212(4)	3378(3)	55(1)
C(5)	541(4)	3351(4)	2620(3)	52(1)
C(6)	517(4)	3962(4)	1629(3)	55(1)
C(7)	1624(4)	3942(4)	145(3)	67(2)
C(8)	2795(5)	3418(4)	−294(3)	67(2)
C(9)	1726(5)	1154(5)	−509(3)	69(2)
C(10)	2030(5)	−234(4)	−365(3)	70(2)
C(11)	2488(4)	−1769(4)	878(3)	61(2)
C(12)	2580(4)	−1993(3)	1953(3)	57(1)
C(13)	4964(4)	−1233(4)	2310(3)	55(1)
C(14)	5991(4)	−65(4)	2636(3)	53(1)
C(15)	6692(3)	2220(3)	2381(3)	50(1)
C(16)	6500(4)	3382(3)	1731(3)	51(1)
C(17)	4893(4)	4769(3)	2276(3)	58(1)
C(18)	4931(4)	4491(4)	3346(3)	59(1)
C(19)	3959(4)	2943(4)	4482(3)	57(1)
C(20)	3558(4)	1451(4)	4582(3)	55(1)
Ow	8140(5)	4005(7)	5046(5)	174(3)

[a] Equivalent isotropic U defined as one third of the trace of the orthogonalised U_{ij} tensor.

ligand is preorganized in such a way that it contains 'preorganized convergent endodentate donor atoms' [13]. As shown in Table IV, the torsion angles of X—C—C—X (X = O or N) are about 60° which supports this idea. The C—X—C—C torsion angles also have low energy values. There are a total of 26 such torsion angles in the structure, 16 of which are within 20° of 180° (*trans*) and 10 are closer to 60° (*gauche*). Of the 10 near-*gauche* angles, two are in the 15-crown-5 ring both involving the same donor oxygen and two are in the 18-crown-6 ring involving two oxygens joined by a two carbon bridge. The remaining six *gauche* angles involve angles about C—N bonds. Since there are three carbons bonded to each nitrogen there are two torsion angles involving each C—C—N three atom group and only one of those can be *trans*. This accounts for the six other *gauche* C—C—N—C torsion angles.

Table III. K^+–Donor atom interatomic distances (Å) and donor atom K^+–donor atom angles (deg.).

1	2	3	1—2 (Å)	1—2—3 (deg)
N(1)	K	N(2)	2.910(3)	64.4(1)
N(2)	K	O(3)	2.926(3)	62.5(1)
O(3)	K	O(4)	2.898(3)	58.7(1)
O(4)	K	O(5)	2.817(3)	59.6(1)
O(5)	K	O(6)	2.820(2)	60.4(1)
O(6)	K	N(1)	2.784(2)	63.0(1)
O(7)	K	O(8)	2.834(2)	60.5(1)
O(8)	K	O(9)	2.810(2)	59.2(1)
O(9)	K	N(2)	2.772(2)	59.6(1)
O(7)	K	N(1)	—	62.3(1)

Table IV. Torsion angles about C—C bonds in the complex K^+–diptychand–15C5–18C6.

Atoms	Torsion Angle (deg.)
N(1)—C(1)—C(2)—N(2)	−63.2(4)
N(2)—C(3)—C(4)—O(3)	−62.5(4)
O(3)—C(5)—C(6)—O(4)	−65.3(3)
O(4)—C(7)—C(8)—O(5)	54.5(4)
O(5)—C(9)—C(10)—O(6)	67.5(4)
O(6)—C(11)—C(12)—N(1)	−62.9(4)
N(1)—C(13)—C(14)—O(7)	−67.4(4)
O(7)—C(15)—C(16)—O(8)	56.6(4)
O(8)—C(17)—C(18)—O(9)	−48.1(4)
O(9)—C(19)—C(20)—N(2)	−51.3(4)

Table V. Thermodynamic binding properties of diptychand–15C5–18C6 with K^+ and Na^+ ions in 90% MeOH/10% H_2O (v/v) solvent at 25°C.

Cation	Stoichiometry	Log K	ΔH(kJ/mol)	ΔS(J/mol K^{-1})
K^+	1 : 1	4.35(4)	−63.2(2)	−129
Na^+	1 : 1	3.56(3)	−35.4(3)	−51

As expected from the X-ray study, the thermodynamic values of complexation obtained from calorimetry experiments indicate the formation of stable 1 : 1 complexes of diptychand with either K^+ or Na^+ in solution. The calorimetric titrations also show a selectivity for K^+ over Na^+. The data given in Table V indicate that the complexation for both alkali metal ions is an enthalpy-driven process. The selectivity for K^+ over Na^+ can be explained by the fact that the larger of the two cations, K^+, can interact with more donor atoms than Na^+ without costing energy because of donor atom-donor atom repulsion. This is in agreement with the known trend [14] that there is a decreased coordination number with a decrease in metal ion radius. The entropy changes for the interactions of the bicyclic ligand with K^+ and Na^+ are also consistent with the trends [15] of the entropy changes of the corresponding reactions with monocyclic crown ethers and bicyclic cryptands. The negative entropy changes for these interactions may result from the combination of (1) the increased molecular orderliness in solution, (2) the restricted conformational tumbling of the ligand molecule, and (3) the minor contribution from the desolvation of both ligand and metal ion during complexation. The Na^+–ligand interaction leads to a more favorable entropy change compared to that for the K^+–ligand interaction because of the difference in the metal ion desolvation processes [16]. It should be noted that, although the selectivity of K^+ over Na^+ observed in this study is not as high as in the case of 18-crown-6, the moderate metal ion selectivity coupled with a pH complexation dependency and the effortless syntheses make this ligand a possible candidate for practical use in metal separations chemistry.

Acknowledgement

We thank the Brigham Young University Department of Chemistry for the internship for W. Jiang and the Centers of Excellence Program of the State of Utah for financial support.

References

1. C. J. Pedersen: *J. Am. Chem. Soc.* **89**, 7017 (1967).
2. H. Taukube, K. Yamashita, T. Iwachido, and M. Zenki: *Tetrahedron Lett.* **30**, 3983 (1989).
3. H. Tsukube, H. Adachi, and S. Morosawa: *J. Chem. Soc., Perkin Trans. I.* 1537 (1989).
4. R. A. Bartsch, G. S. Heo, S. I. Kang, Y. Liu, and J. Strzelbicki: *J. Org. Chem.* **47**, 457 (1982).
5. G. W. Gokel, D. M. Goli, C. Minganti, and L. Echegoyen: *J. Am. Chem. Soc.* **105**, 6787 (1983).
6. M. J. Van der Merwe, J. C. A. Boeyens, and R. D. Hancock: *Inorg. Chem.* **24**, 1208 (1985).
7. R. W. Hay and D. M. S. Clark: *Inorg. Chem. Acta* **83**, 123 (1984).
8. J. S. Bradshaw, H. An, K. E. Krakowiak, G. Wu, and R. M. Izatt: *Tetrahedron* **46**, 6985, (1990).
9. R. M. Izatt and G. Wu: *Thermochim. Acta* **154**, 161 (1989).
10. J. D. Lamb, R. M. Izatt, C. M. Swain, and J. J. Christensen: *J. Am. Chem. Soc.* **102**, 475 (1980).
11. G. M. Sheldrick: *SHELXTL*. An Integrated System for Solving, Refining and Displaying Crystal Structures from Diffraction Data, University of Gottingen, Federal Republic of Germany, 4th revision, 1983.
12. *International Tables for X-ray Crystallography*, Vol. 4, J. A. Ibers and W. C. Hamilton (eds.), Kynoch Press, Birmingham, England, 1974, p. 99. (Distributed by Kluwer Academic Publishers Group, Dordrecht and Boston.)
13. D. J. Cram, T. Kaneda, R. C. Helgeson, S. B. Brown, C. B. Knobler, E. Maverick, and K. N. Trueblood: *J. Am. Chem. Soc.* **107**, 3645 (1985).
14. R. D. Shannon: *Acta. Crystallogr., Sect. A* **32**, 751 (1976).
15. R. M. Izatt, J. S. Bradshaw, S. A. Nielson, J. D. Lamb, and J. J. Christensen: *Chem. Rev.* **85**, 271 (1985).
16. I. Persson: *Pure Appl. Chem.* **58**, 1153 (1986).

Journal of Inclusion Phenomena and Molecular Recognition in Chemistry 12: 341–360, 1992.

The Crystal and Molecular Structures of Bridged Spherands: The Host $C_{52}H_{52}O_8$, Its Complex $C_{52}H_{52}O_8 \cdot Li^+$, and the Complexes $C_{50}H_{48}O_6 \cdot Li^+$ and $C_{52}H_{52}O_6 \cdot Li^+$

CAROLYN B. KNOBLER, EMILY MAVERICK and KENNETH N. TRUEBLOOD*
J. D. McCullough Laboratory for X-ray Crystallography. Department of Chemistry and Biochemistry, University of California, Los Angeles, California 90024, U.S.A.

(Received 24 September 1990; in final form: 7 January 1991)

Abstract. The structures of the title novel hosts and host–guest complexes, synthesized by D. J. Cram and his colleagues, were determined by direct methods from diffractometer data and refined by least-squares. The hosts $C_{52}H_{52}O_8$, **I**, $C_{50}H_{48}O_6$, **II**, and $C_{52}H_{52}O_6$, **III**, are di-bridged derivatives of the prototype spherand $C_{48}H_{48}O_6$, **IV**. In **IV**, six 4-methylanisole units are joined in the 2,6 positions. In **I** two $—CH_2CH_2OCH_2CH_2—$ bridges replace four of the methoxy methyl groups, joining pseudo-*meta* oxygen atoms; these two bridges are on the same face of the macrocycle (*syn*). In **II** and **III**, trimethylene and tetramethylene bridges, respectively, are similarly situated, again replacing four methoxy methyls of **IV**. Compounds **Ia, IIa** and **IIIa** are Li^+ complexes of these hosts. The lithium ion is essentially hepta-coordinated in **Ia**, penta-coordinated in **IIa** and hexa-coordinated in **IIIa**. These structures are extremely rigid, and the aromatic rings in them are greatly deformed. There are intramolecular oxygen contacts in each complex only slightly greater than 2.5 Å, about 10 percent shorter than the sum of the usual van der Waals radii for oxygen atoms. Thus the structures provide useful data for parameterizing potential functions for molecular mechanics calculations.

Key words. Intramolecular overcrowding, crystal structures, spherand, Li^+ complexes, host, guest.

Supplementary Data relating to this article are deposited with the British Library as supplementary publications No. SUP 82121 (154 pages) and SUP 82122 (36 pages).

1. Introduction

Charles Pedersen's two epoch-making and inspirational papers that created and named the field of crown chemistry [1, 2] had the same title; the first was a communication and the second the promised full paper, which was submitted only three weeks after the first (how often has that been done?). These papers were devoted almost entirely to compounds prepared from aromatic vicinal diols, or substances prepared from such compounds by hydrogenation, the only exception being 18-crown-6, which was prepared by "a special method" in less than 2% yield and was included, although little discussed, "as an example of an unsubstituted crown compound". He reports that "many of these compounds have the unusual property of forming stable complexes with alkali and alkaline earth metal ions", as well as with ammonium ion and transition metal ions, and that the "more effective

This paper is dedicated to the memory of the late Dr C. J. Pedersen.
* Author for correspondence.

ligands are those containing 5 to 10 O atoms, each separated from the next by 2 C atoms". He made Fischer–Hirschfelder–Taylor molecular models, and used them to estimate the sizes of the holes in the rings, corresponding to the diameter of the largest sphere that could pass through. Although Pedersen does not mention explicitly the conformational flexibility of any of the 48 substances he prepared, he was clearly aware of the possibilities and implications, mentioning that "when 7 or more O atoms are present in the polyether ring, they cannot arrange themselves in a coplanar configuration, but can arrange themselves around the surface of a right circular cylinder, with the apices of the —C—O—C— angles pointed toward the center".

Readers of this journal know how quickly and in how many directions this field grew. One of the areas pursued by Cram and his coworkers has been that of the chemistry of what he has called 'spherands', which are synthetic hosts that show minimal conformational reorganization upon complexation [3, 4].

The structures of spherand hosts and their complexes are exceptionally rigid [5], with appreciable deformation from conventional geometry because of intramolecular overcrowding in the inner portion of the molecules. We report here the structures of Li^+ complexes of three bridged spherands and of the host of one of them. Each of these structures shows even more severe distortions than do the spherands themselves. Some of these results[1] have been presented earlier [4, 6].

Since systematic names are lengthy and complex [4], a structural formula is given for each host, together with the Roman numeral that will be used to refer to the compound in the text. Previously reported structural studies establish that indeed for the prototype spherand, **IV**, there is essentially no (net) reorganization upon complexation [5], that is, the conformation of the complex is essentially identical to that of the host. All pure spherand hosts, including those of the present compounds, have been prepared only by expulsion of Li^+ from complexes. In the one

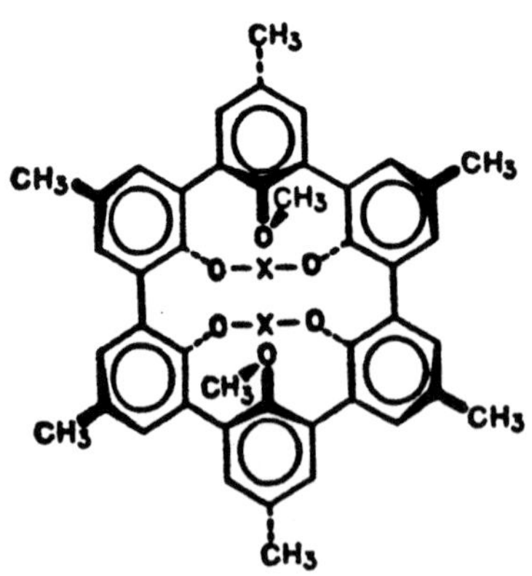

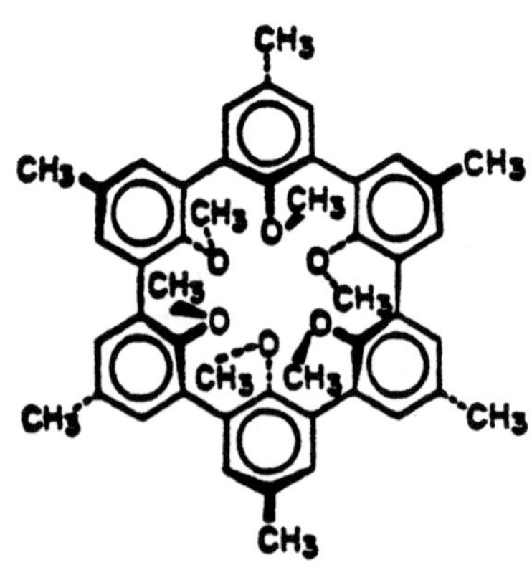

I: X = $-CH_2-CH_2-O-CH_2-CH_2-$

II: X = $-CH_2-CH_2-CH_2-$

III: X = $-CH_2-CH_2-CH_2-CH_2-$

IV

present example in which a comparison of the structure of a host (**I**) and its complex (**Ia**) is possible, there is a small degree of reorganization of one bridge. This implies that with **I**, in contrast to **IV**, after the distortions incident to expulsion of the Li^+, the host does not quite relax back to the exact conformation it had in the complex.

2. Experimental

The macrocycles **I–IV** were all first isolated [4] as crystalline $LiFeCl_4$ complexes, and then, after anion-exchange, as LiCl complexes. The pure hosts were prepared by heating the LiCl complexes for varying periods in $CH_3OH:H_2O$ at 125–130° (**I**, **IV**) or in H_2O:pyridine at 100° (**II**).

Most of the relevant crystallographic information is summarized in Table I. Crystals of **I**, $C_{52}H_{52}O_8$; **Ia**, $C_{52}H_{52}O_8 \cdot LiCl \cdot 3H_2O$; **IIa**, $C_{50}H_{48}O_6 \cdot LiFeCl_4 \cdot 0.5\,CH_2Cl_2$; and **IIIa**,[1] $C_{52}H_{52}O_6 \cdot LiCl \cdot 3C_6H_6$, were grown by slow evaporation from the indicated solvents; crystals of **Ia** were grown from non-aqueous solvents, but the complex had earlier been heated in water solutions (see above), and the solvent in the crystals was best characterized as H_2O. Monoclinic crystals of **IIb**, the LiCl complex of **II**, were grown from a mixture of toluene and methylene chloride.[2] Although this structure also was solved, there was unresolved disordered solvent present, and after refinement of the structure of **IIa**, we did not pursue the analysis of **IIb**.

2.1. REFINEMENT

All four structures were readily solved: **I**, **IIa** and **IIIa** with MULTAN [11], and **Ia** with SHELX76 [8]. Most H atoms appeared on difference maps in early stages of refinement. For **Ia** and **IIa**, H positions were refined in early cycles; in the final refinement, H atoms were treated as parts of rigid groups (methyl) or riding (methylene and aromatic) for all four structures [8]. C—H distances were fixed at 1.08 Å, and displacement parameters for H were set to about 0.01 $Å^2$ more than the equivalent isotropic U value for the attached C atom.

For **I**, after H atoms had been added and the refinement converged, peaks in the difference map of about 0.5 $e^-Å^{-3}$ suggested disorder in four of the *para* methyl groups. The final model contains two partial-occupancy sets of positions for the methyl H atoms for these groups. The largest peaks in the final difference map are near the C and H atoms of one of the bridges.

For **Ia**, peaks in the difference map after convergence suggested solvent molecules, perhaps water or molecules of another (disordered) solvent. The present model contains one fairly well characterized water molecule (H atoms were refined with O1S; one of them is 2.12 Å from the Cl^- ion), another reasonable O atom, and two rather poorly defined O atoms at 0.5 occupancy. One of the latter is 2.3 Å from its counterpart related by the symmetry operation $1-x$, $-y$, $-z$, with large displacement parameters suggesting disorder. The observed room-temperature density is more than 5% greater than that calculated for the room-temperature cell with the contents we find at low temperature (a room-temperatue data set was collected first, about six months before the low-temperature one). We suspect that this deficiency of some 5% was caused by the loss of water from the crystals over

Table I. Crystal, data collection and refinement information[a].

Compound	I	Ia	IIa	IIIa
Formula	$C_{52}H_{52}O_8$	$C_{52}H_{52}O_8 \cdot LiCl \cdot 3H_2O$	$C_{50}H_{48}O_6 \cdot LiFeCl_4 \cdot 0.5CH_2Cl_2$	$C_{52}H_{52}O_6 \cdot LiCl \cdot 3C_6H_6$
Solvent[b]	E-C	M-T	M-T	B
M_r(g mol^{-1})	805	901	992	1050
Space Group	$P\bar{1}$	$P\bar{1}$	$P2_1/c$	$C2/c$
T(K)	128	115	115	295
Crystal dim. (mm[c])	0.27 × 0.24 × 0.51	0.15 × 0.20 × 0.46	0.20 × 0.38 × 0.41	0.08 × 0.04 × 0.50
2θ(°), latt. par	9–20	23–27	24–29	16–32
(# reflns)	15	15	15	35
a(Å)	11.964(9)	12.773(3)	12.251(3)	24.581(3)
b(Å)	12.717(10)	14.125(3)	22.582(7)	16.458(2)
c(Å)	15.534(12)	15.147(4)	17.236(4)	16.049(2)
α(°)	67.05(2)	79.99(2)	90	90
β(°)	87.27(2)	67.50(2)	101.33(3)	115.036(4)
γ(°)	75.22(2)	67.72(2)	90	90
V(Å^3)	2101	2335	4676	5883
Z	2	2	4	4
D_x(g cm^{-3})	1.273	1.282	1.409	1.184
D_m(g cm^{-3}) (295 K)	—	1.312(5) (D_x = 1.250)	1.354(5)	—
Radiation, λ (Å)	MoK_α, 0.7107	MoK_α, 0.7107	MoK_α, 0.7107	CuK_α, 1.5418
Max. sin θ/λ	0.595	0.650	0.704	0.639

Range of h	−13 to 14	0 to 16	0 to 17	0 to 24
k	0 to 15	−17 to 17	0 to 31	0 to 16
l	−16 to 18	−18 to 18	−24 to 23	−14 to 14
Scan (deg min^{-1})	4.5	2.0	2.0	3.0
Background scan	No	Yes	Yes	No
$\mu(cm^{-1})$	0.48	1.03	6.0	9.0
$F(000)$	856	956	2056	2232
No. measd. refl	7425	10823	14626	3365
No. unique refl.	7407	10814	12348	3026
Unobs. $< n\sigma(F)$; $n =$	2	3	3	3
No. unobs. refl	1262	2539	1553	927
No. refl. in L.S	6163	7275	10795	2099
No. params. ref.	541 (2 blocks)	595 (2 blocks)	586 (2 blocks)	309
R	0.057	0.083	0.068	0.102
R_w	0.063	0.073	0.074	0.116
Goodness of fit	1.76	1.71	1.99	3.00
Final Peaks ($e^- Å^{-3}$)	0.35, −0.28	0.53, −0.70	1,07, −1.05	0.47, −0.45

[a] All measured densities were determined by flotation; diffractometer Syntex P1bar, graphite-monochromatized radiation; all low-temperature measurements were made with the device of Strouse [7]; the maximum average deviation of a standard reflection for **I**, **Ia**, **IIa** was 2%, for **IIIa** 4%; all final refinements were done with SHELX76 [8]; all least-squares refinements, based on $|F|$, with weighting inversely proportional to $\sigma^2(F)$; scattering factors were taken from Cromer and Mann [9], and Stewart, Davidson and Simpson [10].

[b] Abbrevations for solvents: **B**, benzene; E, ethanol, C, chloroform; **M**, methylene chloride; T, toluene.

[c] Absorption corrections were not made. For the needle-shaped crystal of **IIIa**, transmission factors vary from 0.90 to 0.96.

time (two molecules per asymmetric unit would almost account for the difference). Loss of water is not uncommon from crystals such as these, in which the water is neither strongly hydrogen bonded nor bound to an ion, and it is consistent with the poor definition of some of the solvent molecules. The largest peaks in the final difference map are near the solvent O atoms, or near Cl^-.

The structure of **IIa** refined well, but even before convergence a large solvent peak (about 10 e^- Å^{-3}) appeared. It was near the center of symmetry at the origin, about 2.9 Å from its symmetry-equivalent position. Since the crystals had been grown from methylene chloride, and since a peak appeared in a reasonable position for a C atom, we characterized the solvent as disordered CH_2Cl_2. Thus in this model one CH_2Cl_2 molecule (refined with restrained bond lengths and bond angle [8]) is present for two formula units of the complex, in one of two equally probable positions in each solvent 'hole'. In the final difference map, the largest peaks are near the Cl atoms of the counterion, $FeCl_4^-$; attempts to fit these peaks to a disorder model failed, and the displacement parameters for Cl and Fe can be reasonably interpreted as evidence of motion (see below).

Finally, the structure of **IIIa** (space group *C*2/*c*) contains disorder of two kinds. In contrast to **I, Ia** and **IIa**, the asymmetric unit of **IIIa** is a half-molecule, with the Li^+ and Cl^- ions in special positions. The two halves of the macrocycle are related by a two-fold axis; two of the four bridging —CH_2— groups are disordered, with nearly equal occupancies (0.56 and 0.44, respectively) for C(19A)—C(20A) and C(19B)—C(20B). The solvent, benzene, appeared in two regions: quite clearly near 0.2, 0, 0.3, and more ambiguously near a center of symmetry. In both regions, difference maps after refinement suggested alternate positions for benzene molecules. The final model contains benzene molecules (refined as rigid bodies [8]) as follows: two molecules have a total occupancy 1.00 (0.55, 0.45) near 0.2, 0, 0.3; two molecules near the center have a combined occupancy of 0.50 (0.29, 0.21). There are thus 12 molecules per unit cell or three per formula unit. The largest peaks in the final difference map are near these solvent molecules. Since the systematic absences do not distinguish between the space groups *C*2/*c* and *Cc*, we tried to refine in the latter, even though intensity statistics strongly favored a centrosymmetric structure. We began by phasing with half the macrocycle, adding other atoms as they appeared on Fourier maps; three independent solvent molecules (modeled as benzene) developed. The outcome was unsatisfactory for the macrocyclic host, however, with large correlations, high e.s.d.s, and strange bond lengths. The number of observations is also too small for reasonable anisotropic refinement in *Cc*. We believe that the macrocycle is best described in *C*2/*c*, but that the solvent may fit the lower-symmetry space group, which suggests that the disorder of benzene molecules, and perhaps also of the bridge atoms, is static. In any event, this structure determination is not precise.

All calculations were done on an IBM 3090 or on DEC VAX 11/750, 11/780 or 3100 computers, with the UCLA Crystallographic Package [12], which includes locally edited versions of CARESS and PROFILE, and with SHELX76 [8], PLUTO78 [13], a local molecular geometry program, and THMA11 [14].

The final positional and isotropic displacement parameters (equivalent values for atoms refined anisotropically, H atoms omitted) are given in Tables IIA through IID.[3] PLUTO stereodrawings in Figures 1 through 4 show conformations and atomic numbering in **I** and in the cationic complexes of **Ia**, **IIa**, and **IIIa**.

Table IIA. Position and displacement* parameters for $C_{52}H_{52}O_8$ (**I**).

Atom	x/a	y/b	z/c	$\langle u^2 \rangle$
C(1)	0.19877(19)	0.68783(19)	0.42942(16)	0.024
C(2)	0.31490(20)	0.68974(21)	0.42450(16)	0.028
C(3)	0.37821(20)	0.68358(21)	0.34849(17)	0.029
C(4)	0.50451(22)	0.68275(26)	0.34643(19)	0.039
C(5)	0.32126(20)	0.68349(20)	0.27214(16)	0.027
C(6)	0.20476(19)	0.68203(19)	0.27367(16)	0.023
C(7)	0.14758(19)	0.67736(19)	0.35454(16)	0.022
O(8)	0.03815(13)	0.65806(13)	0.36400(11)	0.024
C(9)	0.04367(21)	0.53652(21)	0.38322(19)	0.034
C(10)	0.14085(19)	0.68675(20)	0.19099(16)	0.024
C(11)	0.18097(20)	0.60521(20)	0.15007(16)	0.026
C(12)	0.11822(20)	0.60848(20)	0.07531(16)	0.026
C(13)	0.16399(22)	0.52245(22)	0.02886(18)	0.033
C(14)	0.00966(20)	0.68777(20)	0.04845(15)	0.024
C(15)	−0.04012(20)	0.76692(19)	0.09144(15)	0.024
C(16)	0.03525(19)	0.77337(19)	0.15548(15)	0.023
O(17)	0.00869(13)	0.86517(13)	0.18525(11)	0.024
C(18)	0.05128(21)	0.96388(21)	0.12300(17)	0.029
C(19)	0.15658(22)	0.97607(22)	0.16357(17)	0.033
O(20)	0.15185(14)	0.95662(14)	0.25986(11)	0.030
C(21)	0.04686(23)	1.02037(22)	0.28337(18)	0.033
C(22)	0.03683(21)	0.95913(20)	0.38703(17)	0.028
O(23)	−0.00779(13)	0.85859(13)	0.39962(10)	0.024
C(24)	0.01750(20)	0.77698(19)	0.49001(15)	0.023
C(25)	−0.06297(20)	0.77318(19)	0.55833(16)	0.023
C(26)	−0.01892(20)	0.70404(20)	0.65099(16)	0.025
C(27)	0.09407(20)	0.63285(20)	0.67380(16)	0.027
C(28)	0.13264(22)	0.55638(23)	0.77483(18)	0.036
C(29)	0.16565(20)	0.62700(20)	0.60209(16)	0.026
C(30)	0.12866(19)	0.69692(19)	0.50934(16)	0.024
C(31)	−0.19278(19)	0.81281(19)	0.53864(15)	0.023
C(32)	−0.25358(20)	0.73282(20)	0.59739(16)	0.025
C(33)	−0.36986(20)	0.74222(20)	0.58364(16)	0.026
C(34)	−0.43009(22)	0.65649(22)	0.65194(18)	0.035
C(35)	−0.42697(20)	0.82951(20)	0.50104(16)	0.025
C(36)	−0.37086(19)	0.90934(20)	0.43832(16)	0.024
C(37)	−0.25844(19)	0.90943(19)	0.46158(15)	0.023
O(38)	−0.20909(13)	0.99809(13)	0.40776(11)	0.027
C(39)	−0.28011(21)	1.11838(20)	0.37364(17)	0.028
C(40)	−0.24807(22)	1.18584(20)	0.27616(17)	0.029
O(41)	−0.27986(15)	1.14069(14)	0.21290(11)	0.031
C(42)	−0.23813(22)	1.18659(20)	0.12298(17)	0.029
C(43)	−0.26067(21)	1.12203(20)	0.06408(16)	0.028
O(44)	−0.19175(13)	1.00089(13)	0.10116(11)	0.027
C(45)	−0.23952(19)	0.91405(19)	0.09980(15)	0.023
C(46)	−0.17100(19)	0.81680(19)	0.08189(15)	0.023
C(47)	−0.23188(20)	0.74286(20)	0.06695(15)	0.025
C(48)	−0.35090(20)	0.75812(20)	0.07119(16)	0.025
C(49)	−0.41068(21)	0.67788(22)	0.05153(18)	0.032
C(50)	−0.41157(20)	0.84510(20)	0.10174(16)	0.025
C(51)	−0.35605(19)	0.91848(20)	0.12108(15)	0.024
C(52)	−0.41731(19)	0.97780(19)	0.18278(16)	0.023

Table IIA. (*Continued.*)

Atom	x/a	y/b	z/c	$\langle u^2 \rangle$
C(53)	−0.52933(20)	1.05127(20)	0.16295(16)	0.025
C(54)	−0.58897(19)	1.08880(20)	0.22934(16)	0.025
C(55)	−0.70702(21)	1.17458(21)	0.20587(17)	0.029
C(56)	−0.53707(20)	1.04538(20)	0.31908(16)	0.025
C(57)	−0.42491(19)	0.97215(19)	0.34085(16)	0.023
C(58)	−0.36434(19)	0.94163(19)	0.27166(16)	0.023
O(59)	−0.25102(13)	0.87523(13)	0.29033(11)	0.025
C(60)	−0.23289(20)	0.74939(21)	0.32863(18)	0.031

Table IIB. Position and displacement* parameters for $C_{52}H_{52}O_8 \cdot LiCl \cdot 3H_2O$ (**Ia**).

Atom	x/a	y/b	z/c	$\langle u^2 \rangle$
C(1)	−0.51660(31)	0.43775(28)	0.28536(25)	0.021
C(2)	−0.58177(31)	0.45625(30)	0.22428(26)	0.024
C(3)	−0.57818(32)	0.53273(31)	0.15271(26)	0.025
C(4)	−0.65636(36)	0.55679(34)	0.09122(29)	0.034
C(5)	−0.50232(31)	0.58759(29)	0.13776(26)	0.023
C(6)	−0.43338(31)	0.56978(28)	0.19629(25)	0.022
C(7)	−0.44794(31)	0.49940(28)	0.27232(25)	0.021
O(8)	−0.39374(21)	0.48968(19)	0.33917(17)	0.022
C(9)	−0.45585(36)	0.57676(31)	0.40023(28)	0.031
C(10)	−0.34656(31)	0.62404(27)	0.17901(25)	0.020
C(11)	−0.38139(33)	0.73021(29)	0.17340(26)	0.024
C(12)	−0.29900(33)	0.77792(27)	0.16136(26)	0.023
C(13)	−0.33625(37)	0.89352(31)	0.15388(31)	0.035
C(14)	−0.18408(32)	0.71774(28)	0.16142(25)	0.022
C(15)	−0.14486(30)	0.60968(27)	0.17053(23)	0.017
C(16)	−0.22554(31)	0.56561(27)	0.17000(24)	0.019
O(17)	−0.18846(21)	0.46172(18)	0.15518(17)	0.019
C(18)	−0.16593(34)	0.45300(29)	0.05477(26)	0.025
C(19)	−0.09847(33)	0.34276(28)	0.02762(25)	0.023
O(20)	−0.13522(21)	0.27481(18)	0.10507(17)	0.021
C(21)	−0.25260(33)	0.27386(29)	0.11764(26)	0.023
C(22)	−0.29625(32)	0.22478(28)	0.21334(26)	0.022
O(23)	−0.30524(21)	0.28682(18)	0.28562(17)	0.020
C(24)	−0.40462(31)	0.28547(27)	0.36575(26)	0.021
C(25)	−0.39693(30)	0.21813(26)	0.44455(25)	0.018
C(26)	−0.50581(32)	0.20481(28)	0.50662(26)	0.021
C(27)	−0.61767(32)	0.26406(30)	0.49825(26)	0.024
C(28)	−0.73175(34)	0.24725(32)	0.56792(29)	0.031
C(29)	−0.62237(32)	0.34299(29)	0.42794(26)	0.023
C(30)	−0.51679(31)	0.35459(28)	0.36130(25)	0.020
C(31)	−0.28895(30)	0.17655(26)	0.47693(26)	0.019
C(32)	−0.31697(31)	0.17552(27)	0.57625(26)	0.021
C(33)	−0.23127(32)	0.15476(26)	0.62028(25)	0.020
C(34)	−0.26742(33)	0.15190(31)	0.72709(27)	0.028
C(35)	−0.11501(31)	0.14681(26)	0.56021(26)	0.020
C(36)	−0.08275(30)	0.15137(26)	0.46182(26)	0.020
C(37)	−0.16574(31)	0.15514(25)	0.41992(25)	0.018

Table IIB. (*Continued.*)

Atom	x/a	y/b	z/c	$\langle u^2 \rangle$
O(38)	−0.12732(21)	0.14559(18)	0.32244(17)	0.020
C(39)	−0.02187(32)	0.05576(27)	0.28243(26)	0.022
C(40)	0.04284(32)	0.08371(27)	0.18117(26)	0.022
O(41)	0.06494(23)	0.17592(19)	0.18313(18)	0.025
C(42)	0.14205(33)	0.20075(28)	0.09363(27)	0.024
C(43)	0.15175(31)	0.30371(28)	0.09890(26)	0.022
O(44)	0.03280(21)	0.37699(18)	0.14766(18)	0.022
C(45)	0.03185(31)	0.44643(27)	0.20310(25)	0.019
C(46)	−0.03372(30)	0.55323(27)	0.19973(24)	0.018
C(47)	−0.00552(31)	0.61550(27)	0.24424(26)	0.021
C(48)	0.07135(31)	0.57872(28)	0.29606(26)	0.022
C(49)	0.09724(36)	0.65231(30)	0.33951(30)	0.030
C(50)	0.11403(32)	0.47337(27)	0.31306(26)	0.023
C(51)	0.09314(30)	0.40779(27)	0.26931(25)	0.019
C(52)	0.11760(32)	0.30013(27)	0.30779(25)	0.020
C(53)	0.23029(31)	0.23616(28)	0.31350(27)	0.022
C(54)	0.24382(32)	0.14192(28)	0.36352(28)	0.024
C(55)	0.36737(34)	0.07328(30)	0.36667(33)	0.034
C(56)	0.14426(32)	0.11139(27)	0.40945(27)	0.022
C(57)	0.03052(31)	0.17300(27)	0.40399(25)	0.019
C(58)	0.02111(31)	0.26533(27)	0.35152(25)	0.019
O(59)	−0.09177(20)	0.32713(17)	0.34421(17)	0.018
C(60)	−0.16731(33)	0.39842(29)	0.41982(27)	0.026
Li	−0.12040(59)	0.30209(53)	0.22896(46)	0.028
Cl	1.05110(13)	−0.15205(10)	0.13025(9)	0.053
O(1S)[a]	0.85078(37)	0.02933(30)	0.06587(28)	0.076
O(2S)	1.31496(48)	−0.12524(46)	0.02833(35)	0.119
O(3S)	0.46301(96)	0.22282(116)	0.08716(82)	0.174
O(4S)	0.56379(195)	0.02081(194)	−0.06501(173)	0.361

Table IIC. Position and displacement* parameters for $C_{50}H_{48}O_6 \cdot LiFeCl_4 \cdot 0.5\ CH_2Cl_2$(**IIa**).

Atom	x/a	y/b	z/c	$\langle u^2 \rangle$
C(1)	0.22917(23)	0.19928(12)	0.58522(16)	0.017
C(2)	0.17949(24)	0.25184(13)	0.60416(17)	0.019
C(3)	0.18226(24)	0.30351(13)	0.56098(17)	0.020
C(4)	0.12240(27)	0.35849(13)	0.57845(19)	0.026
C(5)	0.23432(24)	0.30163(12)	0.49545(17)	0.019
C(6)	0.28438(23)	0.24975(12)	0.47477(16)	0.017
C(7)	0.28581(22)	0.20008(12)	0.52254(16)	0.015
O(8)	0.34028(15)	0.14865(8)	0.50459(11)	0.015
C(9)	0.45727(27)	0.15007(15)	0.53747(25)	0.039
C(10)	0.32690(23)	0.24458(12)	0.40033(16)	0.017
C(11)	0.40990(23)	0.28324(12)	0.38570(17)	0.019
C(12)	0.45568(23)	0.27527(12)	0.31899(17)	0.020
C(13)	0.54051(26)	0.31783(13)	0.29864(20)	0.027
C(14)	0.42501(23)	0.22539(12)	0.27309(17)	0.019
C(15)	0.34307(22)	0.18396(12)	0.28514(15)	0.015
C(16)	0.28680(22)	0.19881(12)	0.34701(16)	0.015

Table IIC. (*Continued.*)

Atom	x/a	y/b	z/c	$\langle u^2 \rangle$
O(17)	0.19719(15)	0.16592(8)	0.36284(11)	0.015
C(18)	0.09322(22)	0.19723(12)	0.36121(16)	0.018
C(19)	0.01266(22)	0.15777(12)	0.39508(16)	0.016
C(20)	0.03934(22)	0.14086(12)	0.48257(16)	0.017
O(21)	0.13772(15)	0.10230(8)	0.49880(11)	0.015
C(22)	0.18314(21)	0.09183(12)	0.57807(15)	0.014
C(23)	0.20362(22)	0.03410(12)	0.61048(16)	0.016
C(24)	0.23454(27)	0.03287(14)	0.69418(17)	0.024
C(25)	0.24956(29)	0.08194(15)	0.74256(17)	0.028
C(26)	0.27711(44)	0.07507(17)	0.83142(20)	0.054
C(27)	0.24498(26)	0.13776(14)	0.70706(17)	0.024
C(28)	0.21678(23)	0.14190(12)	0.62556(16)	0.017
C(29)	0.21237(21)	−0.02510(12)	0.56999(16)	0.015
C(30)	0.23324(22)	−0.07662(13)	0.61698(16)	0.018
C(31)	0.26731(23)	−0.13024(13)	0.59091(17)	0.020
C(32)	0.29184(28)	−0.18267(14)	0.64618(20)	0.028
C(33)	0.28312(23)	−0.13470(12)	0.51345(18)	0.019
C(34)	0.25869(22)	−0.08653(11)	0.46274(16)	0.016
C(35)	0.21829(21)	−0.03393(12)	0.49017(15)	0.014
O(36)	0.18440(15)	0.00902(8)	0.43337(11)	0.013
C(37)	0.07071(21)	0.00062(12)	0.38995(16)	0.015
C(38)	0.06512(22)	0.00338(12)	0.30110(16)	0.016
C(39)	0.11810(22)	0.05627(12)	0.26907(16)	0.016
O(40)	0.23287(15)	0.06527(8)	0.31092(11)	0.014
C(41)	0.30537(22)	0.07141(12)	0.25897(15)	0.015
C(42)	0.34020(22)	0.12716(12)	0.23746(15)	0.015
C(43)	0.39212(24)	0.12633(13)	0.17069(17)	0.020
C(44)	0.41466(25)	0.07510(14)	0.13317(17)	0.022
C(45)	0.46363(32)	0.07785(16)	0.05949(20)	0.034
C(46)	0.39347(24)	0.02026(13)	0.16491(17)	0.020
C(47)	0.33880(23)	0.01833(12)	0.22817(16)	0.017
C(48)	0.30995(23)	−0.03845(12)	0.26322(16)	0.017
C(49)	0.25176(24)	−0.08257(13)	0.21482(17)	0.020
C(50)	0.20467(25)	−0.13098(13)	0.24683(18)	0.022
C(51)	0.14059(30)	−0.17760(14)	0.19402(21)	0.032
C(52)	0.21170(24)	−0.13268(12)	0.32856(18)	0.020
C(53)	0.26899(22)	−0.08959(12)	0.37821(16)	0.016
C(54)	0.32587(22)	−0.04518(12)	0.34570(16)	0.016
O(55)	0.38990(16)	−0.00596(9)	0.39743(11)	0.018
C(56)	0.50209(24)	0.00264(15)	0.38687(19)	0.026
Li	0.24694(40)	0.09060(21)	0.42369(28)	0.019
Cl(S)[b]	1.00278(45)	0.46517(23)	0.41397(32)	0.094
Cl(S)′	1.00162(83)	0.51301(31)	0.56383(47)	0.163
C(s)	1.08191(114)	0.51074(65)	0.48713(77)	0.102
Fe	0.77145(4)	0.17332(2)	0.08558(3)	0.021
Cl(1)	0.63087(7)	0.22174(4)	0.11695(6)	0.038
Cl(2)	0.88373(8)	0.23653(4)	0.04504(7)	0.049
Cl(3)	0.85818(9)	0.12322(6)	0.18813(6)	0.054
Cl(4)	0.71319(9)	0.11090(6)	−0.01030(6)	0.057

Table IID. Position and displacement* parameters for $C_{52}H_{52}O_6 \cdot LiCl \cdot 3C_6H_6$(**IIIa**).

Atom	x/a	y/b	z/c	⟨u^2⟩
C(1)	0.33325(42)	0.16000(53)	0.71559(56)	0.039
C(2)	0.27471(42)	0.13029(56)	0.67639(61)	0.046
C(3)	0.24480(41)	0.11614(55)	0.57956(63)	0.045
C(4)	0.18189(44)	0.08359(63)	0.53727(64)	0.061
C(5)	0.27670(43)	0.13242(57)	0.52666(58)	0.045
C(6)	0.33476(42)	0.16019(54)	0.56386(57)	0.040
C(7)	0.36132(40)	0.17881(54)	0.65629(64)	0.042
O(8)	0.41869(28)	0.21449(39)	0.69603(39)	0.052
C(9)	0.41617(56)	0.30058(74)	0.70857(108)	0.118
C(10)	0.37051(43)	0.17037(55)	0.50949(56)	0.041
C(11)	0.34562(40)	0.20898(54)	0.42482(62)	0.045
C(12)	0.37910(44)	0.21766(55)	0.37281(57)	0.045
C(13)	0.35242(46)	0.25428(66)	0.27721(63)	0.070
C(14)	0.43952(44)	0.19666(54)	0.41361(57)	0.044
C(15)	0.46825(40)	0.16074(50)	0.50369(56)	0.038
C(16)	0.42872(44)	0.14021(53)	0.54358(55)	0.038
O(17)	0.44505(26)	0.08728(37)	0.61876(36)	0.045
C(18)	0.42258(47)	0.00642(63)	0.58680(64)	0.061
C(19A)[c]	0.3865(8)	−0.0309(10)	0.6345(10)	0.039(4)
C(19B)	0.4240(11)	−0.0502(13)	0.6572(15)	0.049(6)
C(20A)	0.4189(9)	−0.0500(12)	0.7396(12)	0.063(6)
C(20B)	0.3852(11)	−0.0321(14)	0.7119(15)	0.050(7)
C(21)	0.41291(49)	0.00595(58)	0.80562(66)	0.061
O(22)	0.44157(25)	0.08431(35)	0.80905(36)	0.043
C(23)	0.42468(43)	0.13928(53)	0.86076(54)	0.038
C(24)	0.46384(39)	0.16243(51)	0.95011(58)	0.036
C(25)	0.43483(45)	0.20486(59)	0.99823(61)	0.053
C(26)	0.37564(48)	0.22723(63)	0.95983(62)	0.058
C(27)	0.34765(51)	0.27229(93)	1.01584(75)	0.112
C(28)	0.34062(42)	0.21427(60)	0.86324(64)	0.053
C(29)	0.36609(42)	0.17042(56)	0.81671(60)	0.042
Li	0.50000	0.12343(140)	0.75000	0.062
Cl	0.00000	0.75945(27)	0.25000	0.083
C(1S)[d]	0.2483(10)	0.0490(14)	0.2986(19)	0.084(8)
C(2S)	0.1900	0.0673	0.2354	0.100(8)
C(3S)	0.1415	0.0243	0.2365	0.117(9)
C(4S)	0.1513	−0.0372	0.3008	0.147(11)
C(5S)	0.2096	−0.0556	0.3640	0.121(9)
C(6S)	0.2581	−0.0125	0.3629	0.117(9)
C(7S)[e]	0.1371(8)	−0.0103(21)	0.2534(22)	0.110
C(8S)	0.1756	−0.0526	0.3314	0.110
C(9S)	0.2375	−0.0416	0.3654	0.110
C(10S)	0.2610	0.0118	0.3213	0.110
C(11S)	0.2224	0.0541	0.2433	0.110
C(12D)	0.1605	0.0431	0.2094	0.110
C(13S)[f]	0.4736(25)	0.5205(34)	0.0734(29)	0.096(19)
C(14S)	0.4910	0.4417	0.0648	0.095(22)
C(15S)	0.5160	0.4247	0.0033	0.076(18)
C(16S)	0.5236	0.4866	−0.0498	0.101(21)
C(17S)	0.5061	0.5654	−0.0412	0.099(20)
C(18S)	0.4811	0.5824	0.0204	0.093(22)

Table IID. (*Continued*.)

Atom	x/a	y/b	z/c	$\langle u^2 \rangle$
C(19S)[g]	0.4907(20)	0.4363(24)	0.1100(27)	0.082
C(20S)	0.4733	0.5157	0.1172	0.082
C(21S)	0.4841	0.5780	0.0673	0.082
C(22S)	0.5124	0.5609	0.0102	0.082
C(23S)	0.5299	0.4816	0.0029	0.082
C(24S)	0.5191	0.4193	0.0529	0.082

* Displacement parameters are commonly called vibration or thermal parameters. Units of $\langle u^2 \rangle$ are Å^2. Units of each e.s.d. in parentheses are those of the least significant digit of the corresponding parameter. E.s.d.'s of displacement parameters are given for atoms refined isotropically.

Equivalent isotropic values are $[1/(8\pi^2)]$ times the 'equivalent B value' for an anisotropic atom as defined by Hamilton [15].

[a] Occupancies for solvent O atoms: O(1S), 1.0, O(2S), 1.0, O(3S), 0.5, O(4S), 0.5.

[b] Occupancy for solvent CH_2Cl_2 atoms: 0.5.

[c] Occupancy for C(19A)–C(20A): 0.56; for C(19B)–C(20B): 0.44.

[d] Pivot atom in rigid group. Occupancy for C(1S)–C(6S): 0.55.

[e] Pivot atom in rigid group. Occupancy for C(7S)–C(12S): 0.45.

[f] Pivot atom in rigid group. Occupancy for C(13S)–C(18S): 0.29.

[g] Pivot atom in rigid group. Occupancy for C(19S)–C(24S): 0.21.

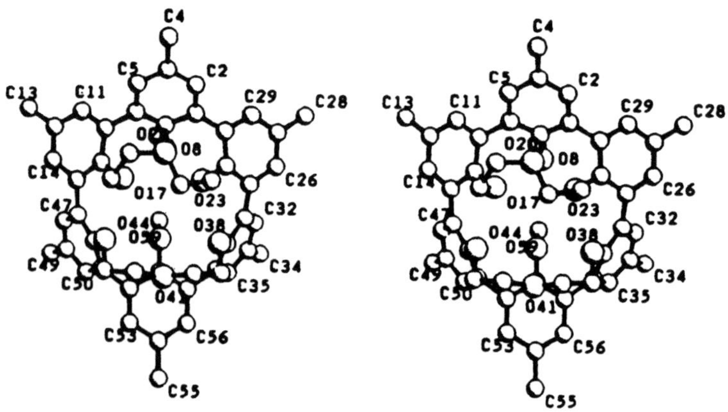

Fig. 1. PLUTO [13] stereoview of **I**. H atoms, and labels for some C atoms, have been omitted to avoid clutter.

3. Results and Discussion

Many of the distinctive properties of these compounds are due to the fact that they have, unexpectedly [4, 6], *syn* bridges. The hosts can only be prepared around Li^+; the pure hosts are then formed by expelling the cation [4]. Calculations [16] indicate that, as models suggest, the isomer of **II** with *anti* bridges should be much more stable, and presumably the same situation obtains for **I** and **III**.

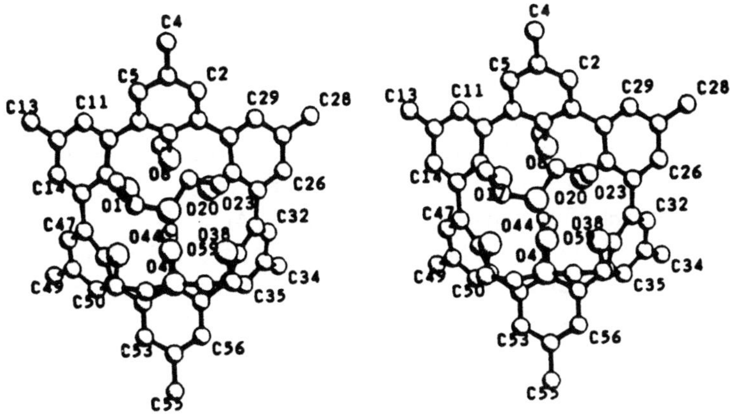

Fig. 2. Stereoview of **Ia**. Solvent and chloride ion have been omitted.

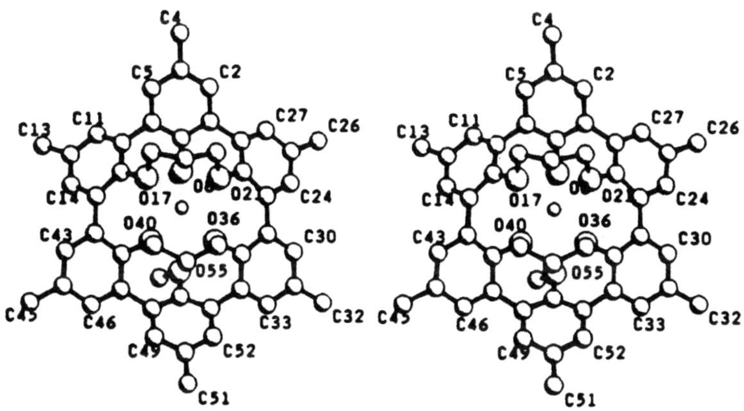

Fig. 3. Stereoview of **IIa**. Solvent and $FeCl_4^-$ ion have been omitted.

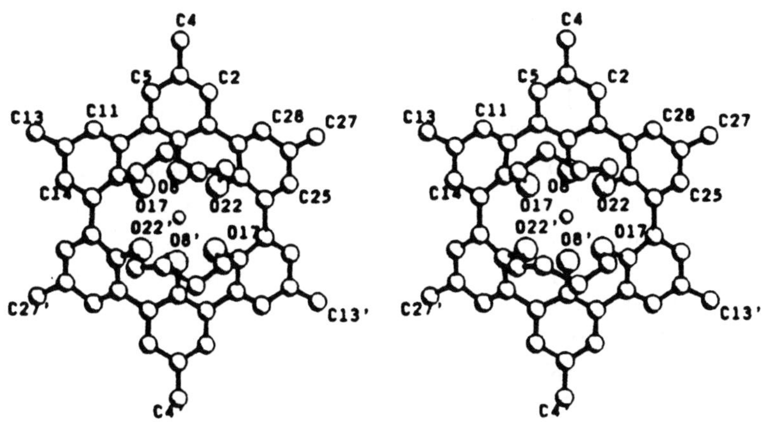

Fig. 4. Stereoview of **IIIa**. The two halves of the macrocycle are related by the symmetry operation $1 - x, y, 3/2 - z$; positions for the disordered bridge atoms are those of C(19B)—C(20B). The solvent, the chloride ion and C(19A)—C(20A) have been omitted.

Among the consequences of the fact that the bridges are *syn* that are most relevant in the present context are: (a) the aromatic rings do not tip alternately up and down relative to the median plane (or to the molecular axis), as they do in **IV** [4, 5]; and (b) the arrangement of the six O atoms attached to the aromatic rings approaches more closely a trigonal prism in each of these structures than it does the approximately regular octahedron characteristic of **IV** [4, 5]. These features are evident in the stereoviews of the structures (Figures 1–4).

A third consequence of the *syn* bridges, not apparent from the Figures, is that there is extreme intramolecular overcrowding in these molecules, and resulting geometrical distortions. Some of the relevant geometrical data are summarized in Tables III and IV, where pertinent properties of **IV** and **IVa** [5] are also shown for comparison; more details are given with the supplementary material. Table III is concerned with the arrangement of the O atoms and the coordination of the Li^+ in the complexes. In **Ia**, seven of the eight O atoms are coordinated to the cation; in **IIa**, five of the six O atoms are, and in **IIIa** (which has a two-fold axis), all six Os are coordinated to Li^+. The free energy of binding [20] might be expected to be related to the coordination number and to the cavity size; generally binding is favored by large total r_i^{-2}, where r_i are the Li^+—O distances. In this respect **IIa** and **IVa** are about the same, while **Ia** is more favored; the fact that **IVa** has much greater binding energy is presumably due chiefly to the absence of short repulsive O···O contacts in **IVa**. The effect of the small cation, with its strong field, in contracting the cavity lined with O atoms can be seen by comparing the number of O···O contacts shorter than the sum of the van der Waals radii (2.8 Å) in **I** (only three) and in **Ia** (seven). On the other hand, the numbers of distances below 3.0 Å (those between 2.8 and 3.0 Å involving little strain) are almost the same (ten and eleven respectively). The O···O distances below 2.8 Å involving atoms coordinated to Li^+ in **Ia** are shorter, by an average of 0.14 Å, than the corresponding distances in **I**.

Comparison of Figures 1 and 2, as well as a detailed study of the respective geometries, shows that the host structure in the pure host (**I**) is very similar to that in the complex (**Ia**). The major difference is in the position of O(20), which is more than 2 Å farther from O(41) in the pure host (5.06 Å vs. 2.99 Å), and also 1.3–1.4 Å farther from O(38) and O(44). It appears that when the Li^+ is expelled from the complex, it escapes by a path that involves moving O(20) (and its adjoining bridge) out of the way. In **I**, the conformation allows short H···O distances, for example O(38)···H(22A) and O(44)···H(18B), both about 2.3 Å.

The fact that some O···O distances, even in the pure host **I**, are distinctly shorter than the sum of the van der Waals radii is an indication of the strain built into the structures. The final ring-closing step in the synthesis of the complexes involves what is probably [4] a Li^+-templated coupling of two aryl radicals with the release of sufficient energy to allow the considerable molecular deformations that are found. Some information about these deformations is summarized in Table IV. In an unperturbed aromatic system, the atoms of the ring and the atoms immediately attached to the ring are all coplanar. In the present molecules, as in **IV** and related unbridged spherands [5], the aromatic rings are folded or buckled, and the immediately attached atoms are often far out of the mean aromatic plane. The distortions in the *anisyl* rings in each of the present

Table III. Short intramolecular O···O and Li^+···O distances and complexation energies.**FFb**

		Ia	**IIa I**	**IIIa**	**IV**	**IVa**
O···O						
No. of distances						
smaller than 2.8Å	3	7[a]	7	6	0	6
smaller than 3.0Å	10	11	8	6	6	6
Shortest O···O distance	2.62Å	2.57Å	2.51Å	2.53Å	2.92Å	2.79Å
Li^+···O						
No. of atoms within 2.5Å of Li^+	—	7[b]	5[c]	6	—	6
Ave. distance, r	—	2.26Å	2.04Å	2.17Å	—	2.14Å
Range[d] of distances	—	2.03–2.43	2.00–2.08	2.05–2.35	—	—
Effective Li^+ radius[e]	—	0.86	0.64	0.77	—	0.74
Sum[f] of r_i^{-2}	—	1.47Å^{-2}	1.33Å^{-2}	1.28Å^{-2}	—	1.31Å^{-2}
$-\Delta G^0$(kJ mol^{-1}) of complexation[g]	—	67	70	—	—	>96

[a] Six of these 7 shortest O···O distances in **Ia** are shorter than the corresponding distances in **I**, by an average of 0.14Å. Only O(8)···O(23) is longer in **Ia** than in **I**, by 0.06Å, and O(8) is not coordinated to Li^+.
[b] O(8) is 3.47Å from Li^+.
[c] O(55) is 2.89Å from Li^+; one aromatic C, C(16), is as close to Li^+ (2.87Å).
[d] This range of distances is in agreement with a survey by Hermansson *et al.* [17], 2.04–2.12Å for 5-coordinate and 2.13–2.17Å for 6-coordinate Li^+; Reiss *et al.* [18] find a range of 1.99–2.20Å, mean 2.06Å for a 5-coordinate Li^+ complex.
[e] This radius is calculated by subtracting an oxygen radius of 1.40Å from the average Li^+···O distance. Shannon [19] gives 0.76Å as the radius for 6-coordination; he also calculates the radius as 0.92Å for 8-coordination, so the value here for 7-coordination is reasonable.
[f] Over all O atoms.
[g] From Ref. [20].

Table IV. Comparison of some features of the molecular geometry[a].

		I	Ia	IIa	IIIa	IV	IVa
Ave. fold angle of aryl ring along CH_3—C···C—O line	Anisyl	5°	4°	6°	4°	6°	3°
	Bridge[b]	9	10	9	10	—	—
Ave. abs. dev. of aryl C atoms from aryl plane	Anisyl	0.023Å	0.018Å	0.026Å	0.020Å	0.032Å	0.014Å
	Bridge	0.042	0.045	0.040	0.046	—	—
Ave. abs. dev. of attached O atom from aryl plane	Anisyl	0.17Å	0.12Å	0.11Å	0.20Å	0.21Å	0.06Å
	Bridge	0.29	0.32	0.24	0.46	—	—
Ave. abs. dev. of attached aryl C from aryl plane	Anisyl	0.19Å	0.17Å	0.28Å	0.14Å	0.17Å	0.13Å
	Bridge	0.43	0.40	0.29	0.33	—	—
Interplanar angle between adjacent rings[c]	Average	52°	50°	43°	40°	52°	56°
	Range	43–66	40–61	28–51	33–46	—	—
Tilt of aryl rings relative to mean plane of the four attached bridge O atoms[d]	Average	37°	35°	28°	26°	30°	33°
	Range	2–67	1–61	12–43	21–31	—	

[a] Values for **IV** and **IVa** are taken from Ref. [5].
[b] 'Bridge' means the aryl (aromatic) rings attached to the O atoms leading to the bridges.
[c] The angle between the normals to the least-squares planes of adjacent aryl rings.
[d] The angle between the normal to the least-squares plane of the aryl ring and the normal to the least-squares plane of the four bridge O atoms that are attached to aryl rings.

molecules, given in Table IV, are comparable to those in **IV** and its complexes [5] (which contain anisyl rings but no bridges), and are, on the average, about half those in the aromatic rings directly linked to the bridges. This is not surprising, since the source of much of the strain in each structure is the unusually short O···O contacts, which always involve the bridge O atoms attached to the aromatic rings. Thus, while the evidence in Table IV shows that complexation relieves some of the strain present in **I** and **IV** as pure hosts, **Ia**, **IIa** and **IIIa** all show more geometrical distortion than **IVa**, presumably because of the short O···O contacts in **Ia**, **IIa** and **IIIa** (Table III).

The dihedral angles between adjacent rings are comparable to those in **IV** and related molecules [5], but the latter have $\bar{3}$ symmetry, or approximate it, with rings tipped up and down alternately, while the present molecules do not. The considerable variation in the degree of tip of the different rings, and especially the extreme tip of two of the bridging rings in **I** and **Ia** relative to the molecular axis (the viewing direction in Figures 1 and 2), has no counterpart in **IV** and related species.

The bond distances and bond angles in the present structures are all quite normal; e.g. the average aromatic C—C distances in **I**, **Ia**, **IIa**, and **IIIa** are 1.399, 1.399, 1.398 and 1.40 Å. (The results for **IIIa** are much less precise than for the other three, as can be seen from Tables I and II.) The fact that these and other distances are not foreshortened is in accord with the lack of both overall and internal molecular motion (except for some wagging of the anisyl methyl groups, and possibly of the bridge atoms in **IIIa**).

The *overall* libration of the host molecules, estimated from the U^{ij} values for the individual atoms, is very small in each of these structures: the *maximum* rms librational amplitudes are about 1° for **I**, **Ia**, and **IIa** and 2° for **IIIa** (**IIIa** was done at room temperature, the others at low temperature). There is, however, evidence of significant torsional oscillation (libration) of the methoxy methyl groups about the C—O bonds in each compound, the rms librational amplitudes being in the range 5 to 10° for the low temperature structures, and 18° for **IIIa** (Table V). Because the overall libration is so small, the corrections to most bond distances for such motion are negligible, only a fraction of the e.s.d. of the distance. On the other hand, some corrections to the O—CH_3 distances are appreciable, an order of magnitude larger (up to as much as 0.06 Å). These distances have been corrected in the supplementary tables.

For bonds of the kinds present in the hosts of these structures, Hirshfeld has pointed out that the difference in mean square displacement amplitudes (MSDAs) of the bonded atoms along the bond direction should be of the order of 0.0010 Å^2 or less [21] if the data are of good quality. The quality of U values, as estimated from the mean e.s.d. of U^{ij}, is reasonably good for **I** and **IIa**, fair for **Ia**, and poor for **IIIa** (Table V). The data for **I**, **Ia** and **IIa** fit the Hirshfeld criterion as well as can be expected, given the quality of the data; for **IIIa**, the U values are so imprecise that the test is hardly meaningful, but nonetheless, the 'fit' is in accord with the mean e.s.d. of U.

Study of the data in Table V (and the more extensive tables of MSDA differences with the supplementary material) shows that the hosts in these structures are indeed very rigid, except for the appreciable librational motion of the methoxy methyl groups about the C—O bonds and a small degree of wagging of

Table V. Some results of analysis[a] of the anisotropic displacement parameters.

	I	Ia	IIa	IIIa	IV	IVa
Rms libration amplitude of CH_3 about the C_{Ar}—O bond	6.6°	6.2°	10.3°[b]	17.8°	10.5°	10.9°
	5.0	5.2	5.5			
Mean MSDA differences[c]:						
(1) between bonded atoms along bonded direction	16 pm^2	20 pm^2	11 pm^2	85 pm^2	9 pm^2	8 pm^2
(2) between all atoms within the individual aryl rings	17	21	13	72	13	13
(3) between non-bonded atoms in different aryl rings	23	34	18	83	17	17
(4) between methoxy methyl C atoms and atoms in aryl rings						
same rings	23	29	34	177	126	60
other rings	186	166	214	1344	437	360
Mean e.s.d.[d] of U^{ij}	11	17	12	63	13	12
T	128 K	115 K	115 K	295 K	295 K	295 K

[a] All calculations were made with THMA11 [14]. Numbers in parentheses represent e.s.d.s, given in units of the least significant digit of the corresponding quantity. See discussion in text concerning the significance of MSDA differences, Data for **IV** and **IVa** taken from Ref. [5]. Detailed tables of these quantities for each structure are included with the supplementary material.

[b] For **IIa**, this larger libration amplitude is, somewhat surprisingly, for C(9), on the methoxy O closer to Li^+.

[c] Units of MSDA differences are $Å^2 \times 10^{-4}$, or pm^2.

[d] The mean e.s.d. of a difference in MSDA values would be expected to be about 1.4 times the mean e.s.d. of an individual U^{ij} term.

the *para* methyl groups. The MSDA differences between non-bonded atoms within each aromatic ring are scarcely larger than those between bonded atoms, and the differences between atoms in *different* aromatic rings are only slightly larger. A much more common situation in compounds containing many aromatic rings is that the differences in MSDA values for atoms in different rings are much larger than those within a ring, indicating significant relative motion [22]. The fact that the MSDA differences for the methoxy methyl groups are much larger relative to atoms in other aromatic rings than relative to those in rings to which they are attached is a manifestation of the fact that if atomic motion is approximately normal to the interatomic vector in question, the MSDA difference may be small even though appreciable internal motion is present. In other words, large MSDA differences are a sufficient but not a necessary condition for the presence of internal motion.

The $FeCl_4^-$ counterion in **IIa** has tetrahedral geometry (bond angles 108.2–111.0°, average 109.5°) with an average Fe—Cl distance (corrected for libration) of 2.209 Å (range 2.205 to 2.215 Å), quite in accord with the average of 17 precise structures in the Cambridge Structural Data Base [23], 2.217 (14) Å. The Cl atoms have rather large displacement parameters; the apparent motion of the anion fits the rigid-body approximation extremely well.

4. Conclusions

Compared to the prototype spherand **IV**, the 'augmented' spherand **I** provides more cation···O contacts and a greater cavity size (which should favor Na^+), and the 'diminished' spherand **II** provides a smaller cavity size (which should favor Li^+). In fact, **IV** is most strongly bonded to each of these cations; for Na^+ the order of binding is **IV** > **I** > **II**, and for Li^+ the order is **IV** > **II** > **I** [20]. The present structural study shows that a high degree of strain remains in the complexes of **I** and **II**, in the form of short O···O distances and out-of-plane deformations, and that, in addition, at least for **I**, minor rearrangement occurs upon complexation. These results also suggest that **III** will not bind Li^+ as strongly as **IV**, for although the average Li^+—O distance is only 0.03 Å longer in **IIIa** than in **IVa**, and there are two very short Li^+—O distances of 2.05 Å in **IIIa** the short O···O contacts necessitated by the *syn* configuration will render this complex less stable relative to the pure host than **IVa** is relative to **IV**.

Acknowledgments

We are indebted to Drs. D. J. Cram, R. C. Helgeson, G. M. Lein, and T. Takeda, for samples of the crystals studied, to the staffs of the UCLA Office of Academic Computing and of the departmental VAX computer for facilitating this work, and to the National Science Foundation for support of much of it.

Notes

[1] The new complex **IIIa** was prepared by R. C. Helgeson; its properties will be published elsewhere.

[2] Crystal data: $a = 22.791(5)$, $b = 16.716(4)$, $c = 16.138(3)$ Å, $\beta = 104.32(2)°$ at 296 K, $V = 5957$ Å^3, space group $I2/a$, $Z = 4$, $D_m = 1.192(3)$ g cm^{-3}, $D_x = 1.186$ g cm^{-3} for the host with LiCl and three molecules of toluene.

[3] Lists of structure factors (Tables 1S through 4S) have been deposited with the British Library at Boston Spa, Wetherby, West Yorkshire, U.K., as Supplementary Publication No. SUP 82121 (154 pages); atomic positions with anisotropic displacement parameters, (Tables 5S through 8S), bond distances, bond angles and torsion angles (9S through 12S), and tables of differences of mean square displacement amplitudes (Tables 13S through 16S) have been deposited as Supplementary Publication No. SUP 82122 (36 pages).

References

1. C. J. Pedersen: *J. Am. Chem. Soc.* **89,** 2495–2496 (1967).
2. C. J. Pedersen: *J. Am. Chem. Soc.* **89,** 7017–7036 (1967).
3. D. J. Cram, T. Kaneda, R. C. Helgeson, and G. M. Lein: *J. Am. Chem. Soc.* **101,** 6752–6754 (1979).
4. D. J. Cram, T. Kaneda, R. C. Helgeson, S. B. Brown, C. B. Knobler, E. Maverick and K. N. Trueblood: *J. Am. Chem. Soc.* **107,** 3645–3657 (1985).
5. K. N. Trueblood, E. F. Maverick and C. B. Knobler: *Acta Crystallogr.* **B47,** 389–398 (1991).

6. D. J. Cram, G. M. Lein, T. Kaneda, R. C. Helgeson, C. B. Knobler, E. Maverick, and K. N. Trueblood: *J. Am. Chem. Soc.* **103,** 6228–6232 (1981).
7. C. E. Strouse: *Rev. Sci. Instr.* **47,** 871–876 (1976).
8. G. M. Sheldrick: *SHELX76*, Program for crystal structure determination. University of Cambridge, England (1976).
9. D. T. Cromer and J. B. Mann: *Acta Crystallogr.* **A24,** 321–324 (1968).
10. R. F. Stewart, E. R. Davidson, and W. T. Simpson: *J. Chem. Phys.* **42,** 3175–3187 (1965), as employed in *SHELX76* [8].
11. P. Main, S. E. Hull, L. Lessinger, G. Germain, J. P. Declercq, and M. M. Woolfson: *MULTAN78*, A System of Computer Programs for the Automatic Solution of Crystal Structures from X-ray Diffraction Data. Univ. of York, England (1978).
12. *UCLA Crystallographic Package*: J. D. McCullough Laboratory of X-ray Crystallography, Univ. of California, Los Angeles, (1984).
13. W. D. S. Motherwell and W. Clegg: *PLUTO78*. A program for drawing crystal and molecular structures. Univ. of Cambridge, England (1978).
14. K. N. Trueblood: *Acta Crystallogr.* **A34,** 950–954 (1978); J. D. Dunitz, V. Schomaker and K. N. Trueblood: *J. Phys. Chem.* **92,** 856–867 (1988).
15. W. C. Hamilton: *Acta Crystallogr.* **12,** 609–610 (1959).
16. P. A. Kollman, G. Wipff and U. C. Singh: *J. Am. Chem. Soc.* **107,** 2212–2219 (1985).
17. K. Hermansson, J. O. Thomas, and I. Olovsson: *Acta Crystallogr.* **B33,** 2857–2861 (1977).
18. C. A. Reiss, K. Goubitz and D. Heijdenrijk: *Acta Crystallogr.* **C46,** 465–467 (1990).
19. R. D. Shannon: *Acta Crystallogr.* **A32,** 751–767 (1976).
20. D. J. Cram and G. M. Lein: *J. Am. Chem. Soc.* **107,** 3657–3668 (1985).
21. F. L. Hirshfeld: *Acta Crystallogr.* **A32,** 239–244 (1976).
22. R. E. Rosenfield, K. N. Trueblood, and J. D. Dunitz: *Acta Crystallogr.* **A34,** 828–829 (1978).
23. F. H. Allen, O. Kennard, D. G. Watson, L. Brammer, A. G. Orpen and R. Taylor: *J. Chem. Soc. Perkin Trans. 2*, S1–S19 (1987).

Journal of Inclusion Phenomena and Molecular Recognition in Chemistry **12**: 361–375, 1992.

Complexes between Disubstituted Benzo-15-Crown-5 Ligands and Sodium or Potassium Bromides#

MARTIN P. PAYNE and MARY R. TRUTER*
Molecular Structures Department, Rothamstead Experimental Station, Harpenden, Herts., AL5 2JQ and Chemistry Department, University College, 20 Gordon Street, London, WC1H OAJ, U.K.*

(Received: 13 September 1990; in final form: 31 January 1991)

Abstract. Stability constants have been determined with ion selective electrodes for complexes between sodium or potassium bromide in methanol with each of four crown ethers, benzo-15-crown-5 (**Ia**), dibromobenzo-15-crown-5 (**Ib**), dimethoxybenzo-15-crown-5 (**Ic**) and di-*n*-butoxybenzo-15-crown-5 (**Id**). Those for (**Ib**) were significantly lower than the others. The stability constants for complexes between sodium bromide and (**Ia**) and (**Ib**) in dimethylformamide (DMF) were found to be about one fifth of the corresponding values in methanol. The conductivity method was used to measure the ion pairing in methanol of sodium bromide alone and in the presence of (**Ia**), (**Ib**), or (**Ic**). Ion pairing is increased on complexation, the association constants being 3.3 $mol^{-1}\,dm^3$ for $Na^+\,Br^-$ and 20–23 $mol^{-1}\,dm^3$ for $Na(\mathbf{Ia{-}c})^+\,Br^-$. The syntheses of compounds (**Ic**) and (**Id**) are described.

Key words. Crown ethers, stability constants, ion pairing.

1. Retrospect

See following page

This paper is dedicated to the memory of the late Dr C. J. Pedersen.
* Author and address for correspondence.

THE GUARDIAN
Friday November 3 1989

Obituary: Charles Pedersen

Nobel prize for a crowning achievement in chemistry

CHARLES. J. Pedersen, who has died in New Jersey at the age of 85, was awarded a Nobel Prize for Chemistry in 1987; the twenty-year delay between his first seminal publications and the award is an indication of the time required for something really new to be appreciated.

His father was Norwegian, his mother Japanese, and he was born in Korea. He studied chemistry in the University of Dayton, Ohio and at M.I.T., and he then joined du Pont at Wilmington, Delaware. He stayed there, becoming an American citizen in 1953, until his retirement in 1969. His status there from 1946 was that of research associate, a position of prestige which allowed him to do research free from managerial responsibility.

In 1961, after successful work in other fields, Pedersen started on the project which was to lead to his Nobel Prize. But his discoveries were entirely different from the intended objective. He set out to make compounds containing vanadium to act as catalysts for polymerization. He actually found, by mid-1962, that he had made a new kind of neutral organic compound capable of forming complexes with inorganic salts, such as sodium chloride, and extracting them into organic solvents.

During the next five years he worked with ever-increasing excitement and made sixty of these new compounds. He called them "crowns" because the systematic chemical names are very cumbersome and because with them the positive ions in the salts "could be crowned and uncrowned without physical damage to either, just like the heads of royalty". The surrounding of the positive ion by the organic crown molecule conferred new properties on it and also left the negative ion more reactive. Not all of the crowns were effective complexing agents. This allowed Pedersen to deduce the factors that were important, one of which was that the crown must be the correct size to encircle the ion. He found similar interactions (that is, without the formation or breaking of chemical bonds) to be possible between crowns and many metal salts and ammonium salts and also small organic molecules.

While individuals of vision in du Pont allowed him to continue his "blue skies" research, they did not allow it to be made public until 1967. Then he gave a talk at a meeting in Japan and as sole author he published one short and one full paper in the Journal of the American Chemical Society. The full paper described 49 new compounds which are difficult to make, and their interactions with various salts.

In all he published fewer than ten papers on crowns but these papers show amazing insights into future developments. The significance of his work was first realised only by a few people with prepared minds, notably biophysicists who were studying the transport of ions through biological membranes.

After his 1967 talk in Japan only two people came up to speak to him, one from a local pharmaceutical firm and the late Sir Ronald Nyholm, head of the chemistry department at University College, London. The latter was honorary director of the recently formed Agricultural Research Council's Unit of Structural Chemistry where the fundamental chemistry behind such "transport" was being investigated.

He invited Pedersen to work in the Unit for three months in 1969 and also arranged for samples of Pedersen's compounds to be studied by X-ray crystallography in the Unit. Results of the X-ray "pictures" showed that Pedersen's ideas about the crown sitting round the spherical positive ion were correct and also, that the uncomplexed crown molecule was a different shape in the absence of the ion so that the fit was more like that of a hood than of a rigid crown.

Two of the chemists who were among the first to appreciate Pedersen's work shared the Nobel Prize with him: Professors J-M Lehn of Strasbourg and D.J.Cram in California. They had in common that each led a large team, published several hundred papers and acknowledged a debt to Pedersen. They provided different, and brilliant, contributions to the field variously described as "molecular recognition" and "host-guest" chemistry. This — the ability deliberately to synthesise a larger host molecule to enclose selectively a chosen smaller guest — gives a new measure of control in a range of applications from pharmaceuticals to catalysts.

Charlie Pedersen was a most delightful person, very modest and extremely kind and helpful to young people. He liked to work at the bench himself, always wearing his black beret, and chose not to have assistants. He was a keen fisherman and gardener with a happy home life. He and his wife, Sue, greatly enjoyed their stay in London, partly because they felt it was safe to walk about at night.

His later years were saddened by her death in 1983 and his own increasing physical disablement. By the time his belated award of the Nobel Prize arrived he was so frail that he was able to go to Stockholm only because du Pont provided a private aeroplane for him. But his physical disabilities did not affect his brain or personality. In a letter to me, dictated eight days before he died, he said, "Have you been doing any new work in crystallography? Please keep me informed!"

Mary R. Truter

Charles John Pedersen, born Pusan, March 19, 1904; died Salem, New Jersey, October 26, 1989.

2. Introduction

There is a delicate balance between ion-pairing, solvation and complex formation in the isolation of crystalline derivatives of alkali metal salts and crown ethers. Pedersen's [1] original samples were dried before being analysed or having their infrared spectra recorded. He was intrigued to learn that some of his crystalline complexes had been shown by crystal structure analysis to contain water, e.g. the complex between sodium iodide and benzo-15-crown-5, (**Ia**) [2]. After his retirement he worked for three months in the Agricultural Research Council Unit of Structural Chemistry (University College London). There, he ran his own infrared spectra, modestly refuting a suggestion that this was a waste of his talent by saying "I see things when I do it myself."

Sodium bromide reacts with (**Ia**) to give a hydrated complex isomorphous with the iodide [3] i.e. containing $(NaH_2O(\textbf{Ia}))^+$ whereas with the dimethoxy derivative (**Ic**) it gives an unsolvated complexed ion pair, (NaBr(**Ic**)) with an Na—Br bond length of 2.763(3) Å. We wondered if there might be a detectable difference in ion pairing in solution. We tried field desorption mass spectrometry [4]. From acetone solution the molecular ions $(Na\textbf{Ia})^+$ and $(Na(\textbf{Ic}))^+$ were obtained but not $(NaH_2O(\textbf{Ia}))^+$ or $(NaBr(\textbf{Ic}))^+$; however, the complex (NaBr(**Ic**)) gave Na_2Br^+ which might prove diagnostic for ion pairing.

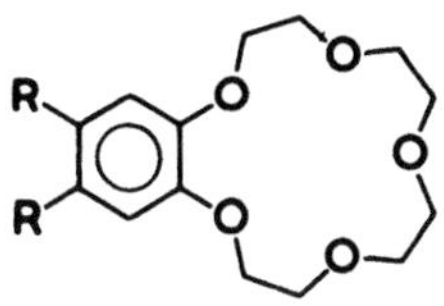

R = H **Ia**
R = Br **Ib**
R = OCH_3 **Ic**
R = OC_4H_9 **Id**

We now report investigations of solutions of sodium bromide or potassium bromide with benzo-15-crown-5 and three disubstituted compounds, (**Ib**), (**Ic**), and (**Id**). Stability constants were measured with ion selective electrodes in a system described previously [5]; our experiments using constant ionic strength conditions and earlier ones with tert-butylammonium bromide as an inert medium [6] had shown that it was reasonable to approximate ionic activities to concentrations so the stability constants are defined as:

$$K_1 = c(ML^+)/c(M^+) \cdot c(L) \qquad (1)$$

$$K_2 = c(ML_2^+)/c(ML^+) \cdot c(L) \qquad (2)$$

If ion pairing can be assumed to be negligible, e.g. for sodium tetraphenylborate, changes in conductivity may be used to determine stability constants. This technique was used by Smid and coworkers [7] to study complexes of benzo-15-crown-5 and its monosubstituted derivatives in acetone. In the present work we have used changes in conductivity to study ion pairing of $Na^+ Br^-$ or NaL^+Br^- and complex dissociation is an important complication. If K_1 is about 10^3 the complex will be dissociated to the extent of 30% at 10 mM and 60% at 1 mM. To reduce dissociation to less than 5% a 15–20 mM excess of ligand had to be used in conductivity experiments which were designed to determine the association constant of ion pairing defined as

$$K_A = c(MX)f_{mx}/c(M^+) \cdot c(X^-)f_m^+ \cdot f_x^- \qquad (3)$$

$$= xf_{mx}/((1-x)^2 \cdot f_m^+ \cdot f_x^-) \qquad (4)$$

Where c is the concentration, f_1 are the activity coefficients and x is the fraction of ion associated.

3. Experimental

3.1. MATERIALS

Sodium bromide and potassium bromide were of A.R. quality (Aldrich), dried in a vacuum at 150°C for 2–3 h and stored in a desiccator.

For potentiometry, reagent grade DMF (Lancaster Synthesis) was purified by distillation at reduced pressure under a nitrogen atmosphere and A.R. grade methanol (Fisons) was used without further purification. For conductivity measurements fresh A.R. methanol (Fisons) was deionized by running through a 1 : 1 mixture of freshly activated cationic (IR 120) and anionic (IR A400) ion-exchange resins and stored in specially cleaned Pyrex distillation flasks. The specific conductivity of the deionized methanol was in the range $0.6–2.0 \times 10^{-6}$ ohm^{-1} cm^{-1}. The cleaning procedure consisted of coating the surfaces with freshly prepared chromic acid ($K_2Cr_2O_7$–60%H_2SO_4) for 30 min–1 h, followed by rinsing many times with distilled water, A.R. methanol and deionized methanol in that order until a negligible increase in the conductivity of the deionized methanol was observed. All glassware used in the titration, including the conductivity cell, was cleaned in a similar manner.

Benzo-15-crown-5(**Ia**), 4,5-dimethoxybenzo-15-crown-5 (**Ic**) and 4,5-dibromobenzo-15-crown-5 (**Ib**) had been prepared by D. G. Parsons [8]. Benzo-15-crown-5 was formed by standard routes, for example coupling of the α,β-ditosylate of 1,11-dihydoxy-3,6,9-trioxyundecane with catechol, and the 4,5-dibromo derivative by the direct bromination of the unsubstituted crown. A similar synthesis of the dibromo compound has recently been published by other workers [9]. It was necessary to synthesize (**1c**) and dibutoxybenzo-15-crown-5 (**Id**) for the present work, success being achieved after considerable adaption from Ref. [8]. Details are given below.

3.2. SYNTHESIS OF DIMETHOXYBENZO-15-CROWN-5 **Ic**

There are four principal reaction steps, summarized in Scheme 1. Steps (i) and (ii) are based on the work of Dallacker and Lohnert [10] with minor modifications to reduce side-reactions. In step (i) the catechol in methanol was added over a period of about 2 hours, with vigorous stirring to keep the temperature below 20°C. Step (ii) was carried out under nitrogen to prevent aerial oxidation of the product.

(iii) 51.5 g, 1,11-dihydroxy-3,6,9-trioxyundecane (Aldrich) was dissolved in *ca.* 160 mL pyridine and *ca.* 130 g *p*-toluenesulphonyl chloride added with shaking over a period of 40 min and left at 0–5° for 3 h. The precipitated pyridinium hydrochloride was dissolved by stirring with 400 mL water and the pale yellow viscous oil which separated collected and washed with water (4 × 150 mL) and dried by addition of a little toluene and its subsequent removal as a toluene–water azeotrope under reduced pressure in a rotary evaporator. In the I.R. spectrum the strong doublet at 1190 cm^{-1} and 1175 cm^{-1} characteristic of *p*-toluenesulphonates and only a very weak OH stretch in the 3000–3500 cm^{-1} indicated the reaction to be practically complete. The yield of dried tosylate was approx. 50 g (38%).

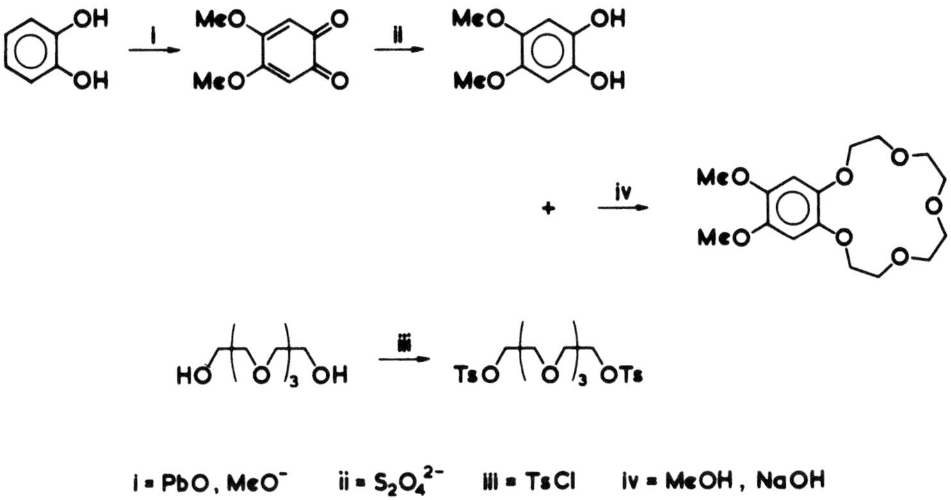

Scheme 1

(iv) A solution of 4,5-dimethoxycatechol (7.0 g, 0.0412 moles) in methanol (200 mL) was brought to reflux under a N_2 atmosphere, and 1,11-ditosyloxy-3,6,9-trioxaundecane (24.8 g, 0.0493 moles) added in one batch, and the mixture treated with sodium hydroxide pellets (4.9 g). Reaction occurred rapidly and the solution became dark green in colour. The addition of 10–20 mg sodium borohydride to prevent oxidation of the catechol resulted in the solution becoming pale orange-brown and it remained in this state throughout the refluxing period of $3\frac{1}{2}$ h.

The solution was allowed to cool to room temperature under the nitrogen atmosphere and a further 10 mg of sodium borohydride added to lessen the extent of oxidation during the subsequent workup. 20 mL aqueous 2M hydrochloric acid was added to neutralize the excess sodium hydroxide and the solution concentrated to *ca.* 50 mL by evaporation of solvent at reduced pressure, a solid (sodium tosylate) separated out. 200 mL H_2O was added and the solution extracted with CH_2Cl_2 (3 × 100 mL) and combined organic extracts washed with water (3 × 125 mL). Removal of the solvent under reduced pressure gave about 15 g of an oil dried by azeotropic distillation of added toluene (2 × 50 mL). This product was dissolved in ethyl acetate (40 mL) and a solution of lithium perchlorate (4 g) in ethanol (40 mL) added. On cooling to 0–5°C a fine white crystalline solid separated out, filtration, followed by drying using suction on a glass sinter gave 7.9 g of product.

To recover the free crown, 7.8 g lithium perchlorate complex were dissolved in 50 mL water and the solution extracted with dichloromethane (3 × 50 mL), the combined organic extract was washed with water (4 × 50 mL). The aqueous washings were extracted with CH_2Cl_2 (2 × 50 mL) and the combined organic extracts evaporated to dryness, producing a buff coloured oil which on cooling and seeding with a few crystals of the previously prepared crown crystallized to give 4.8 g of a buff coloured solid.

A 85% petroleum ether (b.p. 80–100°), 15% ethyl acetate mixture proved suitable for recrystallization giving 2.0 g of creamy white crystals, m.p. 80–81°C, *found*: C 58.48, H 7.45% *calc.* for $C_{16}H_{24}O_7$, C 58.52, H 7.37%. The IR spectrum of this product was identical with that of the sample used to make the 4,5-dimethoxybenzo-15-crown-5 sodium bromide complex [3]. The IR active C—O—C stretching bands give rise to strong bands at 1220, 1200 and 1140 cm^{-1}. The ^1H NMR spectrum in $CDCl_3$ showed multiplets due to methylene resonances at 4.14(4H), 3.92(4H) and 3.75(8H) ppm in addition to a singlet due to aromatic protons at 6.60 ppm and the methoxy methyl groups at 3.83 ppm, fully consistent with the above formulation.

3.3. SYNTHESIS OF 4,5-DIBUTOXYBENZO-15-CROWN-5 (**Id**)

Our first attempts to make 4,5-dimethoxybenzo-15-crown-5 were modelled on the conditions employed for the original synthesis of benzo-15-crown-5 itself (Pedersen [1]), coupling of 1,11-dichloro-3,6,9-trioxaundecane in *n*-butanol with 4,5-dimethoxycatechol. This method, however, resulted in exchange of aromatic ring substituent methoxy groups with butoxy groups to give 4,5-dibutoxybenzo-15-crown-5, as shown in Scheme 2. The mechanism probably involves oxidation to the quinone followed by nucleophilic attack by the butanol oxygen with simultaneous or subsequent elimination of methoxy groups.

(iii′) 1,11-dichloro-3,6,9-trioxyundecane was prepared by the chlorination of the dihydroxy compound with purified thionyl chloride using pyridine to remove the HCl produced. After work up the product was purified by extraction with benzene and vacuum distillation b.p. 130–135°C, 0.2 mm Hg.

(iv′) This dichloro compound and 4,5-dimethoxycatechol were dissolved in equimolar proportions (38 g and 28 g respectively) in 400 mL *n*-BuOH. The mixture was heated to reflux (110°C) and 10% excess of KOH added. The reaction mixture turned dark brown in colour. After 3 h gently refluxing, a small sample was

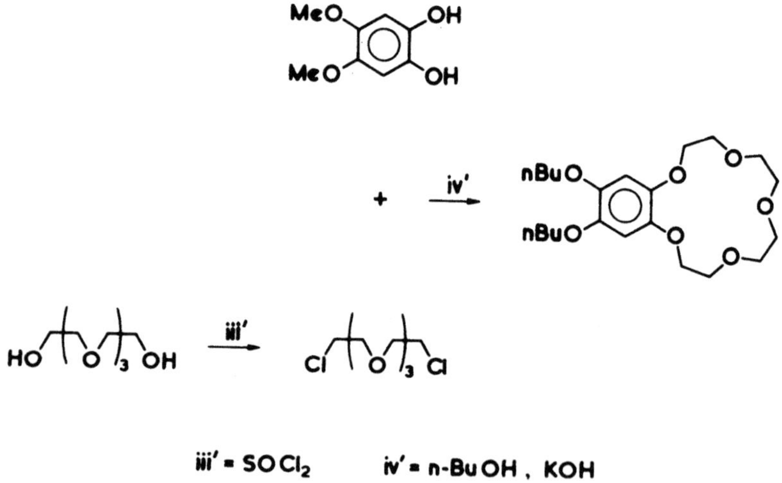

Scheme 2

worked up as described below, and an IR spectrum run on the resulting dark brown oil. The absence of bands at 660 cm^{-1} and 740 cm^{-1} due to C—Cl stretch indicated that all the dichloro compound had reacted. On cooling the reaction mixture was extracted with chloroform (3 × 150 mL), washed with water (3 × 100 mL), dried with anhydrous magnesium sulphate and the solvent removed by distillation under reduced pressure giving approx. 40 g of a dark brown oil. In an attempt to remove some highly oxidized dark materials, the oil was dissolved in ethyl acetate and passed through a layer of alumina on a Grade 2 glass sinter funnel. Only a small proportion of the dark material was retained by the alumina. The ethyl acetate solution was evaporated to dryness, the brown oil redissolved in *ca.* 100 mL methanol and the solution left overnight at 0–5° in a refrigerator. Successively over a few days two white solids separated, the first melting sharply at 132–134°C. and the second over the range 106°–116°C. Neither could be identified from their IR and NMR spectra. The filtrate was evaporated to dryness and the resulting dark brown viscous oil (35–40 g) was distilled under reduced pressure. The main product was obtained at 210–240°C and redistilled, the fraction b.p. 210–240°C (0.2 mm Hg) was collected. Its ^{1}H NMR spectrum was consistent with 4,5-di-*n*-butoxybenzo-15-crown-5 as the major product. In $CDCl_3$: δ0.96(*t*), 1.48(*m*), 1.73(*m*), 3.88(*m*), 3.93(*t*), 4.10(*m*), 6.57(*s*) ppm. Recrystallization of *ca.* 5.2 g of this crude product in 40 mL petrol (b.p. 80–100°) gave 1.1 g of almost colourless (very pale creamy yellow-brown) crystals, m.p. 44–47°C.

3.4. MEASUREMENTS OF STABILITY CONSTANTS

Stability constants were measured by potentiometry using the apparatus and methods described previously [5]. The titration vessel was kept at 25.0 ± 0.2°C, the ion selective electrodes were ORION for sodium and Electronic Instruments Ltd for potassium cations. The reference electrode was an Ag/AgCl type filled with a solution of the alkali metal chloride in the current solvent. Before use in methanol or DMF the electrodes were immersed in *ca.* 10 mM solutions of NaBr or KBr in the solvent for at least 48 h then rinsed with the solvent.

The electrodes were calibrated with solutions of known concentration. Decade shifts in methanol were close to 59.2 mV as predicted for Nernstian response; they were checked before each tritration and the actual values used. For DMF, the anomalously high value of 65 ± 1 mV was found, for two titrations for each of Na^+, K^+, Rb^+ and Cs^+.

For methanol, titrations were carried out with salt solution (1–5 mM) in the 'pot' and 25 to 30 aliquots of crown solution were added the e.m.f. being measured when it had steadied after each addition. The titrations were repeated several times with different concentrations of solutions. For DMF, the solid addition method was used; it required 250 mg of benzo-15-crown-5 (**Ia**) to produce a change of 24 mV.

3.5. CONDUCTIVITY MEASUREMENTS

Electrical conductances were measured using a WPA CMD 400 digital conductance meter. The conductance cell was of traditional design, made of Pyrex glass with brightly platinized electrodes.

The extent of ion-pairing of simple 1 : 1 salts with single charged ions in methanol is small at the millimolar concentration level, requiring an accuracy of better than *ca.* 1% for a reliable comparison of association constants.

The conductivity cell was in a water bath thermostatically regulated to *ca.* 0.2° between 24.5 and 25.5°C. The cell was fitted with a cork top with two bored holes into which the tapered tip of the precision burette (10 mL, 0.005 mL graduations, Pyrex Grade A) and a soft glass air leak were inserted. The air leak was made with the narrowest possible capillary consistent with a reasonable rate of delivery of titrant (*ca.* 1–2 mL/minute). The temperature of the water bath was recorded using a thermometer accurate to 0.1°.

In the titrations *ca.* 10 mM NaBr in 20 mM crown solution was employed as titrant and *ca.* 20 mM crown solution as titrand. The former was generally made up from *ca.* 10 mM salt solution and a known mass of crown. The crown compounds despite recrystallization were slightly contaminated with salt producing a small but significant contribution (*ca.* 5×10^{6} ohm^{-1}) to the conductance throughout the titration. Correction for this 'background' was carried out by subtraction of the initial conductance of titrand from the subsequent values. To allow the accurate determination of initial titrand volumes and estimates of evaporation losses from weighings the densities of both solutions were determined using a pyknometer previously calibrated with distilled water.

At the beginning of each titration the conductance cell was weighed in air and filled to a volume of *ca.* 7.5–8.0 mL with titrand and re-weighed allowing the initial volume to be precisely determined. The initial volume employed was *ca.* 50% greater than that required to submerge the electrodes completely.

The cell was left in the water bath for 30 min–1 h for thermal equilibrium to be established, the conductivity was recorded and the first addition of titrant (0.1–0.5 mL) made. The cell was gently agitated in the bath allowing mixing and thermal equilibrium to be established and the steady value of the conductivity obtained after 3–5 min recorded. The additions of titrant and conductivity readings were continued until the cell was nearly full and the concentration had risen to *ca.* 4 mM. Between twenty and thirty additions were made in total. The cell was then re-weighed allowing the volume of evaporation occurring over the course of the titration (DV) to be estimated by subtraction of the addition volume indicated from the readings of a calibrated burette. An approximate volume correction was applied on the assumption that the total volume of evaporation increased linearly with the volume of titrant added.

The effective cell constant was measured over a range of concentrations by means of a conductometric titration solution as titrant at *ca.* 25°C using a method essentially the same as the main experiment. The molar conductance of aqueous KCl at 25°C has been shown [11] to follow the equation:

$$\Lambda = 149.873 - 95.01c^{1/2} + 38.48c \log c + 183.1c - 176.4c^{3/2} \qquad (5)$$

where c is the molality (mol/kg). Coefficients of this equation for conductivites at 18°, 10° and 0° were also given in the same paper.

Since the temperature of the bath could not be easily adjusted to exactly 25°C small corrections were required. An interpolation procedure was employed.

Since viscosity variation is the main source of the temperature dependence of conductivities the coefficients at the above four temperatures were fitted to simple functions of the viscosity. E.g.

$$\Lambda_o = 135.454\,(10^3\eta)^{-0.8708} \quad (R^2 = 0.9999988) \tag{6}$$

Viscosities were estimated from the empirical relation [12],

$$\log(\eta T/\eta 20°) = (1.3272(20\text{-}T) - (0.001053(T\text{-}20)^2/(T + 105)$$

with $\eta 20°(H_2O) = 1.002 \times 10^{-3}\ \mathrm{N\,s\,m^{-2}}$. The error in the conductivity produced by this method is negligible compared with that resulting from the uncertainty in the temperature ($\pm 0.1°$).

3.6. ANALYSIS OF CONDUCTANCE DATA

The development of conductivity theory for electrolytes in solution has been reviewed extensively [e.g. 13, 14]. Although a wide variety of equations are available only the most theoretical treatments, requiring considerable computing power, yield reasonably consistent K_A values when association is small, e.g. in methanol or water. In view of this and the comparative nature of the present study, we have used the simpler and more convenient Shedlovksy analysis [15]. It is based on a modification of the 1933 Fuoss Equation [16], i.e.

$$\Lambda = y(\Lambda_0 - (A + B\Lambda_0)(yc)^{1/2}) \tag{7}$$

in which the second term in the fraction dissociated, $y = 1 - x$, is replaced by the experimental factor Λ/Λ_0:

$$\Lambda = y\Lambda_0 - (A + B\Lambda_0)\Lambda/\Lambda_0(yc)^{1/2} \tag{8}$$

A and B are functions of the solvent viscosity (η_0), dielectric constant (D) and temperature (T) ($A = 82.5/\eta_0(DT)^{1/2}$, $B = 8.204 \times 10^5/(DT)^{3/2}$ [17]. Rearrangement of (8) gives:

$$\Lambda = y\Lambda_0/(1 + (A + B\Lambda_0)(yc)^{1/2}/\Lambda_0) \tag{9}$$

Solution of the Shedlovsky equation proceeds as follows. Equation (9) is a quadratic in $y^{1/2}$, giving:

$$y = \Lambda/\Lambda_0(z/2 + (1 + (z/2)^2)^{1/2})^2 = S(z)\Lambda/\Lambda_0 \tag{10}$$

where $z = (A + B\Lambda_0)\,(\Lambda c)^{1/2}/\Lambda_0^{3/2}$. $1/S(z)$ is the fractional decrease in conductivity predicted due to ionic (asymmetry and electrophoretic) effects in the absence of association. Considering the ion-pair equilibrium:

$$M^+ + X^- \leftrightharpoons MX$$

$$K_A = a(MX)/a(M^+)\cdot a(X^-) = (1 - y)\cdot f/f_\pm^2 \cdot y^2 \cdot c \tag{4}$$

where c = molar concentration of MX and f = activity coefficient. The mean ionic activity coefficients ($f_\pm$) were evaluated in the present work using the extended Debye–Hükel Equation [18].

$$\ln(f_\pm) = \alpha c^{1/2}/(1 + \beta Rc) \tag{11}$$

where

$$\alpha = e^3(2.10^3 N_A)^{1/2}/(2.303.8\pi(\varepsilon_0 DK_b T)^{3/2}) \tag{12}$$

$$\beta = (2.10^3\, N_A e^2/\varepsilon_0 Dk_b T)^{1/2} \tag{13}$$

and R = the 'ion-size parameter'. Following the recommendations of Justice [19] we employ $R = q$ = the Bjerrum distance ($e^2/4\pi\varepsilon_0 Dk_b$, 858 pm in methanol). Substitution of Equation (10) into (4) gives:

$$1/\Lambda S(z) = 1/\Lambda_0 + K_A \Lambda S(z)\, cf_{\pm}^2/\Lambda_0^2 \tag{14}$$

A plot of $1/\Lambda S(z)$ vs. $\Delta S(z)cf_{\pm}^2/\Lambda_0^2$ gives an intercept $1/\Lambda_0$ and slope K_A. A first estimate of $\Lambda_0 = 100$ was employed for the initial evaluation of $S(z)$ and in the first Shedlovsky plot. From the latter a more accurate value of Λ_0 was obtained for use in subsequent iterations. Only four or five cycles of refinement were required for successive Λ_0 values to differ by less than 0.1%. The calculation was easily handled by a small computer or programmable calculator. For the present purposes a programm 'KIPFIT' was written [20] in BASIC for operation on a Tektronix 4051 including corrections for complex dissociation and viscosity changes described below.

3.6.1. *Viscosity Corrections*

Since conductance measurements on the complexes have been carried out in solutions approx. 20 mM in crown ligand the viscosity of the medium is significantly greater than that of the pure solvent and the conductances accordingly reduced.

The relative viscosity of electrolyte solutions as a function of concentration is given to a first approximation by the Jones–Dole equation [21]:

$$\eta/\eta_0^0 = 1 + A'c^{1/2} + B'c \tag{15}$$

Where A' = Debye–Falkenhagen constant, an ion–ion interaction term expressible as a function of solvent properties, and B' = a factor directly proportional to the molecular volume of the solute in solution.

Flow times in an Ostwald viscometer were measured for the pure solvent, 20 mM crown solutions and solutions 20 mM in crown and *ca.* 10 mM in salt. Determinations of the densities of these solutions using a pyknometer precalibrated with distilled water allowed the conversion of the kinematic viscosities thus derived into absolute viscosities. The results may be expressed by the following approximate relations in which the viscosity is assumed to be a linear function of total ligand and salt concentrations:

$$\eta/\eta_0 = 1 + 0.41[\mathbf{Ia}]_{TOT} + 0.69[NaBr]_{TOT} \tag{16}$$

$$\eta/\eta_0 = 1 + 0.70[\mathbf{Ic}]_{TOT} + 0.49[NaBr]_{TOT} \tag{17}$$

$$\eta/\eta_0 = 1 + 0.77[\mathbf{Ib}]_{TOT} + 0.98[NaBr]_{TOT} \tag{18}$$

The molar conductivities derived for each titration point were corrected for viscosity effects by applying Walden's rule [22].

$$\Lambda_2 = \Lambda_1(\eta/\eta_0) \tag{19}$$

in conjunction with Equations (16)–(18).

3.6.2. *Temperature Corrections*

The temperature of the water bath was set as close as possible to 25°C. Small deviations (up to 0.3°) were observed. The factors A and B in Equation (8) were corrected for direct temperature dependence and for the effect of temperature on the viscosity and dielectric constant of methanol from the empirical relations [12]:

$$\eta T = \eta 25°C - 8.7 \times 10^{-3}(T\text{-}25) + 0.26 \times 10^{-3}(T\text{-}25)^2 \quad (20)$$

$$\log D_{\mathrm{T}} = \log D^0_{25} - 0.0026(T\text{-}25) \quad (21)$$

The effect of these corrections on derived association constants was very minor ($\ll 1 K_{\mathrm{A}}$ unit).

From expressions for the ionic mobility based on a spherical ion model the temperature dependence of Λ resides principally in the variation of the viscosity of the solvent. Thus:

$$\Lambda_0 \propto 1/\eta_{\mathrm{T}}$$

Use of this relation allowed approximate correction of derived Λ_0 values for the small temperature variations. Such corrections were in no case greater than 0.5% of the total conductivity.

3.6.3. *Complex Dissociation Corrections*

Further corrections to the conductivities were made to allow for complex dissociation by the calculation of the percentage dissociation and assuming additivity of the conductivities of free and complexed ions. Stability constants for each complex are in Table I and the extent of dissociation (x') was estimated from the equation:

$$K_1 \approx c(\mathrm{ML}^+)/c(\mathrm{M}^+) \cdot c(\mathrm{L}) = (1 - x')/x' \cdot (c°(\mathrm{L}) - (1 - x')c°(\mathrm{M})) \quad (1')$$

Table I. Mean stability constants from potentiometric titrations in methanol and DMF at 25.0 ± 0.2°.

Ligand	n	MX	$\log K_1$	(±)	n	MX	$\log K_1$	(±)	$\log K_2$	(±)
In MeOH		25–30 data points per titration								
Ia	3	NaBr	2.94	(1)	5	KBr	2.88	(4)	3.15	(4)
Ib	3	NaBr	2.32	(2)	3	KBr	2.39	(3)	3.01	(1)
Ic	6	NaBr	2.87	(3)	3	KBr	2.80	(2)	3.23	(2)
Id	3	NaBr	2.81	(3)	4	KBr	2.76	(9)	3.35	(2)
In DMF		15–20 data points per titration								
Ia	2	NaBr	1.38	(5)						
Ib	2	NaBr	0.82	(3)						

n number of titrations.
(±) error in the least significant digit.

The superscript ° denotes total concentrations of crown and metal ions, free or complexed. Rearrangement and solution of the resulting quadratic in x' gives Equation (23):

$$x' = [-(K_1(c^\circ(L) - c^\circ(\mathrm{Na})) + 1) + (K_1(c^\circ(\mathrm{L}) - c^\circ(\mathrm{Na})) + 1)^2 + 4c^\circ(\mathrm{Na}) \cdot K_1)^{1/2}]/2 \cdot c^\circ(\mathrm{Na}) \cdot K_1 \tag{23}$$

Since to a very good approximation

$$\Lambda_2 = (1 - x')\Lambda\mathrm{NaL^+Br^-} + x'\Lambda\mathrm{Na^+Br^-} \tag{24}$$

$$\Lambda_3 = \Lambda\mathrm{NaL^+Br^-} = (\Lambda_2 - x' \cdot \Lambda\mathrm{Na^+Br^-})/(1 - x') \tag{25}$$

where Λ_3 is now the molar conductivity corrected for both viscosity and complex dissociation effects. $\Lambda\mathrm{Na^+Br^-}$ was estimated from the Shedlovsky Equation (9).

4. Results

From the e.m.f. titrations the stability constants and mass balances were calculated with the program MINIQUAD 75 [23] modified by Dr. J. D. Owen to run interactively on a Prime 750 computer. Rapid convergence was obtained with the stoichiometries 1 : 1 for sodium complexes and 1 : 1 plus 1 : 2 for potassium complexes. Results are in Table I, the estimates of error are derived from the independent experiments not the standard deviations calculated by MINIQUAD for individual values.

As found previously [24] the initial addition of crown ether did not produce as great a change in e.m.f as that calculated from the overall stability constant. We think this may be a surface effect on the electrode.

The Shedlovsky plots of corrected Δ_3 vs. $c^{1/2}$ were linear over the concentration range 0.6 to 2.5 mM and gave the results in Table II. For $\mathrm{Na^+Br^-}$ the mean association constant is in good agreement with that [$K_\mathrm{A} = 3.7$ mol dm^{-3}] calculated by our method from the conductivity measurements given by Jervis *et al.* [25] but our Λ_0 is slightly lower than theirs [101.5 ohm^{-1} mol^{-1} dm^3]. Results of our comparative study are clear cut. The mobility of the sodium ion in methanol solution is reduced on complexation with benzo-15-crown-5 and to a greater degree with the dibromo- and dimethoxy-crown compounds as expected from their greater molar volumes.

Although there is considerable variation in the association constants for different runs, it is clear that K_A is significantly higher for crown complexed than for uncomplexed sodium bromide. At the highest concentrations reached in the titrations (*ca* 2.5 mM) the complexes are estimated to be present as ion-pairs to the extent of 5%, cf. less than 1% for uncomplexed NaBr.

5. Discussion

Our value of log K_1 for NaBr with benzo-15-crown-5, 2.94, is intermediate between that obtained previously by the same method in the same laboratory [26], 3.05, and in a different laboratory [27], 2.87. Even larger variations are recorded if different methods are used [28], e.g. 3.37 by the conductivity method for this system and 4.03

Table II. Analysis of conductivity measurements for NaBr in methanol, alone and complexed with crown ethers.

Expt.	n^a	$\Lambda_0(25°)$ ohm^{-1} mol^{-1} dm^3	K_A mol^{-1} dm^3	$\delta(\Lambda)^b$
Na^+Br^-				
1	18	98.48	1.7	0.08
2	18	98.61	3.6	0.13
3	18	99.35	4.7	0.29
Mean		98.9 ± 0.4	3 ± 2	
Na(Ia)$^+$Br$^-$				
1	25	94.82	19.2	0.19
2	21	95.44	22.2	0.16
3	24	95.25	18.8	0.13
Mean		95.1 ± 0.3	20 ± 1	
Na(Ib)$^+$Br$^-$				
1	22	90.84	22.6	0.13
2	21	90.42	21.9	0.15
3	18	91.04	26.6	0.14
Mean		90.8 ± 0.3	24 ± 2	
Na(Ic)$^+$Br$^-$				
1	18	90.17	19.4	0.12
2	26	91.10	23.0	0.17
3	19	89.96	24.0	0.15
4	23	89.58	20.4	0.24
5	12	89.02	15.0	0.16
Mean		90.0 ± 0.7	20 ± 3	

[a]No. of points.
[b]Standard deviation from calculated and experimental conductivites $(\Sigma(\Lambda_i^{obs.} - \Lambda_i^{calc.})^2(n-1))^{1/2}$.

to 4.50 for Na with benzo-18-crown-6 in methanol. Our value of log K_1 for NaBr with benzo-15-crown-5 in DMF is in reasonable agreement with that, 1.6, obtained by NMR methods [29]. Lower values in DMF than methanol are to be expected from the donor numbers of the solvents, 26 and 18, respectively.

Comparison of **Ib** and **Ia** shows that for Na^+ and K^+, the former in methanol or DMF, there is a significant reduction in log K_1, 0.6 for Na^+ and ~0.5 for K^+ on dibromination; this is compatible with lower electron density on the aromatic oxygen atoms with electron-withdrawing substituents. Smid and coworkers [7] found a similar effect for monosubstituted bromobenzo-15-crown-5 in acetone. The effect of dibromination on K_1 is less for potassium than for sodium and very small for K_2. This is probably because the $K^+ \cdots 0$ distances are greater in the 1 : 2 complexes than in the 1 : 1 and $K^+ \cdots 0$ is longer than $Na^+ \cdots 0$. The dimethoxy and di-*n*-butoxy substitutions give reductions in K_1 on the margin of significance by this method and further discussion is deferred until the results of NMR studies are presented [30].

Previous investigation of ion pairing in complexes of simple salts with various 18-crown-6 derivatives in coordinating solvents by Kolthoff and coworkers [31] indicated that the extent of ion-pairing in the complex is either similar to or less than that in the free salt; increases in ionic association on complexation were found only for potassium picrate and potassium tetraphenylborate with di-*tert*-butyldibenzo-18-crown-6 in methanol. Unfortunately crystal structures corresponding exactly to those MXL systems are not yet available. Measurements of ionic association for cryptate-222 complexes with sodium or potassium halides show increased ion-association in the complexes in solution in methanol [32] while there is no ion-pairing in the crystals of the corresponding iodides [33]. We have not found significantly greater association for the system which gives crystals containing complexed ion pairs [3]. We do not consider that measurement of conductivities, however carefully carried out, would enable us to predict the nature of the entities giving crystalline products.

Acknowledgements

We are grateful to Drs. D. G. Parsons and J. N. Wingfield for helpful discussions, Dr. P. Gans for the program 'MINIQUAD 75' and the Agricultural Research Council for a Postgraduate Studentship (to M. P. P.).

References

1. C. J. Pedersen: *J. Am. Chem. Soc.* **89**, 7017 (1967).
2. M. A. Bush and M. R. Truter: *J. Chem. Soc. Perkin Trans. 2*, 341 (1972).
3. I. R. Hanson, D. G. Parsons, M. R. Truter: *Acta Crystallogr.* **B38**, 448 (1982).
4. D. E. Games, L. A. P. Kane-Maguire, D. G. Parsons and M. R. Truter: *Inorg. Chim. Acta* **49**, 213 (1981).
5. E. J. Harris, B. Zaba, M. R. Truter, D. G. Parsons, and J. N. Wingfield: *Arch. Biochem. Biophys.* **182**, 311 (1977).
6. J. G. Sapsford: Sandwich Course report, Polytechnic of North London, Holloway, London. (1974).
7. R. Ungaro, B. El Haj and J. Smid: *J. Am. Chem. Soc.* **98**, 5198, (1976).
8. D. G. Parsons: personal communication 1980.
9. V. Ahsen, E. Yilmazer, M. Ertas and O. Bekaroglu: *J. Chem. Soc. Dalton Trans.* 401 (1988).
10. F. Dallacker and G. Lohnert: *Chem. Ber.* **105**, 1586 (1972).
11. J. Barthel, F. Feuerlein, R. Neueder and R. Wachter: *J. Solution Chem.* **9**, 209 (1980).
12. *Chemical Rubber Company Handbook* 61st Ed., p. F51, C.R.C. Press, Boca, Raton, Florida (1980).
13. R. M. Fuoss: *Proc. Natl. Acad. Sci.* **77**, 34 (1980).
14. R. M. Fuoss: *J. Solution Chem.* **7**, 771 (1978).
15. T. Sheldlovsky: *J. Franklin Inst.* **225**, 739 (1938).
16. R. M. Fuoss and C. A. Kraus: *J. Am. Chem. Soc.* **55**, 476 (1933).
17. W. J. Moore: *Physical Chemistry*, p. 458, 5th Edition, Longman, Harlow, Essex (1972).
18. J. Robbins: *Ions in Solution*, Oxford University Press, Oxford (1972).
19. J. C. Justice: *Electrochim. Acta* **16**, 701 (1971).
20. M. P. Payne: Ph.D thesis, University of London 1989.
21. G. Jones and M. Dole: *J. Am. Chem. Soc.* **51**, 2950 (1929).
22. J. O'M. Bockris and A. K. N. Reddy: *Modern Electrochemistry*, Vol. 1, Plenum Press, New York 1970.
23. A. Sabatini, A. Vacca and P. Gans: *Talanta* **21** 53 (1974); P. Gans, A. Sabatini and A. Vacca: *Inorg. Chim. Acta* **18**, 237 (1976).
24. D. G. Parsons: *J. Chem. Soc. Perkins Trans I* 1193 (1984).

25. R. E. Jervis, D. R. Muir, J. P. Butler and A. R. Gordon: *J. Am. Chem. Soc.* **75**, 2855 (1953).
26. D. G. Parsons, M. R. Truter and J. N. Wingfield: *Inorg. Chim. Acta* **47**, 81 (1980).
27. D. E. Fenton, D. Parkin and R. F. Newton: *J. Chem. Soc. Perkin Trans I* 449 (1981).
28. R. M. Izatt, J. S. Bradshaw, S. A. Nielsen, J. D. Lamb, J. J. Christensen, and D. Sen; *Chem. Rev.* **85**, 271 (1985).
29. J. D. Lin and A. I. Popov: *J. Am. Chem. Soc.* **103**, 3773 (1981).
30. M. P. Payne: in preparation, 1990.
31. M. K. Chantooni Jr., I. M. Kolthoff and G. Roland: *Aust. J. Chem.* **36**, 1753 (1983); I. M. Kolthoff and M. K. Chantooni: *J. Anal. Chem.* **52**, 1039 (1980).
32. M. H. Abraham, A. F. D. De Namor, W. H. Lee and R. J. Wheaton: *Acta Chem. Scand.* **A34,** 621 (1980).
33. D. Moras, B. Metz and R. Weiss: *Acta Crystallogr.* **B29**, 8383 (1973); D. Moras and R. Weiss: *Acta Crystallogr.* **B29**, 396 (1973).

Journal of Inclusion Phenomena and Molecular Recognition in Chemistry **12**: 377–387, 1992.

X-Ray Structural Study of a Zinc(II) Inclusion Complex of a Phenolate-pendant Cyclam[#]

EIICHI KIMURA,*[a,b] HIROMASA KUROSAKI,[b] TOHRU KOIKE,[a] and KOSHIRO TORIUMI[b]
[a] *Department of Pharmaceutical Chemistry, Hiroshima University School of Medicine, Kasumi 1-2-3, Minami-ku, Hiroshima 734, Japan.*
[b] *Coordination Chemistry Laboratories, Institute for Molecular Science, Okazaki National Research Institutes, Nishigonaka 38, Myodaiji, Okazaki 444, Japan.*

(Received: 7 August 1990; in final form: 17 October 1990)

Abstract. The molecular structure of phenol-pendant cyclam-zinc(II) complex, **4a**, has been determined by X-ray structure analysis. Crystals of **4a** · ClO_4 · CH_3OH ($C_{16}H_{27}N_4OZn$ · ClO_4 · CH_3OH) are monoclinic, space group $P2_1/n$, with four molecules in the unit cell of dimensions $a = 31.198(2)$ Å, $b = 8.426(1)$ Å, $c = 8.214(1)$ Å, and $\beta = 93.96(1)°$. The structure was solved by the heavy atom method and refined anisotropically to $R = 0.044$, $R_w = 0.062$ for 1551 independent reflections. The complex assumes a five-coordinate, square pyramidal geometry, where zinc(II) is surrounded by the cyclam moiety in a planar fashion with the pendant phenolate anion occupying an axial position. An extremely short Zn—O(phenolate) bond distance of 1.983(5) Å, in conjunction with the 0.288 Å displacement of Zn(II) above the cyclam N_4 plane toward the phenolate, accounts for the extremely low pK_a value of 5.8 for the pendant phenol. These facts about **4a**, in comparison with the previous findings for the Ni(II) (**4b**) and Cu(II) complexes (**4c**) with the same ligand, illustrate well the characteristics of zinc(II) ion coordination properties.

Key words. Charles Pedersen, crown ethers, phenol-pendant cyclam-zinc(II) complex, crystal structure.

Supplementary Data relating to this article are deposited with the British Library as Supplementary Publication No. SUPS211g (4 pages).

1. Retrospect

From July 20–23, 1987 I had the honor to organize the 12th International Symposium on Macrocyclic Chemistry at Hiroshima. That year marked the 20th anniversary of Crown Ethers. In 1967, Charlie orally presented his discovery in Nikko, Japan. We therefore decided to dedicate this symposium to him, and to place a citation to Charlie on the first page of the abstracts and to send a copy to him before the meeting. The citation is reproduced here.

The following is a response letter from him, which reached me just a day before the opening. I thought it my duty to read this letter at the opening ceremony. I believe that Charlie would allow me to exhibit the original copy to all the readers of this journal.

[#] This paper is dedicated to the memory of the late Dr C. J. Pedersen.
* Author for correspondence.

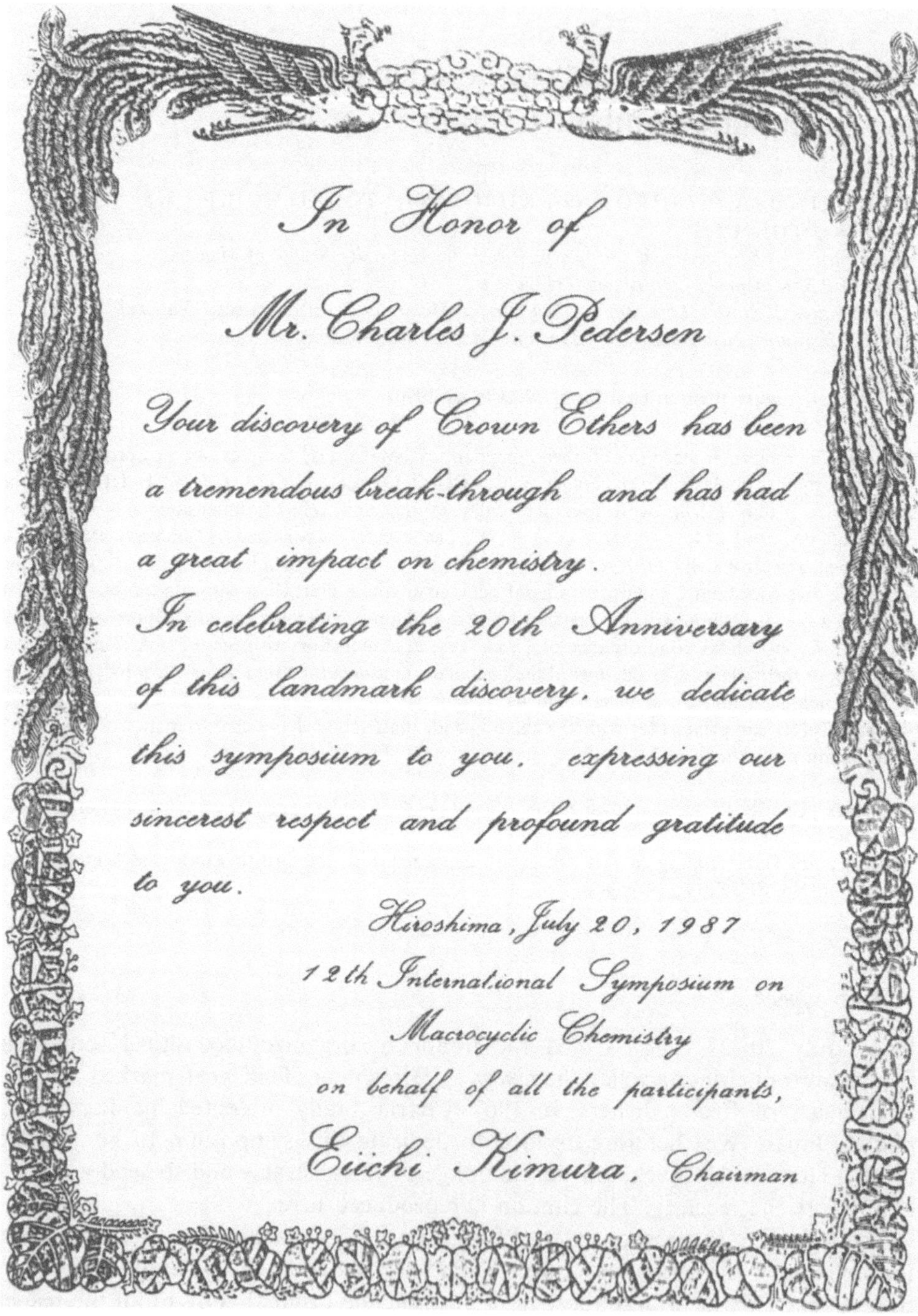

In Honor of

Mr. Charles J. Pedersen

Your discovery of Crown Ethers has been a tremendous break-through and has had a great impact on chemistry.

In celebrating the 20th Anniversary of this landmark discovery, we dedicate this symposium to you, expressing our sincerest respect and profound gratitude to you.

Hiroshima, July 20, 1987

12th International Symposium on Macrocyclic Chemistry

on behalf of all the participants,

Eiichi Kimura, Chairman

Charles J. Pedersen
57 Market Street
Salem, N. J. 08079

Professor Eiichi Kimura
Chairman, XII ISMC
Hiroshima University
School of Medicine
1-2-3 Kasumi, Minami-ku
Hiroshima 734, Japan

July 10, 1987

Dear Professor Kimura:

Thank you for expediting the printing of the abstract so that I might see it before the opening of the Symposium.

I must confess that tears welled up in my eyes when I saw the beautiful page with your kind and eloquent dedication. I will always treasure it and I thank you and your collaborators with all my heart.

Most men achieve "Immortality" through their progeny. I have no child of my own. Possibly, the Crown Ethers will serve, in a small way, to mark my footstep on earth.

I am impressed more than ever by the scope of the intrusion of Crown Ethers into chemistry. I will enjoy reading the abstracts more carefully at my leisure.

Can you doubt that I greatly regret not being with you on July 20, 1987?

Dr. Herman E. Schroeder brings to all of you my cordial greetings.

With my best personal regards,

Sincerely yours,

2. Introduction

Although coordination properties of zinc(II) ion have constantly attracted interest, particularly in connection with the bioinorganic chemistry of zinc enzymes, only a few ligands have been designed that permit us to gain deep insight into these properties [1]. Recently, we have discovered that zinc(II) complexes of a macrocyclic triamine complex **1** is a good mimic of carbonic anhydrase [2] and that its phenol-pendant derivative **2** offers an anion (i.e. phenolate) inhibition model of carbonic anhydrase [3]. In the latter complex, intramolecular coordination of the phenolate anion oxygen as the fourth donor is more facile with a consequence of the low pK_a value of 6.8 compared to the proton dissociation of the H_2O bound as the fifth donor which occurs at a much higher pK_a value of 10.7 [3]. In the absence of the phenol-pendant arm, the H_2O bound as the fourth donor in **1** dissociates with a pK_a value of 7.3 at 25°C [2], a value almost equal to that for carbonic anhydrases [4]. The X-ray structure of **2** discloses a five coordinate, trigonal bipyramidal structure, where the phenolate donor is at an equatorial position with the very short Zn—O(phenolate) distance of 1.930(4) Å [3]. From these studies with macrocyclic triamine ligands, we have concluded that Zn(II) ion, when bonded to three neutral amine donors, appears to prefer to interact with an anion (especially an anion with available oxygen donor atoms) as the fourth donor, a fact possibly related to the behavior of Zn(II) ions in carbonic anhydrases which are also bound to three imidazole nitrogens.

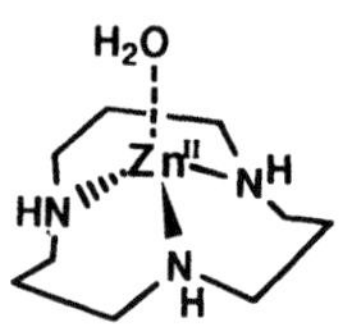

1

Zn^{II} - [12]aneN_3

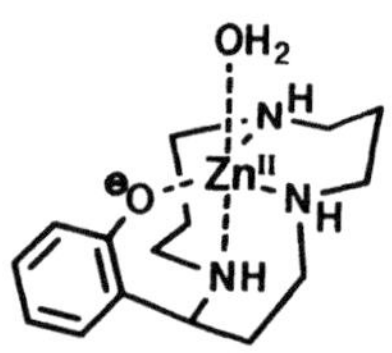

2

We then became interested in whether such a propensity of Zn(II) is present in different environments (e.g. cyclam–Zn(II) complex **3a** [5, 6]). We have thus turned to the phenol-pendant cyclam–Zn(II) complex **4a**, in which Zn(II) is placed in

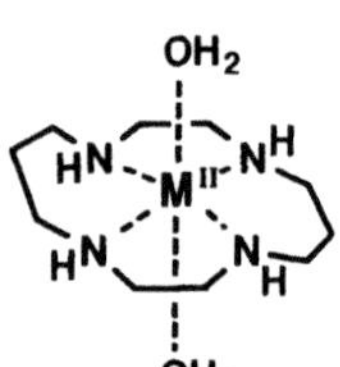

3a; M=Zn
b; M=Ni
c; M=Cu

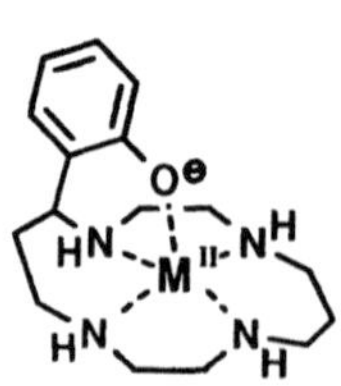

4a; M=Zn
b; M=Ni
c; M=Cu

square planar N_4 environment formed by the cyclam. It would be of interest to see how the activity and ligation of the pendant phenol is affected by the basal N_4 donors. We therefore have investigated the X-ray structure and chemical properties of **4a** with reference to **2** and to the pendantless cyclam complex **3a**. Earlier, we had reported X-ray structures of **4b** with Ni(II) [7] and of **4c** with Cu(II) [8]. A comparison of the Zn(II) complex (**4a**) with **4b** and **4c** should reveal characteristic features of Zn(II) coordination properties. It should be emphasized that one of the advantages of the use of macrocylic ligands is that they can allow selected ligand environments for the encapsulated metal ions, which allow one to detect inherent coordination properties of these metal ions.

3. Experimental Section

3.1. GENERAL METHODS

IR and UV spectra were recorded on a Shimadzu FTIR-4200 and a Hitachi U-3200 spectrophotometer, respectively. All reagents and solvents used were analytical grade. Phenol-pendant macrocyclic tetraamine **6** was synthesized using our reported procedure [8, 9]. The pH measurement procedure was the same as that previously reported for the complexation of **6** with Cu(II) and Ni(II) ions [8, 10, 11].

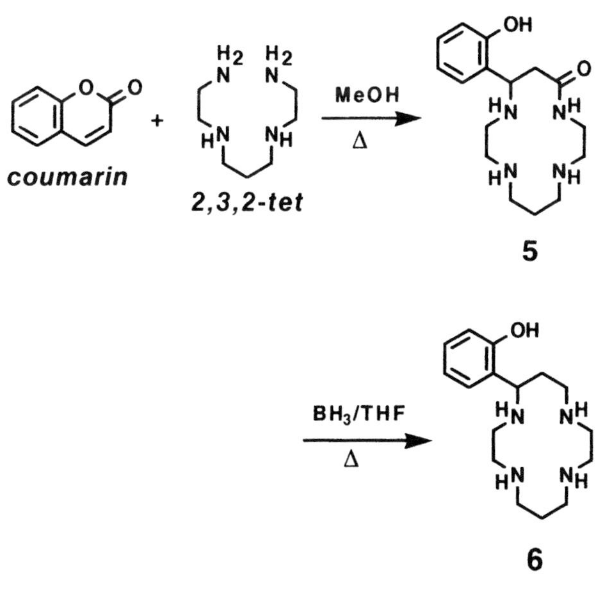

3.2. SYNTHESIS OF (Zn(II)-PHENOLATE-PENDANT CYCLAM) · ClO_4 · CH_3OH, **4a** · ClO_4 · CH_3OH

A solution containing **6** (58 mg, 0.2 mmol) and NaOH (8 mg, 0.2 mmol) in MeOH (5 mL) was added to a solution of $Zn(ClO_4)_2 \cdot 6H_2O$ (74 mg, 0.2 mmol) in MeOH (5 mL) at room temperature. After slow evaporation of the solvent, colorless crystals of **4a** · ClO_4 · CH_3OH were obtained in *ca.* 40% yield. IR (KBr pellet);

3600, 3400 (br), 3250, 2870 (s), 1593 (s), 1478 (s), 1445, 1307 (s), 1100 (br), 952, 885, 864, 775 (s) cm^{-1}. *Anal. Calcd.* for $C_{16}H_{27}N_4OZn \cdot ClO_4 \cdot CH_3OH$: C, 41.82; H, 6.40; N, 11.48. *Found*: C, 41.79; H, 6.42; N, 11.34.

3.3. CRYSTALLOGRAPHIC STUDY

A colorless crystal with dimensions 0.47 × 0.36 × 0.09 mm of **4a** $\cdot ClO_4 \cdot CH_3OH$ was used for data collection. The lattice parameters and intensity data were measured on a Rigaku AFC-5 diffractometer with graphite monochromated MoK_α radiation. The structure was solved by the heavy atom method and refined anisotropically using data corrected for absorption to give $R = 0.044$, $R_w = 0.062$ for 1551 independent observed reflections. All the hydrogen atoms are located in a difference electron density map and refined with isotropic temperature factors. The molecular structure is illustrated in Figure 1. The summary of crystal data and data collection parameters, selected bond lengths and bond angles in comparison to those for **4b** and **4c** are presented in Tables I, II and III, respectively. The atomic positional parameters and distance to atoms from a plane defined by four cyclam nitrogens are listed in Tables IV and V, respectively.

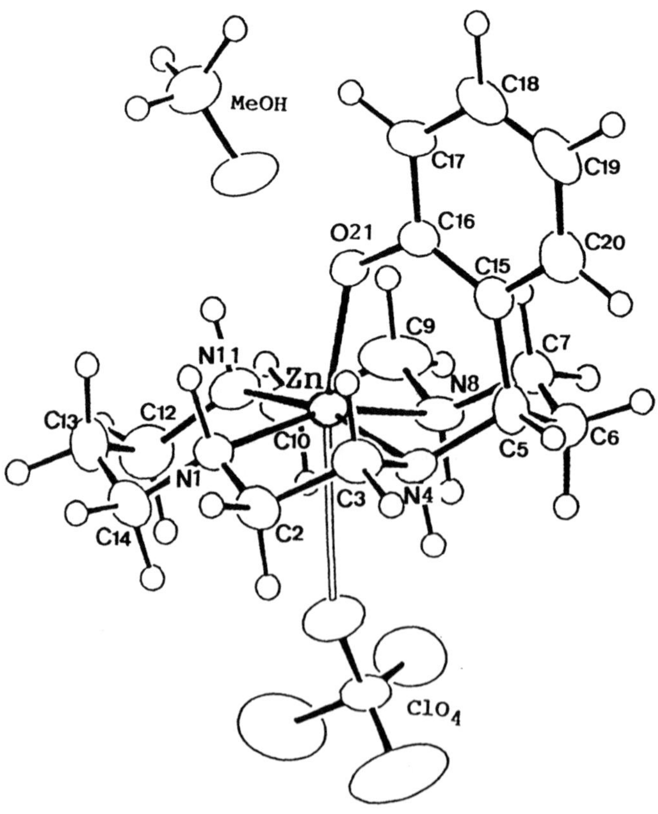

Fig. 1. Molecular structure of **4a** $\cdot ClO_4 \cdot CH_3OH$. Atoms are drawn with 30% probability ellipsoids.

Table I. Summary of crystal data and data-collection parameters for **4a** · ClO_4 · CH_3OH.

formula	$C_{16}H_{27}N_4OZn \cdot ClO_4 \cdot CH_3OH$
formula wt.	488.3
crystal system	monoclinic
space group	$P2_1/n$
cell dimensions	
a (Å)	31.198(2)
b (Å)	8.426(1)
c (Å)	8.214(1)
β (deg)	93.96(1)
V (Å^3)	2154.0(2)
Z	4
density calc. (g cm^{-3})	1.506
crystal color	colorless
crystal dimensions (mm)	0.47 × 0.36 × 0.09
radiation	MoK_α (graphite monochromated)
μ (mm^{-1})	1.340
temperature (°C)	25
2θ range, deg	2–50
index range	$(\pm h, -k, +l)$
diffractometer	Rigaku AFC-5
scan width, deg	$1.1 + 0.5 \tan \theta$
phasing	heavy atom method
refinement	block-diagonal least-squares method
no. of measd refl.	4353
no. of indep refl.	
with $\lvert F_0 \rvert > 3\sigma(F_0)$	1551
final R	0.044
final R_w	0.062
weighting scheme	$w = [(\sigma_{count})^2 + (0.03 \times \lvert F_0 \rvert)^2]^{-1}$

Table II. Comparison of selected bond distances (Å) for **4**.

	4a[a]	**4b**[b]	**4c**[c]
M—N(1)	2.115(7)	2.072(6)	2.035(4)
M—N(4)	2.067(7)	2.051(5)	2.003(4)
M—N(8)	2.120(7)	2.085(6)	2.037(5)
M—N(11)	2.110(7)	2.078(5)	2.018(5)
M—O(21)	1.983(5)	2.015(5)	2.145(4)
M—O(1C1)	2.765(11)	2.402(7)	3.135(6)

[a] Estimated standard deviation in parentheses.
[b] From Ref. [7].
[c] From Ref. [8].

Table III. Comparison of selected bond angles (deg) for **4**.

	4a[a]	**4b**[b]	**4c**[c]
N(1)—M—N(4)	83.9(3)	85.6(2)	86.0(2)
N(1)—M—N(11)	94.4(3)	92.7(2)	93.3(2)
N(4)—M—N(8)	93.6(3)	96.6(2)	94.3(2)
N(8)—M—N(11)	83.8(3)	84.9(2)	85.5(2)
N(4)—M—O(21)	91.6(2)	88.8(2)	87.5(2)
N(8)—M—O(21)	96.5(3)	94.1(2)	98.0(2)
M—O(21)—C(16)	126.1(5)	126.9(4)	127.4(3)

[a] Estimated standard deviation in parentheses.
[b] From Ref. [7].
[c] From Ref. [8].

Table IV. Atomic positional parameters ($\times 10^4$) and equivalent isotropic temperature factors for **4a** · ClO_4 · CH_3OH.

Atom	x/a	y/b	z/c	B_{eq} (Å^2)
Zn	1186.4(3)	2580.8(12)	1873.4(10)	3.4
N(1)	1228(3)	1246(8)	−291(8)	4.7
C(2)	1652(3)	1555(12)	−852(10)	4.9
C(3)	1982(3)	1707(12)	612(11)	4.8
N(4)	1830(2)	3024(8)	1618(8)	3.8
C(5)	2082(3)	3193(12)	3253(11)	4.6
C(6)	1918(4)	4635(12)	4112(12)	6.3
C(7)	1495(3)	4520(11)	4806(11)	5.1
N(8)	1122(2)	4489(8)	3521(9)	4.5
C(9)	693(4)	4226(13)	4122(14)	7.8
C(10)	388(4)	4035(14)	2685(15)	8.1
N(11)	508(2)	2600(10)	1727(10)	5.3
C(12)	279(3)	2531(14)	70(16)	7.8
C(13)	451(4)	1125(14)	−931(14)	7.9
C(14)	876(4)	1416(13)	−1583(11)	6.8
C(15)	2067(3)	1655(11)	4238(10)	4.4
C(16)	1716(3)	671(10)	4320(9)	3.7
C(17)	1759(3)	−735(10)	5267(10)	4.2
C(18)	2141(4)	−1061(11)	6159(11)	5.7
C(19)	2478(3)	−78(14)	6126(11)	6.1
C(20)	2451(3)	1265(12)	5130(12)	5.5
O(21)	1327(2)	933(6)	3550(6)	3.8
Cl	1138(1)	6625(3)	−588(3)	5.0
O(1Cl)	1243(4)	5004(13)	−371(15)	7.7
O(2Cl)	957(5)	7308(15)	710(15)	9.5
O(3Cl)	785(5)	6631(26)	−1675(24)	16.6
O(4Cl)	1424(5)	7396(19)	−1422(24)	14.6
C(MeOH)	461(4)	−1502(14)	4497(15)	7.4
O(MeOH)	568(3)	69(10)	4159(11)	8.6

Table V. Distance (Å) to atoms of **4a** from a plane defined by four cyclam nitrogens.

Atoms	Distance	Atoms	Distance
Zn	0.288	C(6)	0.387
N(1)	−0.037	C(7)	0.735
N(4)	0.027	C(9)	0.357
N(8)	−0.030	C(10)	−0.373
N(11)	0.035	C(12)	−0.872
C(2)	−0.432	C(13)	−0.493
C(3)	0.343	C(14)	−0.926
C(5)	0.865		

4. Results and Discussion

4.1. CRYSTAL STRUCTURE OF **4a** · ClO_4 · CH_3OH

The side view of the complex is shown in Figure 1. The five-coordinate square-pyramidal coordination geometry around Zn(II) ion is similar to that of the high-spin Ni(II) [7] and Cu(II) [8] complexes. The configurations ('*trans*-III' form) [12] of the macrocyclic ligand (cyclam) are fundamentally idenical for the Zn(II), Cu(II), and Ni(II) complexes. The phenolate oxygen O(21) is almost at the apex of the pyramid. The structure of these complexes are similar.

The four equatorial Zn—N bond lengths of 2.115(7), 2.067(7), 2.120(7) and 2.110(7) Å are slightly longer than the corresponding Cu—N and Ni(high spin)—N bond lengths. This increase in length is caused by the slightly larger ionic radius of Zn(II). The Zn(II) is too large to lie in the plane of the four nitrogens and so is 0.288 Å above the plane (see Figure 2). By contrast, Cu(II) and Ni(II) are coplanar with the four nitrogens of the ligand due to the strong ligand field of cyclam [7, 8]. As a consequence, the Zn—O(21) bond length of 1.983(5) Å is significantly shorter

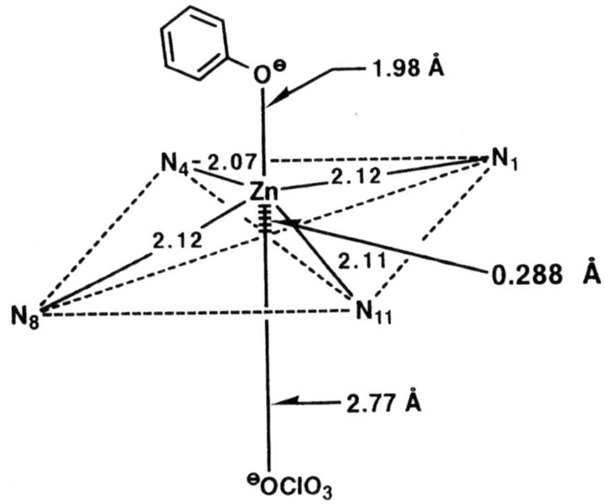

Fig. 2. Coordination geometries around the zinc(II) atom in **4a** · ClO_4 · CH_3OH.

than those in the Cu(II) (2.145(4) Å) and Ni(II) (2.015(5) Å) complexes. Moreover, the Zn—O(21) bond is much shorter than Zn—N bond lengths. These facts lead us to conclude that *zinc ion shifts toward the phenolate anion for the maximum stabilization.* If zinc(II) ion were in the cyclam N_4 plane, it would not be possible for the phenolate group to be brought into a position to bind Zn(II) so strongly. The shortest attainable Zn—O distance in such a structure would be *ca.* 2.0 Å, as estimated by the Dreiding model. The Ni(II) complex assumes almost an ideal square-pyramidal structure [7]. These facts are compatible with the strong tendency of Zn(II) to bind with an anionic ligand. Being a d^{10} ion, the Zn(II) ion is not restricted to directional ligand fields and it can come closer to the donor it most favors at the sacrifice of bonding with other neutral nitrogen donors.

Comparing the N_4 complex **4a** with its N_3 counterpart, **2**, the Zn—O^- and average Zn—N bond distances in **4a** (1.983(5) Å and 2.10 Å) are longer than those in **2** (1.930(4) Å and 2.07 Å). This is understood in light of its lower coordination number.

4.2. PHENOL–PHENOLATE EQUILIBRIUM IN METAL COMPLEXES

The ease of the axial phenol dissociation (**7** → **4**) was determined using pH-metric and spectrophotometric data and found to vary for each metal ion. These pK_a values are directly related to the M(II)—O^- bond distances and are as follows: Zn(II) ($pK_a = 5.8$, 25°C and $I = 0.1$ ($NaClO_4$)) > Ni(II) ($pK_a = 6.3$) > Cu(II) ($pK_a = 9.2$). For comparison, the pK_a value of the diprotonated ligand (**6** · $2H^+$) is 8.9 under the same conditions. It is thus concluded that with the Zn(II) complex the pendant phenol is most acidic, which is related to the Zn(II) coordinating properties.

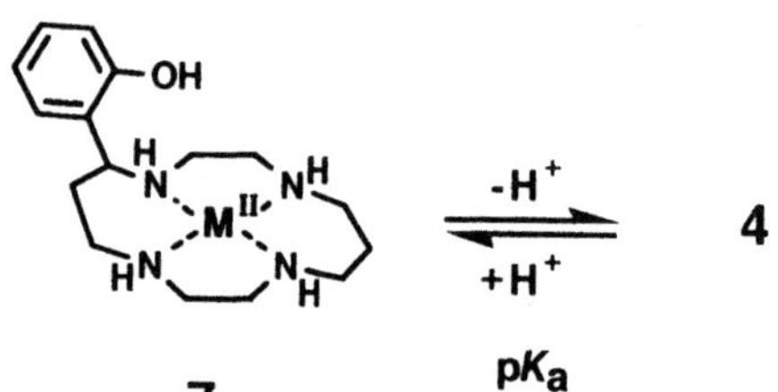

7

It should be added that the UV spectra of the phenolate bands of the Zn(II) complex (**4a**) at λ_{max} 288 nm (ε2600) and of the Ni(II) complex (**4b**) at λ_{max} 293 nm (ε3700) [8] are almost the same as that of the uncoordinate phenolate anion at λ_{max} 292 (ε4000). These facts indicate that there is little mixing between the orbitals of those M(II) ions and the phenolate orbital. Namely, there is mainly an ionic interaction between these M(II) ions and the apical O^-. On the other hand, in the Cu(II) complex (**4c**), the phenolate absorption band is significantly broader with no peak at *ca.* 290 nm [8].

4.3. ACIDITY OF H_2O BOUND TO M(II)-CYCLAM 3

It is appropriate here to compare the above findings with the pK_a values for the H_2O which most likely occupies the axial site and is bound to Zn(II) (**3a**), Ni(II)

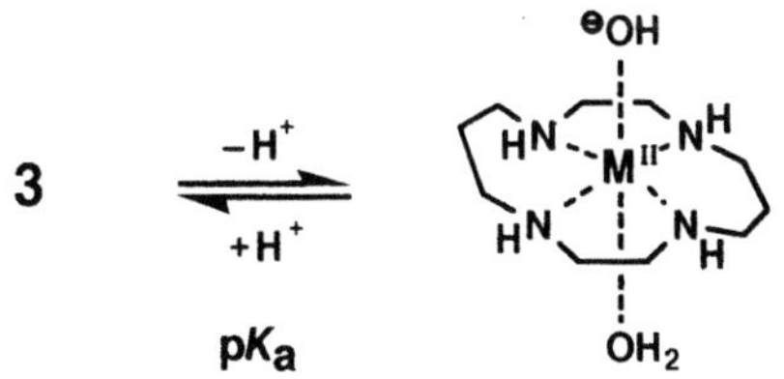

8

(**3b**), and Cu(II) (**3c**). The pK_a value of 9.8 (25°C, $I = 0.10$ ($NaClO_4$)) for Zn(II) was determined pH-metrically. For Ni(II) and Cu(II) complexes, the proton dissociation did not occur until pH *ca.* 11 and we therefore conclude that their pK_a values should be higher than 11. The lowest pK_a for a water in the Zn(II) complex (**3a**) is reasonable when compared to the complexes of **4**.

5. Conclusion

It has been shown that Zn(II) ion shows a strong affinity to bind with an anionic oxygen when the Zn(II) ion was complexed with three nitrogen atoms of a macrocyclic triamine. This study shows that this same tendency is present when Zn(II) ion is coordinated by four nitrogens. In both situations the Zn—O^- bond distances are rather short indicating a strong interaction between Zn(II) and O^-. Thus, the tailored macrocyclic polyamines [13] prove to be a useful tool in studying otherwise undetectable significant properties of the Zn(II) ion and assisting in elucidation of its biochemical significance.

References

1. P. Woolley: *Nature (London)* **258**, 677 (1975). J. T. Groves, and R. R. Chambers, Jr: *J. Am. Chem. Soc.* **106**, 630 (1984).
2. E. Kimura, T. Shiota, T. Koike, M. Shiro, and M. Kodama: *J. Am. Chem. Soc.* **112**, 5805 (1990).
3. E. Kimura, T. Koike, and K. Toriumi: *Inorg. Chem.* **27**, 3687 (1988).
4. Y. Pocker and S. Sarkanen: *Adv. Enzymol.* **47**, 149 (1987). Y. Pocker, J. T. Stone: *J. Am. Chem. Soc.* **87**, 5497 (1965).
5. M. Kodama and E. Kimura: *J. Chem. Soc., Dalton Trans.* 2269 (1977).
6. T. Ito, M. Kato, and H. Ito: *Bull. Chem. Soc. Jpn.* **57**, 2634 (1984). M. Kato and T. Ito: *Inorg. Chem.* **24**, 509 (1985).
7. Y. Iitaka, T. Koike, and E. Kimura: *Inorg. Chem.* **25**, 402 (1986).
8. E. Kimura, T. Koike, K. Uenishi, M. Hediger, M. Kuramoto, S. Joko, Y. Arai, M. Kodama, and Y. Iitaka: *Inorg. Chem.* **26**, 2975 (1987).
9. E. Kimura, T. Koike, and M. Takahashi: *J. Chem. Soc., Chem. Commun.* 385 (1985).
10. E. Kimura, M. Yamaoka, M. Morioka, and T. Koike: *Inorg. Chem.* **25**, 3883 (1986).
11. E. Kimura, T. Koike, H. Nada, and Y. Iitaka: *Inorg. Chem.* **27**, 1036 (1988).
12. V. J. Thöm, C. C. Fox, J. C. A. Boeyens, and R. D. Hancock: *J. Am. Chem. Soc.* **106**, 5947 (1984).
13. Our recent papers about Zn(II) chemistry: E. Kimura, T. Koike, T. Shiota, and Y. Iitaka: *Inorg. Chem.* **29**, 1460 (1990). E. Kimura, and T. Koike: *Comments on Inorg. Chem.* Gordon and Breach Science Publishers: New York, Vol. 11, 285 (1991).

Journal of Inclusion Phenomena and Molecular Recognition in Chemistry **12**: 389–395, 1992.

Structural Definition of Chemical Templates and the Prediction of New and Unusual Materials[#]

DARYLE H. BUSCH
Department of Chemistry, University of Kansas, Lawrence, Kansas 66045, U.S.A.

(Received: 31 October 1991; in final form: 22 January 1991)

Abstract. Synthetic applications of chemical templates have yielded, in succession, macrocycles, cage compounds, interlocking rings and a trefoil knot. The known kinds of templates are analyzed and used to provide a broad definition of a chemical template, which, in turn, leads to the prediction of exciting new forms of matter based on orderly molecular entanglements.

Key words. Chemical templates, molecular entanglements, molecular organization, template definition.

1. Retrospect

On the basis of the dates of a number of events, I believe I first met Charles Pedersen either late in 1965 or in 1966. At the time I was spending a week at the Experimental Station of du Pont for the purpose of consulting with chemists in various departments. I was asked to talk with Charlie and I clearly recall certain aspects of that rather special visit. He sought me out because we were both interested in macrocyclic ligands and he enthusiastically described his new family of crown ethers and their interactions with alkali metal ions. I recall how he used the old Fischer–Hirschfelder models in presenting his work; it was clear that he recognized the role of complementarity in the metal ion/ligand interaction. He wanted to know if this was significant and if any other alkali metal complexes were known. I explained to him that alkali metal ion complexes were indeed rare but not unknown, and, in fact, his cesium complexes were probably the first claimed for that element. I quickly became enthusiastic about his work and encouraged him to publish promptly. Further, I realized that the Nikko meeting of the International Conference on Coordination Chemistry was being planned and I volunteered to contact the organizing committee and suggest they invite him. That committee was receptive and he made his celebrated disclosure at Nikko. I met with Charlie and with his collaborator in solution studies, Dr. Frensdorf, on a number of subsequent consulting visits. Eventually I helped his supervisor nominate Charlie for the Delaware Section Award which I believe was the first recognition he received for his seminal works on alkali metal complexes of macrocyclic polyethers.

2. Introduction

Chemical templates are the basis of genetics and lie at the foundation of biology and, in recent decades, they have been used extensively in both inorganic and organic

[#] This paper is dedicated to the memory of the late Dr C. J. Pedersen.

chemistry, particularly in synthesis [1, 2, 3]. Chemical templates were first conceived and demonstrated for the field of synthetic chemistry many years ago [4], and they played a major role in the development of macrocylic ligand and inclusion chemistry [1–3, 5]. More recently extraordinary new examples have been devised, producing interlocking rings and a molecular knot [2]. Despite the extensive activity in this area, the term *template* has not been adequately defined in the past and various processes have been dubbed template reactions without sharing common critical characteristics. By analyzing the structural relationship associated with molecular organization [3], it is now possible to provide a precise definition of a *chemical template* that brings order to the subject and gives promise of opening new vistas in chemistry, including the possibility of entirely new kinds of molecular materials.

3. Discussion

Chemical templates are defined in terms of the basic elements of molecular organization [6].

> *A chemical template organizes an assembly of atoms, with respect to one or more geometric loci, in order to achieve a particular linking of atoms [7].*

In synthetic applications, the goals have been topological until relatively recently [8], although the most obvious natural templates have sequencing as their goal.

The simplest goal of a chemical template was and remains the synthesis of macrocyclic rings, and this has been accomplished by the closing of individual chelate rings by reactions carried out on ligands that are bound to metal ions. In such a case, it is necessary to organize the reacting ligands with respect to one or more edges of a coordination polygon or polyhedron, depending on the nature of the reactive ligand. As Scheme 1 shows, when a single tetradentate ligand is the

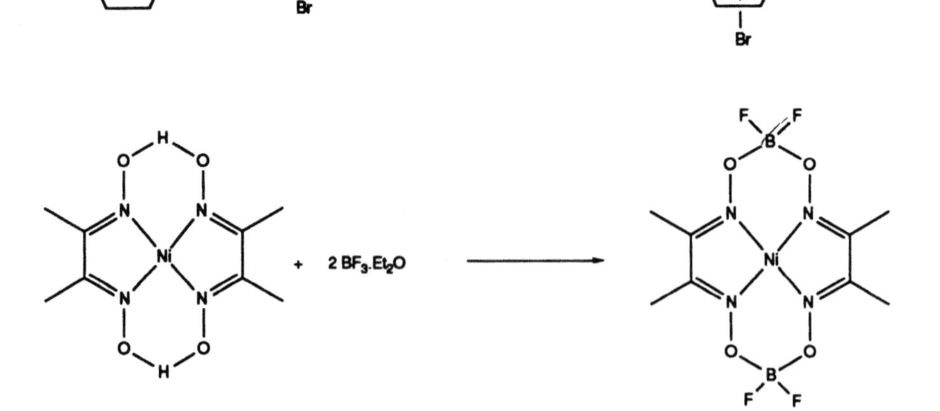

Scheme 1.

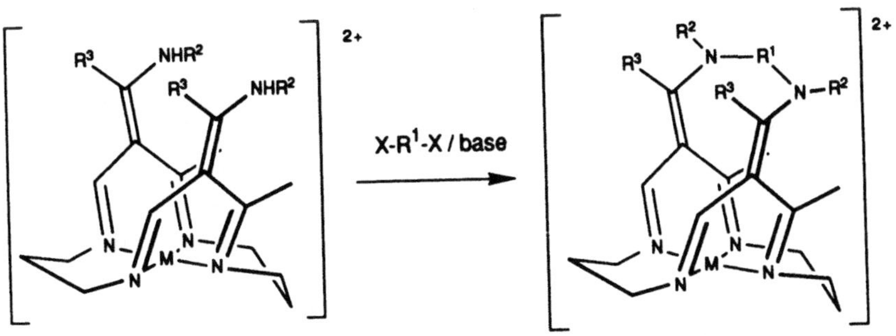

Scheme 2.

reactant, the organization is with respect to a single polyhedral edge, but when the reacting system is a pair of bidentate ligands then organization is required with respect to two edges [9]. Scheme 2 shows that the simple edge in question need not be a part of a *coordination polyhedron*, so long as it conforms to a sufficiently rigid orientation that is equivalent to the edge of a hypothetical polyhedron.

Macrobicyclization, and the concomitant enclosure of metal ions in bicyclic cages, was the second topological class of chemical templates to be recognized and exploited [10]. As Scheme 3A (in which methyl substituents and double bonds

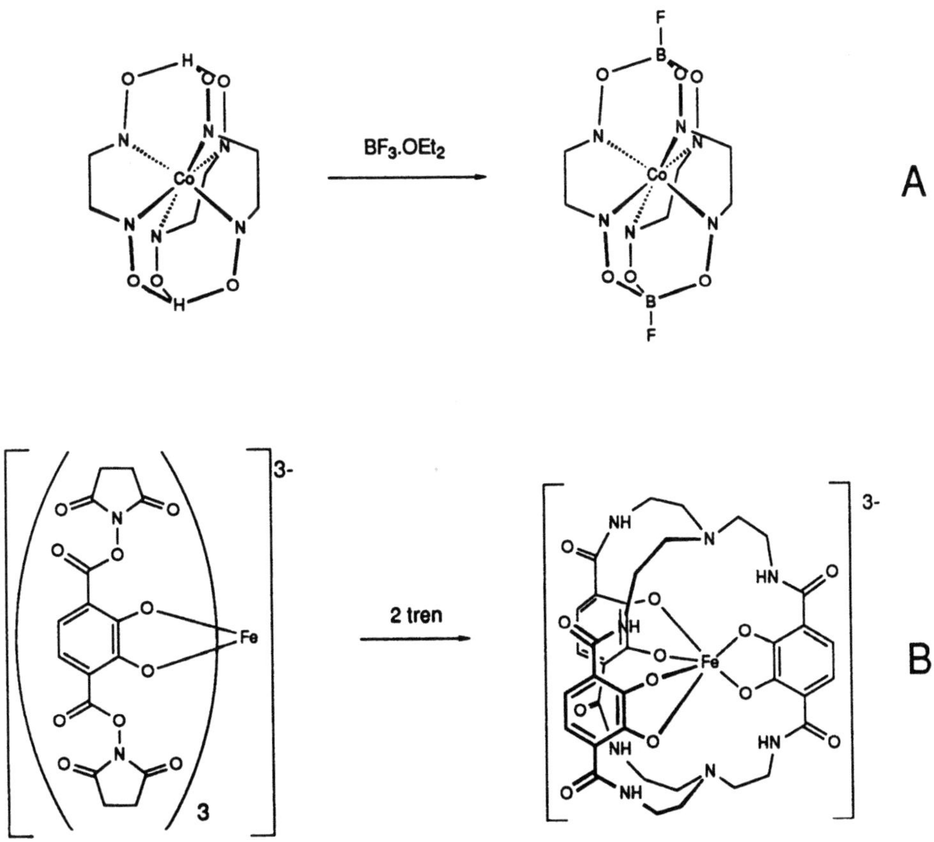

Scheme 3.

omitted for clarity) shows, this is accomplished by organization of the reacting bound ligands with respect to one face of a coordination polyhedron. As with simple macrocyclization, the atoms ligated to the metal ion may be those involved in the bicyclization reaction [11], but this need not be the case. The first example of a template bicyclization reaction used atoms α to the donor atoms [12] and studies on siderophore models [13] have used functional groups quite remote from the metal atoms (Scheme 3B). The key feature remains the organization of those reaction centers with respect to the polyhedral face. For these examples the polyhedral face has three corners and three strands are organized in the template process. Higher order faces, e.g. a square or pentagon, would organize corresponding numbers of strands.

Understanding the chemical templates in macrocyclization and macrobicyclization reactions is a simple exercise, but it is the analysis of the new topological types of chemical templates leading to catenanes and trefoil knots that makes the formal definition of a chemical template necessary and significant. In the template synthesis of interlocking molecular rings (catenanes) Dietrich–Buchecker and Sauvage [3] utilized two very similar templates. The simplest involves the positioning of the organized moiety within the compass of a previously formed and simultaneously coordinated ring. As Strategy A of Scheme 4 shows, the reactive ligand is organized with respect to a single tetrahedral edge that is orthogonal to the edge to which the preformed ring is bound.

It is simpler to state the case for the second and more basic template reaction used by those investigators to produce catenanes (Strategy B, Scheme 4). In this case the two reactive moieties that are precursors to the interlocking rings are organized with respect to mutually orthogonal tetrahedral edges. In general the four points involved in this template could be associated with structures other than a tetrahedral one and the edges need not be precisely orthogonal. For example two tridentate ligands, each of which occupies a pair of mutually planar octahedral edges (Figure 1), organize their ligands with respect to two mutually orthogonal octahedral four fold axes and this is a suitable template for the synthesis of

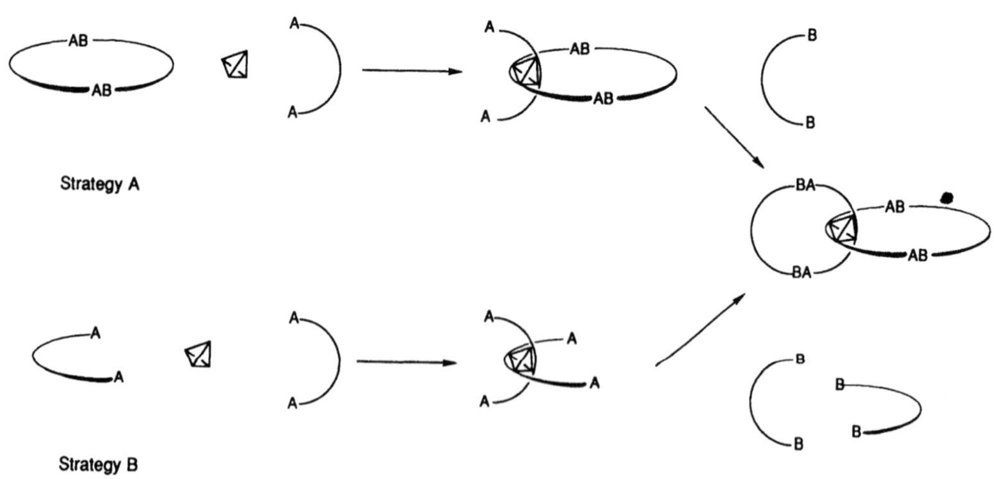

Scheme 4.

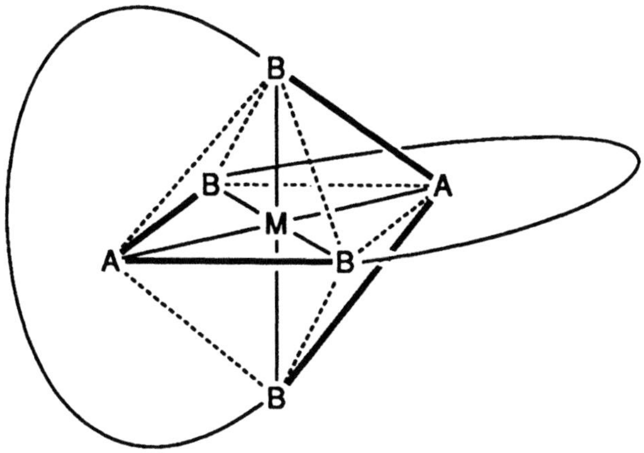

Fig. 1. An octahedral template designed to produce catenanes.

interlocking rings, although it has not yet been used. Similarly, trans-spanning pairs of bidentate ligands might be used in a square planar coordination sphere for the same topological purpose.

The template reaction leading to the trefoil knot (Scheme 5) breaks new ground as it extends organization to two metal centers and the 'assembly of atoms'

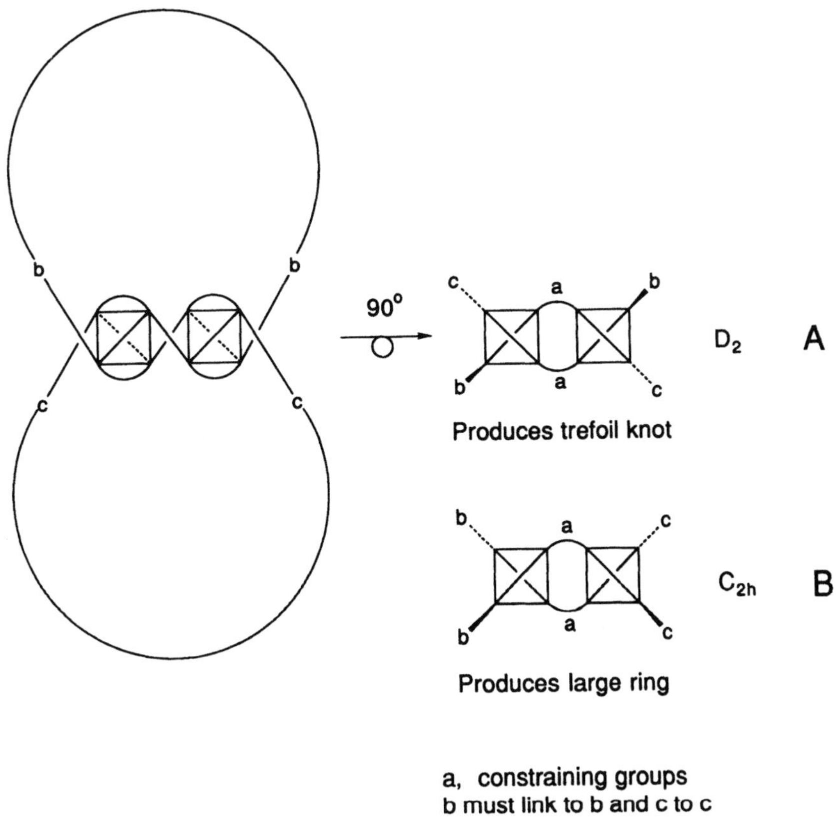

Scheme 5.

undergoing organization includes pairs of donors that chelate to each of the two templating centers [14]. Like the catenane template, two pairs of bridging bis(didentate) ligands are simultaneously bound with didentate groups from the different ligands occupying orthogonal tetrahedral edges. Further, the two templating centers must exhibit the same chiralities; otherwise the product of the template reaction would be a simple ring having no knot (Scheme 5B). This can be concisely stated:

for the trefoil knot, the 'assembly of atoms' is organized with respect to pairs of orthogonal edges on two mutually constrained and cochiral (tetrahedral) centers.

It should be noted that the template illustrated at the right in Scheme 5A is related to the structure shown (as its left) for the trefoil knot by rotation through 90° about an axis lying in the plane of the picture and passing through both metal atoms. Ring closure requires the joining of the two points marked C to each other below the plane of the paper and the two points marked B, above the plane of the paper.

4. Conclusions

Whereas the enormous impact of the processes leading to simple macrocyclization and to macrobicyclization is well known, the ultimate significance of the ingenious new template types devised by Sauvage and associates remains unclear. Here we predict that complicated templates may eventually produce entirely new forms of matter based on *orderly entanglements*. Further, these novel materials may have exceptional properties. Elastomers illustrate that *random molecular entanglements* have properties of great value. It is intriguing to reflect on the possibility of producing highly ordered molecular entanglements.

If template reactions can produce chains and knots, then they should be capable of producing molecular braids. And, if braiding is possible then so should be molecular knitting and even molecular weaving. Consider the properties of such intermolecular entanglements. Unusual strengths, uncommmon flexibilities and fascinating anisotropies are obvious possibilities. The successful applications of the more obvious, but still subtle *chiral* templates in organometallic chemistry [15] gives promise that the further application of chemical templates to transition metal catalysis of polymerization processes can be expected to produce a variety of such ordered molecular entanglements, ultimately yielding the components of '*a molecular macrame*'.

Acknowledgement

The persistent support of the U.S. National Science Foundation and the U.S. National Institutes of Health has made our studies of chemical templates possible.

References and notes

1. T. J. McMurry, K. N. Raymond and P. H. Smith: *Science* **24**, 938 (1989).
2. D. H. Busch and N. A. Stephenson: *J. Coord. Chem.* **100**, 119 (1990).
3. C. O. Dietrich-Bucheker and J. P. Sauvage: *Chem. Revs.* **87**, 795 (1987).
4. M. C. Thompson and D. H. Busch: *J. Am. Chem. Soc.* **86**, 3642 (1964).

5. Template macrocyclization and macrobicyclization reactions contributed heavily to the founding and developing of the fields of macrocyclic ligand chemistry and inclusion chemistry. Many symposia, journals, meetings and books have been devoted to these subjects.
6. The five fundamental factors that underlie molecular organization are so basic that they seem almost trivial; however, their proper consideration is essential in all molecular designs focused on problems involving molecular organization. Indeed this is probably why their essential mutual interdependence has only recently been articulated and demonstrated in analyzing such familiar 'effects' as the chelate effect, the macrocyclic effect and the template effect. They are geometry, size, topology, rigidity and complementarity.
7. The word 'organizes' (with respect to a geometric element; i.e. an edge or face) is used rather than the word 'orientation' because the group must be both propitiously located and capable of reaction. For example, if the atom or group of atoms in question is acting as a donor and is to serve as a nucleophile, then it must have the requisite potent lone pair of electrons. If the reactive site undergoing organization is a remote acyl carbon, then it must be sufficiently flexible to reach the critical transition state. Further, the definition says '*... with respect to ...*' (a geometric locus) instead of 'in conformity with' because template relationships at the collection of points describing a geometric element require a vector sense as well a positional one. This will be apparent as the discussion proceeds.
8. (a) T. R. Kelly, C. Zhao and G. J. Bridger: *J. Am. Chem. Soc.* **111**, 3744 (1989); (b) T. Tjivkua, P. Ballester and J. Rebek, Jr.: *J. Am. Chem. Soc.* **112**, 1249 (1990).
9. In the latter case it is particularly obvious that misorganization leads to failure. If the two bidentate ligands are chelated in an octahedral structure so that two of their ends are *trans* rather than *cis* to each other, then template ring closure is disfavored.
10. The first example of such a process was due to Boston and Rose and involved the case chosen as Scheme 3 (Ref. [12]). This was followed by examples from Parks, Wagner, and Holm (*Inorg. Chem.* **10**, 2472 (1971)) and from Goedken and Peng (*J. Chem. Soc., Chem. Commun.* 62 (1973)).
11. A. M. Sargeson: *Pure Appl. Chem.* **58**, 1511 (1986), and references therein.
12. D. R. Boston and N. J. Rose: *J. Am. Chem. Soc.* **90**, 6859 (1968).
13. T. J. McMurry, S. J. Rodgers and K. N. Raymond: *J. Am. Chem. Soc.* **109**, 3451 (1987).
14. C. O. Dietrich-Buchecker and J. P. Sauvage: *Angew. Chem. Int. Ed. Engl.* **28**, 189 (1989).
15. K. E. Koenig, M. J. Sabacky, G. L. Bachman, W. C. Christopfel, H. D. Barnstorff, R. B. Friedman, W. S. Knowles, B. R. Stults, B. D. Vineyard and D. J. Weinkauff: *Ann. N.Y. Acad. Sci.* **333**, 16–22 (Eng.) (1980).

ANNOUNCEMENTS

SIXTH INTERNATIONAL MEETING ON LITHIUM BATTERIES

Münster, Germany, 10–15 May, 1992.

Further information from: Prof. Dr. J. O. Besenhard, Department of Inorganic Chemistry, University of Münster, Wilhelm-Klemm-Strasse 8, W-4400 Münster, Germany. Fax: (251) 83-3169.

SEVENTH INTERNATIONAL SYMPOSIUM ON MOLECULAR RECOGNITION AND INCLUSION

Kyoto, Japan, 26–31 July, 1992.

Further information from: Prof. H. Ogoshi, Department of Synthetic Chemistry, Kyoto University, Sakyo-ku, Kyoto 606, Japan. Fax: 75-753-5917.

SEVENTEENTH INTERNATIONAL SYMPOSIUM ON MACROCYCLIC CHEMISTRY

Provo, Utah, USA, 9–14 August, 1992.

Further information from: Dr R. M. Izatt, Chemistry Department: ISMC, 226 ESC, Brigham Young University, Provo, Utah 84602, USA. Fax: 801-378-5474.

MACROCYCLIC LIGANDS FOR THE DESIGN OF NEW MATERIALS

Guildford, U.K., 14–16 September, 1992.

Further information from: Dr A. F. Danil de Namor, Chemistry Department, University of Surrey, Guildford, Surrey, GU2 5XH, U.K.

Author Index

Subject Index

GPSR Compliance

The European Union's (EU) General Product Safety Regulation (GPSR) is a set of rules that requires consumer products to be safe and our obligations to ensure this.

If you have any concerns about our products, you can contact us on ProductSafety@springernature.com

In case Publisher is established outside the EU, the EU authorized representative is:

Springer Nature Customer Service Center GmbH
Europaplatz 3
69115 Heidelberg, Germany

Batch number: 10607320

Printed by Printforce, the Netherlands